Geoarchaeology of Lebanon's Ancient Harbours

Nick Marriner

BAR International Series 1953
2009

Published in 2016 by
BAR Publishing, Oxford

BAR International Series 1953

Geoarchaeology of Lebanon's Ancient Harbours

ISBN 978 1 4073 0436 6

BAR Publishing is the trading name of British Archaeological Reports (Oxford) Ltd. British Archaeological Reports was first incorporated in 1974 to publish the BAR Series, International and British. In 1992 Hadrian Books Ltd became part of the BAR group. This volume was originally published by Archaeopress in conjunction with British Archaeological Reports (Oxford) Ltd / Hadrian Books Ltd, the Series principal publisher, in 2009. This present volume is published by BAR Publishing, 2016.

Printed in England

BAR
PUBLISHING

BAR titles are available from:

BAR Publishing
122 Banbury Rd, Oxford, OX2 7BP, UK
EMAIL info@barpublishing.com
PHONE +44 (0)1865 310431
FAX +44 (0)1865 316916
www.barpublishing.com

Contents

Acknowledgements

I wish to thank the following people for their assistance with this project: E. Anthony, Y. Battiau-Queney, M. Bourcier, J.-Y. Bottero, M. Boudagher-Fadel, H. Brückner, H. Bruneton, N. Carayon, P. Carbonel, M. Chalabi, B. Devillers, C. Doumet-Serhal, H. Frost, J.-P. Goiran, F. Husseini, E. Marcus, S. Meulé, B. Millet, C. Morhange, G. Noujaim-Clark, P. Pirazzoli, M. Provansal, A. Rosen, M. Saghieh-Beydoun, A. Sandler, A. Véron et G. Willcox.

I would also like to express my gratitude to the following institutions:
Association Internationale pour la Sauvegarde de Tyr
Bourses Entente Cordiale
The British Museum
CEDRE (programme franco-libanais)
Centre Européen de Recherche et d'Enseignement des Géosciences de l'Environnement (CEREGE)
Centre National de Recherche Scientifique (CNRS)
Conservatoire du littoral
Direction Générale des Antiquités du Liban
European Union for Coastal Conservation (EUCC)
Fondation Procter et Gamble pour la protection du littoral
Institut Universitaire de France
Lebanese British Friends of the National Museum
Leverhulme Trust
UNESCO World Heirtage Commission
Université Aix-Marseille

Introduction

Beirut, Sidon and Tyre have been occupied by human societies since the third millennium BC (Katzenstein, 1997; Elayi and Sayegh, 2000; Doumet-Serhal, 2003, 2004a). The sites grew up around easily defendable promontories, for Beirut and Sidon, and an offshore island, as in the case of Tyre. All three possessed natural low energy basins that could be exploited as anchorage havens with little or no need for human artificialisation (Marriner *et al.*, 2005, 2006a-b). In spite of their former maritime glories, however, the evolution of these three important Phoenician city-states has remained largely enigmatic (Aubet, 2001). With reference to the maritime façades of Sidon and Tyre, much of the present literature continues to cite the pioneering 1930s work of Poidebard (Poidebard, 1939; Poidebard and Lauffray, 1951). Although Frost (1971) questioned some of her predecessor's interpretations, notably regarding the southern harbour at Tyre, many of these have failed to filter down to the wider archaeological literature (Katzenstein, 1997; Moscati, 1997; Briquel-Chatonnet and Gubel, 1999; Markoe, 2002; Strong, 2002; Holst and Harb, 2006). In the southern Levant, the Bronze and Iron Age harbours of Israel have been the object of significant scientific endeavours since the early 1980s (Raban, 1981, 1984, 1985a-c, 1995). Geopolitical frictions meant that similar data from the Lebanon had been missing, important in understanding not only the archaeological evolution of the whole Syro-Canaanite coast (Wachsmann, 1998; Marcus, 2002a-b), but also the geomorphological and geological responses of the Levantine seaboard since 6000 BP (Dalongeville, 1975; Sanlaville, 1977; Raban, 1987a; Sanlaville *et al.*, 1997; Stanley, 2002; Morhange *et al.*, 2006b).

Under the auspices of the British Museum (London), the Franco-Lebanese program CEDRE (Beirut and Paris), the Leverhulme Trust (London) and UNESCO World Heritage (Paris), this project has looked to better comprehend 5000 years of human-environment interactions at Beirut, Sidon and Tyre through the application of geoscience techniques developed and refined during the past 15 years (Goiran and Morhange, 2003; Morhange, 2001; Marriner and Morhange, in press).

Chapter 1: Although innumerable studies have addressed the various aspects of ancient harbour geoarchaeology, there is no single monograph that treats the subject in its entirety. The aim of this first chapter is therefore to comprehensively review the present literature, and set ancient harbour geoscience within the wider context of Mediterranean coastal archaeology (Brückner, 1997; Reinhardt and Raban, 1999; Stanley *et al.*, 2001, 2004a-b; Brückner *et al.*, 2002; Kraft *et al.*, 2003; Vött *et al.*, 2006a-b). Ancient harbour archives have been demonstrated to be appropriate for the analysis of many environmental and cultural questions, at multiple historical and geographical scales (Morhange, 2001). Within this context, a number of temporal and spatial patterns are manifest in the literature. Based on how the basins have survived to present, we have looked to order these into a typology of ancient harbour sites. We move on to formulate a chronostratigraphic model of ancient harbours, building on the now rich bio- and lithostratigraphic datasets that exist on the subject (Marriner and Morhange, 2006b). This review of the literature and description of the principles underpinning ancient harbour geoarchaeology set the backdrop for our research undertaken on the Phoenician seaboard (Marriner *et al.*, 2005; Morhange and Saghieh-Beydoun, 2005; Marriner and Morhange, 2006a; Marriner *et al.*, 2006a-b).

Chapter 2: The most pronounced coastal changes of all three sites have been observed at Tyre. This is attributed to three complimentary dynamics: (1) significant coastal artificialisation during the Hellenistic period, with the construction of Alexander the Great's causeway (Katzenstein, 1997); (2) the location of the site at the distal margin of the Litani delta, the Lebanon's largest fluvial system and a major source of sediment at the Holocene timescale (Abd-el-Al, 1948; Soffer, 1994); and (3) rapid relative sea-level changes, attributed to a tectonic subsidence of the Tyrian horst by ~3 m since late Roman times (Morhange *et al.*, 2006b). Chapter 2 analyses the role of these various natural and anthropogenic forcings to reconstruct the Holocene accretion and progradation of Tyre's tombolo, a peculiar sand isthmus linking the former offshore bastion to the continent (Nir, 1996). Despite a rich literature on spit and barrier island evolution (Davis, 1994; Clemmensen *et al.*, 2001; Stapor and Stone, 2004; Stone *et al.*, 2004; Schwartz and Birkemeier, 2004; Otvos and Giardino, 2004; Gardner *et al.*, 2005; Simms *et al.*, 2006), the morphogenesis of tombolos has been largely neglected in the literature (Flinn, 1997; Courtaud, 2000; Goiran *et al.*, 2005). At Alexandria, Goiran (2001) has shown the existence of a proto-tombolo around mean sea level during Hellenistic times. This natural sediment wedge, part of the High Stand Systems Tract, greatly facilitated the construction of Alexander's causeway. In this chapter, we elucidate the stratigraphic data from Tyre's salient to propose a tripartite model of tombolo evolution during the past 8000 years. We subsequently review the data from Alexandria (Goiran, 2001; Goiran *et al.*, 2005) to compare and contrast the salients' evolution (Nir, 1996).

Chapter 3: The exact location of Tyre's ancient anchorages has been a source of archaeological speculation since the sixteenth century (Villamont, 1596; Van Cotvyck, 1620; Stochove, 1650; Besson, 1660; Maundrell, 1703; Arvieux, 1735; Pococke, 1745; Bertou, 1843; Kenrick, 1855; Poulain de Bossay, 1861, 1863; Renan, 1864; Rawlinson, 1889; Poidebard, 1939; Frost, 1971; Katzenstein, 1997). In chapter 3, we review this earlier literature before moving on to precisely relocate the ancient northern harbour, the city's principal transport hub during antiquity, and its phases of evolution (Marriner *et al.*, 2005). While a natural anchorage is inferred during the Bronze Age, we expound the increasing weight of anthropogenic forcings from the Iron Age onwards, culminating in a technological apogee during the Byzantine period (Hohlfelder, 1997). Using coastal stratigraphy and underwater archaeological data (Frost, 1971; El Amouri *et al.*, 2005; Noureddine and Helou, 2005; Descamps, personal communication) we demonstrate that Poidebard's (1939) Egyptian harbour is in reality a drowned quarter of the ancient city. Informed hypotheses are proposed for a possible second anchorage on Tyre, and it is demonstrated that presently drowned portions of the sandstone ridge served as outer harbours during the Bronze and Iron Ages. Stratigraphic data in the vicinity of Tells Mashuk, Chawakir and Rachidiye are also presented and discussed.

Chapter 4: At Sidon, in tandem with British Museum terrestrial excavations (Curtis, 2000; Doumet-Serhal, 2003, 2004a), coastal stratigraphy has been used to reconstruct where, when and how the city's ancient anchorages evolved. During the Bronze Age, the city's southern bay, or *Crique Ronde*, and northern pocket beach served as proto-harbours (Poidebard and Lauffray, 1951; Marriner *et al.*, 2006a-b). It is hypothesised that Zire island was also an integral part of the seaport façade at this time (Frost, 1973; Carayon, 2003). Towards the end of the Bronze Age, archaeological data from Sidon tell attest to increasing levels of Mediterranean (Doumet-Serhal, 2004). The ability of Middle to Late Bronze Age societies to build larger vessels capable of withstanding the rigours of open-sea navigation (Wachsmann, 1998), necessitated a concomitant evolution in seaport technology. At the Bronze/Iron Age transition we present data attesting to an artificialisation of Sidon's northern pocket cove, naturally predisposed to become the city's main transport hub from this period onwards.

Chapter 5: At Beirut, redevelopment of the central business district during the 1990s exposed great tracts of the city's archaeology (Elayi and Sayegh, 2000; Doumet-Serhal, 2004). Often dubbed as the 'largest archaeological dig in the world' (Lefèvre, 1995a), we were called upon by Saghieh-Beydoun to link the historical data with the coastal stratigraphy and reconstruct the ancient harbour's history. Unlike Sidon and Tyre, Beirut's ancient coastline and seaport has been extensively concreted since the nineteenth century, rendering precise reconstructions of the coastal topography and seaport progradation during antiquity very challenging. The chronostratigraphic data are compared and contrasted with Beirut's sister harbours, Sidon and Tyre.

Chapter 6: Draws together the data from all three sites to propose a general model of Phoenician harbour evolution since the Bronze Age. From a technological apogee during the Byzantine period, the asynchronous demise of the three harbours during the sixth to eighth centuries AD leads us to propose four possible forcing factors (1) historical, namely a series of internal and external malaises leading to an erosion of Byzantine hegemony in the Levant (Kennedy, 1985a-b; Bonner, 2005); (2) rapid, tectonically induced, relative sea-level changes (Guidoboni *et al.*, 1994; Darawcheh *et al.*, 2000; Morhange *et al.*, 2006b); (3) tsunamogenic destruction (Soloviev *et al.*, 2000); and (4) climatic deterioration (Schilman *et al.*, 2001; Rosen, 2006; Migowski *et al.*, in press). The geoarchaeological data are subsequently set against research undertaken from the Israeli seaboard (for example Raban, 1995, Marcus 2002a-b; Haggai, 2006) in an attempt to better comprehend the nature of Levantine harbour technology and the mechanisms and directions of cultural flows during the Bronze Age to Islamic period. The subject of sediment sources, imperative to understanding the infilling and progradation of the sites' coastlines since 6000 BP, is also addressed using clay mineralogy. These data are compared and contrasted with previous research (Stanley and Galili, 1996; Stanley and Wingerath, 1996; Stanley *et al.*, 1997; Sandler and Herut, 2000; Ribes *et al.*, 2003) to critically assess the role of local versus regional sediment sources during the Holocene. Next, the occupation histories of Sidon and Tyre are explored using trace metal geochemistry, undertaken in collaboration with Elmaleh, Le Roux and Véron (Le Roux *et al.*, 2002, 2003a-b; Elmaleh, in preparation). Finally, our geoarchaeological datasets allow us to propose a cartography of the most sensitive archaeological zones at Beirut, Sidon and Tyre (Marriner and Morhange, 2005a-b). A program of further research and protection is elucidated.

Chapter 1

Geoscience of ancient Mediterranean harbours

Although much has been written on the subject of ancient Mediterranean harbours, the relatively new area of harbour geoarchaeology remains dispersed in the geoscience and archaeological literature. Over a decade of research has amassed rich and varied datasets of anthropogenically forced coastal evolution, with a remarkable number of between-site analogies. This new research field also shows the rich potential of geoscience to reconcile important archaeological questions. No single publication, however, has yet drawn on these geological patterns to yield a detailed overview suitable for geoscientists and environmental archaeologists. The aim of this review article is to (1) discuss how ancient harbours have come to be preserved in the geological record; (2) expound the basic principles and palaeoenvironmental tools underpinning ancient harbour geoarchaeology; (3) outline some of the most significant research advances made; and (4) discuss a new chrono-stratigraphic model applicable to harbour sequences.

1.1 Introduction

Maritime archaeology has for many decades been dogmatically skewed towards what Breen and Lane (2003) term 'ship-centrism', the study of boats and their architectural aspects, to the detriment of physical harbour landscapes and infrastructures (Muckelroy, 1978; Sherwood Illsley, 1996; Gould, 2000; Green, 2004; Pomey and Rieth, 2005). Indeed, the opening sentence of a 2001 article by Gibbins and Adams emphatically encapsulates this: "Maritime archaeology, the study of the material remains of human activities on the seas and interconnected waterways, is fundamentally focused on shipwrecks. Although the definition encompasses many other types of context such as harbours, submerged land surfaces and coastal settlements, shipwrecks are the most distinctive and numerous type of site studied by maritime archaeologists." Whatever the benevolent aspects of the former, and there are many, we argue that it has yielded a patchy and archaeologically biased picture that largely disregards the port area and adjacent coastal environments. Significantly, Gibbins and Adams (2001) go on to concede that "in the fields of study focused on the maritime past, the ubiquity of shipwrecks as a site type has seemed at times to promote their research profile at the expense of palaeo-landscapes, submerged settlement sites and other structures such as harbours and fish traps."

Ports were for a long time a privileged research object of classical archaeology (Grenier, 1934), focusing namely on the study of harbour infrastructure and the analysis of maritime commerce (Rougé, 1966; Foerster Laures, 1986). Traditionally, reconstruction of ancient palaeogeographies relied heavily on ancient texts such as Strabo and Ptolemy (Ardaillon, 1896). At the start of the 1980s, unprecedented multi-disciplinary excavations were initiated at Caesarea Maritima (Israel) and Marseilles (France), looking to overcome many of these traditional shortcomings and prioritise both large-scale and small-scale archaeological units (Raban, 1985a, 1988; Hesnard, 1994; Raban and Holum, 1996; Hermary *et al.*, 1999; Rothé and Tréziny, 2005). The resultant data are unparalleled in their scope and completeness, with mutually discrete disciplines (archaeology, history, geology, geomorphology and biology) working in tandem to produce a multifaceted basin-wide understanding of the two sites. One important and perhaps most singly pervasive advancements, both in terms of cost effectiveness and data wealth, was the exploitation of the geological record as an aid to the archaeology (Reinhardt *et al.*, 1994; Reinhardt and Raban, 1999; Morhange *et al.*, 2001; Morhange *et al.*, 2003a). Ancient harbour sediments were shown to be rich time-series of human impacts yielding insights into the magnitude, variability and direction of changes during antiquity. Despite this geoarchaeological 'revolution', no single publication has yet given a detailed overview suitable for earth scientists and environmental archaeologists (Goiran and Morhange, 2003).

1.2 Brief history of research

1.2.1 Ancient harbour research in archaeology

1.2.1.1 Early travellers and savants (seventeenth to nineteenth centuries)

Whilst maritime archaeology scarcely predates the invention of underwater breathing apparatus in the early twentieth century, scholarly interest in ancient Mediterranean harbours is a centuries-old field of inquiry stretching back to the Renaissance. Revival of the Classics around this time meant that the Grand Tour of Italy and Greece was an important rite of passage for young aristocrats looking to 'enlighten' themselves with the architecture and archaeology of antiquity's most important sites (Bourguet *et al.*, 1998; Horden and Purcell, 2000). In the Near East meanwhile, pilgrims en route to the Holy Land extensively described the landscapes of numerous Levantine coastal sites (Pococke, Volney etc.).

Whilst interesting from a historiographical standpoint, these early works were dominated by speculative reverie void of any real scientific substratum. Numerous authors attempted to deductively translate the classic written sources and personal observations of relic landscape features into graphic representations of harbour topographies (e.g. the sixteenth century reconstruction of Portus in Braun and Hogenberg's *Civitates Orbis Terrarum*, volume IV, 1588, **Figure 1.1**). This early corpus of antiquarian study significantly influenced scholarly research during the proceeding centuries, and idealised views of many ports continued to be produced with just slight variations (see Paroli, 2005 for Portus). Although this piecing together of the literary jigsaw yielded frail and greatly tentative results it bears testimony to the intrigue surrounding ancient harbours.

Many of these early sixteenth to eighteenth century voyagers were startled by the paradoxically small size of ancient port basins, a fact which appeared greatly enigmatic with their former maritime glories. For example, Maundrell (1703) cannot but help hide the disenchantment of his stay at Tyre: "This city, standing in the sea upon a peninsula, promises at a distance something very magnificent. But when you come to it, you find no similitude of that glory, for which it was so renowned in ancient times." Volney (1792) further compounds his predecessor's words: "Where are those fleets of Tyre, those dock-yards of Arad, those work-shops of Sidon, and that multitude of sailors, of pilots, of merchants, and of soldiers? [...] I sought the ancient inhabitants and their works, and found nothing but a trace, like the foot-prints of a traveller over the sand. The temples are fallen, the palaces overthrown, the ports filled up, the cities destroyed; and the earth, stripped of inhabitants, has become a place of sepulchres." A century later, Renan reiterated his predecessor's words "I think that no other first order city of antiquity has left behind as few traces as Tyre."

During the nineteenth century a clear transition is observed away from the simple landscape descriptions of the antiquarian traditions to the employment of more rigorous scientific techniques by intellectuals. Colonialism and the fight not only for territorial but also cultural supremacy gave rise to a number of important geographical and archaeological works. Napoleon's ambitious military expedition to Egypt (1798-1801) epitomises these new scientific currents. Although the military campaign itself was a debacle, Napoleon took with him 167 of France's leading scholars: mathematicians, astronomers, naturalists, engineers, architects, draftsmen, and men of letters. Their task was to study the country in all its aspects and eventually led, some twenty years later, to the elephantine publication the *Description de l'Egypte*.

After the fall of Napoleon I, the French looked to offset rising British influence in the Mediterranean using culture and science. In 1846, the *Ecole française d'Athènes*, the first research school of its kind was created in Greece and spawned an annex school, the *Ecole française de Rome,* which eventually became autonomous in 1875. The German Archaeological Institute was founded in 1871. These two schools had been preceded at Rome by a joint Franco-German venture, the *Institut de correspondance archéologique* (1829), which aimed to publish new archaeological findings from around the Mediterranean in the institution's journal. At Rome, the model was eventually followed by other European and North American nations.

In addition to the archaeological schools, a number of geographical societies, including the Royal Geographical Society, the Palestine Exploration Fund and the *Société de Géographie de Paris*, also came into being around this time. Archaeology was seen as a means of showcasing political prowess and stronghold, and was very much at the centre stage of the British, French and German colonial movements. The resulting encyclopaedic works, and the supporting institutional frameworks which accompanied them, typify the shift towards precise recording and measurement in all areas of the natural sciences and archaeology. For coastal archaeology, the most important works of this period include Renan (1864), El-Falaki (1872), Ardaillon (1896) and Georgiades (1907). Significantly, it was during this period that many scholars started to draw parallels between coastal progradation and harbour silting (**Figure 1.2**) to explain the reduced size or isolation of many ancient port basins, notably Renan at Tyre (1864) and Canina at Ostia (1830).

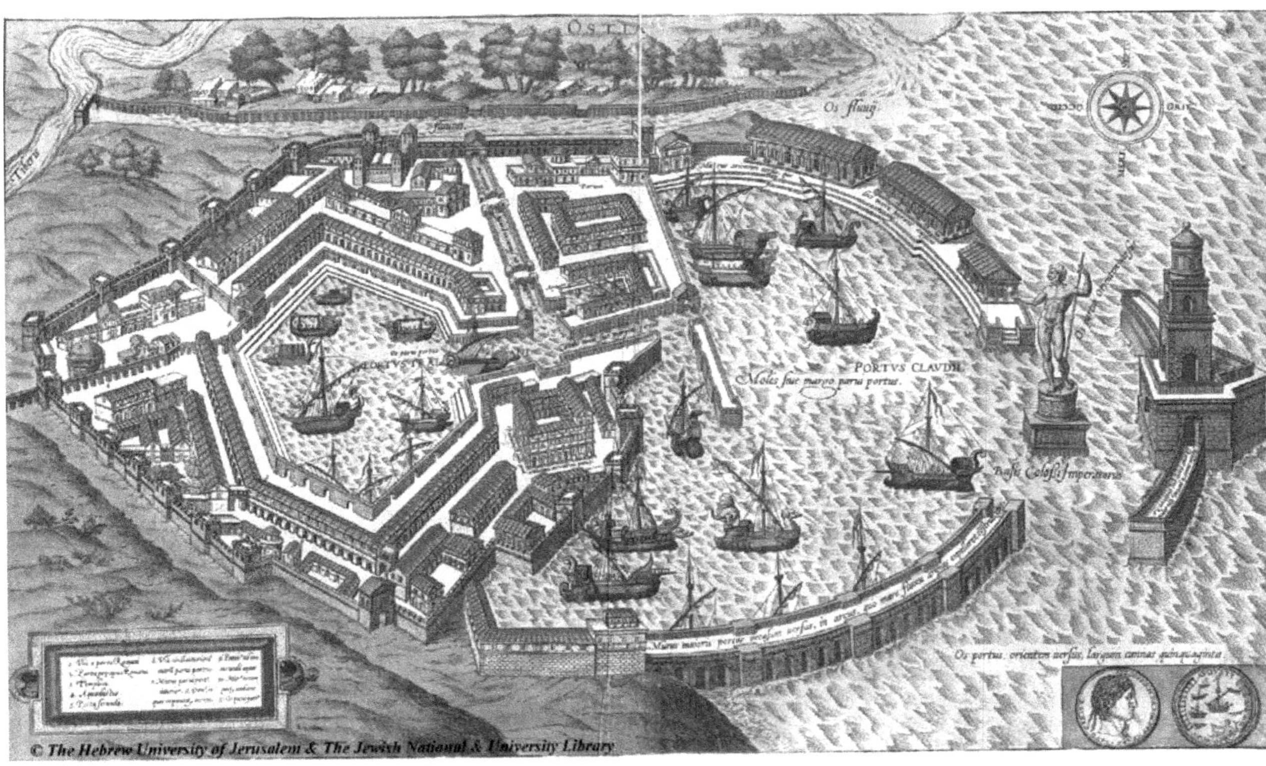

Figure 1.1: Braun and Hogenberg's sixteenth century reconstruction of Portus, near Rome (volume IV, 1588). This representation is typical of the Renaissance antiquarian tradition whereby, in the absence of precise topographical information, speculative graphic reconstructions were reached.

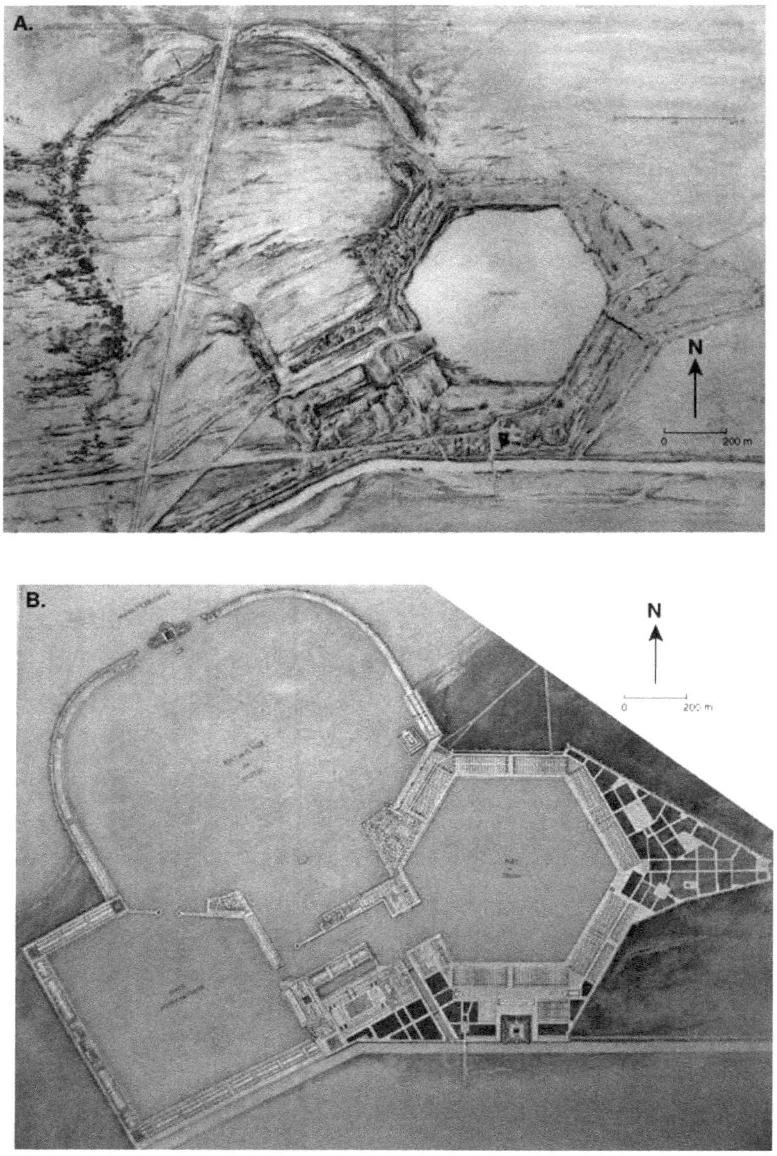

Figure 1.2: A. Silting up of Portus at Ostia as represented by Garrez (1834). **B.** Garrez's reconstruction of the ancient port on the basis of measured topographical features. The plans were inspired by the work of Canini (1827). In general terms, the nineteenth century marked a watershed between the antiquarian tradition of the Renaissance period and modern topography. It was around this time that a number of scholars started to draw the link between harbour silting and coastal progradation to explain the demise of ancient seaports.

1.2.1.2 Pre-World War II – the beginnings of underwater archaeology

While underwater archaeology was given its first major impetus in the 1850s, when low water levels exposed tracts of archaeology in Swiss lakes, it was not until the beginning of the 1900s that the technology had sufficiently caught up with the theory (Keller, 1854; Desor and Favre, 1874; Paret, 1958; Dumont, 2006). Despite cumbersome and rudimentary breathing apparatus, the discovery and partial excavation of shipwrecks at Antikythera (1900-1) and Madhia (1908-13) were archaeological firsts (Pomey and Rieth, 2005). Around the same time, a number of scholars undertook systematic surveys of submerged port structures (for example Negris, 1904a-b; Jondet, 1916; Paris, 1916; Halliday Saville, 1941). In 1923, Lehmann-Hartleben furnished one of the most comprehensive and authoritative early studies on Mediterranean port infrastructures.

During the 1930s, Poidebard pioneeringly transposed aerial photography techniques he had developed working on Syrian limes to the maritime context of the Phoenician coast (Poidebard, 1939; Poidebard and Lauffrey, 1951). Like the British archaeologist Crawford (1886-1957), who also advocated the use of aerial photography in archaeology, Poidebard coupled this aerial reconnaissance with diving surveys to propose harbour reconstructions for numerous Phoenician city-states (Nordiguian and Salles, 2000; Viret, 2000; Denise and Nordiguian, 2004; **Figure 1.3**). The aeroplane, symbol of technical modernity during the 1920s, offered new research avenues in archaeology and even if Poidebard was not the first to realise this, he was a pioneer in developing a clearly defined methodology. Relying on reports from 'hard-hat' divers he surveyed and mapped a number of important coastal sites including Arwad, Sidon, Tyre and Carthage. Although some of his findings have since proved erroneous (Frost, 1971; El Amouri *et al.*, 2005) his work laid the foundations for maritime archaeology as we know it today. His methodical documentation and recording (aerial reconnaissance, cartography and underwater photography) is all the more important for modern scholars working in this region given the destruction of many remains wrought by recent building (Marriner and Morhange, 2005a).

1.2.1.3 Post-war period

Invention of the aqualung revolutionized underwater archaeology after 1945, with a coeval shift in research loci away from the ports themselves to shipwrecks. Over the following decades, many ancient wreck sites were subsequently localized along the coasts of France and Italy (Pomey and Rieth, 2005). A minority of scholars continued to show an interest in ports and harbour infrastructure, notably at Appollonia in Libya (Flemming, 1961, 1965, 1971), Athlit in Israel (Linder, 1967), Carthage in Tunisia (Yorke and Little, 1975; Yorke, 1976; Yorke *et al.*, 1976), Cosa (Lewis, 1973; McCann, 1979; McCann *et al.*, 1987), Pyrgi (Oleson, 1977) and Portus in Italy (Testaguzza, 1964), Phaselis in Turkey (Blackman, 1973a), Kenchreai (Scranton and Ramage, 1967) and Porto Cheli in Greece (Jameson, 1973), and the ancient harbours of Israel (Galili *et al.*, 2002). Underwater archaeologists such as Frost also did much for the development of aquatic survey techniques during the 1960s and 1970s (Frost, 1964, 1966, 1971, 1973). Placing emphasis on precise observation and recording, her surveys were a continuum in the long history of inquiry on the Phoenician coast (Frost, 2000a-b, 2002a-b, 2004, 2005).

In recent decades, progress in coastal archaeology can be attributed to the emergence of new tools (Leveau, 2005). For example, aerial photography, use of the aqualung, underwater robotics, advances in dating and geochemistry have all facilitated a revolution in traditional approaches.

Figure 1.3: Antoine Poidebard was one of the pioneers in underwater and coastal archaeology. He coupled aerial photography and underwater prospecting to propose harbour locations for a number of Phoenicia's coastal sites. The oblique aerial photograph is of the Tyrian peninsula. In 332 BC Alexander the Great linked the offshore island fortress to the continent by means of a causeway. This engineering feat profoundly influenced sediment trapping behind the island and today a tombolo, or sand spit, links the two. Poidebard diagnosed a series of harbourworks on the southern coastal fringe of Tyre. New research suggests that this area is in reality not a harbour basin but rather a drowned urban quarter, which sank into the sea during late Roman times (Photographs from Poidebard [1939] and Denise and Nordiguian [2004]).

1.2.2 History of ancient harbour geoarchaeology

In many ways, geology and archaeology have played a mutually perpetuating role in the development of one another. Stratigraphy and the processes involved in the formation of the sedimentary record have influenced archaeology since the late nineteenth century. In the absence of written documentary evidence, prehistorians turned to earth scientists in an attempt to better understand human-environment interactions. Similarly, archaeology has long been used as a temporal control in the interpretation of the Quaternary geological record (Bridgland, 2000; Westaway *et al.*, in press). In coastal areas it was notably the theme of sea-level changes and archaeological structures that especially focused the attentions of early natural scientists. During the nineteenth century, scholars started to recognise the importance of landscape mobility, which went against the vein of the dominant "fixist" and "catastrophist" theories that had dominated thinking in the preceding centuries (Desjardins, 1876). Lyell (1830), for example, was intrigued by marine borings he observed on pillars in the Roman market of Pozzuoli and ingeniously deduced their potential as a gauge of sea-level changes since antiquity. As attest the drowned remains of Pozzuoli's Portus Julius, this area, which lies in the centre of an active caldera, has been subject to significant rapid sea-level changes during the past two millennia (Morhange *et al.*, 2006a).

Lyell opined that the only way to understand the past was to assume that the earth-modifying processes which occur today also occurred in the past (Ager, 1989, 1995). To clearly illustrate this idea of uniformitarianism Lyell used the three pillars at Pozzuoli as the frontispiece to his *Principles of Geology* (1830) (**Figure 1.4**). Lyell's later text *Geological Evidence of the Antiquity of Man* (Lyell, 1863) was an even more transparent manifestation of the reciprocity between the two disciplines.

The existence of submerged vestiges in Crete had been known since the seventeenth century, but it was not until 1865 that Spratt published a study of coastal movements on the island (Spratt, 1865). Spratt not only identified and measured the altitude of ancient coastlines at various localities around the island, but was also the first to understand, on the basis of his observations of Phalasarna harbour that uplift of the western part of the island dated to the historical period. In 1869, Raulin, having translated Spratt's study, postulated the island had undergone uplift in its western part whilst suffering collapse on its eastern side (Raulin, 1869).

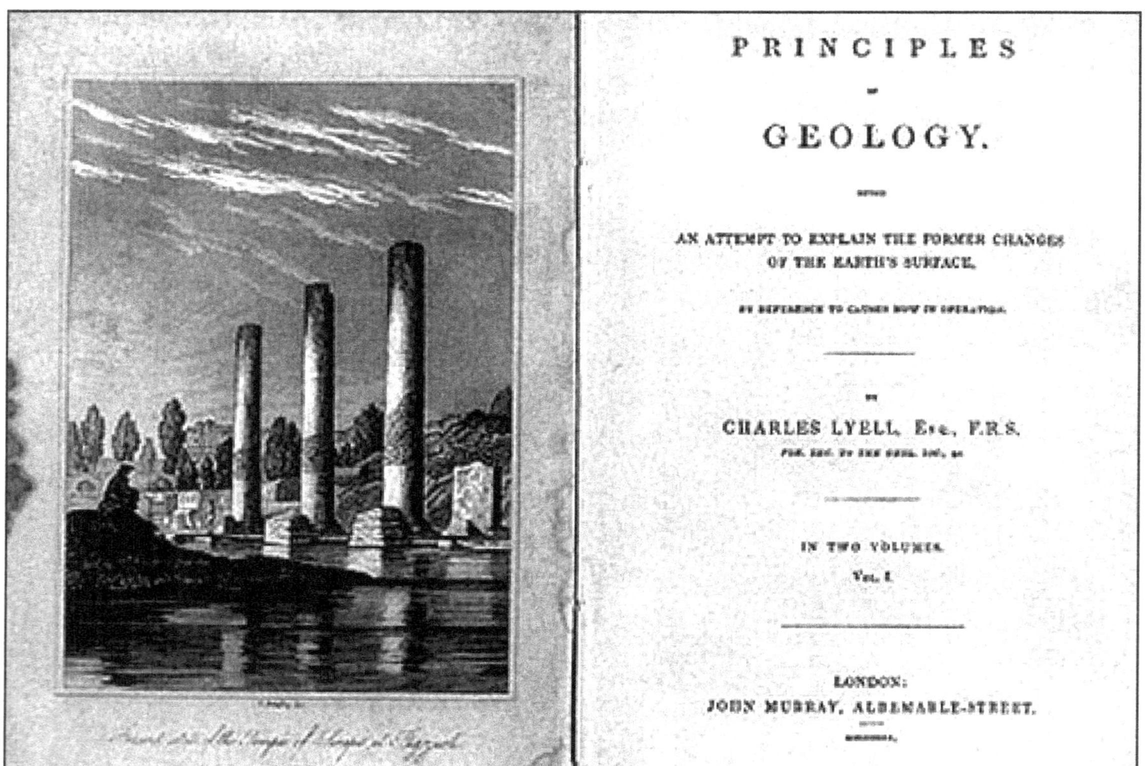

Figure 1.4: The three pillars of the Roman market at Pozzuoli (southern Italy) have become a secular icon of uniformitarianism since their publication in Charles Lyell's Principles of Geology (1830). Lyell argued that the rise and fall of the coastal archaeology showed that the land had undergone significant vertical movements since antiquity.

At the beginning of the twentieth century, debate opposed the Greek Negris (1903a-b, 1904a-b) and the influential French geologist Cayeux (1907, 1914), regarding the position of submerged archaeological vestiges around the coasts of the eastern Mediterranean (Delos, Leucade, Egine, etc.). For Negris observed metric submergence, such as that of Alexandria (Egypt), were not to be linked with localised phenomena but rather a ubiquitous basinwide rise in sea level (Negris, 1921). Cayeux on the other hand, greatly influenced by Suess' fixist postulates for the historical period, affirmed that the general level of the Mediterranean had not varied significantly since antiquity. At Delos, Negris and Cayeux identified archaeological examples of sea-level fixity and instability, and Cayeux interpreted drowned structures as being the result of localised sediment compaction. The latter ignored the well-documented example of Phalasarna's uplifted harbour (western Crete), refusing to concede the possibility of regional scale sea-level changes since antiquity. Cayeux's conclusion is a classic example of scientific rigidness, resulting from the defence of a pre-established theory. Much of this debate subsided from the 1950s onwards, when the advent of radiometric dating techniques ushered in much greater temporal control.

During the 1970s, scholars such as Flemming (1969, 1971, 1978) and Blackman (1973b, 2005), resuscitated ancient harbour research, notably the study of sea-level change as an aid to understanding submerged settlement sites. In a similar vein, Schmiedt (1972) used Roman fish tanks to precisely reconstruct Tyrhennian sea-level changes during the past 2000 years. Following in Poidebard's footsteps, he coupled these data with aerial photographs and geomorphology to propose palaeogeographical reconstructions of numerous ancient harbour sites on Italy's coastline (Schmiedt, 1970, 1975; Acquaro, 1988). These geocentric research currents, influenced by palaeoenvironmental inquiry, served as important precursors to ancient harbour geoarchaeology as it is today practiced. The latter has its origins in two research programs beginning in the early 1990s: the first at Marseilles, France and the second at Caesarea, Israel. Although geoarchaeology was not a new notion, it had rarely been applied to the rich archaeological and sedimentary contexts of ancient harbours. At Caesarea, Reinhardt *et al.* (1994, 1998) and Reinhardt and Raban (1999) undertook micropalaeontological and geochemical studies to investigate the harbour's Roman history. Around the same period, Hesnard (1994), Morhange (1994) and Morhange *et al.* (2001, 2003a) investigated large stratigraphic sections to probe the history of Marseilles' Graeco-Roman basin (e.g. relative sea-level rise, coastal progradation, rapid silting) and anthropogenic impacts since antiquity (**Figure 1.5**).

Figure 1.5: Jules Verne 3, a Roman dredging boat unearthed in Marseilles' ancient harbour. The vessel dates from the first to second centuries AD. The central dredging well measures 255 cm by 50 cm. The wooden hull has been preserved by the enveloping harbour clays (Photograph: Morhange).

The history of geoscience in ancient harbour research is nowhere better documented than at Marseilles, where construction work bordering the *Vieux Port* has revealed sandy beach facies and widespread plastic clays since the nineteenth century (Vasseur, 1911, 1914). Speculative interpretations, void of any systematic field and laboratory analyses, were produced although many of the early authors confused coastline deformation and relative sea-level changes (Vasseur, 1914; Bouchayer, 1931; Duprat, 1935). It was not until after the Second World War that the embryo of a multidisciplinary team came into being, comprising a research quartet of Benoît (an archaeologist), Gouvernet (a sedimentologist), Mars (a malacologist) and Molinier (a botanist). Although much of their work was never fully published, Gouvernet studied beach formations around the city's ancient harbour (Gouvernet, 1948). Notwithstanding the research difficulties they faced, notably presence of the water table and absence of precise chronological constraints, the team was one of the first to recognise the scientific scope of harbour basin geology in reconstructing coastal progradation since antiquity (Morhange, 2001). In the 1960s and 1970s, new excavations led to the major discovery of the Roman harbour quays, although research was marred by fixist dogma and a notable absence of any palaeoenvironmental investigations (with the exception of Pirazzoli and Thommeret's 1973 paper in which they describe and date biological mean sea-level indicators attached to harbour structures). Widespread excavations in the early 1990s set out to remedy many of these earlier ills, with the formulation of a multi-disciplinary research framework (Hesnard, 2004a-b).

From these two early works, a new era of coastal geoscience came to fruition with significant advances in our understanding of ancient harbour sedimentary systems (e.g. Kition-Bamboula, Morhange *et al.*, 2000; Alexandria ad Aegyptum, Goiran, 2001, Stanley and Bernasconi, 2006; Tyre, Marriner *et al.*, 2005; Sidon, Marriner *et al.*, 2006a-b). These multiple studies have shown ancient harbours to be rich geoarchaeological records replete with information on occupation histories, human use and abuse of the Mediterranean environment, natural catastrophes (e.g. tsunami impacts) or even local tectonic mobility (Neev *et al.*, 1987).

At a broader scale, research on deltaic systems has also been an important area of inquiry in understanding anthropogenically forced coastal change, pioneered notably by scholars such as Kraft and Rapp since the 1970s. This work has tended to focus on geomorphological research objects - deltas and their sediments - rather than archaeological layers at a given site, and has been qualified as archaeological geology by Rapp and Hill (1998). Although there is a great deal of overlap, the research can broadly be dissected into two separate *écoles*: (1) A first school has looked to validate the classical sources. For example, in Greece and Ionia Kraft *et al.* (1975, 1977, 1980a-b, 2003, 2005), Bousquet and Pechoux (1980), Bousquet *et al.* (1983), Jing and Rapp (2003) and Pavlopoulos *et al.* (2003) have employed extensive coastal stratigraphy and palaeogeographic reconstructions as tests of the Homeric sources; (2) A second school has focused on the more geocentric themes of delta progradation and sediment failure. These notably include the work of Arteaga *et al.* (1988), Rapp and Kraft (1994), Riedel (1995), Schröder and Bochum (1996), Brückner (1997), Brückner *et al.* (2002, 2005) in Ionia, Fouache *et al.* (2005) and Vött *et al.* (2006a-b) in Greece, and Wunderlich (1988) and Stanley *et al.* (2001, 2004a-b) on the Nile delta. Outside the Mediterranean, ongoing research is looking to better understand progradation of the Mesopotamian plain (Tigris and Euphrates deltas) and mid- to late infilling of the Persian Gulf (Baeteman *et al.*, 2005).

Nowadays, most large-scale coastal archaeological projects look to apply a multi-disciplinary approach at different temporal and spatial scales. Since 1985, harbour archaeology and geoscience workshops have furnished important scientific arenas for multidisciplinary discussion and debate, and attest to a clear growth in this domain as a research focal point (Raban, 1985b; Euzennat, 1987; Raban, 1988; Karageorghis and Michaelides, 1995; Briand and Maldonado, 1996; Pérez Ballester and Berlanga, 1998; Leveau *et al.*, 1999; Morhange, 2000; Zaccaria, 2001; Vermeulen and De Dapper, 2001; Berlanga and Pérez Ballester, 2003; Fouache, 2003; De Maria and Turchetti, 2004a-b; Zevi and Turchetti, 2004a-b; Fouache and Pavlopoulos, 2005; Bochaca *et al.*, 2005; Morhange *et al.*, 2005a).

1.3 Defining ancient harbours

Goiran and Morhange (2003) have formulated a tripartite definition of ancient artificial harbours, which is a practical starting point for geoscientists investigating these rich archives (**Figure 1.6**). Although it may seem paradoxical that the earth sciences should show an interest in these peculiar archives, it is in many ways not surprising that ports have focused the attention of physical geographers. In effect, these unique base-level depocentres lie at the intersection between the natural environment and human societies. This new harbour geoarchaeology exploits the originality of these unique depocentres, whereby ancient environmental protection has ensured their preservation in the geological record.

1.3.1 The basin container. For a traditional archaeologist, an ancient harbour is defined on the basis of its physical 'man-made' attributes such as harbourworks (quays, piers, moles), shipsheds and the general satellite infrastructure which gravitates around this entity. In reality there is great diversity in the nature of these containers, be they natural low-energy environments (coves, estuaries or lagoons) used since the Bronze Age, or (semi)artificial basins resulting from human enterprise (e.g. Phoenician cothons [Carayon, 2005] or the Caesarea's concrete moles [Brandon, 1996],

Portus, Cosa [Felici and Balderi, 1997] and Anzio [Felici, 1993, 2002]).

For the earth scientist, an ancient harbour incorporates not merely the physical man-made edifices - which in many cases no longer exist in a conspicuous form - but also two important geological objects: (1) the sedimentary contents of the basin; and (2) the water column.

1.3.2 The sedimentary contents. Ancient harbours are base-level depocentres. A harbour basin typically comprises a suite of fine-grained sediment layers which translate the low-energy conditions created by the port container. High-resolution study of this unit's bio- and lithostratigraphical contents gives insights into the degree of harbour protection, confinement and degradation of local ecosystems, in addition to local occupation histories. The unique anoxic specificity of this layer also renders it important in the preservation of otherwise perishable archaeological material (leather artefacts, wood etc.).

1.3.3 The water column lies at the intersection between the terrestrial and marine domains. Relative changes in the position of the waterline, coupled with an understanding of sediment budgets and fluxes, are important in comprehending aggradation of the harbour bottom and coastal progradation or deformation. For human societies two reference levels, relative sea level and the harbour bottom, are therefore important in dictating the viability of a harbour basin. These reference levels are dynamic, moving up or down (more rarely) as a function of sea-level changes, tectonic isostasy, sediment fluxes and sediment compaction to dictate the total accommodation space. In the face of rapid accumulation rates, maintaining a navigable water column, or draught depth, engendered clear management strategies (Marriner and Morhange, 2006a).

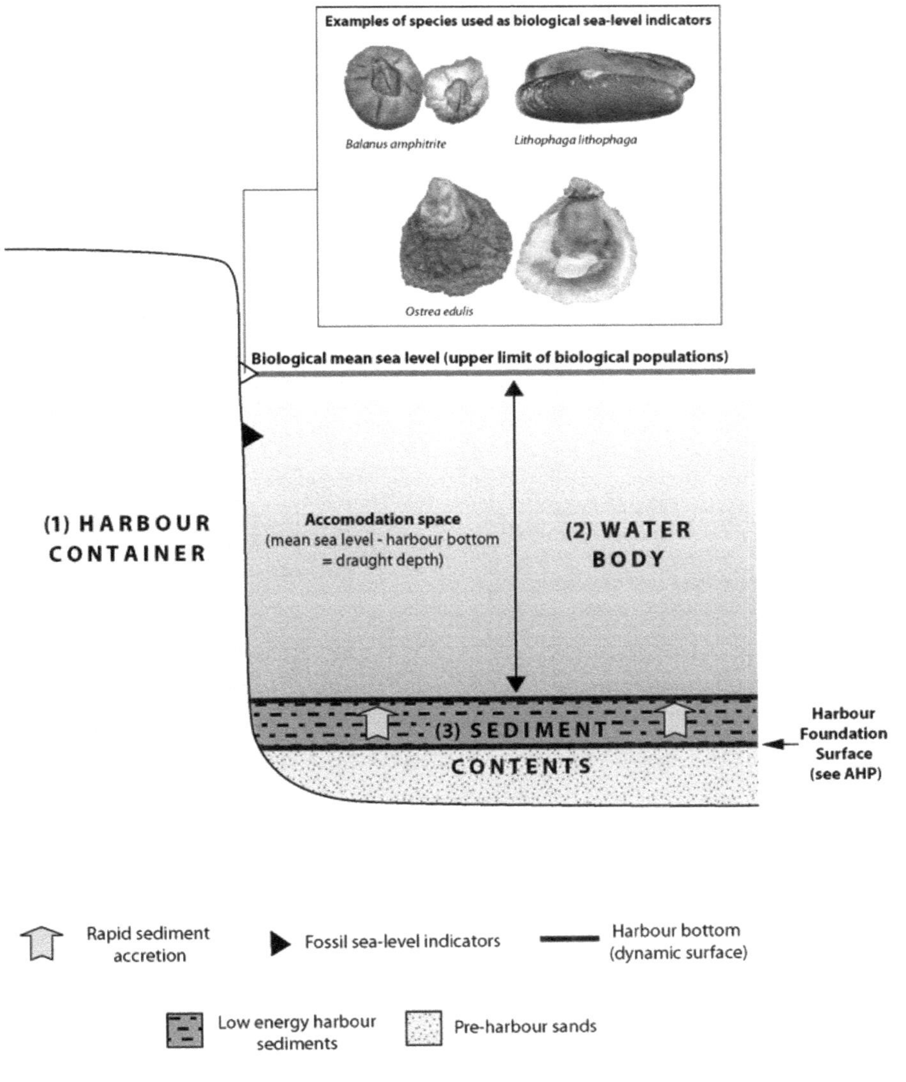

Figure 1.6: Schematic representation of an ancient harbour depositional context with its three defining entities (1) the harbour container; (2) the water body and (3) the sediment contents.

1.4 Ancient harbour typology

Rougé (1966) was the first to propose an archaeological typology on the basis of harbour technology (natural and artificial harbours). A geoarchaeological typology has, however, never been devised. Four main aspects are important in dictating harbour location and design. (1) **Situation**. A port is not built simply anywhere. It forms an interface between the hinterland and sea and its location depends on traffic in these two areas. The margins of large deltas were often attractive locations, e.g. Marseilles for the Rhone valley and Alexandria for the Nile. (2) **Site conditions**. Two types of geological contexts, rocky and clastic coasts, were exploited. During the Iron Age, Greek and Phoenician colonists established harbour complexes in protected rocky bays and coves around the Black Sea and circum Mediterranean. Ports on clastic coasts were generally later, founded by much larger political superpowers with developed harbour technology and engineering *savoir faire* (e.g. Rome and Carthage). When large urban areas lay in inland areas, these types of coastal harbours served as *avant-ports* and operated in tandem with a fluvial port further upriver (e.g. the Ostia complex for Rome). The discovery of hydraulic concrete in the second century BC meant that the Romans were not significantly hindered by environmental constraints, as was typically the case during the Bronze and Iron Ages (Oleson, 1988; Brandon, 1996). (3) **Overall layout**. The layout of a port depends on navigation conditions (winds and waves) and on the types of ship that use it. The size of the ships defines the acceptable wave-induced disturbance and the possible need to build a breakwater providing protection against swell and storms. The number of ships using the port dictates the length of quays and the area of the basins required. (4) **Harbour structures**. The ships' draught defines the depth at the quayside and thus the height and structure of the quay. Locally available materials (wood, stone and mortar) and construction methods define the specific structures for a region and historical period.

In light of these determining factors, there is great variety in harbour types and how these have come to be preserved in the geological record. Four variables determine our chosen harbour typology: (1) distance from the present coastline; (2) position relative to present sea level; (3) geomorphology, and its role in influencing the choice of harbours; and finally (4) taphonomy, or how these ancient ports have come to be fossilised in the sedimentary record (**Figure 1.7**). Seven groups can broadly be discerned (**Figures 1.8 and 1.9**).

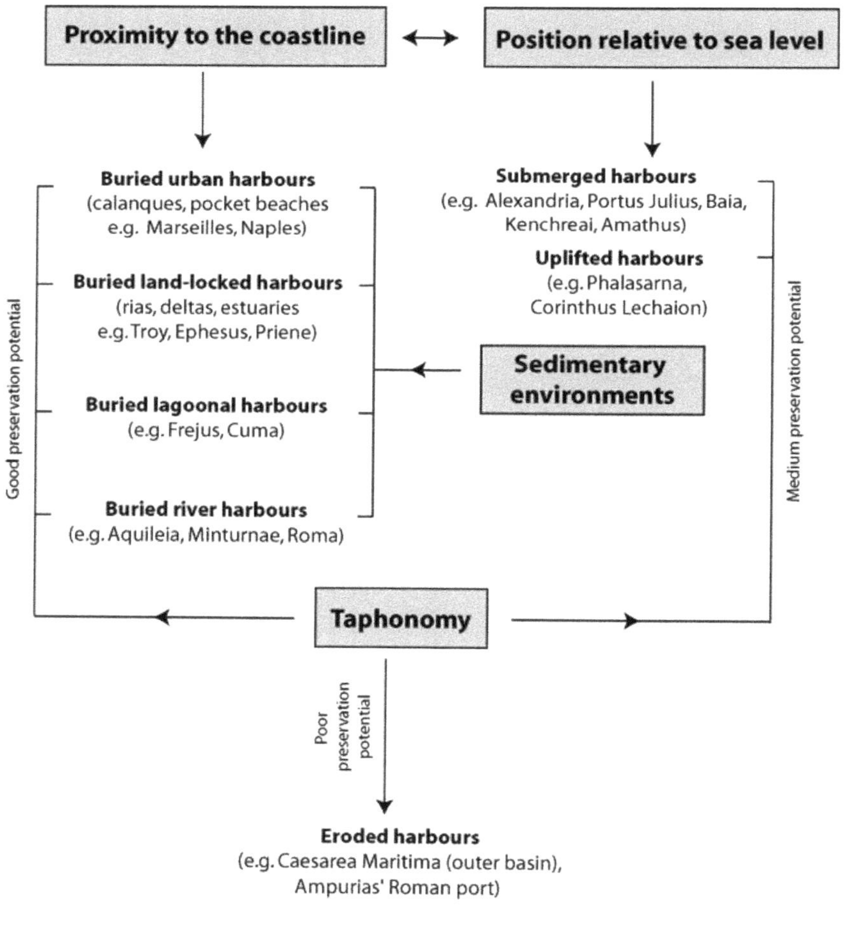

Figure 1.7: Ancient harbour classification based on four variables (1) proximity to the coastline; (2) position relative to sea level; (3) sedimentary environments; and (4) taphonomy.

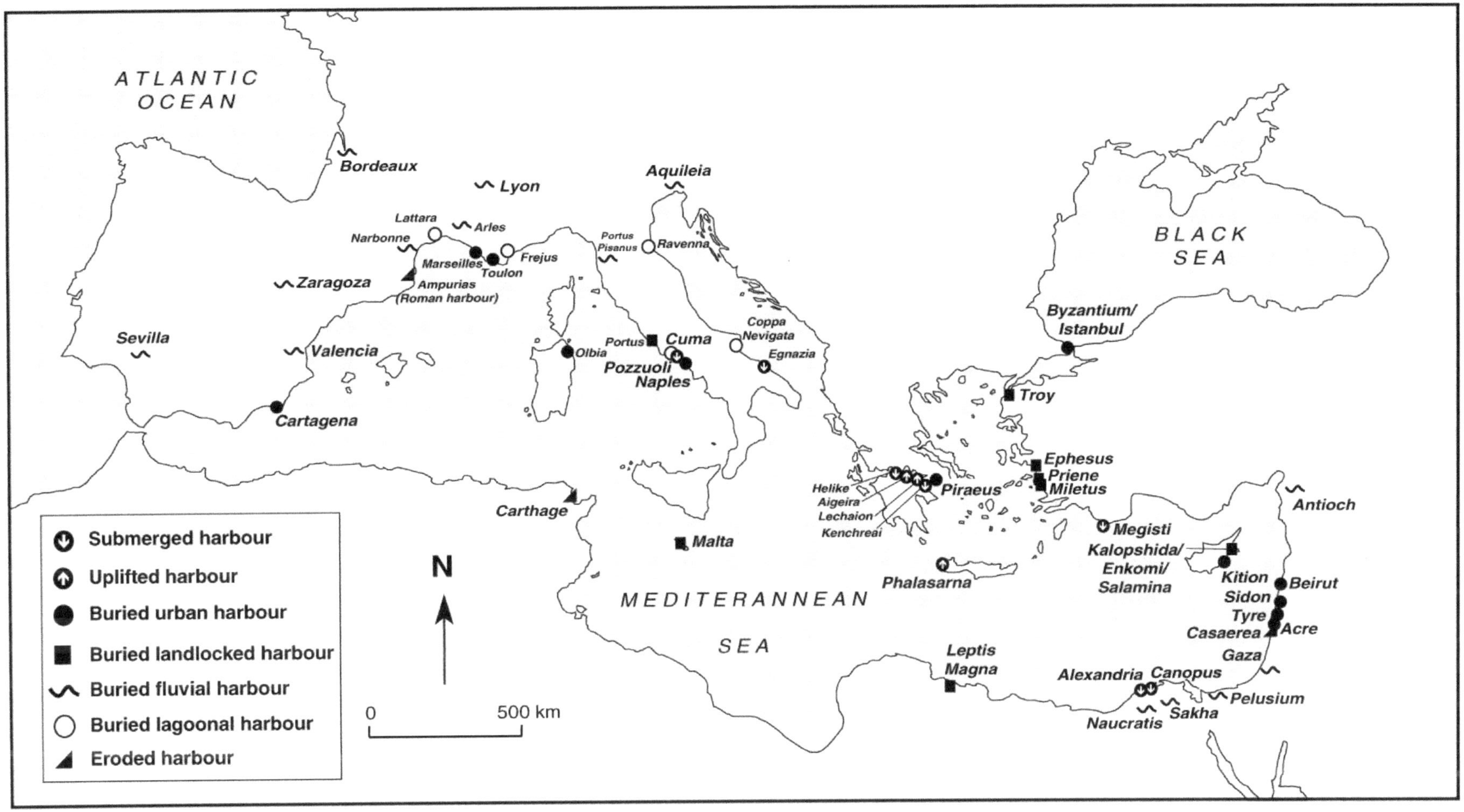

Figure 1.8: Non-exhaustive map of the Mediterranean's ancient habours, grouped according to how they have been preserved in the geological record.

Unstable coasts		Stable coasts				
Submerged harbours	Uplifted harbours	Buried urban harbours	Buried landlocked harbours	Buried fluvial harbours	Buried lagoonal harbours	Eroded harbours
- Alexandria - Baia - Eastern Canopus - Egnazia - Helike - Herakleion - Megisti - Miseno - Pozzuoli	- Aigeira - Kenchreai - Lechaion - Phalasarna - Seleucia Pierea	- Acre - Beirut - Byzantium/Istanbul - Cartagena - Kition Bamboula - Marseilles - Naples - Olbia - Piraeus - Sidon - Toulon - Tyre	- Enkomi - Ephesus - Kalopsidha - Leptis Magna - Miletos - Malta - Priene - Troy - Salamina	- Antioch - Aquileia - Gaza - Minturnae - Narbonne - Naucratis - Ostia (Sardinia) - Pelusium - Rome - Sevilla - Schedia - Thebes - Valencia - Zaragoza	- Coppa Nevigata - Cuma - Frejus - Lattara	- Ampurias - Caesarea (outer harbour)

Figure 1.9: Non-exhaustive list of harbours classified into seven groups.

1.4.1 Unstable coasts

1.4.1.1 Submerged harbours

Freak of the greatly mediatised legend of Atlantis (Collina-Girard, 2001; Gutscher, 2005), drowned cities and harbours have long captured the public imagination (Frost, 1963; Flemming, 1971). Since 18,000 years BP, glacio-eustatic sea-level rise of ~120 m has transgressed important coastal areas of the Mediterranean, submerging numerous Palaeolithic and Mesolithic sites (Shepard, 1964; Masters and Flemming, 1983; Kraft *et al.*, 1983; Shackleton *et al.*, 1984; Galili *et al.*, 1993a-b; Flemming, 1996, 1998; Collina-Girard, 1998; Petit-Maire and Vrielinck, 2005; **Figure 1.10**). In the Aegean, many islands, such as Kerkira, Euboea and the northern Sporadhes, were connected with the mainland and most of the Cycladic islands were joined together to form a Cycladic semi-peninsula (van Andel and Shackleton, 1982). In southern France, the partially drowned Palaeolithic cave of Cosquer is one of the best Mediterranean examples of human occupation of the continental margin and post-glacial sea-level rise (**Figure 1.11**). Painted horses dated ~18,000 BP have been partially eroded at current sea level testifying the absence of any sea-level oscillation higher than present. Transgression of the continental platform gradually displaced coastal populations landwards until, around 6000 BP, broad sea-level stability meant that human societies started to sedentarise along present coastlines (van Andel, 1989). Since this time, coastal site and port submersion can be linked to two different geological factors: (1) tectonic mobility (e.g. subsidence in eastern Crete); and/or (2) sediment failure (e.g. Alexandria, Menouthis and Herakleon at the margin of the Nile delta).

At Alexandria, public interest was roused in the mid-1990s with the spectacular images of J.-Y. Empereur's team surfacing statuary and megalithic blocks, many of them attributed to Pharos' celebrated lighthouse (Empereur and Grimal, 1997; Empereur, 1998; Hairy, 2006). Concomitant underwater research by Goddio *et al.* (1998) and Goddio and Bernand (2004) brought to light drowned harbourworks in the city's eastern harbour, ~5 m below present MSL and covered by a thin layer of sand (**Figure 1.12**). The coastal instability of the Alexandria sector has been attributed to seismic movements (Guidoboni *et al.*, 1994), destructive tsunami waves (Soloviev *et al.*, 2000; Goiran, 2001) and Nile delta sediment loading (Stanley *et al.*, 2001; Stanley and Bernasconi, 2006). Following an opulent Graeco-Roman apogee, recent research suggests that the demise of Alexandria's ports was centred on the eighth to ninth centuries AD, during which time tsunami impacts and dramatic sea-level rise severely damaged harbour infrastructure (Goiran, *et al.*, 2005).

To the east of Alexandria, in the Abu Qir Bay area, Stanley *et al.* (2001, 2004a) have elucidated the submergence of two ancient Greek cities, Herakleion and Eastern Canopus, after the eighth century AD (Stanley *et al.*, 2004a-b). The two seaport cities, which lay at river mouths on the delta coast, have been submerged and drowned by ~8 m during the past 2500 years. The team attribute this subsidence to (1) 4-5 m of RSL rise (eustasy and sediment compaction), and (2) 3-4 m of episodic land failure by sediment loading.

The southwestern coast of the Gulf of Corinth, Greece, lies in a region of rapid tectonic uplift and extension. In 373 BC, the Greek city of Helike and its harbour, built on a Gilbert-type fan delta, were destroyed by an earthquake and submerged (Kiskyras, 1988; Soter and Katsonopoulou, 1998). Using bore-hole datings, Soter (1998) estimates that Helike delta subsided by at least 3 m during the earthquake. The opposition between gradual regional uplift (the Gulf of Corinth rift) and local co-seismic subsidence apparently resulted in a relatively small absolute displacement of the delta during the Holocene.

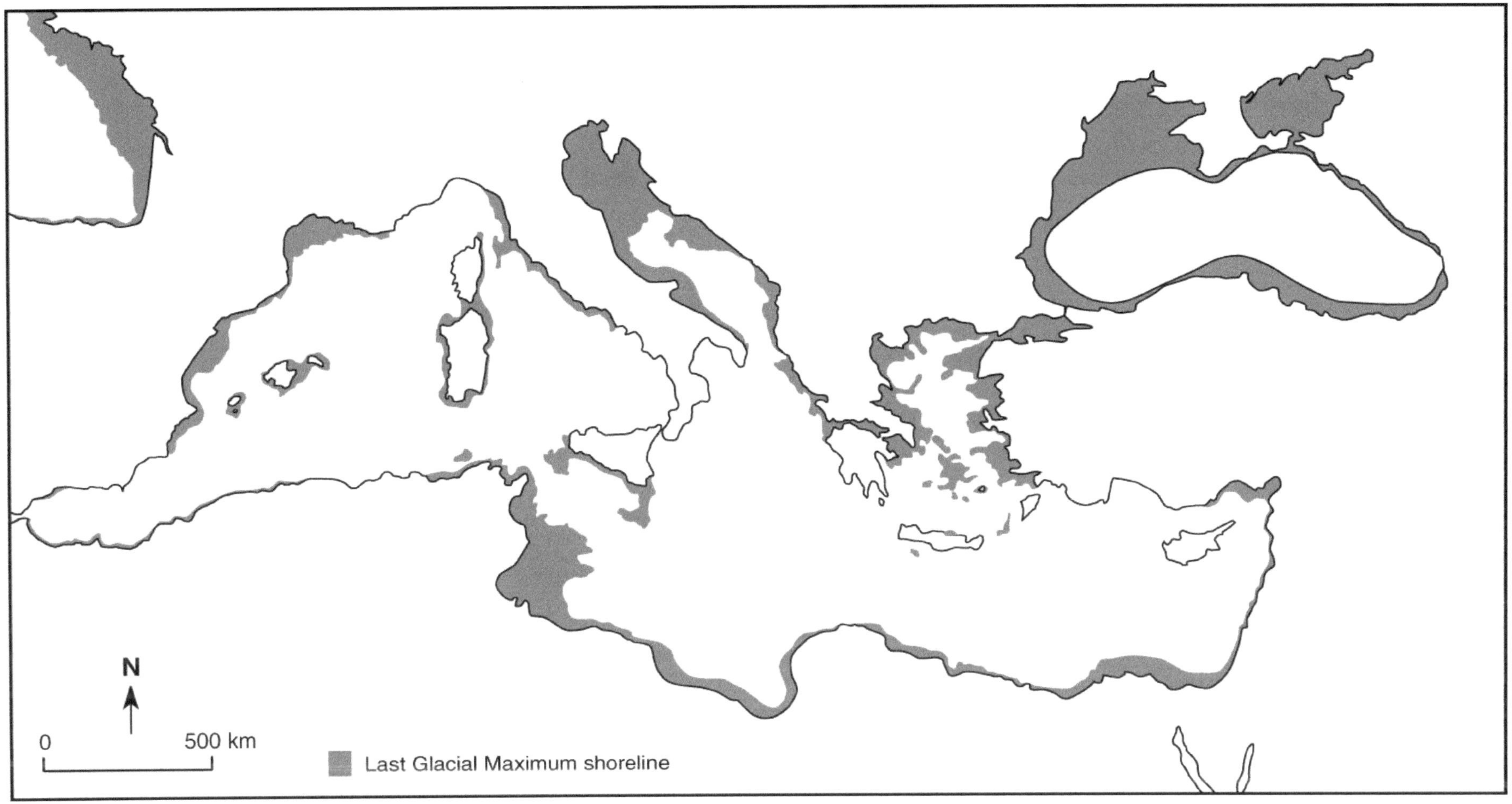

Figure 1.10: Last Glacial Maximum shoreline and transgression of the Mediterranean's coastal margin since 18000 BP, when sea level was around -120 m below present (after Bracco, 2005). The drowning of a great number of Palaeolithic sites means that our understanding of coastal prehistoric human groups in the Mediterranean is relatively poor.

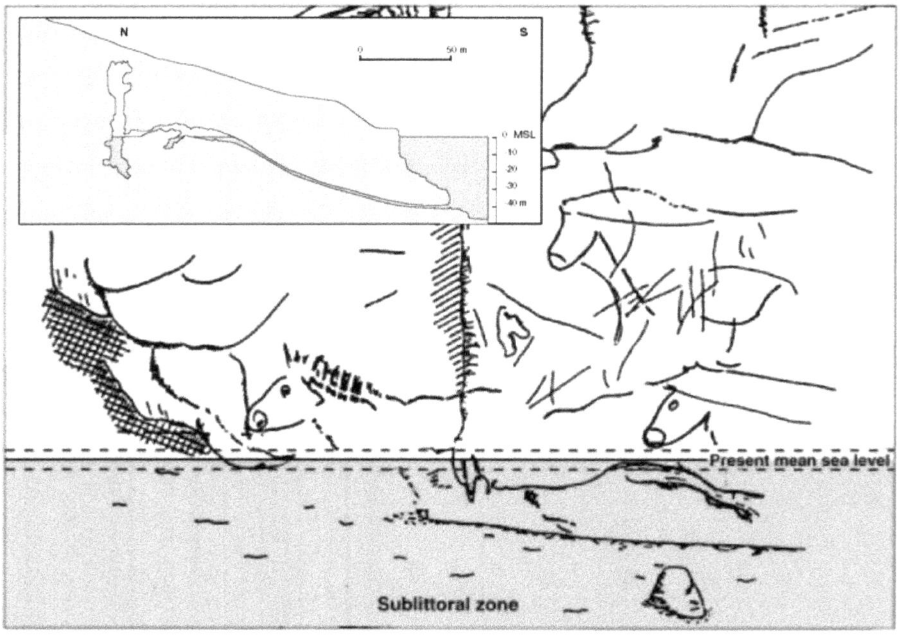

Figure 1.11: The Cosquer Cave is a French Palaeolithic painted and engraved cave (27,000-18,500 BP). The partially drowned cave has formed in the Urgonian limestones of Cap Morgiou, near Marseilles. The entrance, 37 m below sea level, was submerged around 7000 BP (Sartoretto *et al.*, 1995). Archaeological studies indicate that the cave was used as a refuge around 27,000 and 18,000 BP. The partially eroded cave paintings indicate that there has been no sea level higher than present during the Holocene.

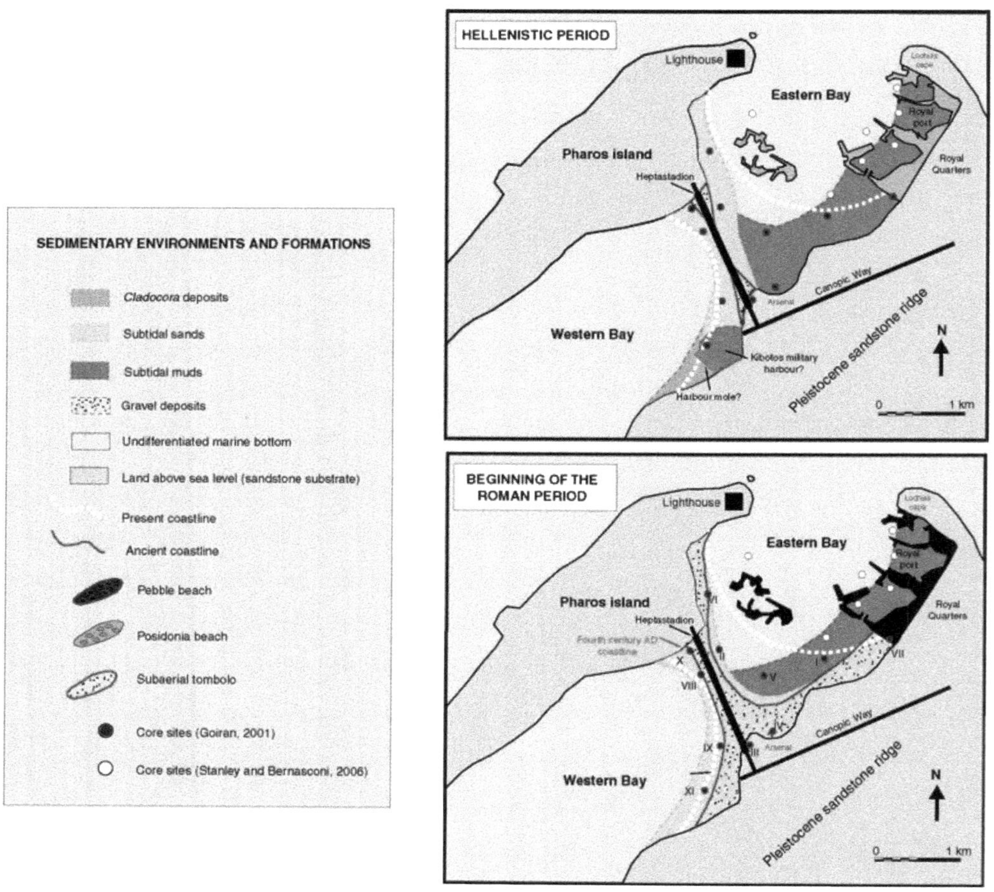

Figure 1.12: Geomorphological reconstruction of Alexandria's ancient harbour by Goiran *et al.* (2005). The Heptastadion, constructed by Alexander the Great in 332 BC, separated Alexandria's bay into two separate coves. The Eastern Bay contained a number of separate port basins. The Royal port is presently drowned some 5 m below present sea level. Goiran (2001) has reconstructed the Holocene history of Alexandria's ancient harbours using coastal stratigraphy and numerous geoscience tools. Corings by Stanley and Bernasconi (2006) from the centre of the Eastern Bay show no diagnostic harbour sediments, suggesting that this area lay outside the main artificial basins.

The Phlaegrean Fields volcanic complex in southern Italy presents a very different crustal context. The ancient ports of Miseno, Baia and Portus Julius (Pozzuoli) are presently drowned ~10 m below mean sea level (Dubois, 1907; Gianfrotta, 1996; Scognamiglio, 1997; **Figure 1.13**). These sites are located inside a caldera and are good examples of shoreline mobility attributed to volcanism and faulting. Recently, scholars have reconstructed a complex history of post-Roman relative sea-level changes evidencing three metric scale crustal oscillations at both Pozzuoli (Morhange *et al.*, 2006a) and Miseno (Cinque *et al.*, 1991) between the fifth and fifteenth centuries AD. Research suggests that inflation-deflation cycles are linked to a complex interplay of deep magma inputs, fluid exsolution and degassing (Todesco *et al.*, 2004) that have profoundly influenced the position of the coastal archaeology. The Roman harbour of Egnazia is another good example of a drowned seaport on the south-eastern coast of Italy. During the past 2600 years a RSL rise of 6 m has been measured using the archaeology, although the exact mechanisms behind this subsidence remain unclear (Auriemma, 2004).

On Castellorizo island, Greece, the ancient city and harbour remains of Megisti have been drowned -2.5 to -3 m below present MSL. Pirazzoli (1987a) has linked this to a gradual 1.5 to 2.0 mm/yr subsidence of the Lycian coast since antiquity. This trend continues today as attest recently abandoned houses on the fringe of the Mandraki basin.

1.4.1.2 Uplifted sites

By contrast, uplifted harbours resulting from crustal movements are much rarer. The best geoarchaeological evidence for this harbour type derives from the Hellenic arc, an area long affected by the complex tectonic interplay of the African and Anatolian plates (Stiros, 2005).

In western Crete, Pirazzoli *et al.* (1992) have ascribed a 9 m uplift of Phalasarna harbour, founded in the fourth century BC, to high seismic activity in the eastern Mediterranean between the fourth to sixth centuries AD (Stiros, 2001). Sometimes referred to as the Early Byzantine Tectonic Paroxysm (Pirazzoli, 1986; Pirazzoli *et al.*, 1996), this episode is concurrent with a phase of Hellenic arc plate adjustment. Synchronous uplift (1-2 m) has been observed in Turkey (the uplifted harbour of Seleucia Pierea, Pirazzoli *et al.*, 1991), Syria (Sanlaville *et al.*, 1997) and sectors of the Lebanese coastline (Pirazzoli, 2005; Morhange *et al.*, 2006b). Phalasarna's ancient harbour sediment archive is of particular interest for the geosciences as it has trapped and archived tsunami deposits inside the basin (Dominey-Howes *et al.*, 1998).

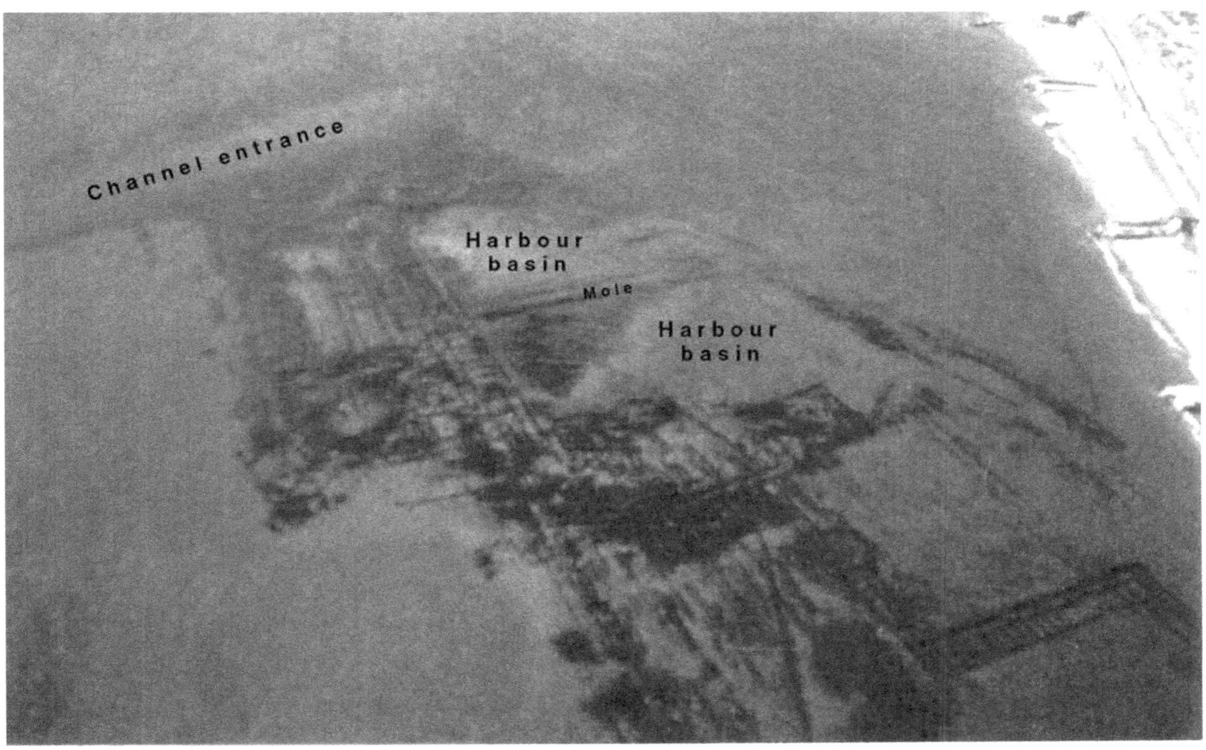

Figure 1.13: Pozzuoli's drowned harbour remains, presently drowned ~10 m below mean sea level. The site lies inside a caldera, where shoreline mobility is attributed to volcanism and faulting. Recently, scholars have reconstructed a complex history of post-Roman relative sea-level changes evidencing three metric crustal oscillations at both Pozzuoli (Morhange *et al.*, 2006) and Miseno (Cinque *et al.*, 1991) between the fifth and fifteenth centuries AD. Research suggests that inflation-deflation cycles are linked to a complex interplay of deep magma inputs, fluid exsolution and degassing.

The Gulf of Corinth is a deep inlet of the Ionian Sea separating the Peloponnese from western mainland Greece. It is one of the most tectonically active and rapidly extending regions in the world (6–15 mm/year), and surface features (uplifted shorelines, reversal of drainage patterns, earthquake-induced landslides) and archaeology are clearly associated with seismic activity (Papadopoulos *et al.*, 2000; Koukouvelas *et al.*, 2001; Kokkalas and Koukouvelas, 2005). Several ancient harbours are known from inside this deep inlet, including Helike (submerged, see above), Aigeira and Lechaion (both uplifted).

At Aigeira, an artificial Roman harbour was built ~100 AD to 250 AD (Papageorgiou *et al.*, 1993). The southern Gulf of Corinth is void of any natural roadstead due to shoreline regularisation by coastal progradation. Stiros (1998, 2005) has described archaeological sea-level indicators attesting to an uplift of ~4 m. Biological and radiometric evidence indicates that this relative sea-level drop was co-seismic in origin and associated with an earthquake around 250 AD.

Interest in Lechaion's harbour organisation and environments, the western seaport of ancient Corinth, dates back to the work of Paris (1915). The harbour was particularly active during Archaic times, a period of Corinthian expansion to the Ionian Sea and southern Italy. Crustal uplift led to the rapid silting up of the harbour basins (Stiros *et al.*, 1996). Bioconstructions (upper limit of *Balanus* sp. or *Lithophaga lithophaga*) on quays at ~1 m above present MSL have been dated to 2500 BP (400-100 cal. BC).

The archaeological data from subsided and uplifted coasts contrast with stable tectonic contexts where harbour infrastructure lie just a few centimetres below present sea level (Pirazzoli, 1976; Morhange *et al.*, 2001). These are the result of relatively minor millennial-scale sea-level rise in response to regional glacio-hydroisostasy. A good example of this derives from the Tyrrhenian coast of Italy, where Roman fish ponds lie ~50 cm below present mean sea level (Schmiedt, 1972; Leoni and Dai Pra, 1997; Antonioli and Leoni, 1998). Lambeck *et al.* (2004) revised these measured data and corrected them on the basis of modelled glacio-hydro isostatic adjusment; they obtain 135 ± 7 cm of sea level change since the Roman period. Compared to direct measurements made on fossil bioindicators, these modelled adjustments appear to overestimate RSL changes during the past 2000 years. For example, well-studied mid-littoral fossil algal ridges in the south of France fit tightly with the 50 cm sea-level rise since Roman times (Pirazzoli, 1976; Laborel *et al.*, 1994; Vella and Provansal, 2000).

1.4.2 Buried harbours

After 6000 BP, slowdown in sea-level rise allied with high sediment supply led to accelerated coastal progradation. Over many millennia, this progradation explains the burial and loss of countless ancient harbours. Human modification of fluvial watersheds (deforestation, agriculture) from the Neolithic period onwards accelerated soil erosion of upland and lowland areas; since the beginning of the Christian era, research has identified three important detritic crises during the Augustean period (~2000 years ago), late Antiquity and the Little Ice Age, translated geologically by pronounced periods of coastal and deltaic progradation (Provansal *et al.*, 1995; Arnaud-Fassetta and Provansal, 1999; Vella *et al.*, 2005). Four different types of buried harbours can be identified.

Figure 1.14: Greek and Roman port remains were discovered in the Bourse area of Marseilles during the 1960s. These vestiges, in the heart of the modern city, have since been converted into museum gardens. The port buildings that have been preserved date from the end of the first century AD but these were preceded by earlier constructions in an area known as "the port's horn". Prone to silting, the area was abandoned from late antiquity onwards. The modern *Vieux Port* is clearly discerned in the background and attests to progradation of the coastline landlocking the ancient seaport.

1.4.2.1 Buried urban harbours

Beirut, Byzantium/Istanbul, Cartagena (García, 1998; Del Carmen Berrocal Caparrós, 1998), Kition Bamboula, Marseilles, Naples, Olbia, Piraeus, and Toulon are examples of buried urban harbours *par excellence*. Whilst the cities' port areas are still fully functional today, the heart of the ancient hubs lay beneath the modern city centres. Rapid rates of sediment accumulation of 10-20 mm/yr during the Roman and late Roman periods have led to the silting-up of the basin fringes, which have been gradually dislocated seawards. Two processes are important in explaining these coastal deformations: (1) high sediment supply linked to fluvial inputs, erosion of clay buildings, urban runoff and use of the basin as a waste deposit (cultural inputs); and (2) human activity has also directly accentuated coastal progradation with a regularisation of coasts since the Bronze Age (**Figures 1.14-1.16**).

On the Phoenician coast, good examples of buried harbours include Sidon and Tyre (Marriner *et al.*, 2005; Marriner *et al.*, 2006a-b). Although the present basins are still in use today, some 5000 years after their foundations, their surface areas have been reduced by almost half. Such a geomorphological evolution offers a great multiplicity of research possibilities in areas where, paradoxically, very little is known about the maritime history.

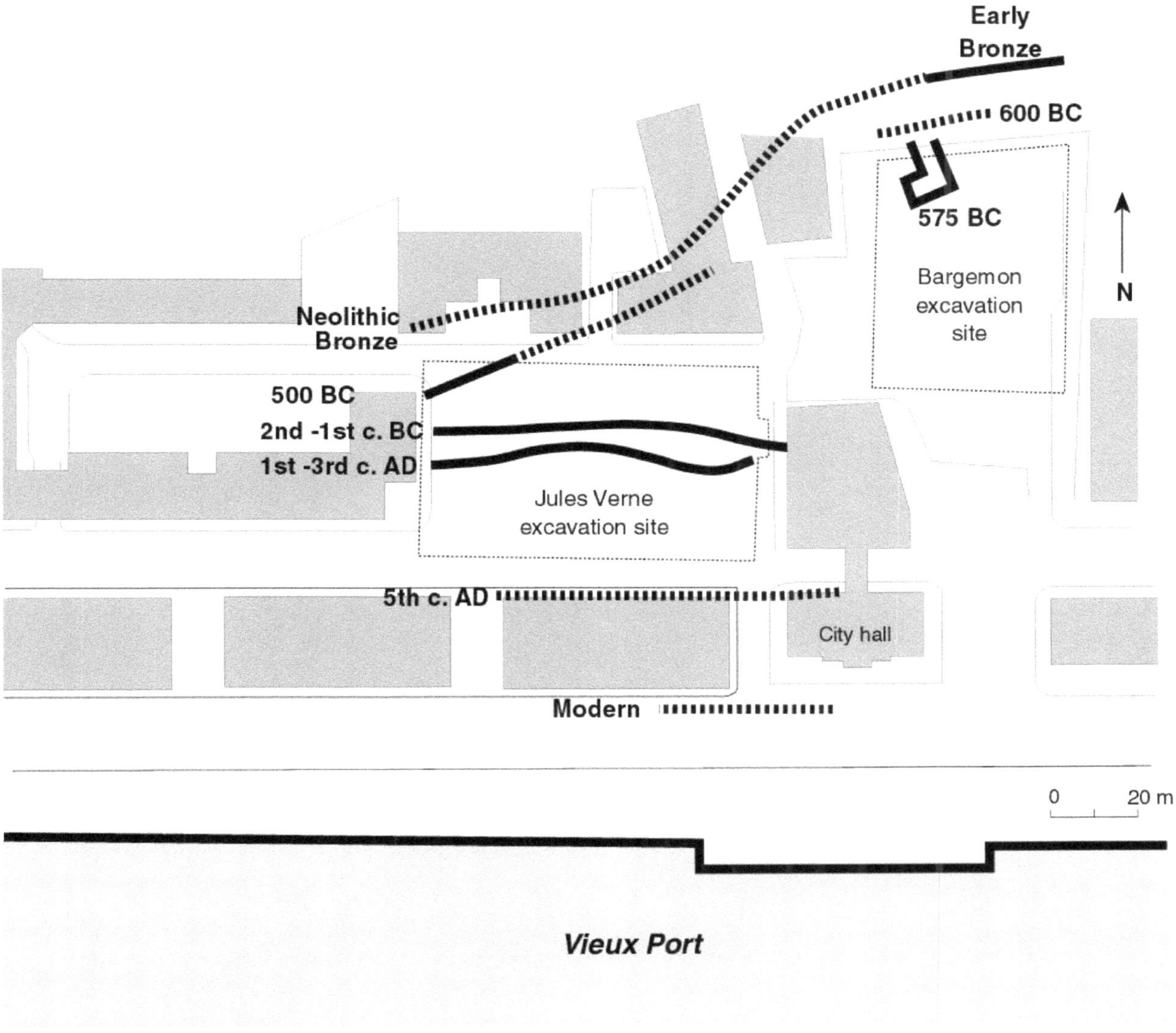

Figure 1.15: Progradation of Marseilles' northern harbour coastline since the Neolithic. The coastlines were studied in section during archaeological excavations and each historic coastline was defined on the basis of its archaeological remains (quays, jetties). Note the progressive regularisation of the coast of this cove. After Morhange *et al.* (2003).

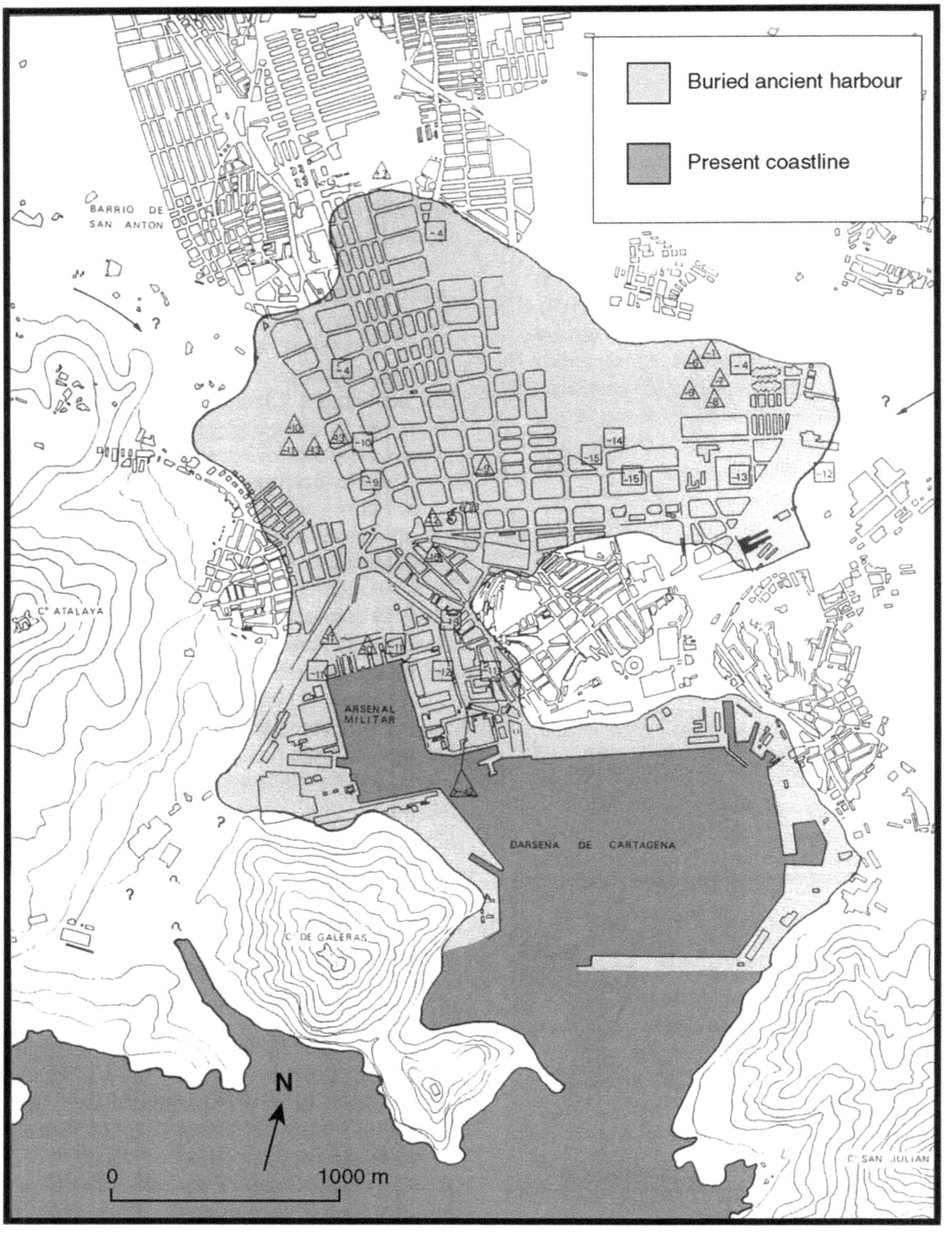

Figure 1.16: Cartagena's buried urban seaport (from García, 1998). Cartagena was founded ~230 BC by Carthaginian general Hasdrubal. Coastal stratigraphy and archaeology indicate that the heart of the ancient seaport is presently located beneath the city centre.

Although not seaports *sensu stricto*, the artificial spits (tombolos) of Tyre and Alexandria are unique geological examples of anthropogenically forced sedimentation (**Figure 1.17**). These rare coastal features are the heritage of a long history of natural morphodynamic forcing and pluri-millennial human impacts. In 332 BC, following a long and protracted seven-month siege of Tyre, Alexander the Great's engineers cleverly exploited the city's atypical geological context to build a causeway and seize the island fortress; this edifice served as a prototype for Alexandria's Heptastadium built a few months later. Both causeways profoundly and pervasively deformed the natural coastline and entrained rapid progradation of the spits (Nir, 1996; Goiran *et al.*, 2005). These two examples underline the impact of human societies in accentuating coastal deformation since the Bronze Age. Other examples of anthropogenically forced tombolos include Clazomenae (Ionia), Apollonia Pontica (Sozopol, Bulgaria) and possibly Orbetello (Italy).

The amount of sediment transported by river systems is important in dictating the scale of coastal progradation. At Marseilles, Toulon or Kition-Bamboula, coastal deformation has been relatively minor, in the order of a few hundred metres. In the Bosporus strait, modest sedimentary inputs have resulted in the long-term viability of Istanbul (Byzantium/Constantinople), an important crossroads between Europe and Asia (Dark, 2004). Modest sediment inputs means that many partially silted up ancient seaports have remained important trade centres since their foundations during antiquity. This harbour type is contrasted by our next group, buried landlocked harbours.

1.4.2.2 Buried landlocked harbours

The buried landlocked harbour type is characterised by kilometric coastal progradation, as epitomised for example by Ionia's ancient ports (Troy, Miletus, Priene or Ephesus; Brückner, 1997). Such rapid deltaic progradation is linked to two factors (1) broad sea-level stabilisation around present level since 6000 BP (**Figure 1.18**) and (2) the unique morphology of these palaeo-rias, which correspond to narrow, transgressed grabens with limited accommodation space (Kayan, 1996, 1999). For example, the delta front of the Menderes ria has prograded by ~60 km since the maximum marine ingression at 7000 BP (Schröder and Bochum, 1996; Müllenhoff *et al.*, 2004; **Figure 1.19**).

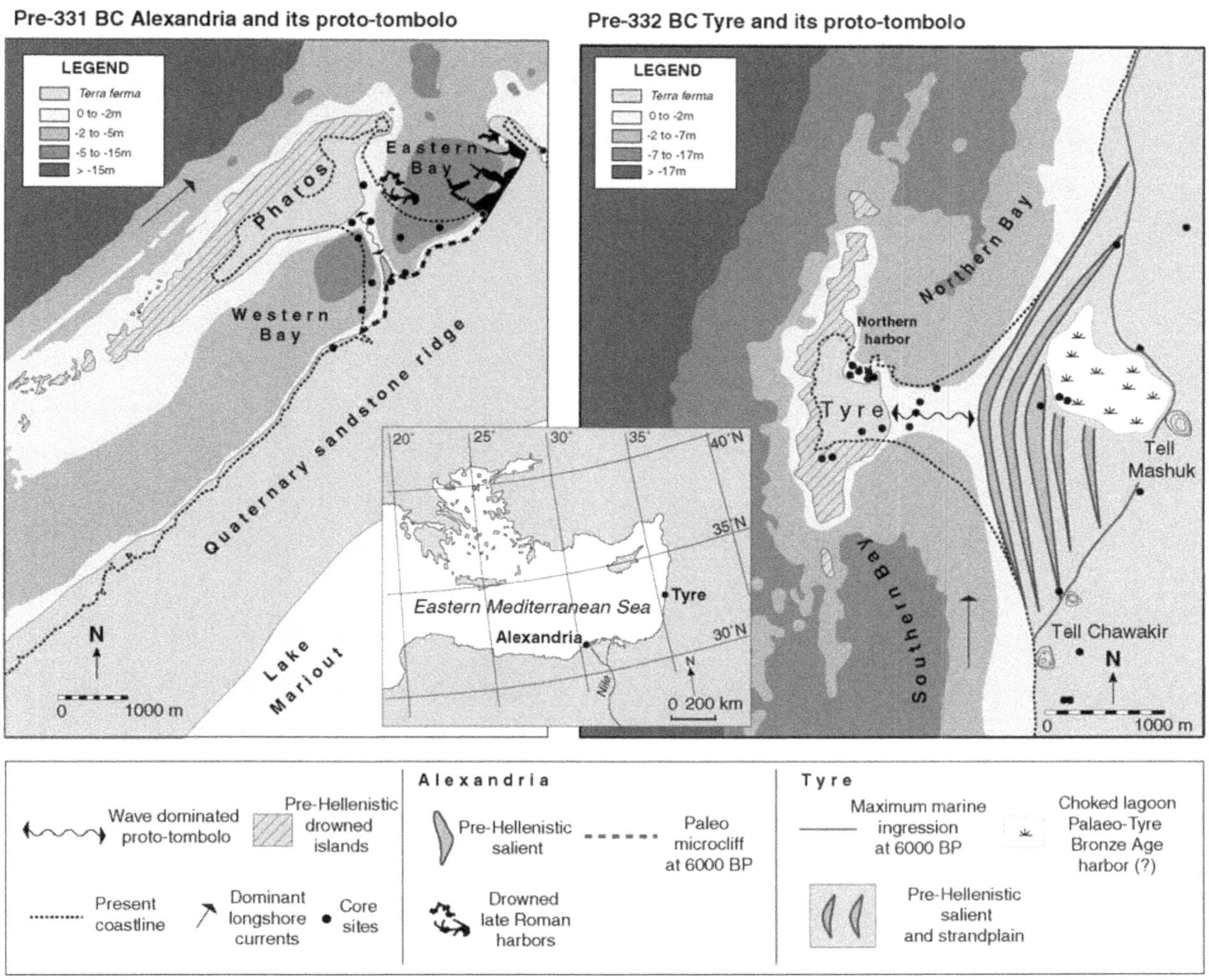

Figure 1.17: Morphodynamic evolution of Alexandria and Tyre's isthmuses since antiquity. Research at both sites has elucidated a proto-tombolo phase within 1-2 m of sea level by Hellensitic times (Goiran, Marriner and Morhange, unpublished data). This proto-tombolo phase greatly facilitated the construction of the two artificial causeways.

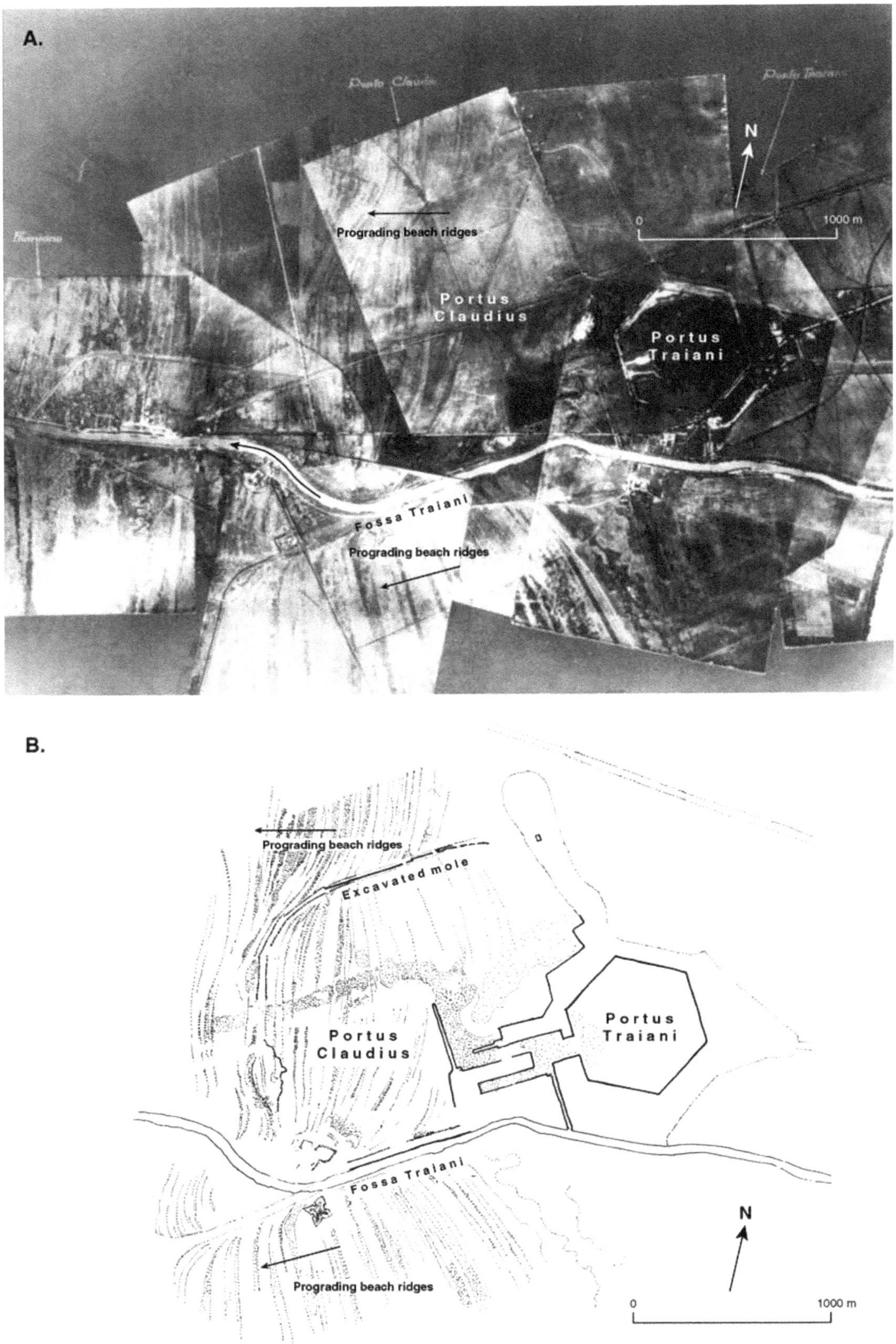

Figure 1.18: A. Early twentieth century photographic mosaic of Portus. B. Geoarchaeological interpretation of the site (after Giuliani, 1996). Portus Claudius has completely silted up since antiquity, whilst the artificial Portus Traiani today comprises a freshwater lake. Progradation of the beach ridges is clearly evidenced by the aerial photography.

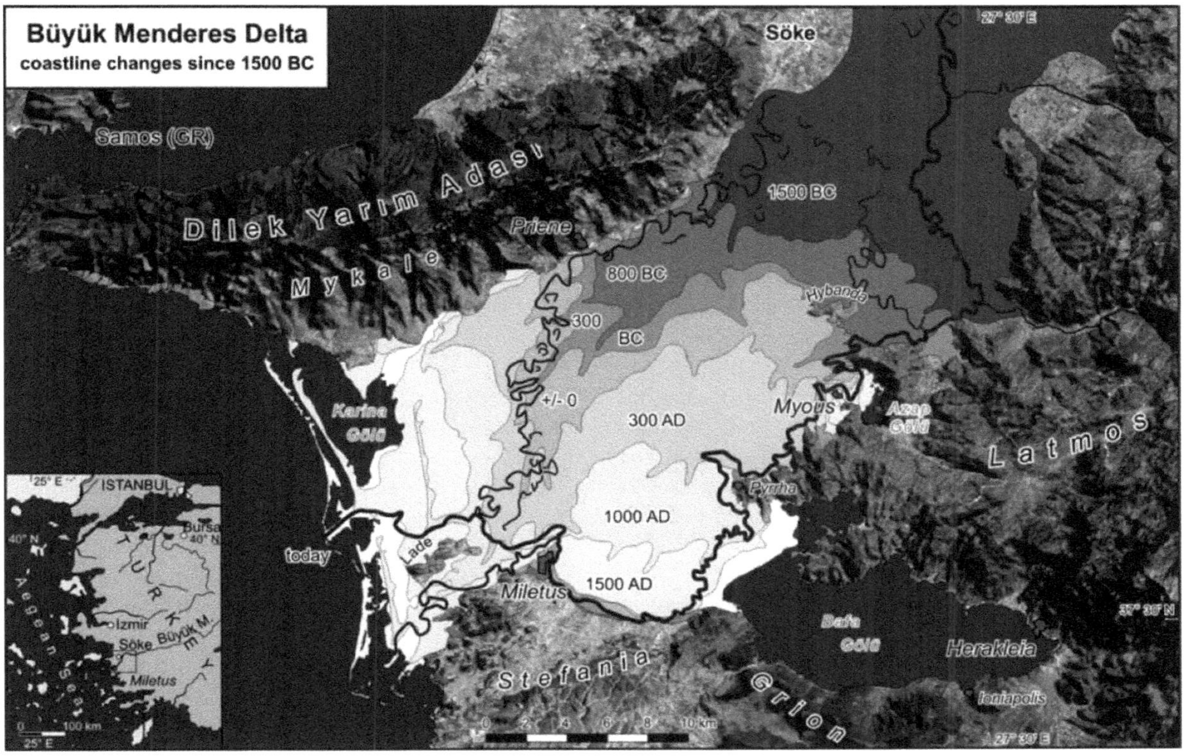

Figure 1.19: Coastal progradation of the Menderes delta and the landlocking of Priene, Miletus and Heracleia's ancient seaports (from Müllenhoff *et al.*, 2004).

The best studied examples include Troy (Kraft *et al.*, 2003), where the harbour areas were landlocked by 2000 BP, and also Ephesus, Priene and Miletos (Brückner, 1997). Brückner's research at Ephesus provides a good illustration of harbour displacement, or 'race to the sea', linked to rapid shoreline progradation. The ancient city's first artificial harbour, near Artemision, silted up as early as the sixth century BC, during a period of rapid delta growth. A second harbour was subsequently built to the west in the fifth century BC, before relocation of the landlocked city at the end of the third century BC.

In Cyprus, Devillers (2005) has elucidated the marine flooding of the Gialias ria around 8000 BP and its ensuing deltaic progradation after ~6000 BP. Using coastal stratigraphy, geomorphology and teledetection, he links a suite of coastal environments (alluvial plain, lagoons, coastal ridges) with an easterly migration of the coastline. Human societies constantly adapted to this changing coastal environment as illustrated by the geographical shift of four ancient harbours: Early/Middle Bronze Age Kalopsidha, Middle/Late Bronze Age Enkomi, Graeco-Roman Salamina and Medieval Famagusta.

Many of the southern Levant's Bronze Age proto-harbours have been investigated by Raban (1987a-b, 1988) and Marcus (2002a-b). Raban elucidated a series of natural estuarine harbours along the presently rectilinear coast of Gaza and Israel. This natural but short-lived protection was eventually compromised by coastal progradation and barrier accretion on the wave-dominated coast to the east of the Nile delta (Raban, 1990). After 2000 BC, many of the coast's Bronze Age tells were no longer viable, eventually abandoned due to silting up of their transport hubs and isolation from maritime trade routes.

The landlocked harbours of the Maltese islands are an example of partially infilled rias (Gambin, 2004). Coastal progradation explains the landlocking and dislocation of many anchorage havens.

Around 6000 BP, the maximum marine ingression created an indented coastal morphology throughout much of the Mediterranean. Over the next few millennia, these indented coastlines were gradually infilled by fluvial sediments reworked by longdrift currents, culminating in a regularised coastal morphology. This process was particularly intense at the margin of major deltas. The coast of Sinai and Palestine, for example, was indented during the Bronze Age before being transformed into a rectilinear shoreline in recent millennia (Stanley, 2002; Morhange *et al.*, 2005b). The attractive proto-historic coves served as anchorage havens for early coastal societies but were rapidly transformed into base-level depocentres and eventually abandoned.

Along the Black Sea shore, peculiar ancient rias known as limans have not progaded. For example in southern Bulgaria, the Roman harbour of Deultum lies inside such a liman. The ria has been transformed into an estuary presently disconnected from the open sea by coastal spits. In eastern Crimea (Ukraine), the harbour of the Greek colony Nymphaion is probably located inside a liman (Ménanteau and Geffray, 2003).

1.4.2.3 Buried fluvial harbours

River harbours are not subject to the same geomorphological and sedimentary processes, and therefore diagnostic harbour sediment signatures are different. Geoarchaeological study of such contexts has been relatively limited until now, but is nonetheless an interesting avenue for future research and provides opportunities with which to compare and contrast the coastal data (Milne and Hobley, 1981; Good, 1991; Izarra, 1993; Bravard and Magny, 2002; Arnaud-Fassetta *et al.*, 2003).

The Egyptians and Mesopotamians were the earliest western civilizations to engage in fluvial transportation and primeval Bronze Age harbourworks are known from the banks of the Nile at Memphis and Giza (Fabre, 2004/2005). These inevitably served as precursors for coastal anchorages, which were much more demanding environments stretching engineering knowledge far beyond primitive river infrastructure (Wachsmann, 1998; Fabre, 2004/2005). Unfortunately, high sediment supply and rapid changes in fluvial geomorphology mean that few conspicuous remains of these early fluvial harbour centres have survived to present. In Mesopotamia, docking basins were excavated and enclosed within the city walls of late third millennium Ur. A small dock dated 700 BC, built with mud-brick and bitumen is also known from alongside a bank of the Euphrates at Assyrian Til-Barsib (Blackman, 1982a-b). In Egypt, the works were many and varied. In the third millennium, for instance, canals were excavated from the Nile to the valley temples of the Giza pyramids so that building materials could be transported. Quays were also commonly established along the Nile, for instance at fourteenth century BC Amarna, boats have been depicted parallel to shoreside quays equipped with bollards. A large basin has also been reported from near fourteenth century Thebes (Blackman, 1982a-b).

Despite extensive excavations at numerous sites on the Nile delta (e.g. Tell El-Daba, Tell el-Fara'in) the exact location of many of the river ports is not known. The port city of Pelusium, located on the delta's northeast margin near the mouth of the former Pelusiac branch, was active during the Late Dynastic Kingdom and is a good example of settlement demise forced by geological changes and sedimentary inputs (Goodfriend and Stanley, 1999). At present, Pelusium is partially buried beneath tracts of deltaic sediments, and lies three kilometres south of the present coast, separated from the sea by extensive beach ridges and salt flats. This site was suddenly cut-off during the ninth century AD from the Nile Pelusiac branch and the sea as a result of floods. These floods induced rapid blockage of the Pelusiac branch and the avulsion of a new branch west of Port Said, most likely the Damietta branch (Stanley, 2005).

Stanley and his team have also worked extensively on the Canopic branch of the Nile delta coast (Stanley and Jorstad, 2006). They have combined geological and archaeological data to show that the Ptolemaic Roman city of Schedia once lay directly on the Canopic channel which was active between the third to second centuries BC until the fifth century AD. Abandonment of the site occurred when Nile waters were displaced to the east via the Bolbitic and later Rosetta branches.

Excavations and geological surveys at Greek Naucratis have revealed a series of active and abandoned channels during antiquity. These channels served as transport pathways for the ancient site, however the site's fluvial port has never been precisely located (Villas, 1996).

Our knowledge of later periods is much better (Berlanga and Pérez Ballester, 2003). The durability of Roman construction works means that well-known buried river harbours are known from Aquileia (Arnaud-Fassetta *et al.*, 2003; Rosada, 2003), Rome (Casson, 1965; Segarra Lagunes, 2004), Portus Ostia (Mannucci, 1996; Keay *et al.*, 2005a), Minturnae (Ruegg, 1988), London (Milne, 1982, 1985; Milne and Bateman, 2003), Bordeaux (Gé *et al.*, 2005), Lyon, Sevilla (Ordóñez Agulla, 2003), Roman and Islamic Valencia (Carmona and Ruiz, 2003) and Zaragoza (Aguarod Otal and Erice Lacabe, 2003). Some studies have yielded less convincing geoarchaeological results as for example at ancient Narbonne (Ambert, 1995).

The main problem of this harbour type is bank instability due to flood erosion and sedimentary accumulation. Two geological processes are important in explaining the archaeology of fluvial ports: (1) **Canalisation and funnelling of the river bed.** In London, for example, Milne (1985) describes a 100 m shift in the port's waterfront between AD 100 and today (**Figure 1.20**). Under a mesotidal fluvial regime, this funnelling of the waterbody has led to a positive increase in tidal amplitude. (2) **The vertical accretion of river banks by flooding.** At Bordeaux, the staircasing of numerous quays and platforms has been described at two sites in the Garonne estuary (Gé *et al.*, 2005). Rapid sediment accumulation during palaeoflood events is indicated by fine-grained organic lithofacies intercalated between the stepped platforms. There is significant scope to translate the waterfront archaeology techniques developed in northern Europe to the rich historical context of the Mediterranean (Milne and Hobley, 1981; Coles, 1984).

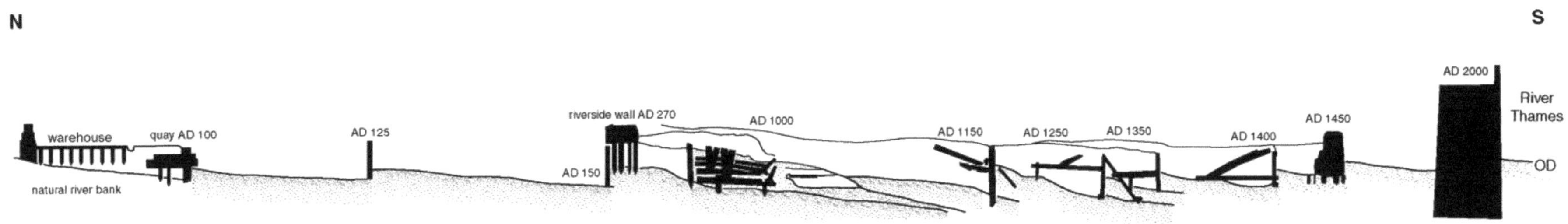

Figure 1.20: London's fluvial harbour, showing a gradual shift in the river bank and quays between the Roman period and today (after Milne, 1985). Marrying archaeological and stratigraphic data facilitates an understanding of sedimentary variations on the fluvial plain linked to the cities installed on the river banks. As in coastal areas, a deformation of the river bank is observed since antiquity.

1.4.2.4 Buried lagoonal harbours

Since 6000 BP, spit accretion on clastic coasts has disconnected a number of palaeo-bays from the open sea. The resulting low-energy depocentres formed lagoons that have gradually infilled since this time to form rich geological archives (Dalongeville *et al.*, 2001; Lespez, 2003a-b; Dalongeville and Müller-Celka, 2003). Lagoons offer natural protection and their use as anchorage havens has been widespread since early antiquity. At Coppa Nevigata, for example, a palaeo-lagoon connected to the sea by several entrances was deep enough for navigation between the third and first millennia BC (Caldara *et al.*, 2002). Caldara *et al.* (2003) have gone on to draw parallels between the nature of the lagoon environment and the development modes of the settlement.

Three main geoarchaeological problems are connected with lagoonal harbour types: (1) Because these anchorages lie at or near river mouths, access to the inlet is hindered by underwater coastal bars which can render navigation treacherous (e.g. the inlet at Licola lagoon near Cuma, Italy [Stefanuik *et al.*, 2003] or Lattara lagoon, southern France [Garcia and Vallet, 2002]). (2) The limited draught depth of the lagoon water body: during the Bronze Age and early Iron Age shallow draught boats meant that lagoons could be widely used (Ambert and Chabal, 1992). The shallow nature of the water column meant, however, that none of the early lagoon ports evolved into large scale seaports. It was not until Roman times that technological advances facilitated artificial remodelling and overdeepening of the basins. Carthage and the basin of Claude at Ostia are the most grandiose examples of these large-scale works. At these two sites, the lagoons had to be repeatedly dredged in order to maintain a navigable water column. (3) Also during Roman times, the ancient lagoons lining the Tyrrhenian and Adriatic seas were used as important intra-lagoonal waterways.

Reconstructing harbour histories using classic biosedimentological techniques (see below) has not been wholly conclusive (Melis, 2000; Lespez *et al.*, 2003a; Caldara *et al.*, 2003; Pasquinucci, 2004; Stefaniuk and Morhange, 2005). Very little human modification of the lagoon is required and therefore precisely discerning the presence of man can be difficult on purely bio- and lithostratigraphical grounds. On the contrary, geochemistry offers interesting avenues for future research.

1.4.2.5 Eroded harbours

Eroded harbours can result from two complementary geological processes: (1) a fall in sediment supply to the coastal zone; and/or (2) the destruction of harbourworks in areas exposed to high energy coastal processes. The best examples of eroded harbours date from the Roman period, when natural low-energy roadsteads were no longer a prerequisite for harbour location. At many high to medium energy coastal sites across the Mediterranean, the Romans constructed large enveloping moles to shelter an anchorage basin on its leeward side. Good examples of eroded ancient harbours include Carthage, Caesarea Maritima and the Roman port of Ampurias (Nieto and Raurich, 1998).

Although the location of Punic Carthage has never been lost to memory, academic speculation has persisted concerning the location, shape and number of its harbours during the Late Punic period (Yorke and Little, 1975; Hurst and Stager, 1978). Gifford *et al.* (1992) cross-correlated brackish-lagoon stratigraphies to reconstruct the coastline and harbour areas of Carthage. The work shows a shallow (4-5 m below MSL) artificial harbour dug down into a Quaternary aeolianite sandstone and flooded by marine waters in the third century BC. The team elucidate two interlinked basins: (1) a commercial port; and (2) a shallower, circular military harbour. Eventual erosion of the harbour is indicated by overlying anthropogenic and high energy beach facies, consistent with the hub's abandonment (Paskoff *et al.*, 1985).

Eroded outer harbour remains at Caesarea Maritima are still clearly visible. During the first century BC, King Herod ordered the construction of a royal harbour of the size of that of Piraeus. Roman engineers employed pozzolan concrete to build free-standing moles and create a closed offshore harbour (Raban, 1992a). Unlike Early Bronze and Iron Age harbours, which exploited natural roadsteads, Caesarea's outer harbour, also known as Sebastos, lay completely exposed to the high energy dynamics of the Levantine coast (Raban and Holum, 1996). Raban and Holum's team have described a short-lived protected outer harbour, eroded and exposed after 200 AD. Coastal erosion of the harbour remains induced Mart and Perecman (1996) into identifying a 5-8 m neotectonic collapse of the western Caesarea area. However, more recent geological research using seismic profiles has found no evidence of any faults, attesting to neotectonic stability for at least the past 2500 years (Gill, 1999). These local data are corroborated by regional sea-level data and coastal archaeology, which indicate the Israeli coast has been stable during the past 6000 years (Sivan *et al.*, 2001, 2004).

Ampurias on the Catalonian coast of Spain had several ancient harbour basins during antiquity. A natural cove/lagoon served as an anchorage during the Greek occupation of the site. Problems of harbour silting mean that this basin was eventually abandoned and superceded by an outer artificial basin during the Roman period. The exposed wave conditions coupled with the avulsion of the Fluvia river to the north have led to a fall in sediment supply and a gradual erosion of the harbour infrastructure during the past two millennia (**Figure 1.21**). Very few remains of the port have survived to the present, rendering a precise reconstruction difficult. Geophysical investigations are presently underway in an attempt to solve many of these geoarchaeological problems (Nieto *et al.*, 2005). Other examples of eroded harbours include Caulonia in Italy (Lena and Medaglia, 2002).

Unlike traditional archaeological typologies that grouped harbours on the basis of chronology and technology, this geoarchaeological classification looks to classify ports using earth science techniques. Sediment supply and RSL changes (neotectonics) are the two most important geological factors in accounting for the position of ancient harbours in relation to the present coastline. Understanding how these two processes have interacted at centennial and millennial timescales is critical to a transparent comprehension of the archaeological record, notably in coastal areas where very little such information is presently available.

1.5. Research methodology

A suite of research tools is available to the geoarchaeologist investigating ancient harbour sequences. These can be ordered into various field and laboratory phases. The onus of recent ancient harbour geoarchaeology has been on the application of a multi-disciplinary research approach to derive accurate palaeoenvironmental and archaeological reconstructions.

1.5.1 Field techniques

Broadly speaking, coastal geoarchaeologists draw their information from two areas, (1) geomorphology; and (2) the sediment archives located within this landscape complex.

1.5.1.1 Geomorphological surveying

In the absence of any conspicuous archaeology, an initial geomorphological prospection and mapping of the coastal landforms aids in identifying those areas of greatest archaeological significance. Geomorphological surveying can yield information on the morphology, genesis and age of key coastal landforms including dune ridges, infilled lagoons, estuaries, fluvial morphology, beach ridges and fossil cliffs. The approach is twofold and consists in (1) identifying key landscape features using a suite of cartographic, iconographic and photographic archives; and (2) verifying and detailing these interpretations in the field.

Figure 1.21: Ampurias' Roman harbour. Exposed wave conditions coupled with avulsion of the river Fluvia to the north have led to a fall in sediment supply and a gradual erosion of the harbour infrastructure during the past two millennia. The ancient Greek harbour had to be abandoned due to silting up of the basin. (Photographs: Marriner).

This terrestrial research can also be coupled with underwater prospecting and mapping looking to better comprehend the evolution of near-shore landscapes and relative sea-level variations, using both morphological (marine notches and erosion benches) and biological indicators. For the latter, precise vertical relationships between species ecology and sea level have been shown to exist for a number of marine taxa, including *Balanus* sp., *Lithophyllum* outgrowths and boring *Lithophaga* (Laborel and Laborel-Deguen, 1994). Understanding these relative sea-level movements is not only critical in comprehending landscape evolution but also human occupation within this landscape complex (Morhange *et al.*, 2001).

Urban contexts are particularly problematic for accurate geomorphological studies because the urban fabric hides many of the most important landscape features. In such instances, a study of the urban tissue and microtopography has been shown to be particularly useful in reconstructing coastal progradation after the Byzantine period at Alexandria, Acre, Sidon and Tyre.

Some degree of caution must however be exercised in the geomorphological approach - hence the importance of multi-disciplinary research - and the dangers of blindly extrapolating present to past geomorphology can create ambiguity and misinterpretation. The technique is also subject to a great deal of subjectivity, which can vary between researchers. For example, at Kition-Bamboula in Cyprus, Morhange *et al.* (2000) questioned landscape interpretations proposed by Nicolaou (1976) and Gifford (1978). Although recent seventeenth to nineteenth century engravings indicate a lagoon environment, the cothon harbour speculated by the latter authors was shown to be erroneous. In reality, Bamboula basin comprised a small cove within a much larger lagoon (**Figure 1.22**).

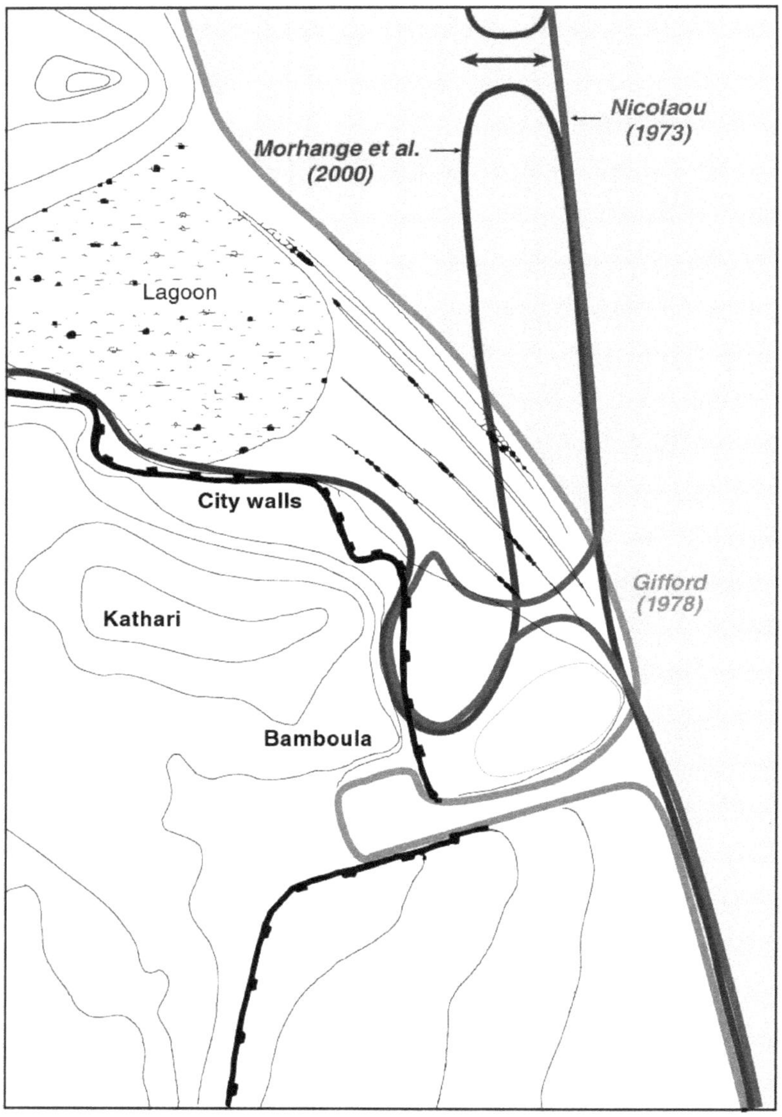

Figure 1.22: Coastal reconstruction of Kition-Bamboula (Cyprus) according to various authors. Nicolaou (1976) and Gifford (1978) speculated the presence of a cothon harbour based on modern engravings showing a lagoon. Morhange *et al.* (2000) demonstrated that Bamboula basin was in fact small cove within a large lagoon. The work reinforces the idea that recent landscape patterns should not be extrapolated to antiquity.

1.5.1.2 Geophysical surveying

Geophysical research techniques also provide a great multiplicity of mapping possibilities, notably where it is difficult to draw clear parallels between the archaeology and certain landscape features (Nishimura, 2001). The method assumes that subterranean conditions can be elicited via differences in physical properties. Geophysical techniques detect subsurface archaeological features that contrast with the surrounding soils in terms of electrical resistance, magnetic, or other properties (Clark, 1990; Scollar *et al.*, 1990; Conyers and Goodman, 1997). Factors that can create a geophysical contrast include soil compaction, particle size, organic content, artefact content, burning, and moisture retention (Neal, 2004).

The non-destructive nature of geophysical techniques means that they have been widely employed in terrestrial archaeology, and increasingly so in coastal geoarchaeology (Hesse, 2000). At Ampurias (Nieto *et al.*, 2005) and Cuma (Stefaniuk *et al.*, 2003) resistance profiles have been processed to produce geophysical profiles that can be interpreted both in stratigraphic and geoarchaeological terms. Significantly, profiles can be grouped together to produce a three dimensional cartography. By searching out the sizes, shapes and extents of differences it has been possible to yield insights into (1) the progradation of the ancient coastlines and the maximal extension of the former water body; and (2) information on coastal infrastructures. At Alexandria, Hesse (1998) has used geophysical prospection to propose a new hypothesis for the location of the ancient city's Heptastadium. Since the nineteenth century and the work of El-Falaki (1872), it was widely accepted that the Heptastadion occupied an axial position on the tombolo. Hesse now suggests that the causeway lay further to the west (**see Figure 1.23**) and was directly linked to the city's grid layout. These findings have since been corroborated by sedimentological data from the area (Goiran, 2001).

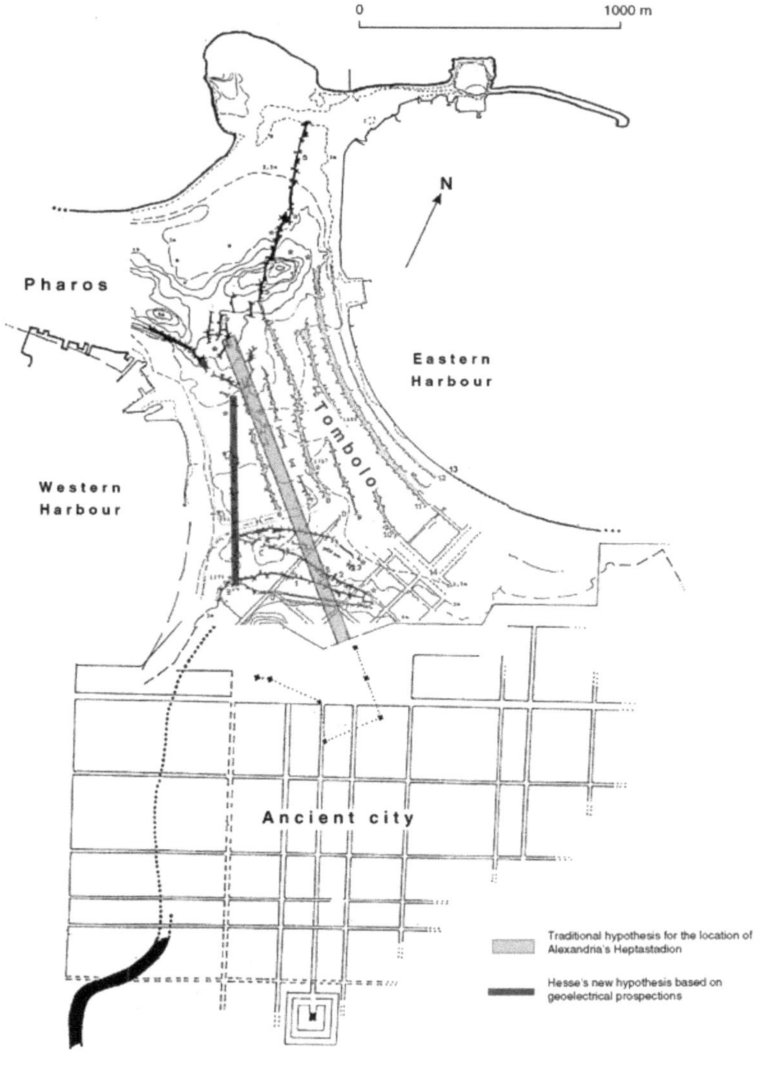

Figure 1.23: Hesse's new hypothesis for the location of Alexandria's causeway, based on geoelectrical prospections (modified from Hesse, 1998).

Since 1997 a geophysical survey has been completed at Portus, covering ~178 ha of the ancient port complex (Keay *et al.*, 2005b). Large areas of the seaport's fringes have been investigated to gain new insights into the harbour's coastal infrastructure and functioning during the Roman period. This technique has also accurately mapped the progradation of coastal ridges on the delta plain.

The main advantages of geophysical techniques are that (1) they are non-destructive (therefore particularly useful at sensitive sites and in urban contexts); and (2) rapid and reliable information can be provided on the location, depth and nature of buried features without the need for excavation. Unfortunately, there are no chronological constraints on the stratigraphy and archaeological features observed. Also, as with geomorphological mapping, there is an element of subjectivity involved in the interpretation of geophysical maps.

1.5.1.3 Coastal stratigraphy

Direct observation of the geological record is one of the key methods used to locate, characterise and reconstruct the physical evidence of past human activity (Kraft *et al.*, 1977). The most ideal means by which to observe coastal stratigraphy, including sedimentary structures, is using excavated sections. In the absence of large scale excavation works, such working conditions are frequently unrealisable and the geoarchaeologist must rely on coring techniques to elucidate the sedimentary column. Cross-correlation and core networking can yield detailed spatial and chrono-stratigraphic information on a given area's sedimentary geometry and key surfaces (e.g. the Harbour Foundation Surface etc.).

For a coastal geoarchaeologist two fundamental scales of analysis can be differentiated: (1) the delta or ria scale (Kraft *et al.*, 1977; Brückner, 1997); and (2) the harbour basin scale (Morhange, 2001). Both sedimentary systems are characterised by a general progradation since 6000 BP, which is critical in comprehending landscape deformation and the changing occupation history of the coastal plain.

The smaller analytical scale of harbour basins means that coastal deformation can be studied more finitely. The work of Morhange *et al.* (2001) at Marseilles, elucidating a rapid shift in shoreline positions from the Bronze Age onwards (**Figure 1.15**), demonstrates the type of spatial resolution that can be obtained when large excavation areas are available for study. Given the relatively limited accommodation space, ancient harbour basins are particularly sensitive to changes in sediment supply. For many years, researchers advocated rapid rates of sedimentation as being conducive to the creation of high-resolution geoarchives of human occupation and impacts. However, recent work from Marseilles and Naples suggests that this premise should be moderated (Hesnard, 1994; Giampaola *et al.*, 2004). Not surprisingly, the dual phenomenon of limited accommodation space and high sediment yields necessitated human management responses to ensure the viability of the harbour basin.

1.5.2 Laboratory techniques

The use of multidisciplinary research techniques is essential in precisely reconstructing coastal palaeoenvironments and shoreline deformation. Integration of multiple proxies ensures the most robust research conclusions possible (**Figure 1.24**). Be the sediments from an excavation section or core sequence, a short sampling interval (<5 cm) facilitates high-resolution reading of the coastal stratigraphy and harbour history. The laboratory techniques employed can be divided into a series of sedimentological, biostratigraphical and geochemical proxies.

1.5.2.1 Sedimentology

After an initial description of the main facies units and sedimentary structures, the section or core is divided into a series of sub-samples. Depending on the nature of the sediments and the coring equipment employed the sampling interval will vary. High-resolution (centimetric) sampling facilitates very precise reconstruction of harbour history and chronology. In reality, very few geoarchaeological teams engage in high-resolution laboratory studies due to their time-consuming nature.

1.5.2.1.1 Colour

Once the sediment samples have been dried they can be described using the Munsell colour scheme. Developed jointly by Munsell and the USDA Soil Conservation Service, these charts were initially used to classify soil colours but are now commonly used in all areas of the earth sciences and archaeology. The Munsell colour system specifies colours based on three dimensions, hue, lightness (called Value by Munsell), and chroma (difference from grey at a given hue and lightness). Although subjective, the charts provide a practical international standard for the communication and specification of sediment colour.

1.5.2.1.2 Sediment texture and granulometry

Particles or 'clasts' are the basic element of any sediment and therefore separating these clasts into discrete fractions is a key *inceptum* for geoarchaeologists. In order to extract the maximum amount of palaeoenvironmental information held within the coastal sediments, the sediment aggregates are wet sieved through two separate meshes (usually 2 mm and 50 μm) to separate out the gravels, sands and silts and clays fractions. The resulting dry fractions are subsequently weighed and data plotted against stratigraphic logs in percentages. The gravels fraction in ancient harbour sediments comprises a suite of interesting material, from marine molluscs, seeds, and grains to ceramic shards. These all attest to the harbour basin being used as a base-level waste dump by human societies.

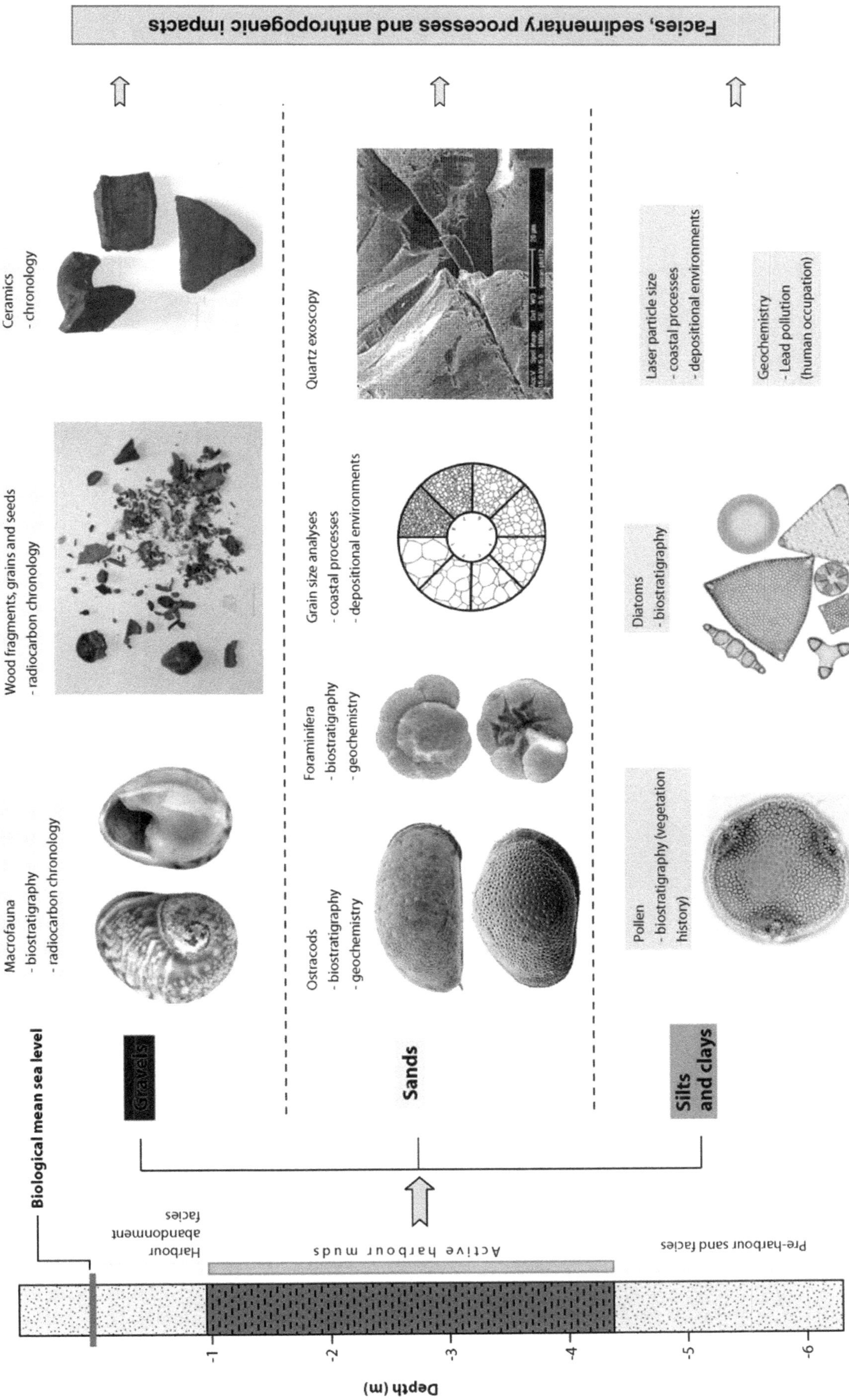

Figure 1.24: Facies, sedimentary processes and anthropogenic impacts: research tools used in the study of ancient harbour sequences.

The sand fraction can be subjected to mechanical sieving to establish various grain size parameters including histograms, fractiles and graphical indices (Folk and Ward, 1957; Folk, 1966). A column of sieves descending in size from 1.6 mm to 0.063 mm is employed and the separated sand fractions accordingly weighed. Results are subsequently statistically analysed, in concordance with various grain size parameters. The silts and clays fraction can also be investigated using laser particle sizing.

The granulometry of port sediments has traditionally interested engineers and not sedimentologists *sensu stricto* (Caldwell, 1939). Breakwaters, groins and jetties are all structures that act in a similar manner in that they impose a physical barrier in the nearshore zone and block the flow of littoral drift. Harbour sediments tend to be characterised by a strong granulometric heterometry, juxtaposing fine-grained silts and sands against coarse grained gravels constituting ceramics, marine macrofauna, wooden fragments etc. Harbour beach faces are generally characterised by a coarser granulometry which contrasts with the predominant silts and clays (50 to 90 %) from within the heart of the basin itself (Morhange, 2001). Unfortunately, although grain size analysis can yield information on the nature of the sedimentary dynamics, marine currents and coastal processes, it is a relatively weak palaeoenvironmental tool when used in isolation.

1.5.2.1.3 Exoscopic indicators

Clastic sediments can also be studied for their grain shape, and traces of surface dissolution and shocks. Sphericity and roundness can yield insights into the nature of the depositional environment and coastal processes. Precise investigation of the sand fraction is now possible using Scanning Electron Microscope (Prone, 2003), and within this context a methodology to characterise ancient harbour quartz grains has been developed by a number of researchers (Georges and Prone, 2000; Goiran, 2001). This particular technique is especially useful in differentiating between sedimentary environments.

SEM photograph of quartz grains evolving in Alexandria's ancient harbour

Alexandria I-26: Evolved Non-Eroded quartz grain from ancient harbour clays at Alexandria.

SEM photographs of palaeo-tsunami traces

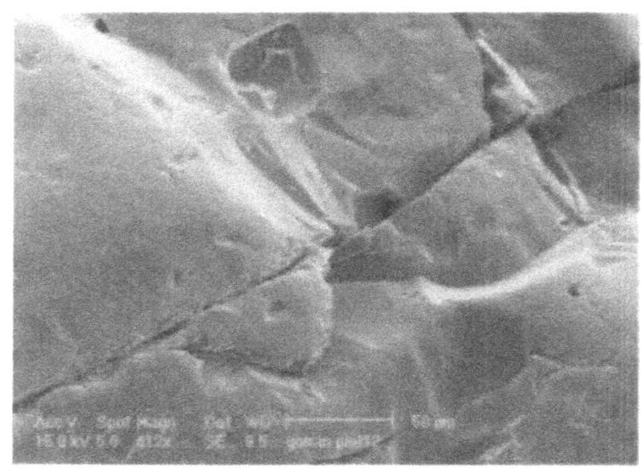

Alexandria I-11: Deep fracture line.

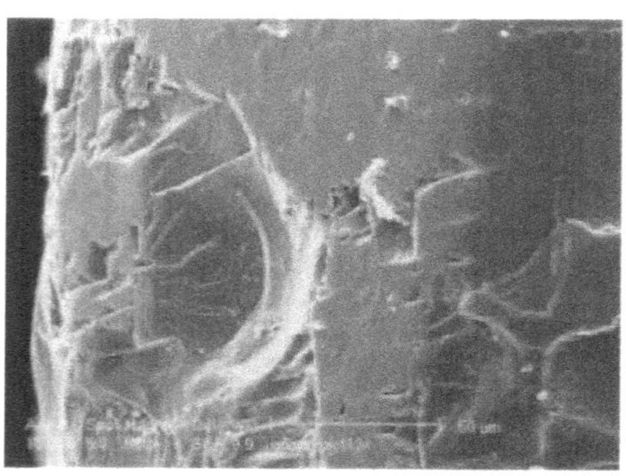

Alexandria I-11: High energy trace shock.

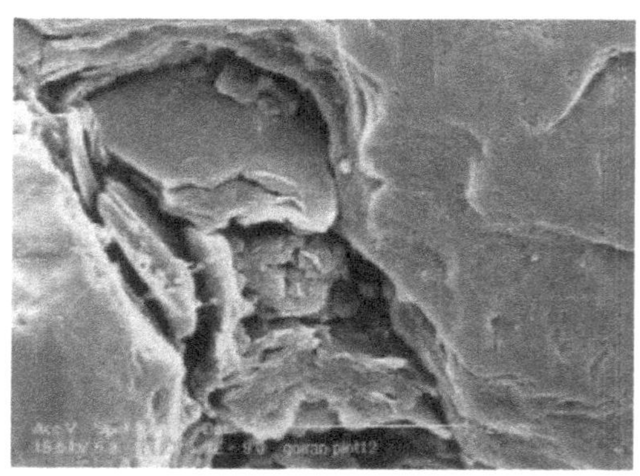

Alexandria I-11: Fracture surface on the face of a quartz grain.

Figure 1.25: Harbour and palaeo-tsunami quartz grains from Alexandria's ancient harbour (from Goiran, 2001). Goiran has shown that tsunami leave traces of high energy shocks on the quartz grains. This technique appears particularly useful in the study of palaeo-tsunami.

In ancient harbour sequences two types of quartz signature can be identified: (1) Quartz grains from low energy harbour muds are characterised by pyritospheres indicative of a confined marine environment. Surface shock marks induced by wave action tend to be polished by low-energy marine processes. Grains can also manifest traces of marine dissolution, characterised by anatomising surface patterns. (2) Quartz grains deriving from midlittoral beaches are marked by shock traces (cupules and croissants) indicative of breaking waves and the higher energy context. The position of the deposits within the intertidal zone is also indicated by traces of aeolian processes, subsequently reworked and polished by marine action.

Palaeo-lagoons are a case apart and research by Georges and Prone (2000) and Georges (2004) has established a typology of lagoon quartz grains which they qualify as Evolved Non-Eroded. Such low energy quartz signatures contrast with recent work at Alexandria, Egypt where paradoxical high-energy deposits present within the harbour muds have been used to establish a typology of tsunami trace shocks (Goiran, 2001; **Figure 1.25**). These have been compared and contrasted with modern analogues from around the globe and provide a preliminary referential with which to corroborate palaeo-tsunami facies. Palaeo-tsunami layers are notoriously difficult to differentiate from storm surge deposits (Dawson, 1994; Dawson, 1999, Dawson *et al.*, 2004).

1.5.2.2 Biostratigraphy

Biostratigraphy is the study of the temporal and spatial distribution of fossil organisms (Jenkins, 1993; Haslett, 2002). In coastal geoarchaeology, it is an important tool in precisely reconstructing depositional environments, the evolution of coastal biocenoses and the impact of human societies upon these ecosystems (pollution, harbourworks etc.). An essential prerequisite for this approach is a good taxonomic framework, and understanding of both the ecology and geographical ranges of species encountered.

1.5.2.2.1 Malacology

Mediterranean malacology has a long history of research and since work on the *Vieux Port* at Marseilles marine molluscs have proved a powerful tool in reconstructing ancient harbour palaeoenvironments, where their fossil shells are found in abundance (Leung Tack, 1971-72). During the 1960s, Péres and Picard (1964) were eminent figures in developing a molluscan classification system that assigned Mediterranean species to well-defined ecological groups (see also Péres, 1982). Refined versions of this classification system are presently used. The most important workable identification frameworks for the Mediterranean include d'Angelo and Gargiullo (1978), Barash and Danin (1992), Poppe and Goto (1991, 1993), Bellan-Santini *et al.* (1994) and Doneddu and Trainito (2005).

Both *in situ* and *extra situ* taxa can be identified on the basis of core lithology and shell taphonomy. This approach can be useful in establishing the degree of harbour confinement/exposure. For example, at Alexandria malacological work on a tsunami layer juxtaposes a great diversity of *in situ* and *extra situ* molluscan tests, consistent with powerful offshore waves reworking biocenoses from deeper bathymetric depths (Goiran, 2001; Stanley and Bernasconi, 2006). Ecological stresses and harbour pollution can also be evidenced by the presence of certain taxa (**Figure 1.26**). Typical ancient harbour taxa include *Parvicardium exiguum*, *Cerastoderma glaucum*, *Loripes lacteus*, *Abra segmentum* and *Cerithium vulgatum*. These species are all consistent with either lagoonal environments or sheltered areas rich in fine-grained sands and silts. Given that marine shells are frequently used to radiocarbon date harbour sequences, it is imperative to understand the ecology of observed taxa and whether or not these are *in situ*.

1.5.2.2.2 Ostracods

Ostracods are microcrustaceans comprising soft body parts enclosed in a low-Mg calcite bivalve (Athersuch *et al.*, 1990). They are typically around 1 mm in size but can vary between 0.2 to 30 mm. As with all crustaceans, ostracods grow by moulting. The carapace has numerous morphological characters which allow taxonomic and phylogenetic studies to be made on living and fossil specimens (Holmes and Chivas, 2002). Ostracods have a long and well-documented fossil record from the Cambrian to the present day, and have been particularly useful for the biozonation of marine strata on a local or regional scale.

Ostracods are excellent indicators of palaeoenvironments because of (1) their ubiquity in both fresh and marine waters; (2) their small size; and (3) their easily-preservable carapaces. Their faunal composition, population density and diversity vary in time and space as a function of numerous environmental factors including water temperature, salinity, water depth, grain size, and anthropogenic impacts (Boomer and Eisenhauer, 2002). In many coastal and nearshore marine areas, human activities can significantly modify the natural coastal system (construction works, pollutants) leading to severe alterations in the different trophic levels of the ecosystems (Ruiz *et al.*, 2005). Ancient anthropogenic activities can affect ostracods in three ways: (1) impact on the densities and diversities of the assemblages; (2) strongly influence the abundance and distribution of selected species; and (3) affect the chemistry of their carapaces.

Although slightly different preparation techniques exist, ostracoda are generally extracted from the dry sand fraction (>150 μm). A minimum of 100 valves is preferred to ensure statistical robustness. Identified taxa are most commonly assigned to five assemblages on the basis of their ecological preferences: freshwater, brackish lagoonal, marine lagoonal, coastal and marine (Müller, 1894; Breman, 1975; Bonaduce *et al.*, 1975; Carbonel, 1980, 1982; Morhange *et al.*, 2000). These clearly discrete ecological groups render ostracods one of the best biostratigraphical markers in ancient harbour sequences. Potential research avenues include the application of ostracod test geochemistry to reconstruct ancient harbour salinity patterns and pollution levels (Ruiz *et al.*, 2005).

Figure 1.26: Biostratigraphic indicators of confined harbour conditions.

1.5.2.2.3 Foraminifera

Foraminifera are single-celled organisms (protists) with tests that have been abundant as fossils for the last 540 million years (Murray, 1991). Foraminiferal tests are commonly divided into chambers that are added during growth, though the simplest forms are open tubes or hollow spheres (Sen Gupta, 2002). Based upon their composition, the tests can be divided into three categories: organic (membranous or tectinous), agglutinated, and calcareous. The agglutinated shells are composed of foreign sediment particles glued together with an organic cement, while the calcareous shells are formed from secreted calcium carbonate. The calcareous shells can be further subdivided into hyaline, porcellaneous, and microgranular types which differ in their calcite crystal arrangement. Fully grown individuals range in size from about 100 μm to almost 20 cm long.

Foraminifera are among the most abundant and ubiquitously spread shelled organisms in marine environments. Their distribution range comprises the intertidal zone to the deepest ocean trenches, and from the tropics to the poles. Although a few species are found in brackish environments, foraminifera do not occur in freshwater environments.

Foraminifera are particularly useful in ancient harbour geoarchaeology because: (1) they are abundant in most marine and marginal marine environments, with a living population density often exceeding one million individuals per square metre; (2) they have a mineralized shell and thus have a high preservation potential in the sedimentary record; (3) only small sediment samples are needed to obtain fossil populations large enough for statistical treatment; (4) being single celled they respond rapidly to environmental changes (Samir, 2000; Pascual *et al.*, 2002); and (5) foraminiferal species characterise very specific environments.

Workable taxonomic frameworks for the Mediterranean region include Cimerman and Langer (1991) and Sgarrella and Moncharmont Zei (1993). In northern Europe and North America, marsh and intertidal foraminifera have been widely used to develop high-resolution sea-level transfer functions (Horton and Edwards, 2006), although studies in the Mediterranean have tended to focus upon establishing broad depth ranges for benthic taxa (Basso and Spezzaferri, 2000). Whilst foraminiferal test geochemistry has long been used in palaeoclimatology (Bard, 1999) to study temperature, salinity, carbonate chemistry, diet, and nutrient conditions, it is only very recently that these principles have been applied to ancient harbour contexts (Reinhardt *et al.*, 1998).

1.5.2.2.4 Diatoms

Diatoms are a major group of eukaryotic algae, and are one of the most common types of phytoplankton (Round and Crawford, 1990; Battarbee *et al.*, 2001). Most diatoms are unicellular, although some form chains or simple colonies. Diatoms cells are contained within an intricate siliceous frustule comprised of two separate valves, the morphology of which forms the basis for their taxonomy. They are a widespread group and can be found in the oceans, freshwater, soils and on damp surfaces. Most live pelagically in open water, although some live as surface films at the water-sediment interface (benthic), or even under damp atmospheric conditions. They are especially important in oceans, where they are estimated to contribute up to 45% of the total oceanic primary production (Mann, 1999).

Unlike many other algal groups, diatoms are readily identifiable to species level and beyond (Round, 1991), and this has made them particularly useful in palaeoenvironmental and palaeoclimatological studies. They manifest a consistent tolerance with a wide range of environmental parameters such as light, moisture, current velocity, pH, salinity, oxygen and inorganic and organic nutrients (Battarbee *et al.*, 2001). Whilst their identification remains difficult, numerous comprehensive taxonomic texts and floras dealing with taxonomy are now readily available.

Their ubiquity in diverse types of waterbodies means that they are also increasingly being used in geoarchaeology (Battarbee, 1988). Although not widely employed in ancient harbour basins until now, recent preliminary studies suggest that diatoms could be rich palaeoenvironmental indicators in these unique archives (Vecchi *et al.*, 2000). Their greatest geoarchaeological scope appears to lie in lagoon and fluvial harbour environments, where coastal geoarchaeological methods are not sensitive enough to accurately reconstruct port history.

1.5.2.2.5 Palynology

Palynology is the science that studies contemporary and fossil palynomorphs, including pollen, spores, dinoflagellate cysts, acritarchs, chitinozoans and scolecodonts (Nair, 1985). Palynomorphs are broadly defined as organic-walled microfossils between 5 and 500 μm in size, and have been widely used in the earth and plant sciences since the 1940s (Moore *et al.*, 1991). They are extracted from rocks and sediments both physically, by wet sieving, often after ultrasonic treatment, and chemically, by using chemical digestion to remove the non-organic fraction.

Pollen are the most commonly employed type of palynormorph fossil in archaeology, where they have been used to reconstruct regional vegetation landscapes and human impacts (Bottema and Woldring, 1990; Atherden *et al.*, 1993; Dumayne-Peaty, 2001). The Mediterranean basin has experienced intensive human development and impact on its ecosystems for thousands of years (Bottema *et al.*, 1990). The greatest impacts of human civilization have been deforestation, intensive grazing and fires, and infrastructure development, especially on the coast. There is a growing body of literature on the use of pollen and plant remains at underwater sites (Weinstein, 1996; Gorham and Bryant, 2001; Gorham, 2001).

The use of pollen in ancient harbour geoarchaeology has been more limited, where abundant and easily-identifiable fossil types such as molluscs, ostracods and foraminifera yield more pertinent information relating to the coastal history. Although fine-grained harbour muds are ideally suited to the preservation of palynomorphs, the most recent studies undertaken upon this type of archive have not proved conclusive (Marriner *et al.*, 2004). Sediments transported and deposited by coastal processes can contain pollen 'contamination'. Differentiating this background noise from 'economic' pollen resulting from local cultural processes at the site can be problematic. On the other hand, plant macrofossils can provide information on (1) the identification of botanical cargoes, provisions, and dunnage; and (2) the identification of plant fibres used to make rope, baskets, matting, and caulking (Gorham and Bryant, 2001). Interesting future avenues of research include the investigation of other plant micro- and macrofossils, notably phytoliths (Rosen, 1999; Gorham, 2001; Pennington and Weber, 2004) and charcoal, wood and seed remains recovered from harbour sediments (Figueiral and Willcox, 1999; Willcox, 1999; Willcox, 2003). Such material is already widely used to establish radiocarbon chronologies in coastal geoarchaeology.

1.5.3 Geochemistry

The association of trace metal anomalies are increasingly being used to reconstruct ancient pollution levels during the historical period and antiquity (Gale and Stos-Gale, 1982; Nriagu, 1983; Sayre *et al.*, 1992; Yener *et al.*, 1991; Renberg *et al.*, 1994; Hong *et al.*, 1994, 1996; Nriagu, 1998; Shotyk *et al.*, 1998, 2005; Martínez-Cortizas *et al.*, 1999; Pyatt *et al.*, 2000; Aberg *et al.*, 2001; Bränvall *et al.*, 2001; Cundy *et al.*, 2006). Within this context, it has been demonstrated that lead (Pb) isotopes in harbour sediments are particularly powerful tools in retracing the history of metallurgic activities at

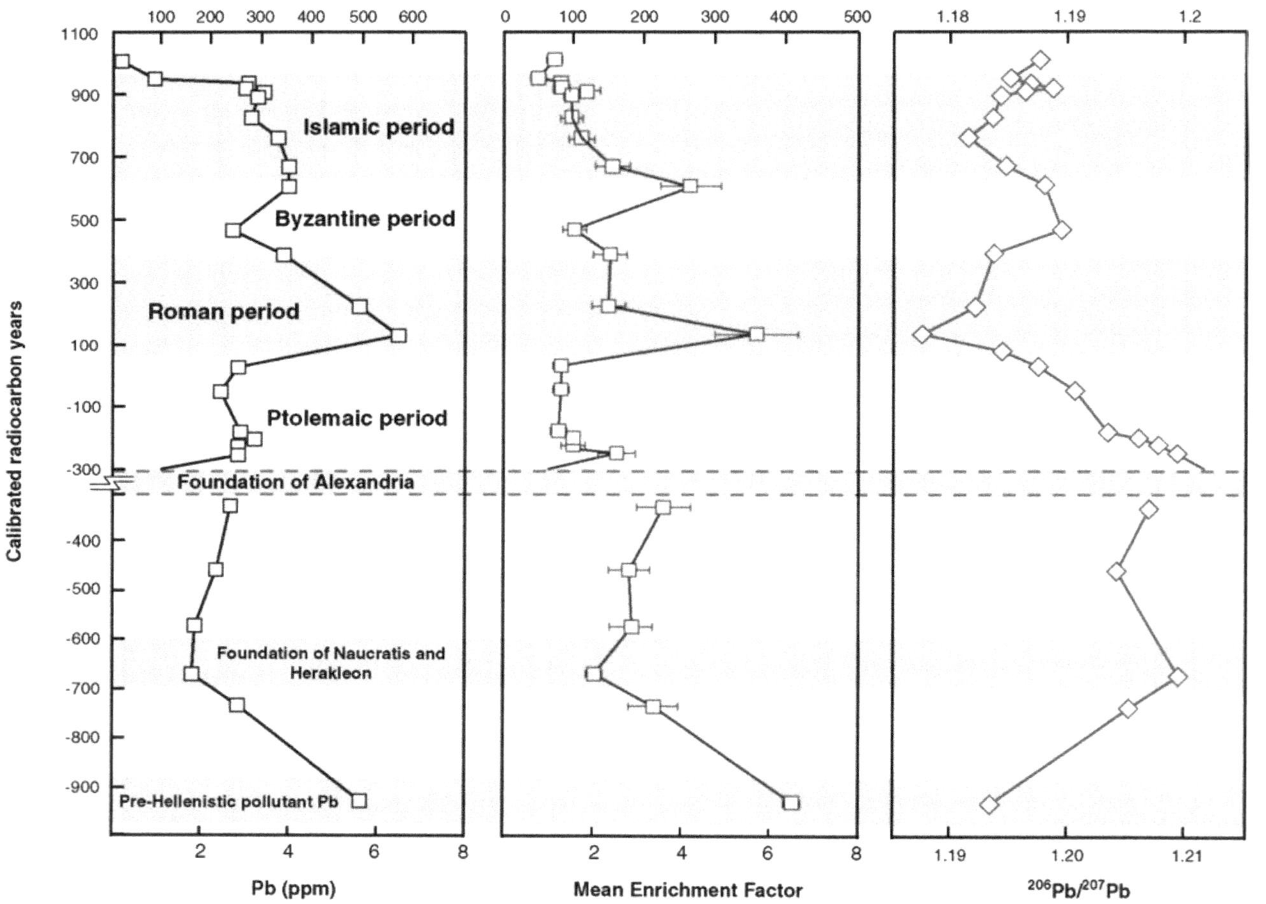

Figure 1.27: Pollutant lead at Alexandria during the past 3000 years (adapted from Véron *et al.*, 2006). The use of lead geochemistry has elucidated pre-Hellenistic pollutants at Alexandria suggesting an advanced settlement at the site before the arrival of Alexander the Great in 331 BC.

coastal sites (Véron *et al.*, 2006). There are a number of reasons why port sediments are particularly conducive to geochemical analyses: (1) ancient seaports comprise low-energy depocentres. Urban and coastal contaminants are thus concentrated in the basins, and have the potential to yield long time series of human-environment interactions; (2) trace metals (e.g. copper, lead), have been used by ancient societies since the end of the Neolithic (Mellaart, 1967). Covariation in the geochemical suites of harbour sediments can therefore be linked to changes in metallurgical *savoir faire* (e.g. copper during the Chalcolithic, copper, tin and lead during the Bronze Age etc.); (3) although small amounts of metals are found in relatively pure form, most must be extracted from more complex ores by removing the 'impurities' (non-metal or other metal) from the combination ore. For example, natural lead is very rare in the earth's crust, therefore when it is found at an archaeological site its presence can invariably be attributed to a mineral fusion. This renders lead and other trace metals particularly powerful tools in reconstructing anthropogenic contaminations (Shotyk *et al.*, 1998, 2005; Martínez-Cortizas *et al.*, 1999); (4) the majority of trace metals are well-preserved in the fossil record, with relatively low levels of degradation over the Holocene timescale; (5) finally, the geochemical composition of some metals means that they can be provenanced. Lead, for example, is a chemical element that possesses three stable radiogenic isotopes ^{206}Pb, ^{207}Pb and ^{208}Pb. These three isotopes vary as a function of the geological source area which means that lead samples can invariably be linked back to a well-constrained mining area.

To date, four ancient harbours have undergone geochemical analyses of this nature. At Alexandria, Veron *et al.* (2006) evidenced Bronze Age human occupation of the site calling into question the Alexandria *ex nihilo* hypothesis (**Figure 1.27**). The Graeco-Roman apogee of the city is indicated by lead pollution levels twice as high as those measured in contemporary ports and estuaries. Graeco-Roman pollution peaks have also been evidenced at Marseilles (Le Roux *et al.*, 2005) and Sidon (Le Roux *et al.*, 2003a; **Figure 1.28**).

1.6. The ancient harbour: stratigraphy and geological principles

Figure 1.29 sets out the type stratigraphy for some of the Mediterranean's best studied examples. Be it in the eastern or western Mediterranean, the stratigraphic similarities are striking, with three distinct facies of note: (1) a middle energy beach sands at the base of each unit, (2) low energy silts, and (3) coarsening up beach sands which cap the sequences. In very general terms this stratigraphic pattern translates a shift from natural coastal environments to anthropogenically modified environments, eventually culminating in a semi- or complete abandonment of the harbour basin.

On the basis of these observed stratigraphic patterns, we have attempted to understand ancient harbour basins within the ordered framework of sequence stratigraphy. Human agency, in addition to modest sea-level variations and overriding sediment supply forcings, are easily incorporated into this conceptual approach that has revolutionised the study of sedimentary formations in geology since the 1970s (Catuneanu, 2002, 2005; Coe, 2003).

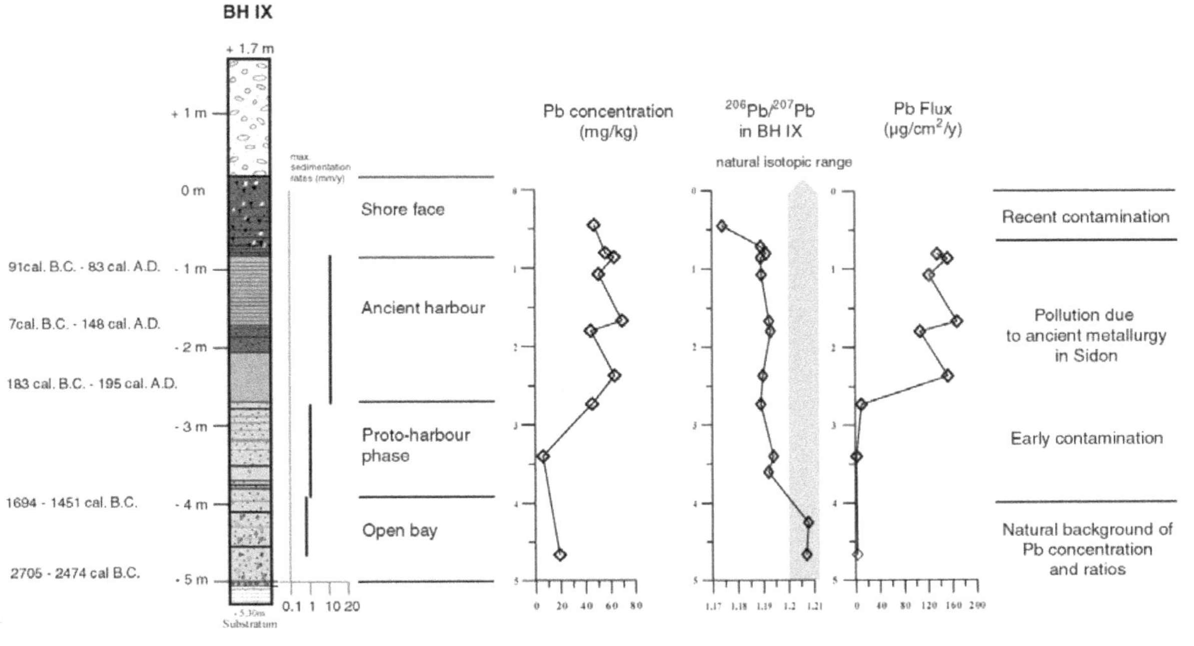

Figure 1.28: Pollutant lead contamination at Sidon's ancient harbour (adapted from Le Roux *et al.*, 2003).

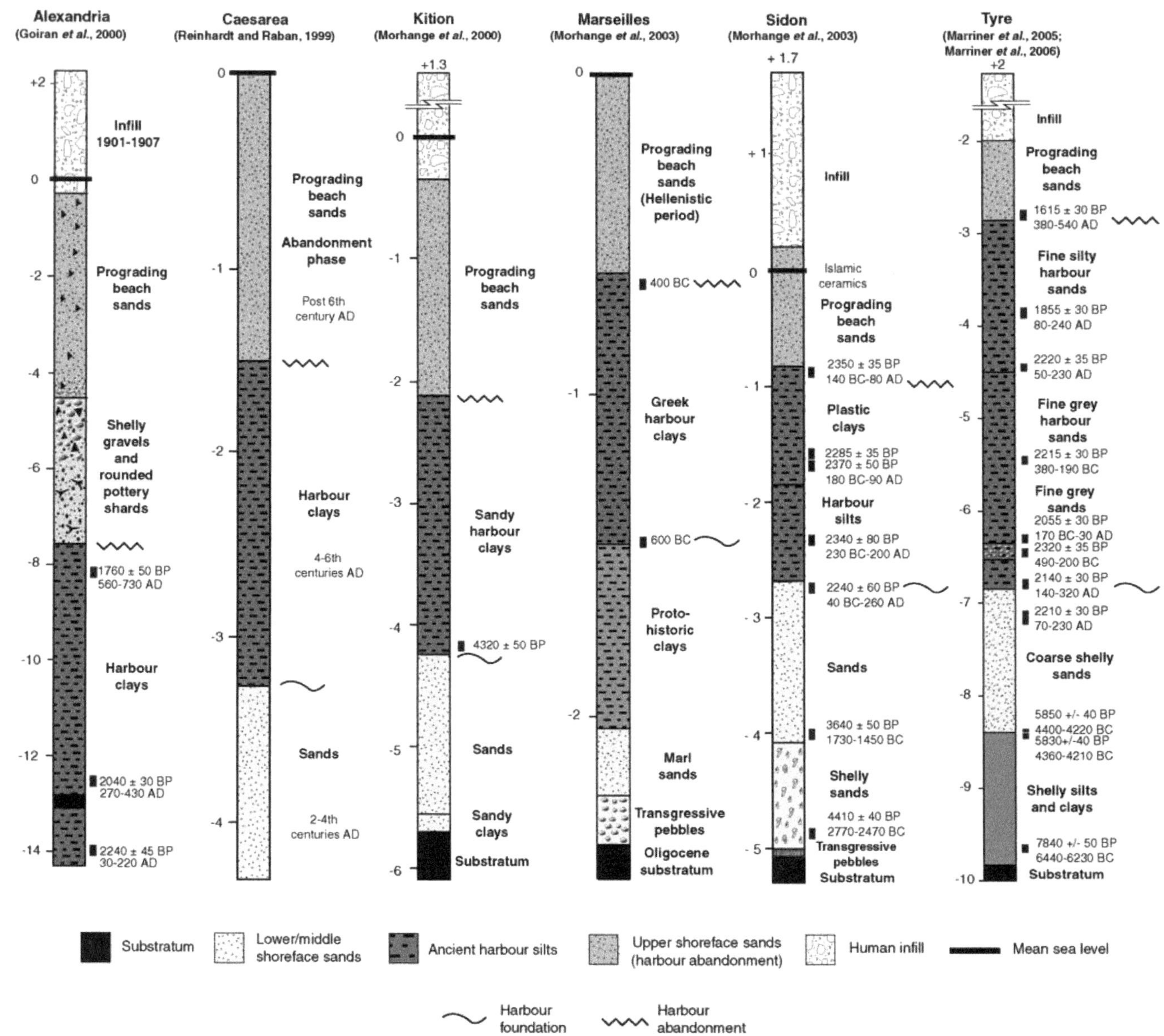

Figure 1.29: Lithostratigraphy of the Mediterranean's best studied ancient harbours. This highstand anthropogenic facies is characterised by uncharacteristically high levels of fine-grained material (Marriner and Morhange, 2006).

There are a number of reasons why sequence stratigraphy provides a robust framework for this new coastal geoarchaeology. (1) Firstly, it is a holistic means of examining the mid- to late Holocene stratigraphic record, in that it considers not only the empirical record but also those parts which are missing. (2) Secondly, evolving palaeogeographies can be studied in their wider contexts and ancient harbours linked into the source to sink sedimentary conveyor. (3) It facilitates different harbour sites to be compared and contrasted with relative ease, whereby relative sea-level variations are correlated with lateral facies changes. (4) Finally, this approach focuses on the mechanisms and processes driving coastal deformation, namely climatic, eustatic, tectono-isostatic and anthropogenic factors (Morhange, 2001). Mankind, as we will go on to demonstrate, has had a greatly distorting role to play within this forcing complex.

1.6.1 Basic principles

Ancient harbours are base-level depocentres that owe their origins to Bronze Age human agency and mid-Holocene sea-level deceleration, stability which brought terrigenous accumulation to certain low-energy littoral zones. The basins form integral components of the Highstand Systems Tract (HST) and their sediment strata comprise aggradational to progradational sets. The Maximum Flooding Surface (MFS) represents the lower boundary of the sediment archive. This surface, dated ~6000 yrs BP, marks the harbour basin's maximum marine incursion and is broadly contemporaneous with the Chalcolithic period and the Bronze Age. Indeed, strong links have been found between the position of the MFS and early coastal settlement patterns along the Levantine coast (Raban, 1987a).

Early Bronze Age societies preferentially concentrated around geomorphologically endowed sites, namely small coves, pocket beaches, estuaries and wadis, which formed natural anchorage havens in little need of human enterprise. It is no coincidence therefore that many of the earliest sea-faring communities originated in the central and eastern Mediterranean. The rocky and convoluted coasts of the Aegean islands were, for example, ideally suited to protect early mariners and their vessels. The presence of man in and around low-energy sediment sinks, allied with relatively continuous rates of sedimentation, has culminated in the formation of rich georchaeological archives.

Although environmental determinism was the rule, societies in the Aegean and the Levant started modifying their natural anchorages during the Middle to Late Bronze Age and Early Iron Age. Two dated examples derive from the Levant coast. Firstly, submerged boulder piles have been evidenced at Yavne-Yam, a Middle Bronze Age site on the coast of Israel; these attest to premeditated human enterprise to improve the quality of the natural anchorage (Marcus, personal communication). Secondly, radiometric dating of wood fragments held between courses of the artificial mole at Atlit constrains this Phoenician structure to the ninth century BC (Haggai, 2006). A similar example is also known from the Syrian coast at Tabbat el-Hammam, where the archaeological evidence supports a ninth/eighth century BC age (Braidwood, 1940).

This artificialisation is translated in the sedimentary record by lower energy facies which chronicles a barring of the anchorage by artificial means. Harbour infrastructure (quays, moles and jetties) accentuated sediment sink properties and this, coupled with important changes in land use and cultural inputs, engendered rapid rates of sedimentation culminating in coastal progradation. By Roman times, harbour engineering had reached its zenith and accumulation rates 10 to 20 times greater than naturally prograding coastlines are recorded. High sediment supply resulted in the deposition of thick beds of fine-grained harbour muds that provide a multiplicity of research possibilities (see below).

1.6.2 Key facies belts and sequence boundaries

Research shows there to be considerable repetition in ancient harbour stratigraphy, both in terms of the facies observed and their temporal envelopes. Much of the data can be correlated on various spatial scales, locally (e.g. Alexandria, Marseilles, Tyre) and regionally (Phoenicia, the Black Sea, Magna Graecia). In light of this, we believe the stratigraphic ensemble of harbour basins to be sufficiently unique as to merit a separate epithet, the Ancient Harbour Parasequence (Marriner and Morhange, 2006b). Although AHPs can vary slightly in thickness and character, a number of key facies belts and sequence boundaries can be identified (**Figures 1.30 and 1.31**).

(1) Around 6000 BP, the MFS marks the basal inception of the AHP and invariably comprises trangressive sands or pebbles. It constitutes the marine flooding of the depocentre and is associated with the most landward position of the shoreline within the basin.
(2) This surface is overlain by naturally aggrading beach sands, a classic feature of clastic coastlines since 6000 BP. After this time, the rate of increase in accommodation space (the space separating the water surface from the basin bottom) became outbalanced by sediment supply. Relative sea-level stability impinged on the creation of new accommodation space, leading to the aggradation of sediment strata. At this time, the coves were completely natural. Where this sedimentation continued unchecked, a coarsening upward of sediment facies is observed, translating high energy wave dynamics in proximity to mean sea level.
(3) The Harbour Foundation Surface (HFS) is one of the most important stratigraphic surfaces in ancient harbour geology. It marks early human modification of the basin, namely the transition from coarse beach sands to fine-grained harbour silts and clays, and corresponds to the construction of protective harbourworks (Goiran and Morhange, 2003).

Facies/surface name	Definition	Diagnostic sedimentology	Diagnostic biostratigraphy	Geochemical imprint
Harbour Abandonment Facies (HAF)	- Degradation of harbourworks and exposure of the basin	- High to middle energy aggradational and progradational sets - Coarsening-up sequence	- Juxtaposition of diverse ecological groups (translate the exposed nature of the depositional environement)	- Weak
Harbour Abandonment Surface (HAS)	- (Semi)abandonment of the basin	- Transition from fine-grained harbour silts and clays to coarse sands and gravels		
Ancient Harbour Facies (AHF)	- Anthropogenically forced low-energy sedimentary environment	- Transition from coarse beach sands to fine-grained harbour silts and clays - Rapid fine-grained sedimentation rates (10-20 mm/yr) - Granulometric heterometry	- Ostracods = *Cyprideis torosa*, *Loxoconcha* spp., *Xestoleberis* spp. - Foraminifera = *Ammonia* spp. - Molluscs = *Parvicardium exiguum*, *Loripes lacteus*, *Cerithium vulgatum*	- Strong
Harbour Foundation Surface (HFS)	- Natural to artificial interface	- Abrupt change from coarse beach sands to fine-grained harbour silts and clays		
Proto-harbour sands / Pre-harbour sands	- Natural beach sediments	- Aggradational coarse to medium grained sands - Coarsening-up sequence	- Coastal and semi-protected sub-tidal taxa	- Weak
Maximum Flooding Surface (MFS)	- Marine flooding of the coastal sequence (ca. 6000 BP)	- Transgressive sands and pebbles	- Coastal and semi-protected sub-tidal taxa	- None

Figure 1.30: Definitions of the AHP's key facies and surfaces.

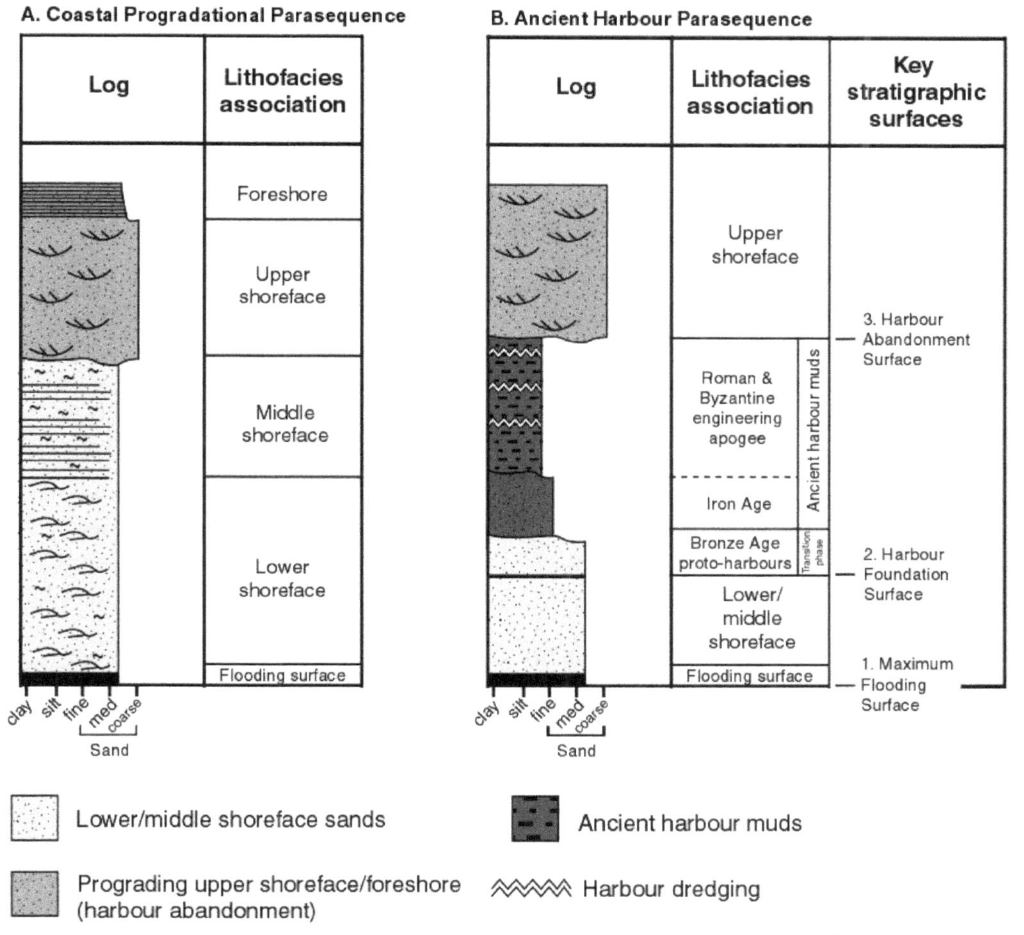

Figure 1.31: Lithofacies and key stratigraphic surfaces of (A) the Coastal Progradational Parasequence and (B) the Ancient Harbour Parasequence.

(4) The Ancient Harbour Facies (AHF) is by no means a uniform unit of plastic clays. Stratigraphic impacts are intrinsically linked to harbour technology. The most moderate signatures of human presence are dated to the Bronze Age, when societies used natural low-energy basins needing little or no human modification. For example, the southern cove of Sidon in the vicinity of Tell Dakerman remained naturally connected and open to the sea throughout antiquity (Poidebard and Lauffray, 1951). The natural to semi-artificial interface is thus by no means transparent on purely granulometric grounds and the astute use of biological indicators, notably molluscan macrofauna and microfauna assemblages, is a much more effective means of establishing this surface. Net transition to fine-grained sands and silts is observed, good examples being those of Tyre and Sidon (Marriner *et al.*, 2006a). During the Late Bronze Age and Early Iron Age, expanding trade forced many Levantine societies to embrace maritime engineering and build artificial harbourworks (Frost, 1995; Raban, 1995). Gradual improvements in harbour engineering are recorded by increasingly fine-grained facies, up until the Roman period when pozzolan concrete marked a revolution in maritime infrastructure (Oleson *et al.*, 2004a-b). Natural roadsteads were no longer a prerequisite to harbour loci and completely artificial ports, enveloped by imposing concrete moles, could be located on open coasts (Hohlfelder, 1997). Plastic clays tend to be the rule for this period.

(5) Harbour Abandonment Surface (HAS). The HAS marks the semi-abandonment of the harbour basin, invariably after the late Roman period. This surface lies at the intersection between the AHF and an exposed beach facies. It corresponds to coastline progradation, culminating in the partial silting or landlocking of the basin heart.

(6) Often, the exposed beach facies comprises coarse grained beach-face sands and is consistent with degradation of harbourworks. This phenomenon can be linked to a number of historical events: (a) settlement demise (e.g. Islamic expansion in the eastern Mediterranean); or (b) natural catastrophes including seismic uplift (e.g. Phalasarna harbour in western Crete), volcano-tectonic subsidence (e.g. Portus Julius in Pozzuoli caldera), or tsunami impacts (e.g. Alexandria). In the absence of clear management strategies, port basins rapidly infilled with thick tracts of coastal and fluvial sediments.

On 'tectonically stable' coasts, sediment supply, whether it be environmentally or anthropogenically forced, has been the chief controlling factor in determining ancient harbour stratigraphy and facies organisation.

1.6.3 Ancient harbour stratigraphy: a geological paradox?

The presence of man has significantly impacted upon the stratigraphic record, to such a degree that we define the AHP as a geological paradox. Anthropogenic forcing of the stratigraphic record is juxtaposed against the classic Coastal Progradational Parasequence (CPP). Under natural conditions, the MFS is overlain by increasingly coarse sediments consistent with an aggradation of the marine bottom. In the AHP model, only the lower portion of this natural stratigraphy is observed; it is overlain by facies presenting a number of stratigraphic aberrations.

(1) **Fining-upward sequence.** The AHP comprises a fining-upward granulometry characterised by coarser-grained transgressive deposits at the base, overlain by increasingly fine-grained deposits up-sequence. The very well protected Roman harbours of Alexandria, Marseilles and Naples all comprise plastic marine muds consisting of >90% silts.

(2) **Accelerated accretion rates,** at least ~10 times greater than nearby naturally prograding coasts, are archetypical of these depocentres (**Figure 1.32**). On natural coasts, sediment was resuspended from the seabed due to energetic wave processes and transported by currents in the water column. In ancient harbour basins, harbour infrastructure attenuated the swell and perturbed marine currents, accounting for a sharp fall in water competence (Inman, 1974). Whilst watershed geomorphology and climate are first-order controls on sediment production and delivery to the ocean, the presence of man also increased the potential for subaerial erosion in coastal watersheds (Horden and Purcell, 2000; Devillers, 2005). Significant increases in the supply term are recorded from the Neolithic period onwards including, for example, anthropogenic changes in the catchments of supplying rivers, erosion of adobe urban constructions (Rosen, 1986) and finally use of the basins as *ad hoc* waste dumps. Even though the qualitative impact of human activity seems obvious, quantitative evaluation of the impact of land use is complicated by coeval variations in other independent environmental factors (e.g. the incidence of intense rainstorms).

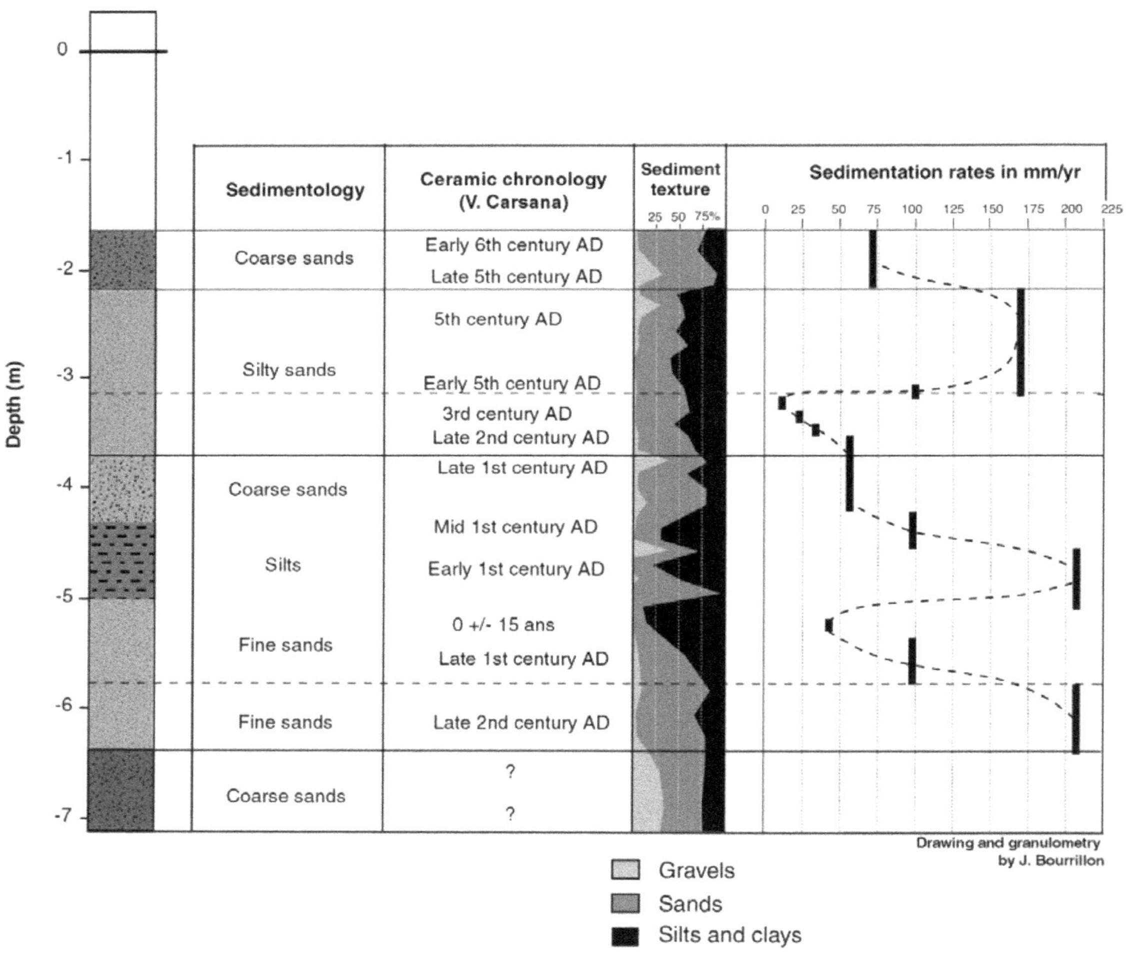

Figure 1.32: Stratigraphic section from Naples harbour (Piazza Municipio) showing the high variability of sedimentary fluxes. Anthropogenic impact is fundamental in explaining the harbour silting up. Excavations directed by D. Giampaola (Archaeological Superintendence of Naples) and assisted by V. Carsana.

(3) **Frequent chrono-stratigraphic inversions** are observed from the Roman period onwards. Rapid aggradation of marine bottoms impinged on the 1 m minimum draught depth required for ancient shipping, a problem which grew so acute that it necessitated a management response (Hesnard, 2004a-b). Harbour dredging phases have been unequivocally evidenced in a number of ancient harbours, characterised by the removal of great tracts of Bronze Age and Iron Age sediments (Marriner and Morhange, 2006a). This can create an important archive loss in certain areas of the basin (**Figure 1.33**). Recognising these hiatuses is significant in estimating the archaeological importance of the archive in question.

(4) **Granulometric paradox.** Roman facies are diagnosed by a granulometric paradox juxtaposing plastic clays, indicative of the prevailing low-energy hydrodynamic processes, and a coarser gravel fraction comprising seeds, charcoal, wood fragments, molluscan shells and a plethora of archaeological artefacts. This latter cultural input attests to the basin being used as a huge waste dump by ancient societies.

A relative decline in harbourworks after the late Roman and Byzantine periods is manifest in a return to 'natural' sedimentary conditions comprising coarse-grained sands and gravels. Following thousands of years of accelerated anthropogenic confinement, reconversion to a natural coastal parasequence is typified by high-energy upper shoreface sands. A change in geometry is also observed with transition from aggradational to progradational strata. This progradation significantly reduced the size of the basins, burying the heart of the anchorages beneath thick tracts of coastal and fluvial sediments.

Figure 1.33: (A-B) Tufa substratum scouring marks resulting from Roman dredging of the harbour bottom at Piazza Municipio, Naples. Excavations directed by D. Giampaola (Archaeological Superintendence of Naples) and assisted by V. Carsana.

1.7 The ancient harbour: an archaeological paradigm?

Traditionally, poor archaeological layers were overlooked as data voids, however an increasing body of literature suggests that the judicious use of litho- and biostratigraphies can yield extensive data in apparently homogeneously sterile strata. In opposition with traditional research approaches to port complexes we posit that the outlined geological processes have created an important archaeological archive that has the potential to significantly advance the study of coastal archaeology.

The latter has traditionally been dominated by architectural approaches to the study of boats, to the detriment of supporting port infrastructure and settlement. Although potentially rich archives in terms of, for example, patterns of consumption, shipwrecks relate to a single event in time and therefore remain relatively modest in spatial and temporal terms. Whilst shipwrecks and their 'sunken treasures' have captured the public imagination (Ballard, 1987; Goddio *et al.*, 1998; Ward and Ballard, 2004) from a holistic perspective they are relatively marginal pieces of the maritime record, given priority over and diverting resources from the local and regional coastal pictures.

Despite the promising beginnings of maritime archaeology in the 1930s and 1940s, our understanding of ancient harbours has actually advanced very little compared to other aspects of the science (Poidebard, 1939). Until the 1990s, and to a great extent even today, it is striking that technological advancements (underwater coring, side-scan sonar, geophysical cartography etc.) are paradoxically contrasted against multi-disciplinary research stagnation. In effect, traditional approaches can only have a limited effect on the future development of this subject whilst the pioneering multidisciplinary projects at Caesarea and Marseilles show the scope of a theoretically informed landscape approach. Areas as culturally complex and dynamic as the coast require flexible multifaceted research strategies. Use of the geological record allows three types of information to be teased out:

(1) **Where?** Geoscience techniques can be used to spatially locate ancient harbour basins. Diagnostic litho- and biostratigraphies, consistent with human modified basins, are clearly translated in the geological record. At locations where the approximate location of the former basins is known, geoarchaeology facilitates a high-resolution reconstruction of the harbour history (e.g. Alexandria, Caesarea and Marseilles). In areas where the

basins are not spatially constrained, reconstruction of the Holocene coastal history can aid in advancing informed hypotheses concerning their location (e.g. Cuma, Tyre's southern harbour). (2) **When?** The transition from natural to anthropogenic environments recorded in the stratigraphy can be dated using either radiometric or ceramic dating techniques. (3) **How?** How did local populations modify their coastal environment? Either (a) indirectly, through the production of increased sediment yields; and (b) directly, by modifying their natural environment to produce low energy basins. Recent work from the Levantine coast indicates that these impacts were relatively moderate during the Bronze Age and reached their apogee during the Roman period.

Harbour basins are also appropriate for the analysis of archaeological data at three scales. (1) Basin scale: an informed geoarchaeological approach can yield insights into the harbour basin topography, its functioning, spatial organisation, and coeval infrastructure through time. (2) Urban scale: information pertaining to the site's occupation history, notably using geochemistry and geophysics, is made possible due to high rates of sedimentation through time. (3) Regional scale: typological data can be derived on how these individual maritime sites interacted as regional economic complexes.

The unique geological specificity of ancient harbours means that former marine environments can be probed over large areas using terrestrial or submarine survey methods. We outline six areas relevant to coastal archaeology at present:

1.7.1 Logistic difficulties of excavating in heavily urban contexts are significant, for example at Alexandria (Goiran *et al.*, 2005), Naples (Giampaola *et al.*, 2004) and London (Milne, 1982, 1985; Milne and Bateman, 1983; see De Maria and Turchetti, 2004a-b; Zevi and Turchetti, 2004a-b for other examples). Evidence is quintessentially fragmentary and difficult to integrate into wider spatial and temporal archaeological pictures.

Coastal archaeology is significantly linked to the whims of the construction industry, whose planning delays are invariably incompatible with scientific agendas. A major challenge is therefore to extract the maximum amount of data over relatively short time periods. Grounding this in a robust intellectual framework is a major challenge. The protected site of Caesarea is an excellent example of meticulous research undertaken over the long-term with an aim not only to expound but also to preserve the site's cultural heritage. At the majority of urban sites such minutiae are not possible and the geosciences have a key role to play in identifying and rapidly delimiting those areas of greatest research potential. At Sidon and Tyre, recent geoarchaeological research has allowed the ancient harbour areas to be precisely delimited, showing that the heart of the ancient basins lie beneath the modern urban centres and constitute areas of immense archaeological importance (Franco, 1996; Marriner and Morhange, 2005a). Future construction must consequently be undertaken in close collaboration with the archaeological authorities. At Marseilles, the precise reconstruction of shoreline progradation is a good example of how urban planning and geoscience can be reconciled (Morhange *et al.*, 2003a).

1.7.2 Difficult working conditions below the water table. Notwithstanding the considerable advances made in coastal archaeology since its beginnings, working conditions remain difficult and often prohibitively expensive. For example, it is not usually possible to excavate vertical cross sections below sea level, or the water table line, because sand and silt do not hold an appreciable wall. This renders the systematic recording of stratigraphy at best difficult and at worst impossible. Consequently, stratigraphy at coastal sites has in the past often been ignored or not properly examined. Casing and coring techniques can overcome these problems and rapidly expound the coastal and archaeological stratigraphies with a view to precisely directing resources.

1.7.3 High-resolution archaeological data. That some objects survive well in certain depositional contexts but not others has been acknowledged by archaeologists since the nineteenth century (Caple, 2001). This paradox has greatly skewed interpretation of the archaeological record and led early antiquarians to define prehistory on the basis of what was preserved: stone, bronze and iron. The only opportunity to gain a holistic view of the past is when we find remains in frozen (e.g. Ötze the ice man), desiccated or waterlogged anoxic conditions which preserve organic and other materials.

Large-scale urban excavations such as those at Marseilles and Naples have shown the considerable research scope of ancient harbour geoarchives. The unique preservation potential of fine-grained silts and sands (low-energy anoxic environment), is one of the key defining elements of the harbour geological paradox and is significant in probing the artefactual record. Enveloping sediments, coupled with the presence of the water table, have the potential to preserve otherwise perishable archaeological artefacts such as wood, leather etc. (Holden *et al.*, 2006). The most impressive archaeological findings to date include numerous shipwrecks (yet again!) and well-preserved wooden harbourworks (Pomey, 1995). The density of both vertical and horizontal information is exceptional.

1.7.4 Sea-level studies. The analysis of harbourworks and the fixed and boring fauna attached to structures has long been recognised as a potential source of sea-level data (Pirazzoli and Thommeret, 1973; Laborel and Laborel-Deguen, 1994; Morhange *et al.* 2000). Such data, fundamental to understanding the vertical distribution of coastal remains, had traditionally derived uniquely from the geological and geomorphological record (note that Lyell, Negris and Cailleux were all exceptions to this). Where precise vertical relationships can be established between archaeological structures and biological palaeo-sea levels it has been possible to accurately reconstruct relative sea-level trends since antiquity at a number of Mediterranean sites (Pirazzoli, 1976, 1979-1980, 1980, 1987b, 1988). Three groups of structures have traditionally been used (Blackman, 1973b; Flemming, 1978, 1979-1980; Flemming and Webb, 1986; van Andel, 1989; Stanley, 1999; Blackman, 2003): (1) emerged vestiges (dwellings, stock houses, walls); (2) partially emerged structures (quays, mooring-stones, slipways, channels; coastal wells can also give insights into sea-level tendencies [Sivan *et al.*, 2001]); and (3) submerged structures (shipwrecks). Unfortunately, the technique tends to be marred by large altitudinal errors, where the envelope of imprecision is often as important as the absolute sea-level change, and dating uncertainties, in addition to the interpretative vagaries of certain structures. Another problem is linked to the geological meaning of these data.

Pioneering work in the 1970s demonstrated that these shortfalls could be overcome using biological fossil remains attached on or in certain interface harbour structures (e.g. quays and jetties). By transposing the techniques developed on rocky coasts (Pirazzoli, 1988; Stiros *et al.*, 1992; Laborel and Laborel-Deguen, 1994) to the ancient harbour context, precise sea-level datasets have become a possibility. Not only do biological remains yield organic material for radiocarbon dating but the biological zonation of certain species (e.g. the upper limit of *Balanus* spp., *Lithophaga lithophaga*, *Vermetus triqueter*, *Chama griphoides* populations) is empirically and precisely linked to biological mean sea level (Péres, 1982; Laborel and Laborel-Deguen, 1994). By measuring the upper altimetric difference between fossil and contemporary populations low vertical error margins of $\pm$ 5 cm can be obtained (**Figures 1. 34-1.35**).

1.7.5 Long-term archives of anthropogenic impacts. Relationships between man and his environment have long been considered in quasi-isolation, either from a human perspective or an environmental stance, rather than a coevolution where both are complimentary. In our opinion, harbours can be used to write the history of man and his interactions with the environment since prehistory. The presence of man is manifested by a number of proxies that cumulatively have significant historical scope. (1) **Granulometric impacts**: the construction of harbourworks is at the origin of a unique fine-grained facies. This lithoclastic signature facilitates a delimitation of the ancient basin topography. (2) **Morphological impacts**: the rapid aggradation of harbour bottoms leads to accelerated coastline progradation. The precise study of accumulation rates and sedimentary fluxes yields insights into the erosional system, notably landscape use at both the urban and watershed scales. (3) **Biological pollution**: modification in faunal assemblages translates local anthropogenic inputs such as increases in turbidity and use of the basin as a waste depocentre over many thousands of years. (4) **Geochemical impacts**: lead has proved to be a powerful tool in recognizing ancient industrial activities (Shotyk *et al.* 1998; Véron *et al.*, 2006).

Figure 1.34: Marseilles excavations, Place Jules Verne (photos by C. Morhange). The upper limit of fixed fauna on stakes from the harbour's Roman quays. By measuring the upper altimetric difference between fossil and contemporary populations it is possible to accurately infer palaeo-sea levels during antiquity (centrimetric precision).

1.7.6 Cultural heritages. It is now widely recognised that archaeological remains are a finite resource endangered by modern encroachment (Galili and Sharvit, 2000). At present, ancient harbours are disappointingly neglected in coastal preservation policy. The paradox of underwater sites today buried on land provides a potentially unique opportunity to showcase these cultural heritages to the general public. Outgrowth ideas for coastal parks have been advanced by a number of authors but very few have in reality come to fruition, be it for financial and/or logistic constraints (Franco, 1996). Unlike their drowned counterparts (e.g. Pozzuoli's Portus Julius, Alexandria's eastern harbour), with their problems of visibility and access, buried harbours are unique in that they can be preserved on land allowing visitors to walk freely amongst the vestiges.

Married with modern forms of museum exhibition and harbour reconstructions the didactical and tourism potentials are far-reaching. Such parks also have a role to play in the preservation of sites.

Very few well-preserved examples actually exist, and Marseilles is at present a rare exception. At Caesarea, an underwater archaeological park for divers was trialled during the 1990s (Raban, 1992b) and in 2006 became the first park of its kind. Visitors are furnished with plastic guidebooks and maps and follow cable guidelines to explore the submerged remains and harbourworks. Divers view some 36 different sign-posted sites along four marked trails in the sunken harbour covering an area of ~73,000 m². For less- experienced divers, one trail is also accessible to snorkelers. The others, ranging from 2 to 9 m below the surface, close to the beach, are appropriate for any beginner diver.

One of the potentially best sites is that of Portus, near Rome Ostia, a site which has long captured the public imagination and scholarly interest. Presently buried beneath thick tracts of coastal and fluvial sediments, the harbour remains are still clearly visible. Although a small museum does presently exist at the site, the rich archaeological potential is not exploited to the full (Mannucci, 1996) and Rome's international airport and large urban agglomeration are pressures on the preservation of the site.

In collaboration with local, national and international (e.g. UNESCO World Heritage) organisations a major challenge facing maritime archaeology is to develop a robust conservation framework at the Mediterranean's most important sites. Many ancient ports raise numerous coastal management issues which need addressing in order to ensure the harmonious development of human activities while protecting the submerged remains of the ancient cities. Issues such as urban expansion and environmental problems have increased significantly in recent years. For example at Alexandria, a site of worldwide significance in terms of its coastal archaeology, problems of wastewater pollution are a major obstacle to an underwater archaeological museum.

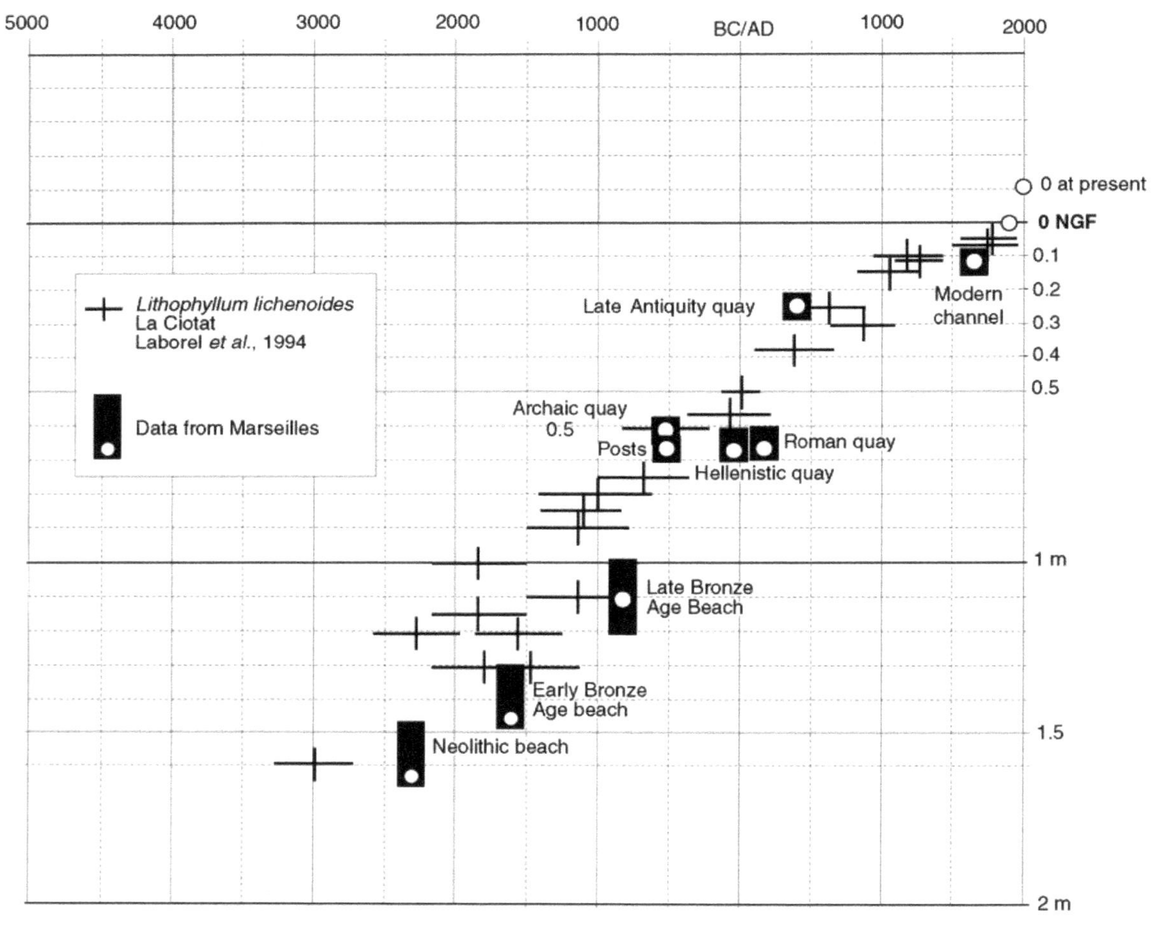

Figure 1.35: In the ancient harbor of Marseilles (southern France), marine fauna fixed upon archaeological structures in addition to bio-sedimentary units document a 1.5 m steady rise in relative sea level during the past 5000 years. A near stable level, at present datum, prevailed from 1500 years AD to the last century. This trend is similar to the one previously documented on the rocky coasts of Provence, southern France. Field observations inside and outside the harbour confirm that no sea-level stand higher than present occurred during the studied period. Since Roman times, relative sea-level has risen by ~50 cm (after Morhange *et al.*, 2001).

1.8 Concluding remarks

It would be wrong to engage in a scientific discourse of the archaeological positives of this new ancient harbour geoscience without exposing some possible shortfalls. Whilst use of the geological record must by no means be seen as a talisman by which to heal all ills, a number of fallibilities remain, often not relating specifically to the overriding discipline itself but rather the tools used. From the perspective of the archaeologist, the new field could be criticised as being overly geocentric. This is to some extent reflected in the propensity for much of the recent literature to feature in specialised geological journals. Thus far, it would appear that a fair *terrain d'entente* has been found whereby geological data is being correctly tied into the archaeological context and vice versa. Progress must be made on two key questions:

(1) Can ancient harbours document historical events? Whilst ancient harbours have been shown to be replete with archaeological information, accurately pinpointing precise historical events can be problematic. Ancient harbours are not annual archives and although considerable advances have been made, discrepancies remain in the relationship between calendrical and radiometric chronologies. At sites where whole sections can be investigated archaeological material can provide good decadal temporal resolution. Where this information is not readily available (e.g. core stratigraphies), radiocarbon chronologies can produce broad bands of dates (centennial resolution), although these are never capable of precisely reconciling certain events, such as harbour construction, with any great degree of precision. This problem is further accentuated in the coastal domain by the vagaries of marine calibration (Reimer and McCormac, 2002). A great deal of recent research has sought to establish local site-specific reservoir ages, although this can be subject to significant variation (Siani *et al.*, 2000). There is therefore a manifest paradox between high-resolution geological data, capable of recording rapid cultural change, and the dating resolutions used to constrain this narration.

(2) Until recently, harbour sediment archives were taken to be relatively continuous records of anthropogenic and coastal change since antiquity. However, a mounting body of evidence elucidating Roman and post-Roman dredging practices suggests that this premise should now be modulated. At sites such as Marseilles, Naples, Sidon and Tyre, significant dating discrepancies coupled with unique sedimentologies provide widespread evidence of repeated dredging from the third century BC to the fifth century AD (Marriner and Morhange, 2006a). Such practices were of course logical management responses to the silting problem and ensured harbour viability. In many ports therefore, significant tracts of Bronze Age and Iron Age strata are missing. Precisely constraining this loss of space and time is fundamental in assessing the archaeological scope of the harbour basin. It also suggests the need for core networking to yield the most complete stratigraphic record possible and identify those areas of missing and complete record.

Although ancient harbour geology is a paradigm thus far little known in archaeology we believe it represents a significant shift from dogmatic to pragmatic research which opens up new interpretative possibilities. It is an innovative means of integrating various scales and types of data with potentially profound and pervasive implications for maritime archaeology as a whole. Traditional disciplinary studies have been shown to be largely inadequate when considered in isolation and a geoarchaeological approach is particularly useful in areas of data paucity.

Whilst paradigm shifts are often slow to reach wider acclaim we believe it is fundamental to link wreck archaeology with the coastal, and notably harbour research, to formulate a more holistic picture of maritime landscapes in antiquity. The benefits of a geoscience approach, both in terms of cost effectiveness and data wealth, are arguably unparalleled. We deem it important to develop new scientific arenas, integrating this new approach with an interpretative archaeology.

Chapter 2

Alexander the Great's isthmus at Tyre

Geological heritage vs. pluri-millennial anthropogenic forcing

*Tyre's coastline is today characterised by a wave-dominated tombolo, a peculiar sand isthmus that links the former island settlement to the adjacent continent (**Figure 2.1**). This rare coastal feature is the heritage of a long history of natural morphodynamic forcing and pluri-millennial human impacts. In 332 BC, following a long and protracted seven-month siege of the city, Alexander the Great's engineers cleverly exploited Tyre's atypical geological context to build a causeway and seize the island fortress; this edifice served as a prototype for Alexandria's Heptastadium built a few months later. Our research shows that both causeways profoundly and pervasively deformed the natural coastline and entrained rapid progradation of the spits.*

The logistics behind Alexander's causeway have been a matter of archaeological speculation for some time. It has been conjectured that the key to explaining how Hellenistic engineers overcame the difficult physical conditions lies in the geological record. A number of authors have proposed informed morphogenetic scenarios based on aerial photography and classical texts (Stewart, 1987; Nir, 1996). However, despite a rich North American literature on spit morphogenesis, this is the first time that anthropogenic tombolo formation and stratigraphy have been precisely studied.

2.1 Introduction

The term tombolo is used to define a spit of sand or shingle linking an island to the adjacent coast. Despite a long history of inquiry (Gulliver, 1896, 1899; Johnson, 1919; Guilcher, 1958; Escoffier, 1954), the literature has traditionally focused upon form-process aspects of their geomorphology with very little attention being paid to the salients' sedimentary history. While the origins of spits and barrier islands have been the object of significant scientific debate, notably with the rich North American literature (see Davis, 1994; Stapor and Stone, 2004; Stone *et al.*, 2004; Schwartz and Birkemeier, 2004; Otvos and Giardino, 2004; Gardner *et al.*, 2005; Simms *et al.*, 2006), papers discussing tombolo evolution - their stratigraphic architecture and stages of accretion - are few and far between. The meagre corpus of recent literature has tended to concentrate on: (1) the role of contemporary processes operating around tombolos (Hine, 1979; Dally and Pope, 1986; Silvester and Hsu, 1993; Flinn, 1997; Courtaud, 2000; Browder and McNinch, 2006); (2) establishing a typology of shoreline salients and tombolos (Zenkovich, 1967; Sanderson and Eliot, 1996); and (3) modelling the geometrical relationships of these depositional forms relative to the breakwater island (Ming and Chiew, 2000). Although the use of detached breakwaters in shoreline protection has renewed interest in salient accretion, understanding of the dynamics and sedimentary history of these forms on longer centennial and millennial timescales remains poor (Clemmensen *et al.*, 2001).

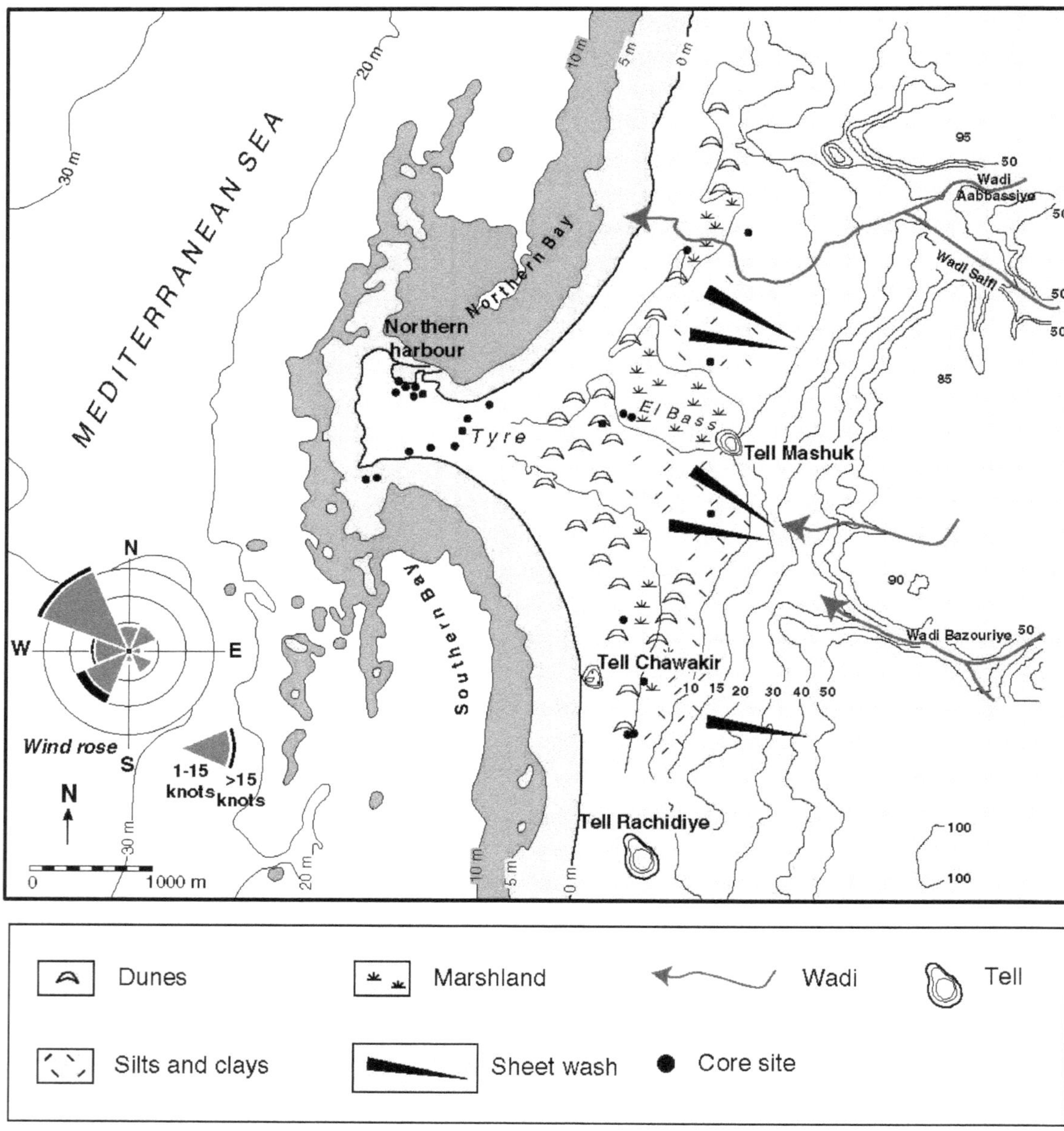

Figure 2.1: Simplified geomorphology of Tyre's tombolo and coastal plain.

Features of an irregular coastline, tombolos have the effect of straightening the littoral zone. The processes determining their formation are similar to those observed in the growth of spits, bars and barrier islands (Anthony and Blivi, 1999; Blivi *et al.*, 2002). Tombolos tend to develop in shallow areas behind island barriers, where sufficient sediment supply coupled with wave and wind action are favourable to beach accretion. Tides, swell and currents serve as the transporting media, interacting with the island to set up a complex pattern of wave refraction and diffraction on the lee of the obstacle. Sunamura and Mizuno (1987) have calculated that a tombolo forms where the ratio of the island's offshore distance to its length is equal or less than 1.5; a salient forms where it is 1.5-3.5, and no protrusion of the coast occurs where it is greater than 3.5. At Tyre, the present ratio is 1.47, at the very limit for tombolo formation. In antiquity, however, this ratio was much lower, for example 0.55 during the Iron Age. By comparison, the ratio for Alexandria's salient is presently 0.35 and 0.21 during the Hellenistic period. The large discrepancies at Tyre are notably due to tectonic collapse of the island bastion during the late Roman period and a reduction in size of the breakwater island by ~470000 m² or ~50 %.

Tombolos are part of the Highstand Systems Tract (Catuneanu, 2002, 2005) and, like deltas and estuaries, the inception of their prograding sediment wedges is dated ~8000 to 6000 BP (Stanley and Warne, 1994; Stefani and Vincenzi, 2005). They comprise triangular forms, widest at the base – where the initial accretion begins – thinning laterally towards the island obstacle. The scales of these landforms can vary considerably from just a few tens of metres behind small obstacles (**Figure 2.2**) up to fifteen kilometres as in the case of Orbetello on the Italian coast (Gosseaume, 1973). Although single tombolo bridges tend to be the rule, there are good examples of tombolo pairs (Giens, France, see Blanc, 1959; Courtaud, 2000) and even triplets (Orbetello, Italy, see Gosseaume, 1973) from the Mediterranean. Double tombolo formation is attributed to waves approaching the island flanks at different angles of incidence, with lagoons developing in the sheltered area between the two tombolo arms (Blanc, 1959).

Tombolo geology is complicated by a multiplicity of depositional environments – including subtidal zone, beach dunes, washover fans, marshes, lagoons and inlets – some of which are laterally continuous, whilst others are not. It is accepted, however, that the overriding requisites for tombolo formation are: (1) high sediment supply; (2) coastal processes conducive to the development and maintenance of the isthmus (e.g. wave diffraction); and (3) a favourable geomorphic setting, most significantly an island obstacle that serves as a physical barrier against the swell.

We have demonstrated that coastal sediments are rich time-series replete with data on the magnitude, variability and direction of natural and anthropogenically forced change since antiquity (Marriner *et al.*, 2005; Marriner *et al.*, 2006a-b). Here, we report stratigraphic and geomorphologic data from Tyre's isthmus, proposing a tripartite chrono-stratigraphic model of tombolo evolution. Juxtaposed against other circum Mediterranean examples (e.g. Orbetello, Giens, Gibraltar and Alexandria [**Figure 2.3**]), Tyre is a small sedimentary system rendering it ideal for scientific inquiry and the formulation of a stratigraphic model. These data are then compared and contrasted with research undertaken on Alexandria's tombolo (Marriner, 2000; Goiran, 2001; Goiran *et al.*, 2005).

Figure 2.2: Accretion of a micro-salient 3 km north of Tyre. The photographs clearly show the outgrowth of regressive beach deposits from the continental edge towards the island breakwater (photograph: N. Marriner; inset: DigitalGlobe, 2006).

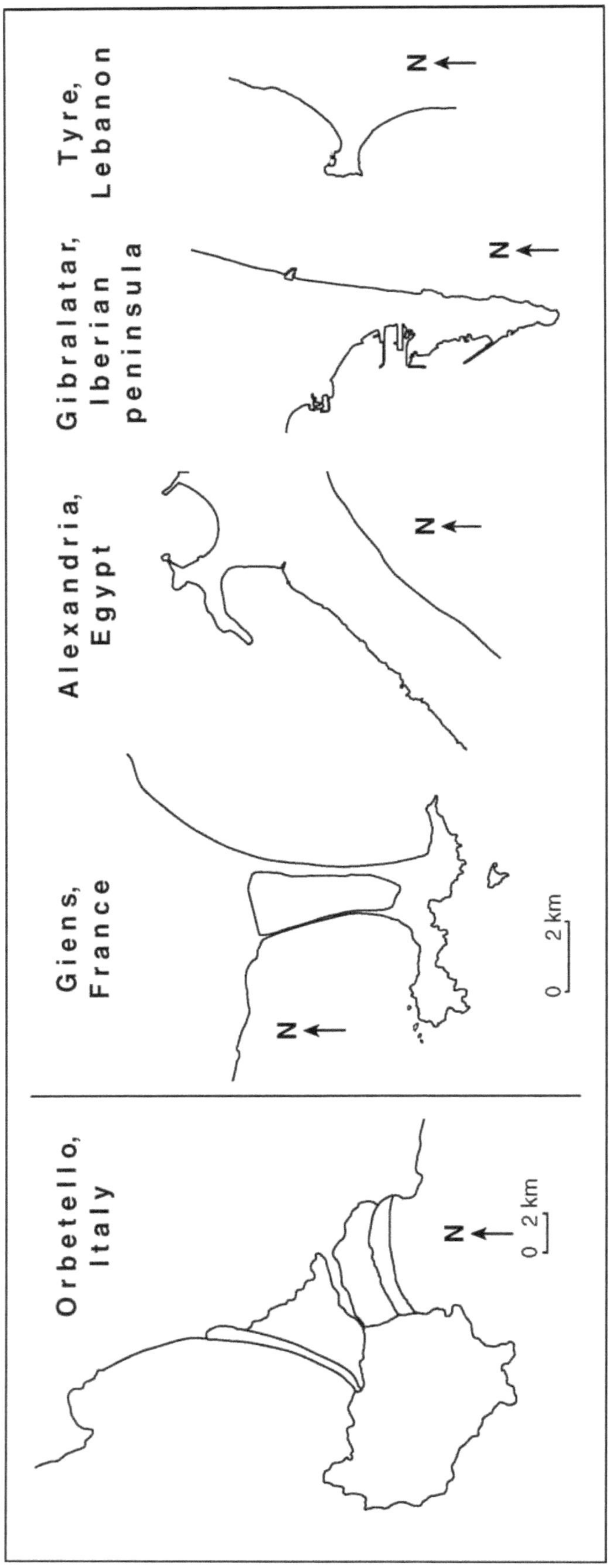

Figure 2.3: Tyre's tombolo set against some of the Mediterranean's better known examples.

2.2 Meteo-marine and geomorphological contexts

South Lebanon is subject to a temperate Mediterranean climate, with humidity of ~70% along the coast (Blanchet, 1976). This humidity is controlled by precipitation from westerlies that bring marine storms during the cold season (Geyh, 1994). The summer climate is influenced by the location of the Inter-Tropical Convergence Zone (ITCZ) over the deserts of the Sahara and Arabia. During autumn and spring, dust storms with hot dry winds govern the weather, sometimes interrupted by sporadic heavy rain storms. The hottest months are July and August (26°C average) and the coldest January and February (13°C average). Generally, the sea is always warmer than coastal air temperatures. The difference is small during the spring (1.4°C in May), but increases during the summer to attain a maximum offset of 5°C during December. During the winter, these differences in temperature lead to high intensity rainfall (Blanchet, 1976). Annual precipitation averages 800 mm, characterised by high intensity rainfall during the winter months with little precipitation for the remainder of the year (Blanchet, 1976; Geyh, 1994; Soffer, 1994).

The coastal plain in the Tyre vicinity reaches a width of 2-3 km. This is atypical of the south Lebanese coastline which is otherwise dominated by a series of Cenomanian and Eocene headlands that protrude into the sea (Dubertret, 1955; Sanlaville, 1977). Tyre lies upon an uplifted horst, bounded to the east by the Roum fault and to the south by the Rosh Hanikra - Ras Nakoura fault (Morhange *et al.*, 2006b). A series of ENE-trending faults intersect the Tyrian coastal plain putting into contact the Cenomanian and Eocene geology. The coastal plain is dominated by both consolidated and clastic Quaternary deposits. Throughout the Holocene, the Tyrian coastline has been protected by a broken chain of sandstone reefs, part of a drowned north south ridge that runs parallel to the coastline (Dubertret, 1955; Sanlaville, 1977). The oblique stratification of these outcrops indicates that they are aeolian in origin. The sandstone ridges are locally referred to as 'ramleh' and are the chronostratigraphic analogues of the 'kurkar' ridges rimming the Israeli coast (Frechen *et al.*, 2001, 2002, 2004; Sivan and Porat, 2004). These Quaternary reefs have dampened the effects of the swell, and aerial photographs show an intricate pattern of wave refraction and diffraction on the lee of Tyre island. The defensive 'impregnability' afforded by these offshore sandstone reefs attracted human societies from the Bronze Age onwards (**Figure 2.4**). Successive cultures, including Phoenician, Persian, Hellenistic, Roman and Byzantine have significantly marked and modified the coastal landscapes (Katzenstein, 1997; Doumet-Serhal, 2004a).

Despite early attempts by Nabuchodonozor II to link Tyre to the mainland during the sixth century BC (Fleming, 1915; Nir, 1996; Katzenstein, 1997), our stratigraphic data demonstrate that the defensive bastion remained an island until 332 BC, separated from the continent by ~800-1000 m of sea. Following stubborn resistance by the Tyrians, Alexander the Great cleverly exploited the shallow bathymetry, induced by Tyre's proto-tombolo, to build a causeway and breach the island's insularity. When the city eventually fell to the Macedonian armies, Alexander strengthened the mole using rubble from destroyed quarters of the city (see Diodorus Siculus, 1967, writing in the 1st century BC). In its most advanced stage, it is speculated the causeway reached an average width of 200 Greek feet, or ~60 m (Nir, 1996). This engineering feat, a prototype to Alexandria's causeway built a year later, significantly artificialised the coastline and accelerated accretion of the salient.

Tyre's tombolo comprises an west-east trending salient, 1500 m long by 3000 m wide (**Figure 2.1**), and lies ~9 km south of the Litani delta. The latter, Lebanon's most important fluvial system, transits 284 x 10^6 m^3 of sediment per year (Abd-el-Al, 1948) and has been one of the primary sediment sources for the isthmus during the Holocene (see chapter 6). Given the microtidal regime (<45 cm, Haifa tide gauge data in Goldshmidt and Gilboa, 1985; **Figure 2.5**), wave processes - both wave impact and longshore currents - and swell diffraction are the key factors driving tombolo formation at Tyre. Dominant coastal winds and swell derive from the south-west, with periodic north-westerlies giving rise to multi-directional long-drift current. Wave climatological data attest to extreme high energy sea-states; wave heights of more than 5 m are measured every ~2 years, and greater than 7 m every ~15 years (Rosen and Kit, 1981; Goldsmith and Sofer, 1983; Carmel *et al.*, 1985a-b). The site is exposed to a wide swell window with a fetch of 743 km from the northwest and 646 km from the southeast (**Figure 2.6**). Although extensively urbanised at present, nineteenth century engravings and aerial photographs indicate important dunefields capping the tombolo until the 1950s (**Figures 2.7 and 2.8**). Rising up to 10 m above present Mean Sea Level (MSL), these dunes have since been partially removed to accommodate urban growth and exhume the tombolo's archaeology.

Today, the tombolo is asymmetric, comprising a better developed southern lobe. Its base lies between Chabriha, a beach rock formation 2.5 km north of the city, and Tell Chawakir, 2.25 km to the south. Wide sandy beaches and mobile dune formations, which cap the Quaternary aeolionite deposits, are characteristic of the El Rachidye area (up to 250 m wide) and run 3 km northwards, through El Chawakir to El Bass (**Figure 2.1**). Dunes up to 9 m in height and containing Bronze Age to Roman ceramics are known from the Tell Chawakir area. Before the subaerial accretion of the salient, this area served as a haulage beach for Bronze Age mariners living on the tells (Marriner and Morhange, 2005a-b).

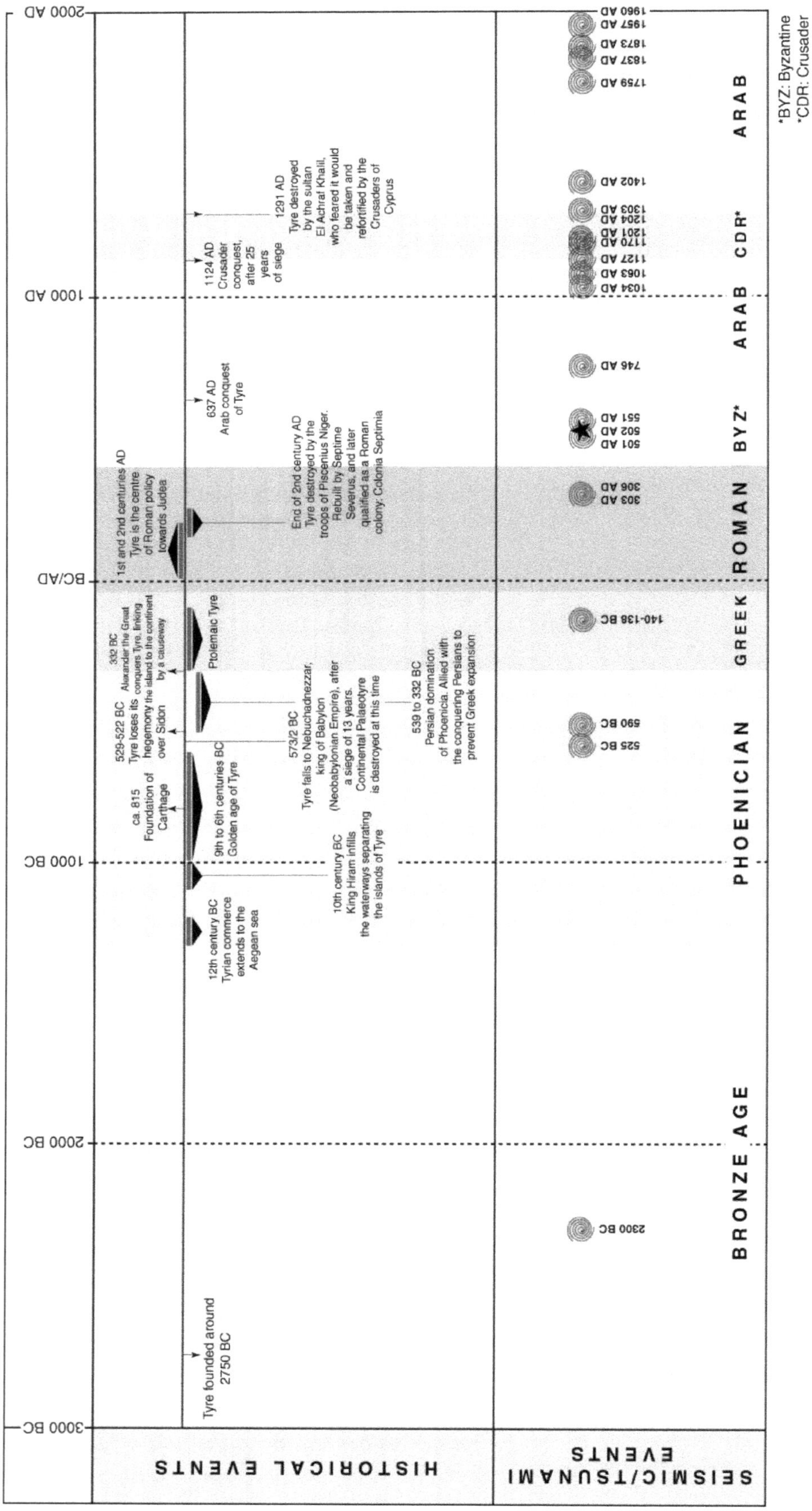

Figure 2.4: Time-line of the history of Tyre.

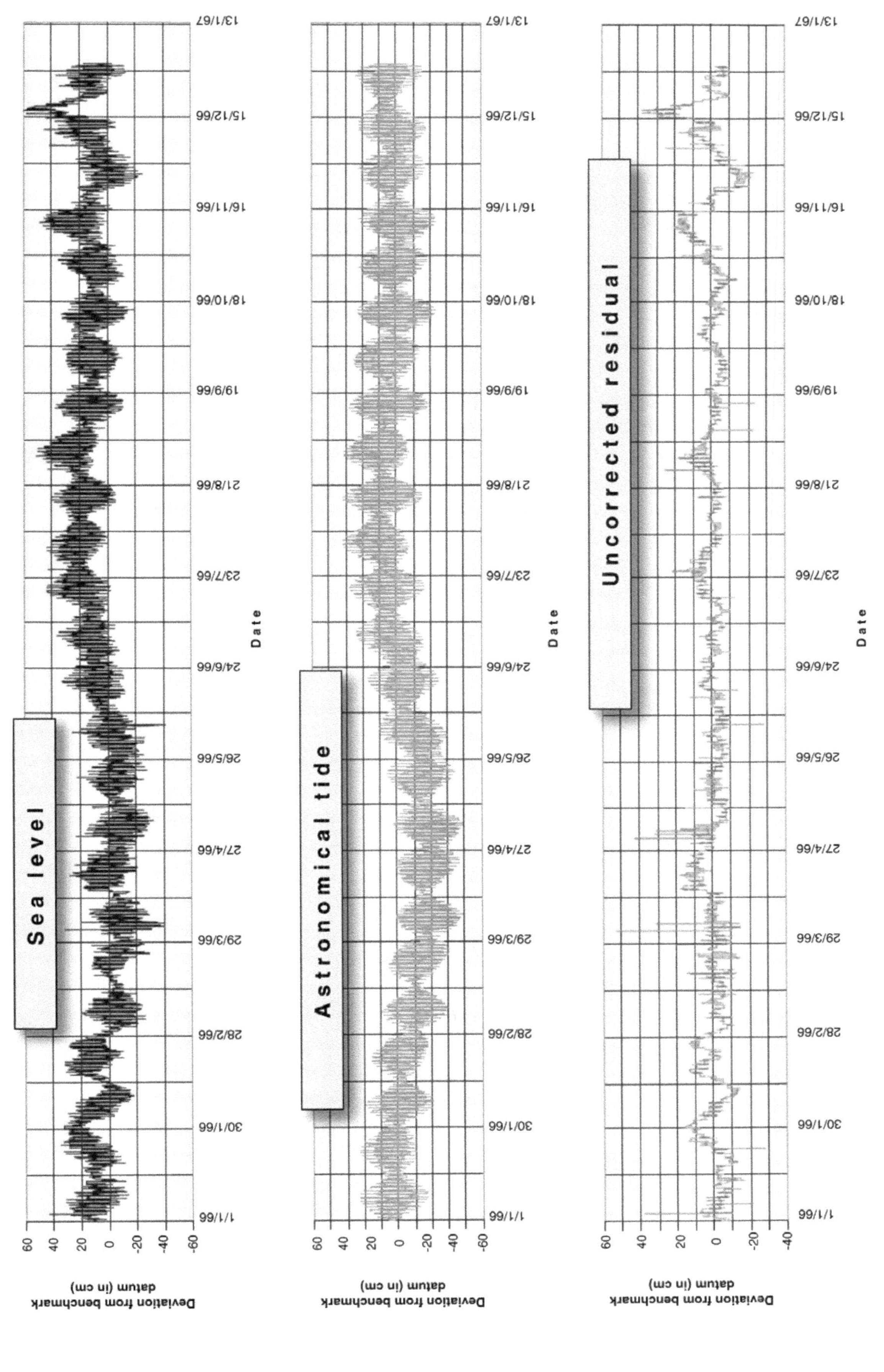

Figure 2.5: Haifa tide gauge data for 1966 showing an average tidal range of ~45-50 cm, consistent with a microtidal regime on the south Lebanese coast (after Goldshmidt and Gilboa, 1985).

Figure 2.6: Tyre's swell window. Tyre is exposed to a wide swell window with fetch distances of over 600 km from the southwest and northeast and ~1890 km from the west.

Figure 2.7: Tyre from the isthmus by David Roberts. The lithograph, dated 1839, looks west from Tell Mashuk towards the tombolo and the former island city. The view reveals the tombolo's morphology before much of its archaeology was exhumed during the twentieth century. Note the silted up northern harbour and an absence of harbourworks in the basin (Roberts, 1839 in Roberts, 2000).

Figure 2.8: Twentieth century evolution of Tyre's tombolo. While the dimensions of the tombolo have remained relatively stable, the sandy salient has been extensively urbanised since the 1970s. This surface construction work has led to a loss of the surficial medieval and modern period geological records.

2.3 Methods and data acquisition

2.3.1 Field data acquisition

At Tyre, a total of 25 cores was obtained (**Figure 2.1**). Sediments were extracted using a 10 cm by 200 cm mechanised corer. Core sections were lined with protective sleeve mechanisms to avoid sediment slumping. All cores were GPS levelled and depths quoted on logs have been benchmarked against biological mean sea level (i.e. the summit of the subtidal zone, Laborel and Laborel-Degeun, 1994). This benchmark was established using the upper limit of *Balanus* populations growing on the modern harbour quay faces. Following extraction, all cores were stored in cold rooms at the CEREGE geoscience facilities in Aix-en-Provence, France.

2.3.2 Stratigraphy and sedimentology

Initial facies descriptions (e.g. colour, petrofacies) were undertaken under standardised laboratory conditions, before sampling to establish general stratigraphy. Samples were oven dried at 40°C and subsequently described using the Munsell colour scheme. Dry sediment aggregates were weighed and washed through two mesh sizes, 2 mm and 50 µm, to separate out the gravels (>2 mm), sands (2 mm to 50 µm) and silts and clays (<50 µm) fractions. In most cases, 100 g of dry aggregate were washed through to ensure the statistical validity of the results. The dried fractions were weighed and data plotted against stratigraphic logs in percentages.

The sand fraction was subjected to mechanical sieving. At least 50 g of dry sands were used. These were mechanically sieved using 15 meshes descending in size from 1.6 mm to 0.063 mm, and accordingly weighed. Statistical analyses were subsequently performed to establish various grain size parameters such as histograms, fractiles and graphical indices (Folk, 1966).

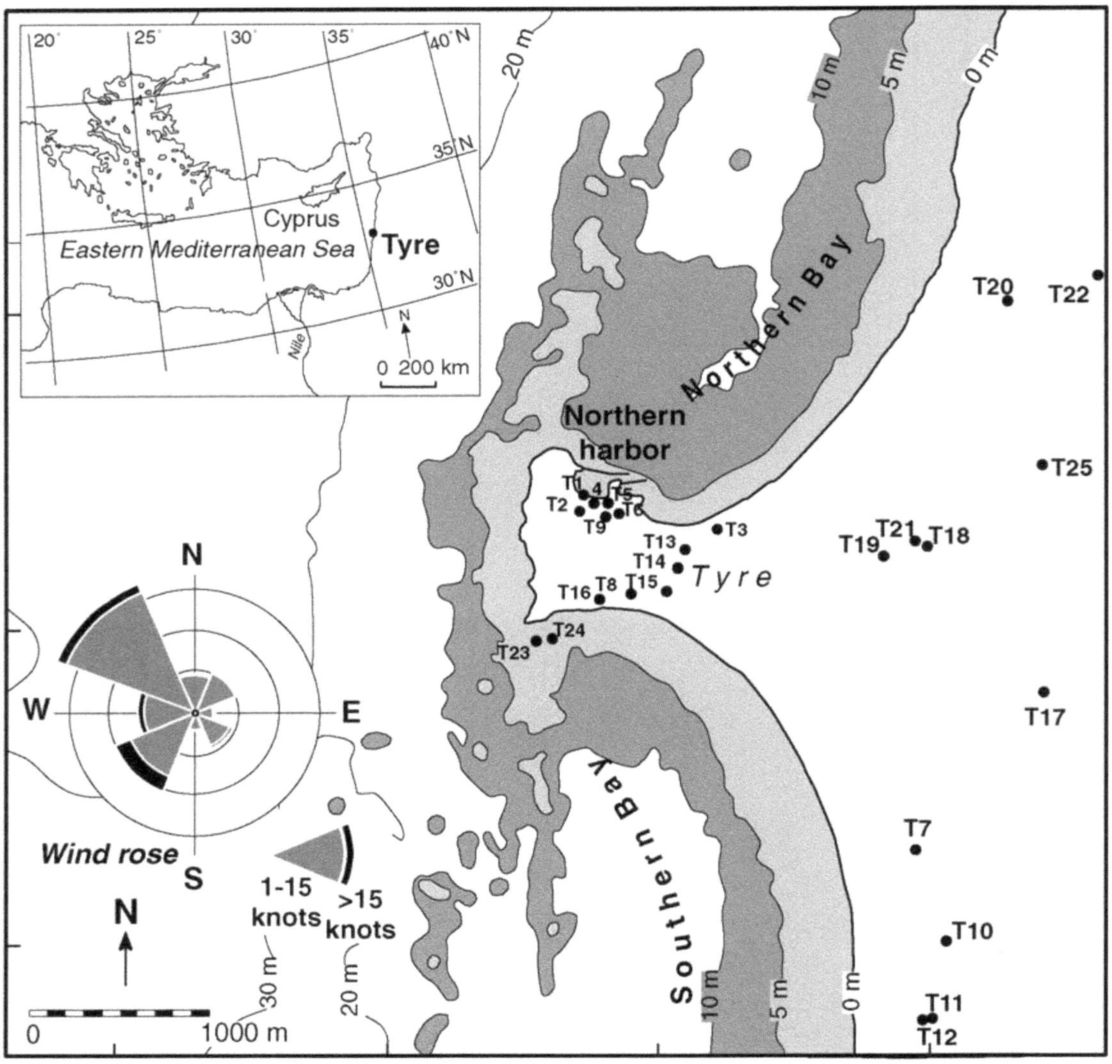

Figure 2.9: Location of core sites on the Tyrian peninsula (black dots denote the sites).

2.3.3 Biostratigraphy

Identification of mollusc shells was undertaken upon the retained gravels fraction and assigned to assemblages according to the Péres and Picard (1964), Péres (1982), Barash and Danin (1992), Poppe and Goto (1991, 1993), Bellan-Santini *et al.* (1994), Bitar and Kouli-Bitar (1998) and Doneddu and Trainito (2005) classification systems. Both *in situ* and *extra situ* taxa were identified.

Ostracoda were extracted from the dry sand fraction (>150 µm). A minimum of 100 valves was identified and assigned to five assemblages on the basis of their ecological preferences: fresh water, brackish lagoonal, marine lagoonal, coastal and marine (Müller, 1894; Breman, 1975; Bonaduce *et al.*, 1975; Carbonel, 1980, 1982).

2.3.4 Chronostratigraphy

More than 40 radiocarbon dates and numerous archaeological data precisely constrain the chronology of the various sedimentary environments observed. The ^{14}C data are summarized in **Table 3.1**, consistent with standard reporting protocols. All dates have been calibrated using OxCal version 3.10, and are quoted to 2σ (Bronk Ramsey, 2001). Material dated included seeds, wood and charcoal remains, and *in situ* molluscan marine shells. Marine samples have been corrected for reservoir effects of 400 years. To test the applicability and robustness of our radiocarbon chronology we dated charcoal (2215 ± 30 BP; Poz-5777) and three *in situ* tests of *Loripes lacteus* (2505 + 30 BP; Poz-5775) from the same stratigraphic layer, TIX 35. These data are consistent with three published ^{14}C ages of known age shells from the Levantine coast, and confirm the absence of any radiocarbon discrepancies at Tyre (Reimer and McCormac, 2002).

2.3.5 Numerical models

For the tombolo, the stratigraphic datasets are based upon a series of 4 cores (**see Figure 2.9**), 8 to 16 m in depth. A north-south transect was obtained with an aim to elucidate the stages of accretion of the sedimentary bodies. Two numerical models were also run to better understand: (1) the role of wind-induced currents in forcing coastal processes; and (2) the refraction of the swell by the breakwater island.

2.3.5.1 Wind induced currents model and sediment resuspension

The wind induced current models were undertaken in collaboration with B. Millet, a modelling specialist from the *Laboratoire d'Océanographie et de Biogéochimie* at the *Université Aix-Marseille II* in Luminy. Similar models have yielded good results for Marseilles and Alexandria's ancient harbours (Millet, 1989; Millet *et al.*, 2000; Goiran, 2001; Millet and Goiran, 2007). Bathymetric data were derived from the 1998 SHOM marine map of the Tyrian coastal area (Service Hydrographique et Océanographique de la Marine, 1998, Ports du Liban, 7514, 1:25000) and coupled with our chronostratigraphic datasets to obtain a representative coastal reconstruction of Tyre at ~5500 BP. Based on our observations of the archaeology and coastal stratigraphy, we factored into this reconstruction a 5 m relative sea-level rise, comprising 2 m of eustasy and 3 m of tectonic subsidence.

For the marine currents, five model runs were performed for wind speeds of 8 m/s from the following angles: 225°, 250°, 270°, 290° and 315°. The 8 m/s is taken as an average value based on local meteorological and oceanographic data (Blanchet, 1976; Goldsmith and Sofer, 1983; Carmel *et al.*, 1985a-b). A series of secondary runs was also performed to calculate the wave heights, bottom tensions and sediment resuspension for the two dominant northeast (315°) and southwest (225°) wind directions. A wind forcing of 8 m/s was equally incorporated into these calculations.

2.3.5.1.1 The wind induced currents model. The numerical model is based on a two dimensional set of equations which calculates the horizontal current speed, averaged out over the vertical water column. It assumes a homogeneous waterbody forced uniquely by the wind, and does not accommodate any swell parameters. The equations are resolved numerically using an Alternating Direction Implicit (ADI) schema. For Tyre, the calculated area comprised a 24 by 54 grid of 50 m squares, a total of 1296 data points. Twenty hour simulations were performed at a time increment of ten seconds to ensure the stability of the numerical calculations. At each time increment, the computer program calculates (1) the horizontal current speeds averaged out vertically over the water column and (2) the height of the water surface.

2.3.5.1.2 The sediment resuspension model employs equations devised by Vlag (1992). These equations are used to calculate sediment resuspension induced by the wind in shallow coastal areas. The formulae used to calculate wave height and period derive from the work of CERC (1975).

2.3.5.2 Swell propagation and diffraction model

In collaboration with S. Meulé of the CEREGE, a swell propagation model was also run for bathymetries at 5500 BP and 2000 AD. The model used equations from the Steady-State Spectral Wave Model (STWAVE) with a JONSWAP swell spectrum. Due to the long nature of these equations, we refer the reader directly to McKee Smith *et al.* (2001). A total of six model runs were completed for swell scenarios at 2000 AD from 225°, 270° and 315 °, and at 5500 BP at 225°, 270° and 315° (dominant swell direction). Wave spectrums were generated for each angle scenario using a standard swell height of 2.6 m with a period of six seconds. Each spectrum was subsequently propagated over the digitised bathymetric area to yield data outputs on swell compass direction (Dir), significant wave height of swell (Hs) and the peak period of the swell in seconds (Tp). Numerical output files comprised a grid area 99 cells n-s by 65 cells e-w, yielding a total of 6425 data points.

2.4 Model results

2.4.1 Wind induced currents and sediment resuspension models

2.4.1.1 Wind-induced currents

Five graphics have been plotted for wind speeds of 8 m/s from the following angles 225°, 250°, 270°, 290° and 315° (**Figure 2.10**). Directional vector arrows are proportional to the speed of the current. For the calculations at 225°, 250°, 290° and 315° the strongest currents generally trend in either a north-south (290° and 315°) or south-north (225° and 250°) direction and attain speeds of 15-20 cm/s. It is interesting to note that none of these four wind directions generates coastal gyres on the leeward side of the island. There are a number of possible explanations for this (1) the regularity in reconstructed bathymetry in this zone; and/or (2) the failure of the model to accommodate swell parameters (e.g. swell wave height, swell direction and swell diffraction). The lowest current speeds of ~5 cm/s are recorded along the continental shoreline and around the drowned reef system.

Westerly 270° winds generate a low current environment on the leeward side of the island, characterised by speeds of <10 cm/s. The model generates two isolated gyres on the down-wind side of the island, one in the northern bay and the other in the southern cove. A low energy zone is represented between these two gyres, an axis which corresponds to the proto-tombolo zone. Low current zones continue to fringe the island, notably in the northern cove and along the southeastern fringe of the drowned ridge. This current pattern emphasises the role of the breakwater island in favouring the early colonisation of Tyre's offshore reefs. According to these model data, westerly winds were most conducive to the initial accretion of the proto-tombolo.

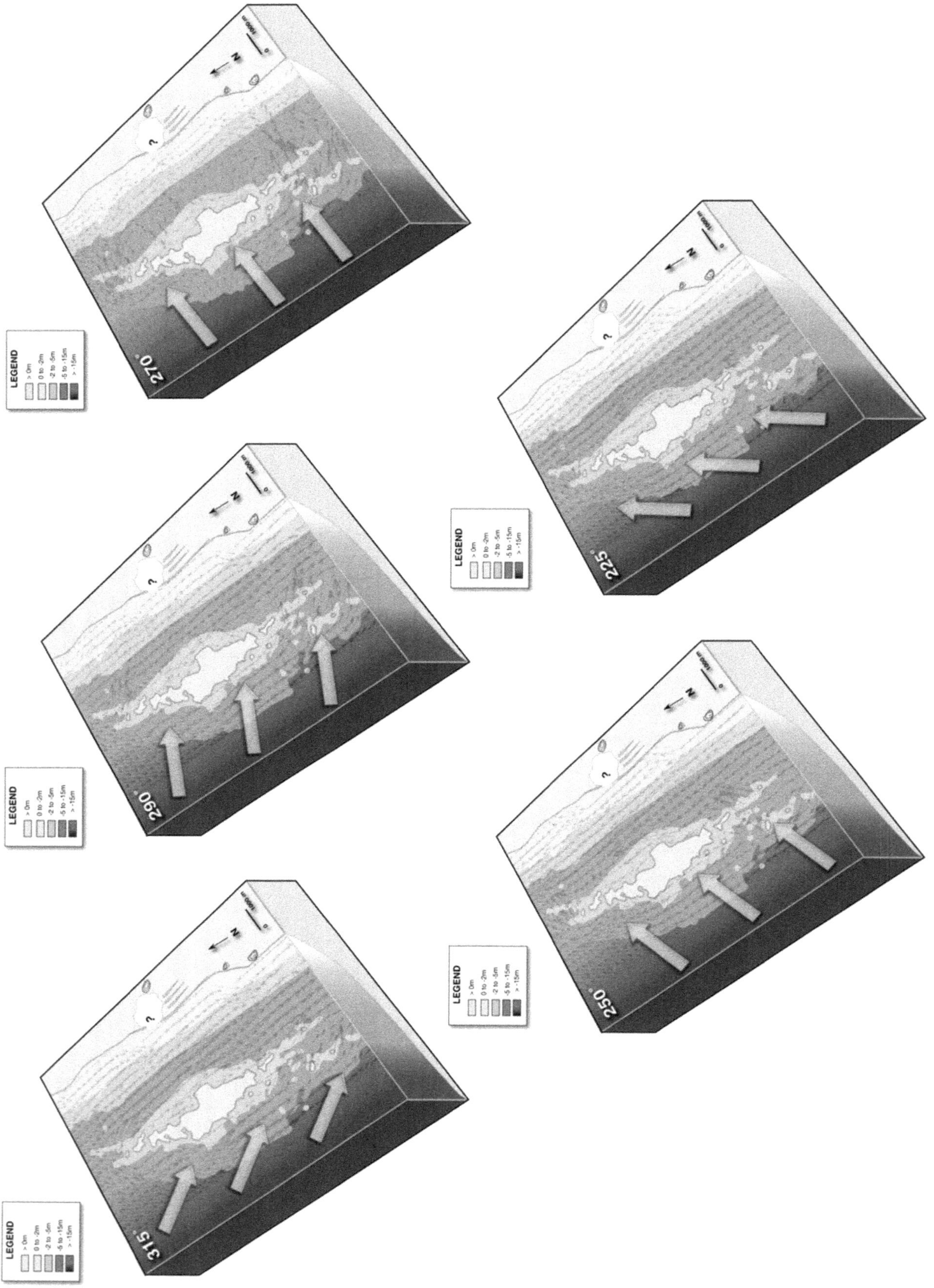

Figure 210: Wind induced currents behind Tyre island at 5500 BP.

2.4.1.2 Wind induced bottom tensions
The graphics of the wind induced bottom tensions correspond to stress imposed by the waterbody on the marine bottom (**Figures 2.11 to 2.14**). These calculations facilitate quantification of the erosive competence of the currents on bottom sediments. They also yield insights into the intensity and direction of sediment transport. Four types of erosional and depositional zones can be discerned: (1) strong, unidirectional currents are diagnostic of areas of high erosion; (2) converging weak currents, are concomitant with zones of sediment deposition; (3) converging strong currents are characterised by localised sediment deposition and relatively turbid waters; and (4) very weak currents, as for example at the centre of a gyre, are typical of zones of sediment accumulation.

For the projections at 225°, 250° and 315° relatively high bottom tension zones are manifest behind the island breakwater, concomitant with the medium to high north-south and south-north trending currents. Such a trend is contradictory with: (1) the accretion of a proto-tomobolo between the island and the continent; and (2) the swell data, which show a significant wave shadow and fall in water energy leeward of the island barrier. This suggests that the wind model is failing to correctly accommodate certain physical parameters. Sediment accretion zones are manifest leeward of the breakwater with the forcing winds at 270° and 290°. This would suggest that westerly winds were important in the initial development of the proto-tombolo. Such a hypothesis is corroborated by the presence of relatively coarse-grained sands around 6000 BP, consistent with these competent north-south and south-north trending currents. After ~4000-3000 BP transition to finer grained sands indicates the establishment of a pronounced topographic barrier.

2.4.1.3 Wave heights
Two wave height maps have been generated for the dominant southwest and northwest wind directions. These maps clearly show the breakwater effect of the island ridge with the creation of low energy zones on the leeward side of the island. The northern harbour and the southeastern fringe of the island are shown to be zones of low wave height (<9 cm for a forcing wind at 8 m/s). Wave heights of 80-99 cm are evidenced west of the drowned ridge and along a narrow north-south strip between Tyre island and the mainland. For a forcing wind at 8 m/s, the model has calculated wave heights of 20-39 cm along the landward margin.

2.4.1.4 Sediment resuspension
Two bottom tension maps have been generated for the combined effects of waves and currents. The model calculations are based upon uniform non-cohesive and non-compacted sediments; cohesive clay sediments cannot be modelled at this present stage. Results are expressed in $kg.m^{-2}.s^{-1}$ and combine the effects of the bathymetry, the fetch, and both wind and wave current bottom tensions.

For the projections at 225° and 315°, the calculations show a north-south trending area of high sediment resuspension (~1.1 $kg.m^{-2}.s^{-1}$) between the island breakwater and the mainland. The lowest areas of sediment resuspension are located in the northern basin, along the southeastern fringe of the island (Renan's southern basin) and on the southern coastal fringe (~0.1 $kg.m^{-2}.s^{-1}$). Low sediment resuspension in this southern area appears consistent with shallow bathymetry and the protective role of sandstone islets around 6000 BP. These data do not satisfactorily account for the initial accretion of the proto-tombolo in deeper areas between the island and the continent. It would therefore appear that either/or (1) westerly winds were one of the most important factors in explaining the initial inception and consolidation of the proto-tombolo; and (2) the present model is not satisfactorily resolving other important physical parameters such as swell refraction and diffraction.

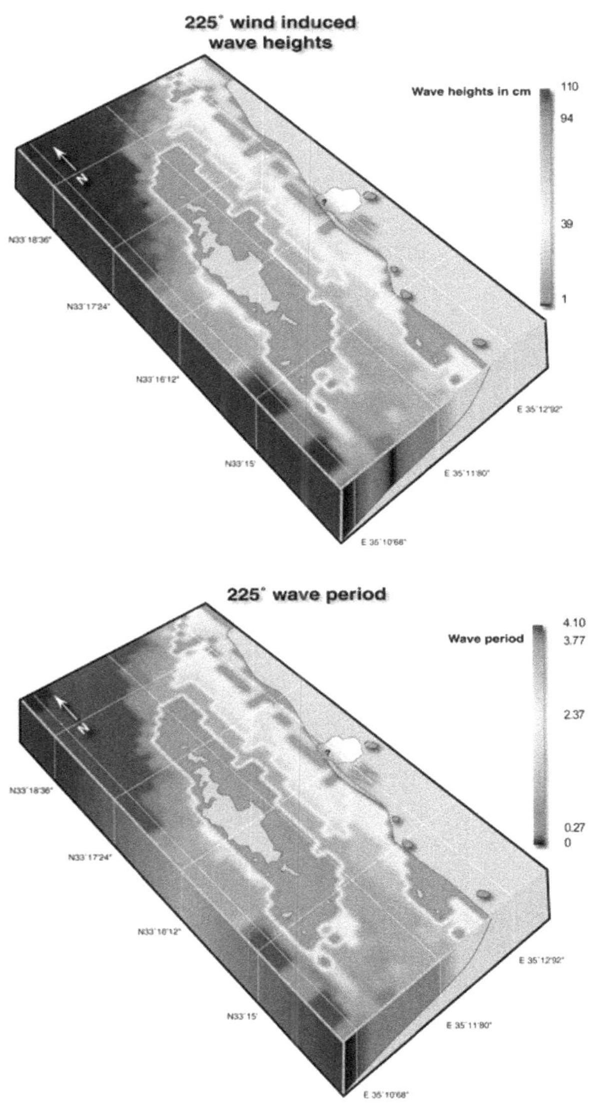

Figure 2.11: Model runs of wind induced wave heights and wave period at 5500 BP from 225°.

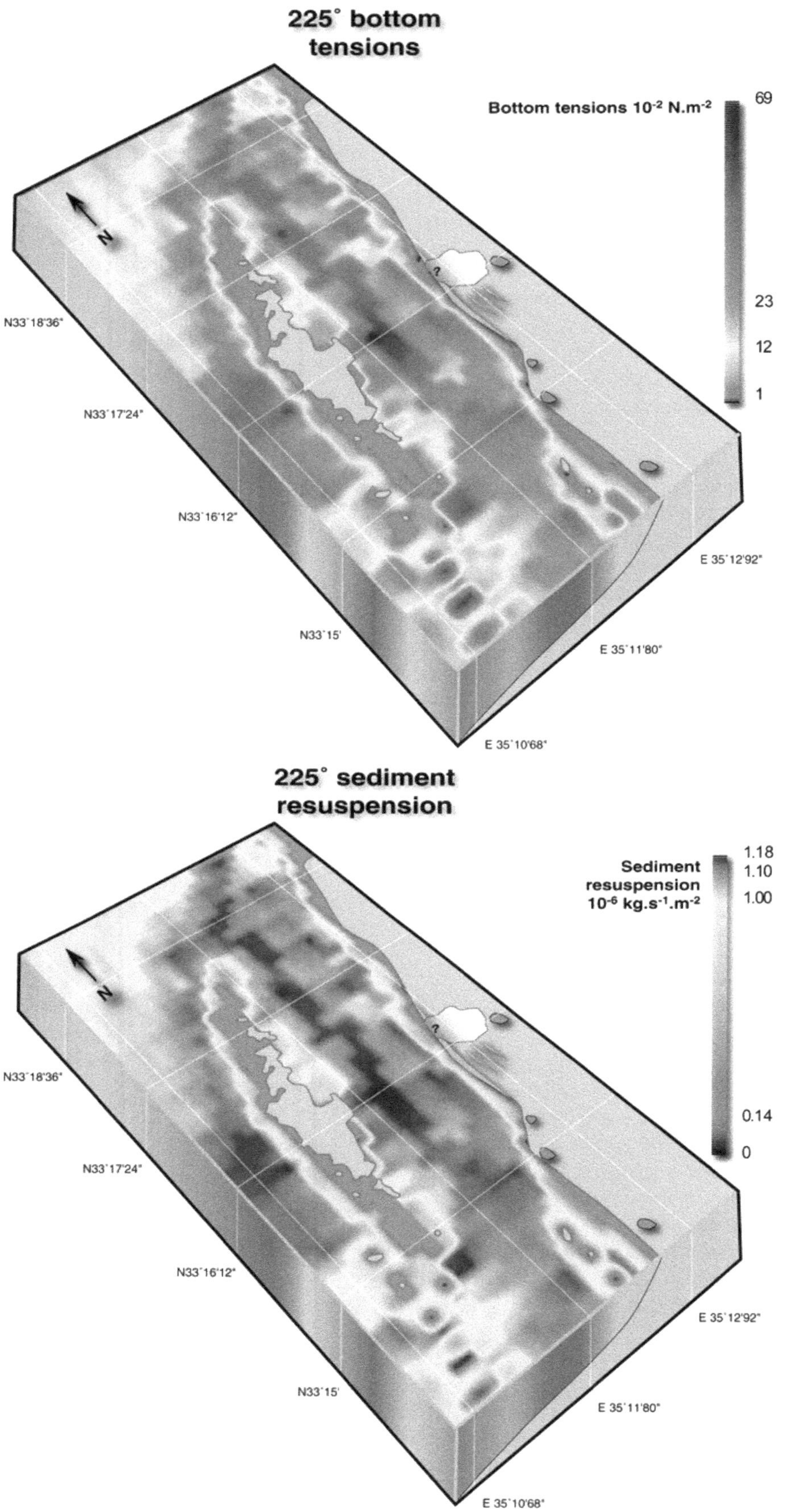

Figure 2.12: Model runs of wind induced bottom tensions and sediment resuspension at 5500 BP from 225°.

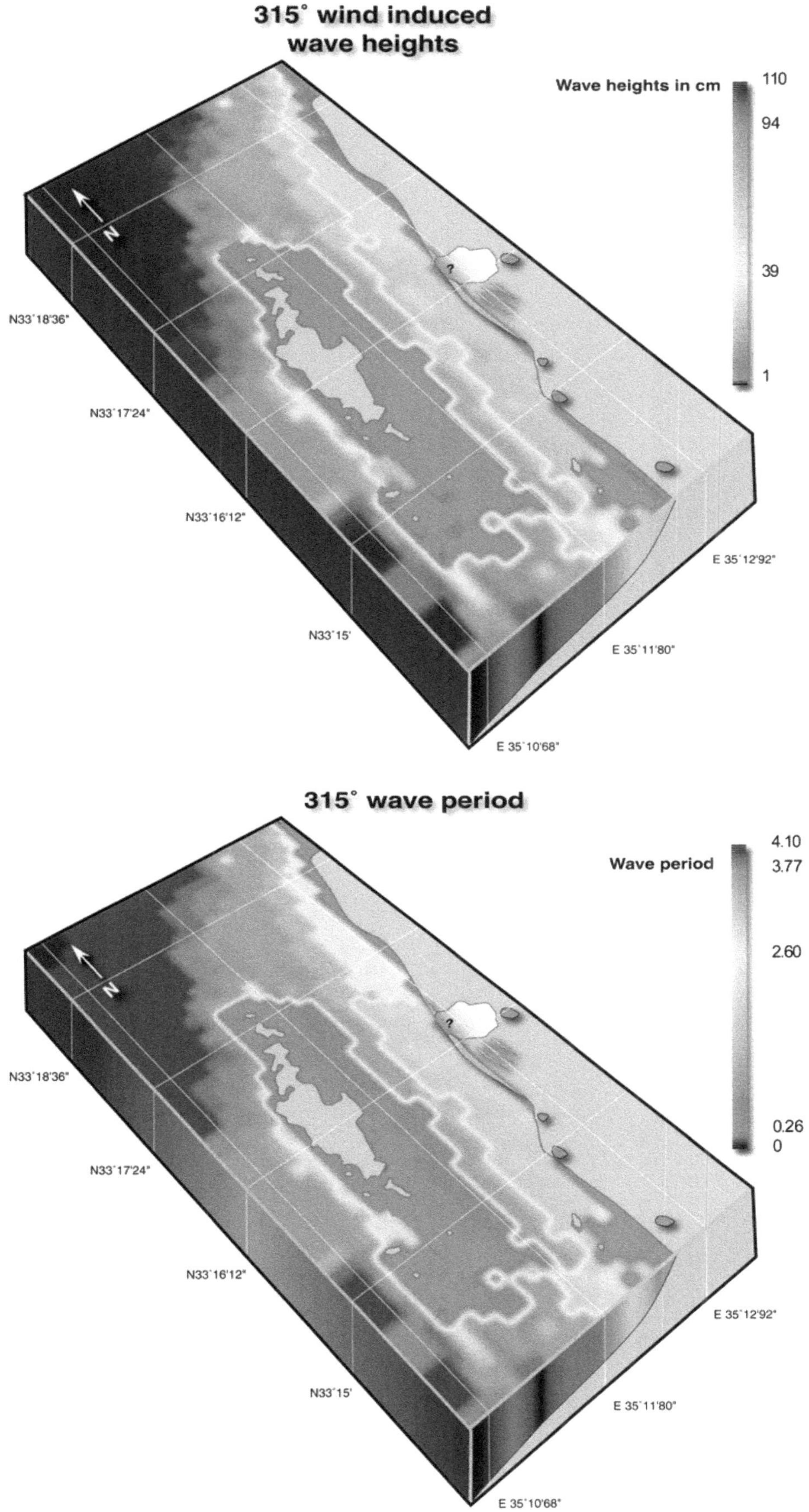

Figure 2.13: Model runs of wind induced wave heights and wave period at 5500 BP from 315°.

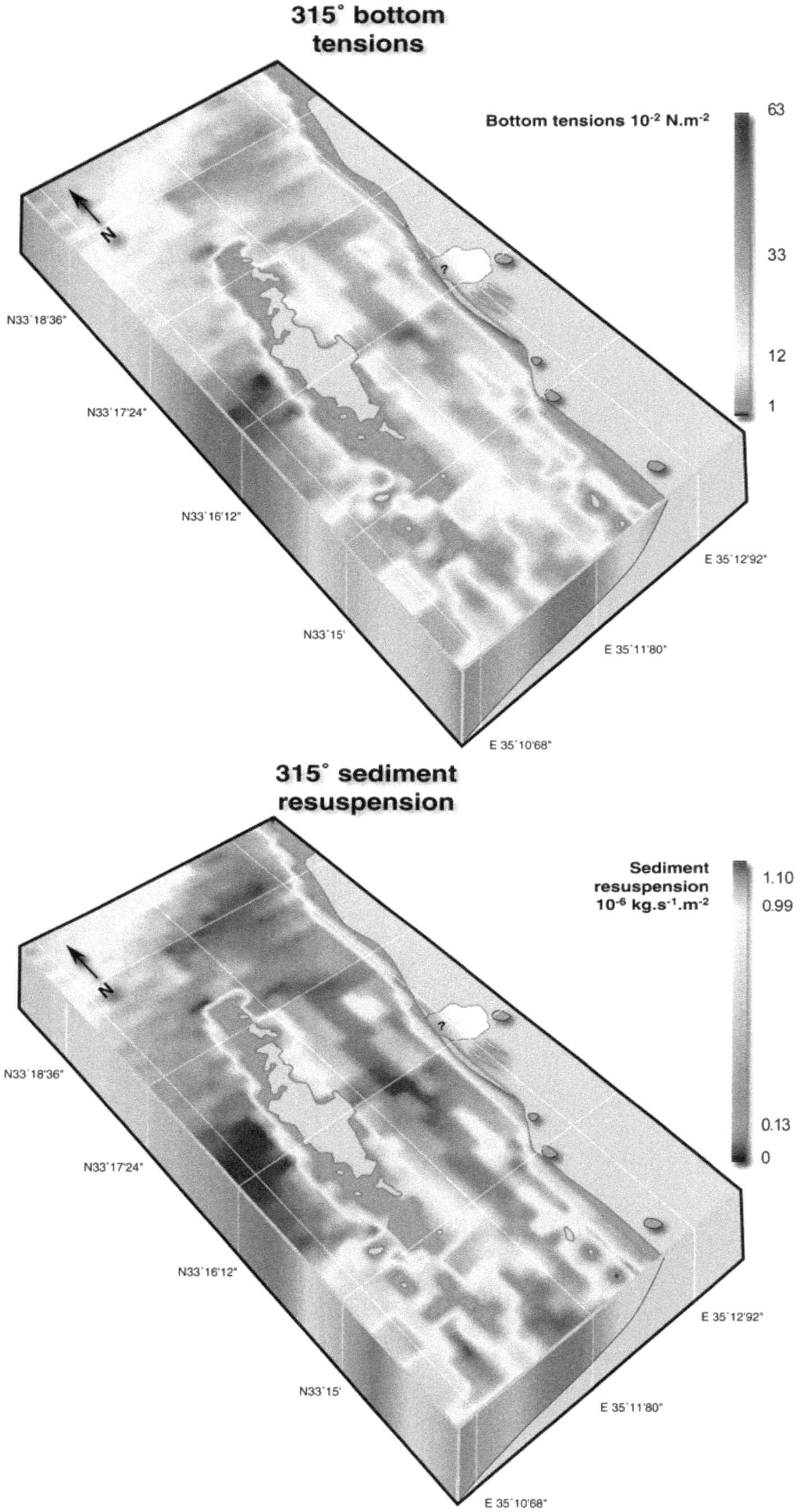

Figure 2.14: Model runs of wind induced bottom tensions and sediment resuspension at 5500 BP from 315°.

2.4.2 Swell propagation and diffraction model

We sought to overcome some of the shortcomings in the wind-induced models by simulating the refraction of the swell around Tyre island. For the projections at 5500 BP, the effects of swell refraction and shoaling are clearly manifest. The incoming swell, which is initially quasi-unidirectional (225°, 270° or 315°), is spread by the lateral movement of energy into the shadowed area behind Tyre island. Projections at 5500 BP indicate that the majority of the swell incident energy is blocked by the natural N-S trending breakwater (**Figures 2.15 - 2.20**). As the wave energy that passes the obstacle spreads laterally into the shadow region, the total energy is distributed over a broader area reducing the swell height leeward of the island from >1.80 m west of the island to <0.75 m on the eastern shadow side (propagated wave height 2.6 m with a period of 6 seconds, **Figure 2.21**). This shadow generated a significant natural shelter zone for the offshore anchoring of boats during periods of fair to medium swell, greatly increasing the anchorage capacity of the city. Our core transect falls in an area of very low wave height (~0.40 m), evoking the rapid accretion of the marine bottom. We conclude that the early accretion of the proto-tombolo took place in two areas: (1) **along the continental margin of Palaeotyrus**, the sedimentological data show that limited accommodation space led to the rapid progradation of a subaerial salient strandplain. This is corroborated by low wave heights along the coastal fringe; (2) **in the wave shadow behind Tyre island**, the models show a sharp fall in energy accompanied by a concomitant marine bottom shallowing. The greater accommodation space means, however, that the tombolo remains a subaqueous form in this zone. By a process of positive feedback, we hypothesise that the shallowing bathymetry altered the pattern of diffraction and refraction leading to the shallowing of deeper distal areas between the island and the continent.

Tectonic subsidence means that the breakwater effects of Tyre island are much less pronounced today than for 5500 BP. A significant wave shadow is generated to the north and south of the tombolo for projections at 225° and 315°. In chapter 6 we demonstrate that the Litani has been one of the dominant sediment sources for accretion along the Tyrian coastline during the Holocene. Meterological data indicate that although the strongest winds and swell derive from the southwest, northwesterlies are also very frequent (33 % of total winds). Northwesterly swell diffraction and refraction appear to be the dominant transport mechanism in translating Litani sediment to the Tyrian peninsula.

For all scenarios, the initial swell energy from all fetch areas has been largely dispersed during the travel distance (i.e., height at destination is significantly lower than the initial value at the fetch exit region). In the blue area behind the island at 5500 BP, the swell direction is essentially reversed. This is the result of refraction, a separate process of bending waves towards shallower regions. At 270° the refracted waves meet in a central area and are translated to strike the shoreline at a perpendicular angle. For the 225° and 315° scenarios the waves meet in areas just north and south of the 270° line. These zones of convergence lie in a west-east trending axis and broadly correspond to the areas occupied by the modern salient.

Much of the previous archaeological work on the Levantine coast has insisted upon the role of dominant south-westerly winds in influencing shipping patterns and the potentiality of seaboard sites. The meteorological and model data, however, indicate that at Tyre north-westerly and westerly winds have been significant in shaping the coastal geomorphology of the pensinsula. The models suggest that westerly winds and swell, though less frequent than the dominant south-westerlies and north-westerlies, were also significant in forcing the initial accretion of the proto-tomobolo.

2.4.3 Modelling data conclusions

The numerical models produce coherent results with regards to the main characteristics of the wave, current and sediment transport fields. This notably includes the blocking of the incident waves by the natural breakwater and circulations in the lee zone of Tyre island at 5500 BP. Comparison of the modelling data with our field and chronostratigraphic data shows satisfactory agreement. At a more general level, the swell model is better at accounting for the initial accretion of the proto-tombolo between the island breakwater and the continent. While the models produce good results for the two time windows analysed, it is important to stress that the hydrodynamics and sediment transport processes operating around a tombolo are highly complex. The resolution of our chronostratigraphic data and the models cannot account for all of this great complexity. At best, the models should be considered as providing a parameterised description of the complex processes taking place in the vicinity of the Tyrian breakwater.

Whilst the numerical models yield a general insight into the wave induced current directions, and zones of sediment deposition and erosion it must be noted that there are a number of possible fallibilities: (1) First, the palaeo-bathymetric data have been interpolated from present day readings. Whilst this does give a general idea of the coastal configuration between 6000-5000 BP, it cannot account for all the spatial variability in littoral topography (differences in accretion/erosion rates). Equally, form-process studies indicate that coastal bathymetry can vary significantly over seasonal, annual, decadal and centennial scales. Unfortunately, the resolution of our chronostratigraphic data does not allow us to precisely model the dynamic nature of these bathymetric changes in time and space. (2) Secondly, a uniform relative sea-level change of 5 m has been assumed since 5500 BP. Due to micro-faulting on the Tyrian horst, there may be some spatial variability (± 0.5

m) in this tectonic parameter. (3) Third, it is important to understand that the model calculations for sediment resuspension are based upon a uniform sediment body throughout the coastal area under investigation, when in reality the spatial distribution of bottom sediments is highly variable. (4) Finally, one of the major downfalls of both models is that neither accommodates both wind induced currents and wave swell propagation.

With reference to the harbours, this model does yield insights into two important areas. (1) It clearly indicates low energy zones in the northern cove and along the southeastern coastal fringe of Tyre island. The numerical models therefore corroborate the environmental potentiality of these areas as anchorage havens. (2) We can better comprehend shipping directions used by ancient mariners during the early Bronze Age.

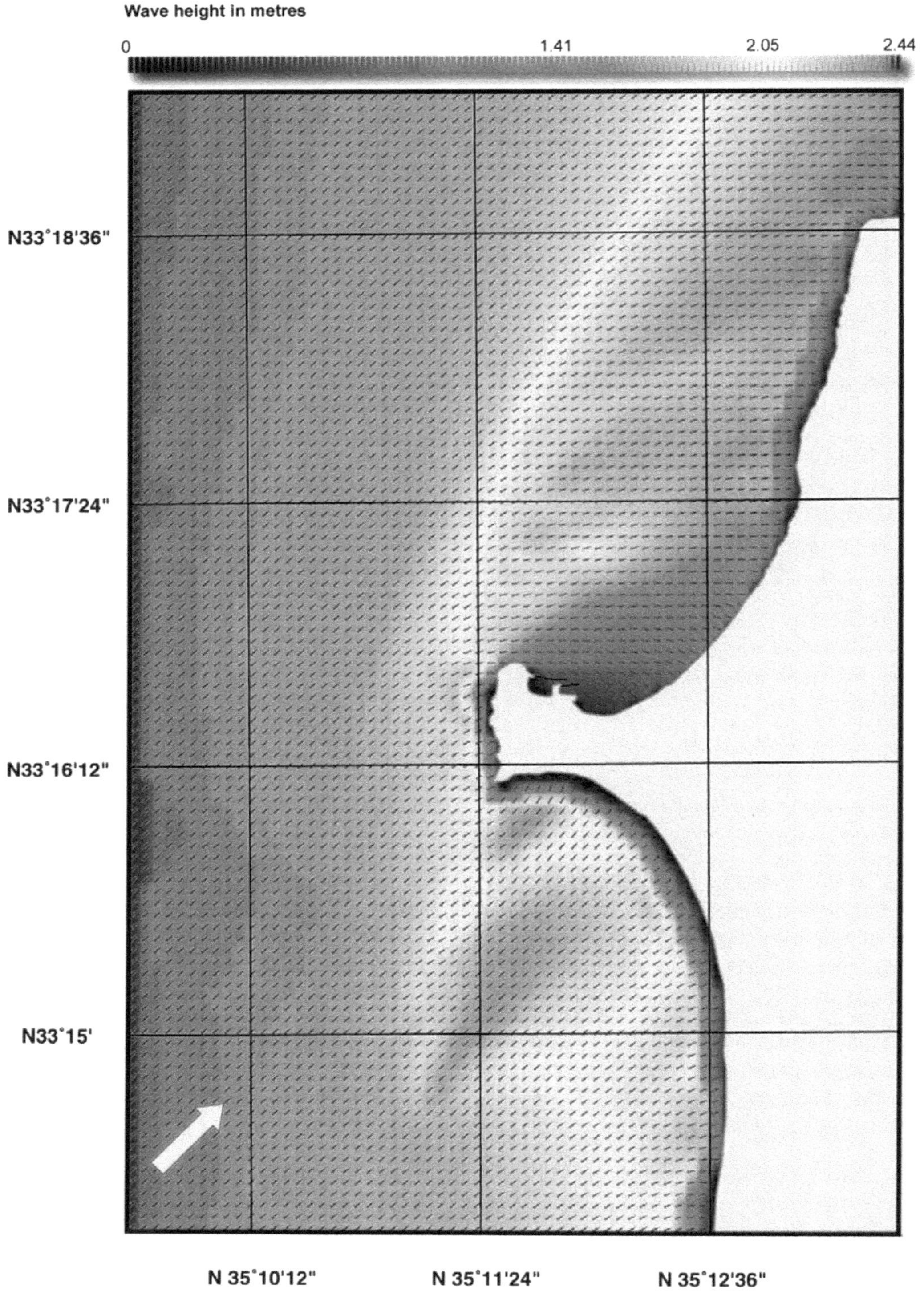

Figure 2.15: Wave height and direction of swell propagated from 225° (present bathymetry).

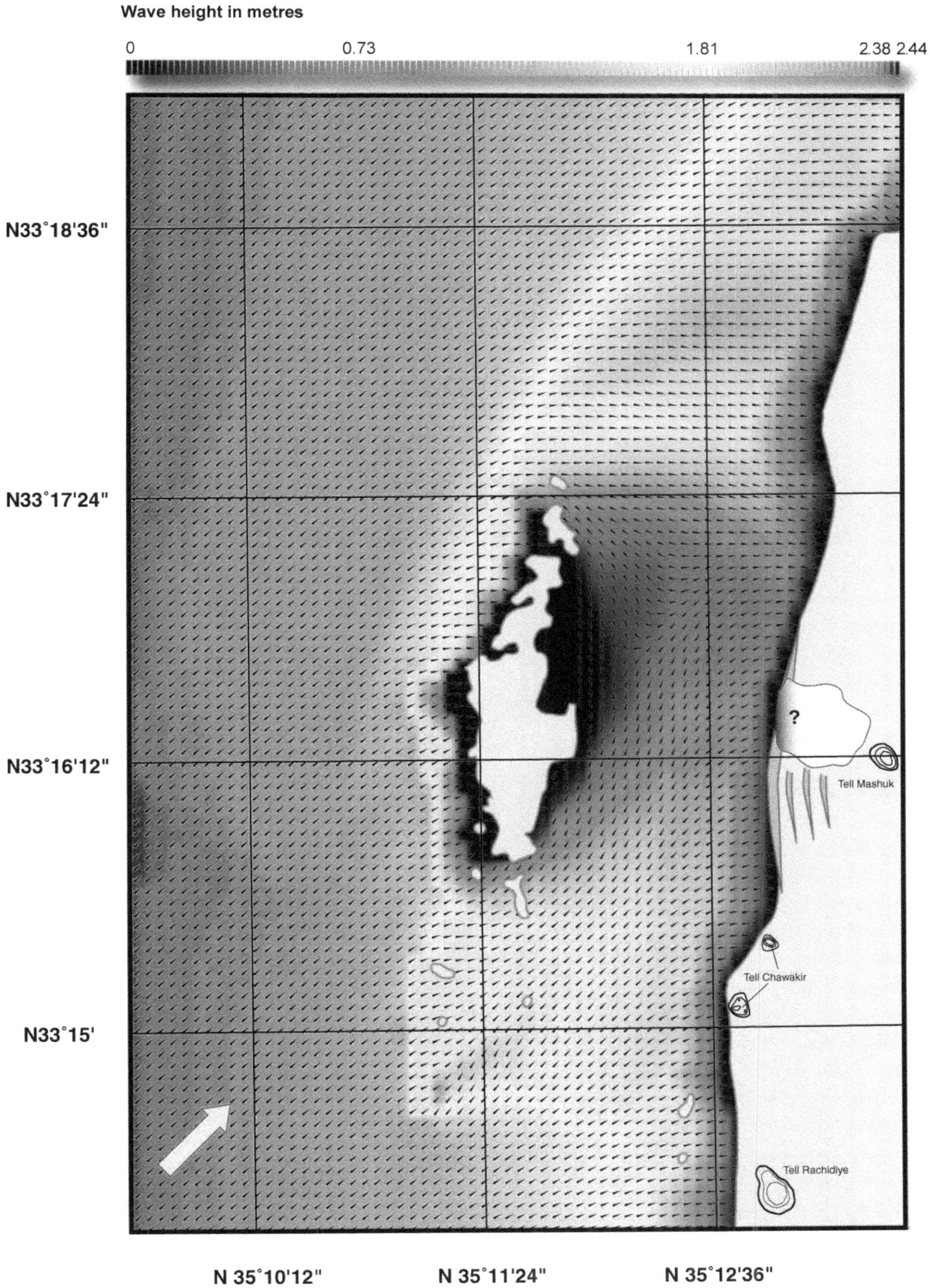

Figure 2.16: Wave height and direction of swell propagated from 225° (bathymetry at 5500 BP).

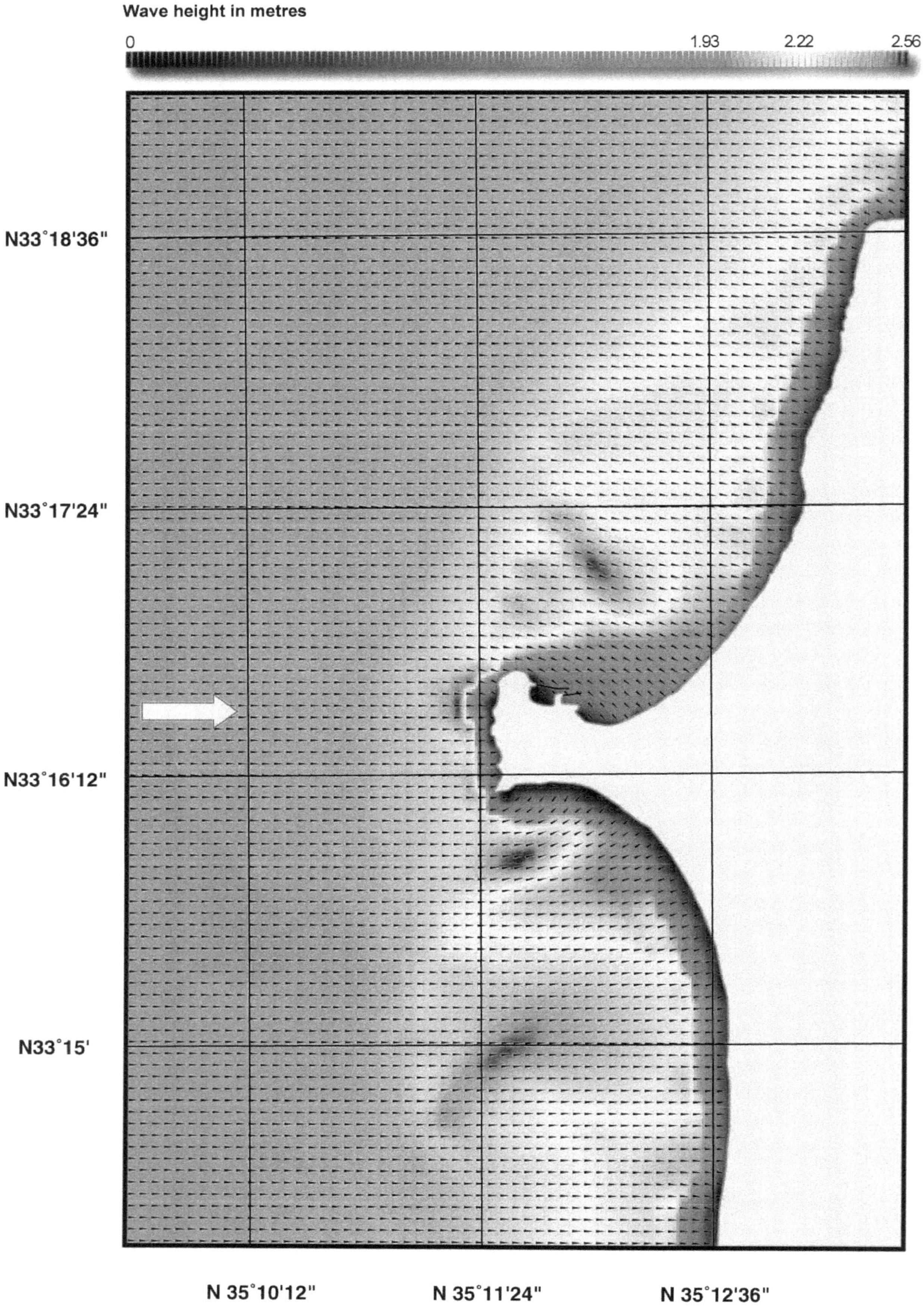

Figure 2.17: Wave height and direction of swell propagated from 270° (present bathymetry).

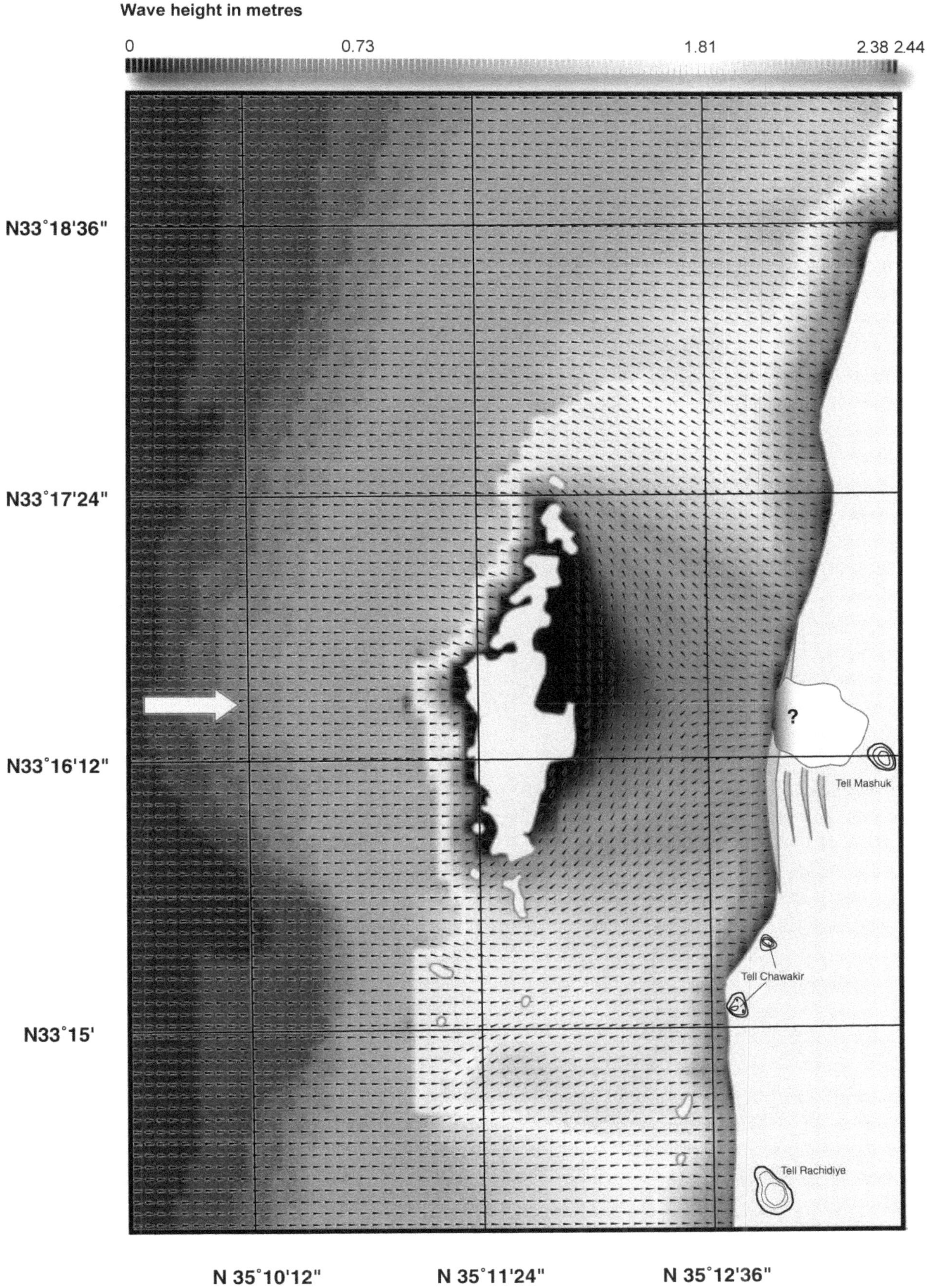

Figure 2.18: Wave height and direction of swell propagated from 270° (bathymetry at 5500 BP).

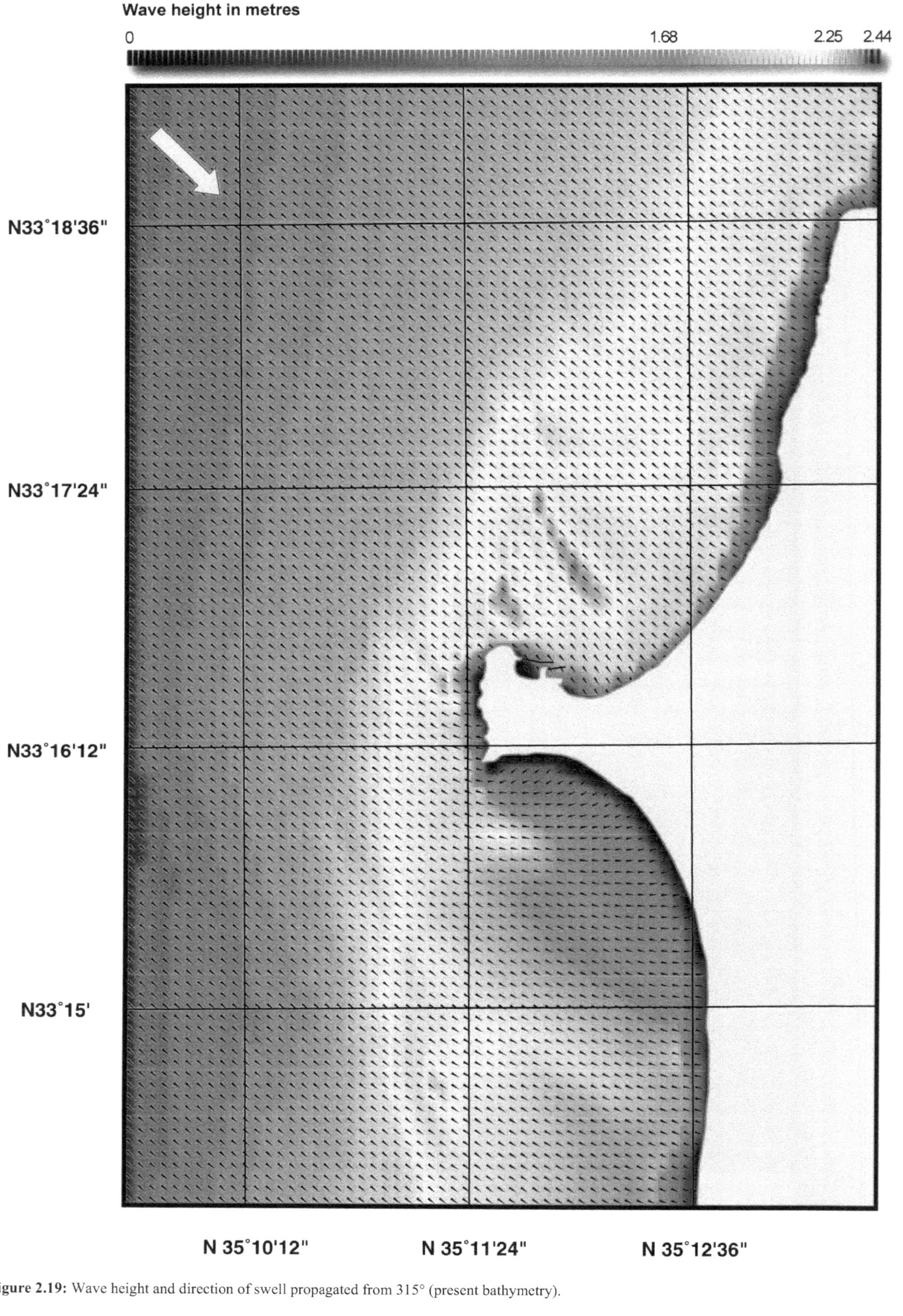

Figure 2.19: Wave height and direction of swell propagated from 315° (present bathymetry).

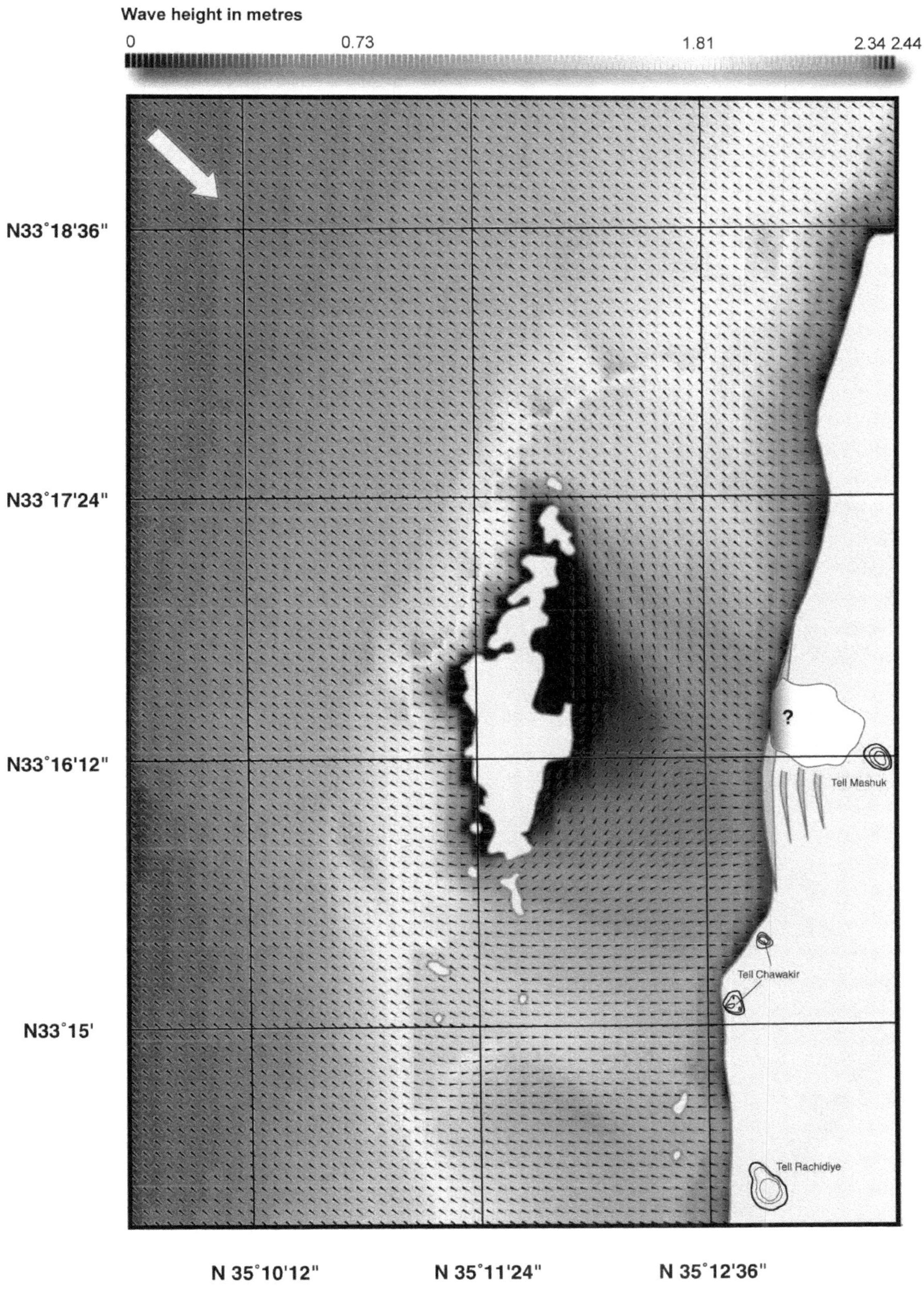

Figure 2.20: Wave height and direction of swell propagated from 315° (bathymetry at 5500 BP).

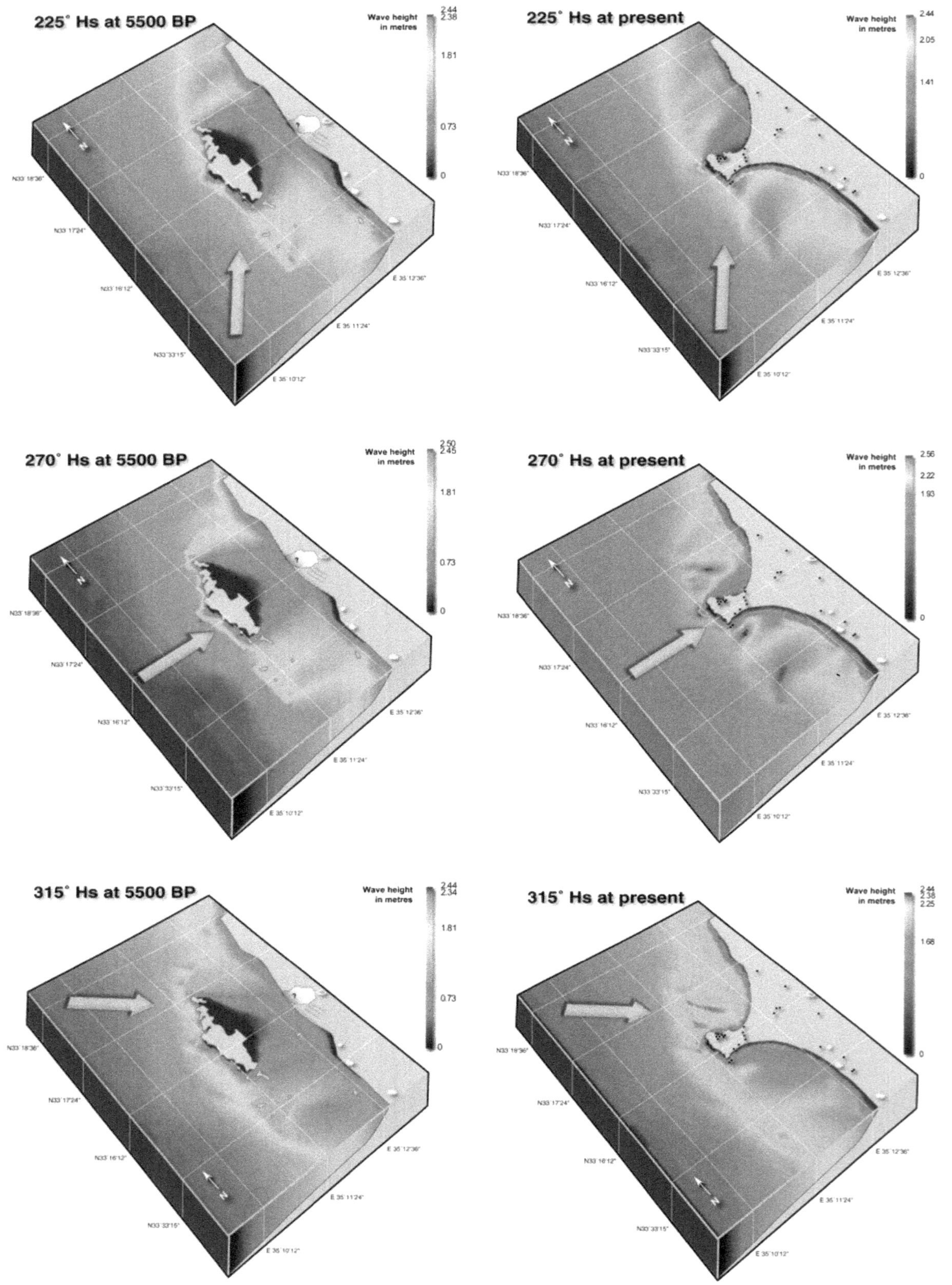

Figure 2.21: Wave heights for 225°, 270° and 315° propagated swells.

2.5 Chronostratigraphic results

Here we describe the stratigraphic data. For a detailed discussion of the tombolo's evolution please see below.

2.5.1 Core TXVIII El Bass/Mashuk salient base: description

Core TXVIII was drilled between Tell Mashuk and El Bass. This area comprises the oldest part of the subaerial salient *sensu stricto*; the sedimentary record attests to the accretion and progradation of the tombolo beach ridges from 6000 BP onwards (**Figures 2.22 and 2.23**). The clay substratum is transgressed by a fine-bedded marine sands (unit A) dated ~6000 BP. Unit A is dominated by molluscs from the subtidal sands (*Bittium* sp., *Tricolia pullus*, *Rissoa lineolata*) and silty or muddy-sand assemblages (*Nassarius pygmaeus*, *Macoma* sp., *Macoma cumana*, *Rissoa monodonta*). Marine-lagoonal (*Loxoconcha* spp., *Xestoleberis* sp.) and coastal (*Aurila convexa*, *Pontocythere* sp., *Urocythereis* sp.) ostracod taxa characterise the unit. The unit records rapid beach ridge accretion and by ~5500 BP, a lagoon environment is indicated by the litho- and biostratigraphical proxies (units B and C).

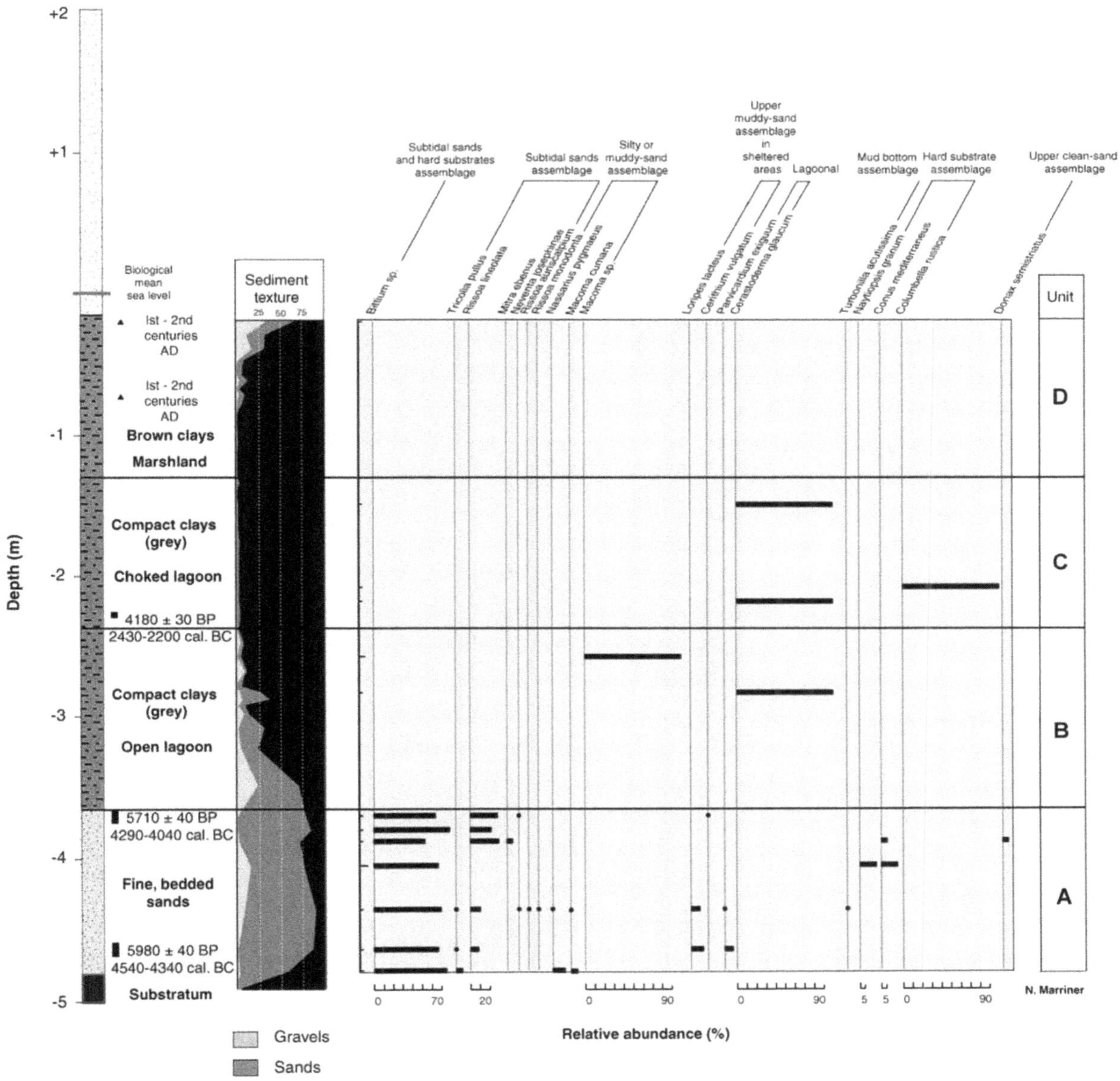

Figure 2.22: TXVIII marine macrofauna. (1) Unit A: The clay substratum is transgressed by a fine-bedded marine sands dated ~6000 BP. It is dominated by molluscs from the subtidal sands (*Bittium* sp., *Tricolia pullus*, *Rissoa lineolata*) and silty or muddy-sand assemblages (*Nassarius pygmaeus*, *Macoma* sp., *Macoma cumana*, *Rissoa monodonta*). The facies records rapid beach ridge accretion until ~5500 BP. (2) Unit B: comprises a dark brown/black plastic clay unit. The sparse molluscan indicators include *Cerastoderma glaucum*, a lagoon tolerant taxa, and *Macoma* sp. (silty or muddy-sand assemblage). (3) Unit C: The silts and clays comprise >95% of the total sediment fraction. The bottom of the facies is dated 4180 ± 30 BP, or 2430-2200 cal. BC. This fine-grained sedimentology and sparse tests of *Cerastoderma glaucum* concur continued lagoonal conditions. We hypothesise the existence of a shallowing-up choked lagoon, which could have served as an anchorage haven during the Bronze Age. (4) Unit D: constitutes a brown clay lithofacies containing plant macrorestes and fossil snails. Ceramics constrain the unit to Roman times (1st-2nd centuries AD). The unit did not yield any marine fauna. The lithology is consistent with a marsh environment.

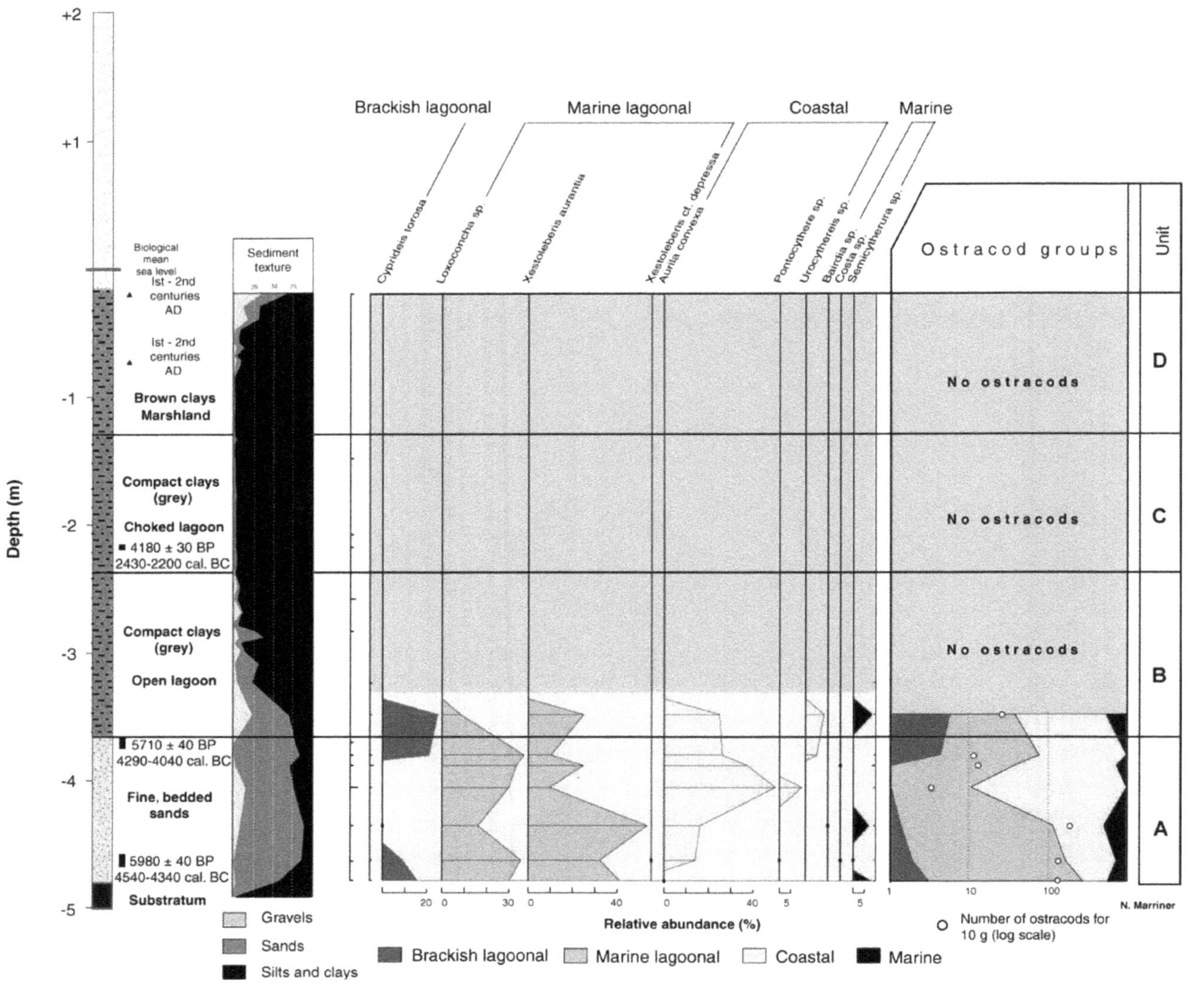

Figure 2.23: TXVIII ostracods. (1) Unit A: is dominated by marine lagoonal (*Loxoconcha* spp., *Xestoleberis* sp.) and coastal (*Aurila convexa*, *Pontocythere* sp., *Urocythereis* sp.) ostracod taxa. Tests of *Cyprideis torosa*, towards the top of the unit, attest to the onset of hyposaline conditions. We infer a low energy beach environment. (2) Unit B: The base of the unit juxtaposes taxa from three ecological groups, brackish lagoonal (*Cyprideis torosa*), marine lagoonal (*Loxoconcha* spp., *Xestoleberis aurantia*) and coastal (*Aurila convexa*) domains. The onset of hyposaline conditions is indicated by a rise in *Cyprideis torosa*. We attribute the absence of ostracods in units B, C and D to a confined anoxic environment marking the onset of a lagoon depositional context.

In unit B, diagnositic beach sedimentology at the base shallows up to a protected lagoon environment. The top of the facies is dated 4180 ± 30 BP, or 2430-2200 cal. BC. The sparse molluscan indicators are exclusively dominated by *Cerastoderma glaucum*, a lagoon tolerant taxa. The reasons for an absence of ostracods are unclear; we evoke an anoxic environment unconducive to their development. In unit C, the silts and clays comprise >95% of the total sediment fraction. This fine-grained sedimentology and sparse tests of *Cerastoderma glaucum* concur continued lagoonal conditions. We hypothesise the existence of a shallowing-up choked lagoon, which could have served as an anchorage haven during the Bronze Age and possibly during the Iron Age. Unit D comprises a brown clay lithofacies dated to Roman times. The unit did not yield any marine fauna; plant macrorestes and fossil snail tests are concurrent with a marsh environment that persisted in the El Bass area until 1864, after which time it was drained for agricultural purposes (Carmona and Ruiz, 2004).

2.5.2 Core TXIII: description

Core TXIII derives from the septentrional lobe of the tombolo, just north of the ancient causeway. Around 10 m of Holocene fill record the marine transgression and progradation of the northern flank of the tombolo. The transgression is dated ~8000 BP and is onlapped by a poorly sorted shelly silts and fine sands unit (**Figure 2.24**). Unit A is lithodependent, with reworking of the underlying clay substratum (Cohen-Seffer *et al.*, 2005). The ostracod fauna is dominated by the brackish-lagoonal species *Cyprideis torosa*, with small peaks of marine lagoonal (*Loxoconcha* spp.) and coastal (*Aurila convexa*, *Aurila woodwardii*) taxa (**Figure 2.25**). Existence of a protected marginal marine environment is corroborated by the macrofauna suites comprising taxa from the upper muddy-sand assemblage in sheltered areas (*Cerithium vulgatum*, *Loripes lacteus*), subtidal sands assemblage (*Rissoa lineolata*, *Rissoa scurra*, *Mitra ebenus*) and the lagoonal assemblage (*Parvicardium exiguum*, *Hydrobia acuta*, *Cerastoderma glaucum*, **Figure 2.26**).

After ~7000 BP, landward-driven, transgressive ridges eventually breached the area, with an onshore movement of coarse sand and gravel (unit B). The unit is characterised by the following macrofauna assemblages: subtidal sands, upper muddy-sand assemblage in sheltered areas and the hard substrate assemblage. A sharp decline in *Cyprideis torosa* is countered by peaks of marine-lagoonal and coastal ostracod species. Outer marine species such as *Semicytherura* spp., *Callistocythere* spp. and *Neocytherideis* sp. are also drifted in.

Around 3000 BP, against a backdrop of decelerating late Holocene sea-level rise, we observe the rapid accretion of a well-sorted fine sand lithofacies (unit C). The subtidal sand and hard substrate taxa *Bittium reticulatum* and *Turboella similis* dominate the unit. High relative abundances of marine lagoonal and coastal ostracod taxa are consistent with a middle to low energy shoreface environment. The top of unit D is dated 2650 ± 35 BP. Factoring in ~3 m of RSL change since antiquity, this suggests a proto-tombolo surface within 1-2 m of the water surface on this portion of the northern lobe by the fourth century BC. Unit E comprises a poorly sorted, upward coarsening sediment devoid of macro- and microfossils. We ascribe this to a prograding Byzantine shoreface.

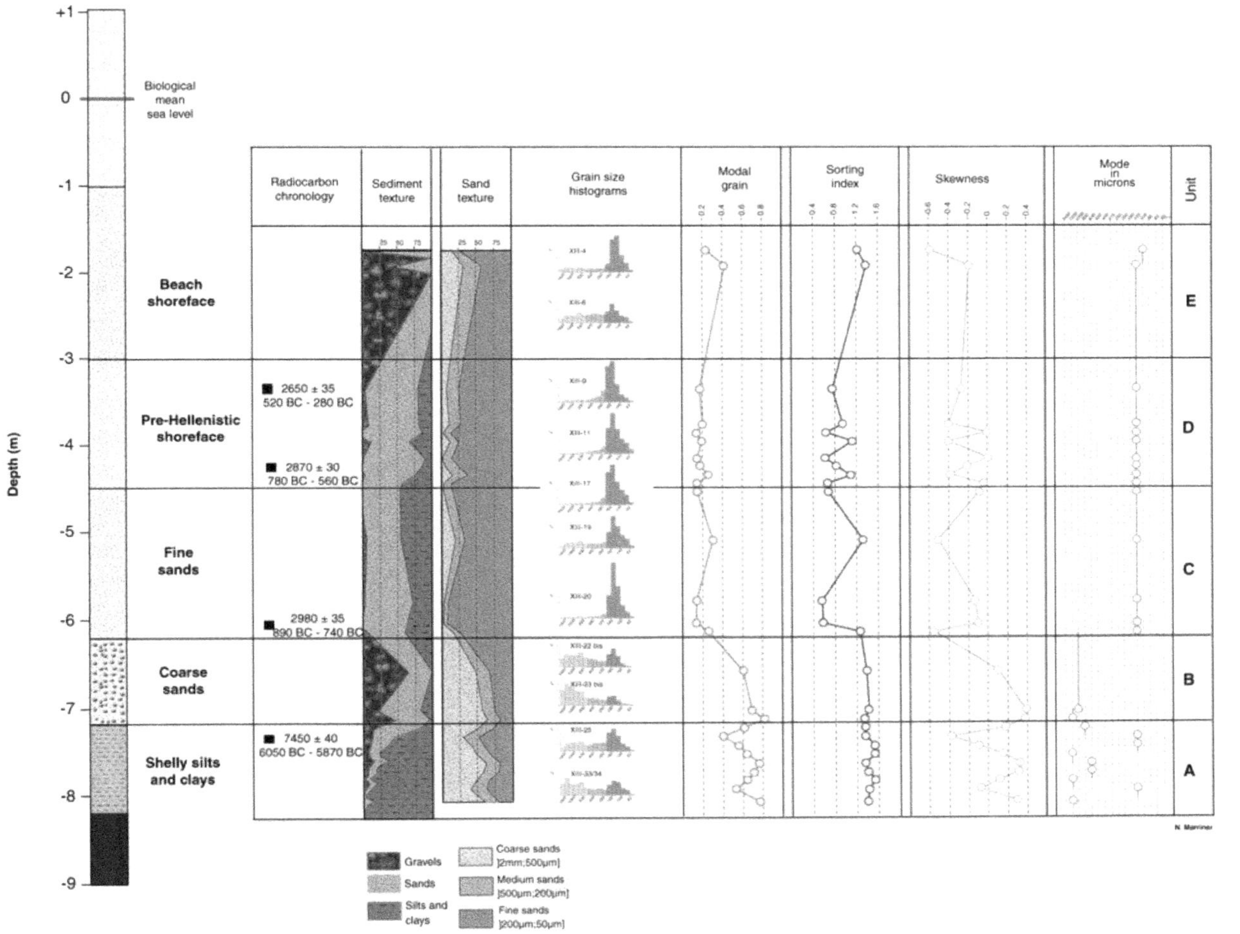

Figure 2.24: Core TXIII lithostratigraphy and grain size analyses. (1) Unit A: is lithodependent, with reworking of the underlying clay substratum. It is dominated by the silts and clays fractions which comprises >75% of the sediment texture. (2) Unit B: constitutes a coarse sand and gravel unit. After ~7000 BP, we infer that landward-driven transgressive ridges eventually breached the area, with an onshore movement of coarse material. (3) Unit C: comprises a well-sorted fine sands (>75 %) facies, which began accreting after ~3000 BP. It attests to a rapidly accreting medium to low energy marine bottom. (4) Unit D: comprises a medium-sorted (0.5 to 1.2) fine sands facies. Variable skewness values of between -0.4 and 0 are consistent with the swash zone and translate a shoreface sedimentary environment. The top of unit D is dated 2650 ± 35 BP. Factoring in ~3 m of RSL change since antiquity, this suggests a proto-tombolo surface within 1-2 m of the water surface on this portion of the northern lobe by the fourth century BC. (5) Unit E: comprises a poorly sorted (>1.2), upward coarsening sediment devoid of macro- and microfossils. We ascribe this to a prograding Byzantine shoreface.

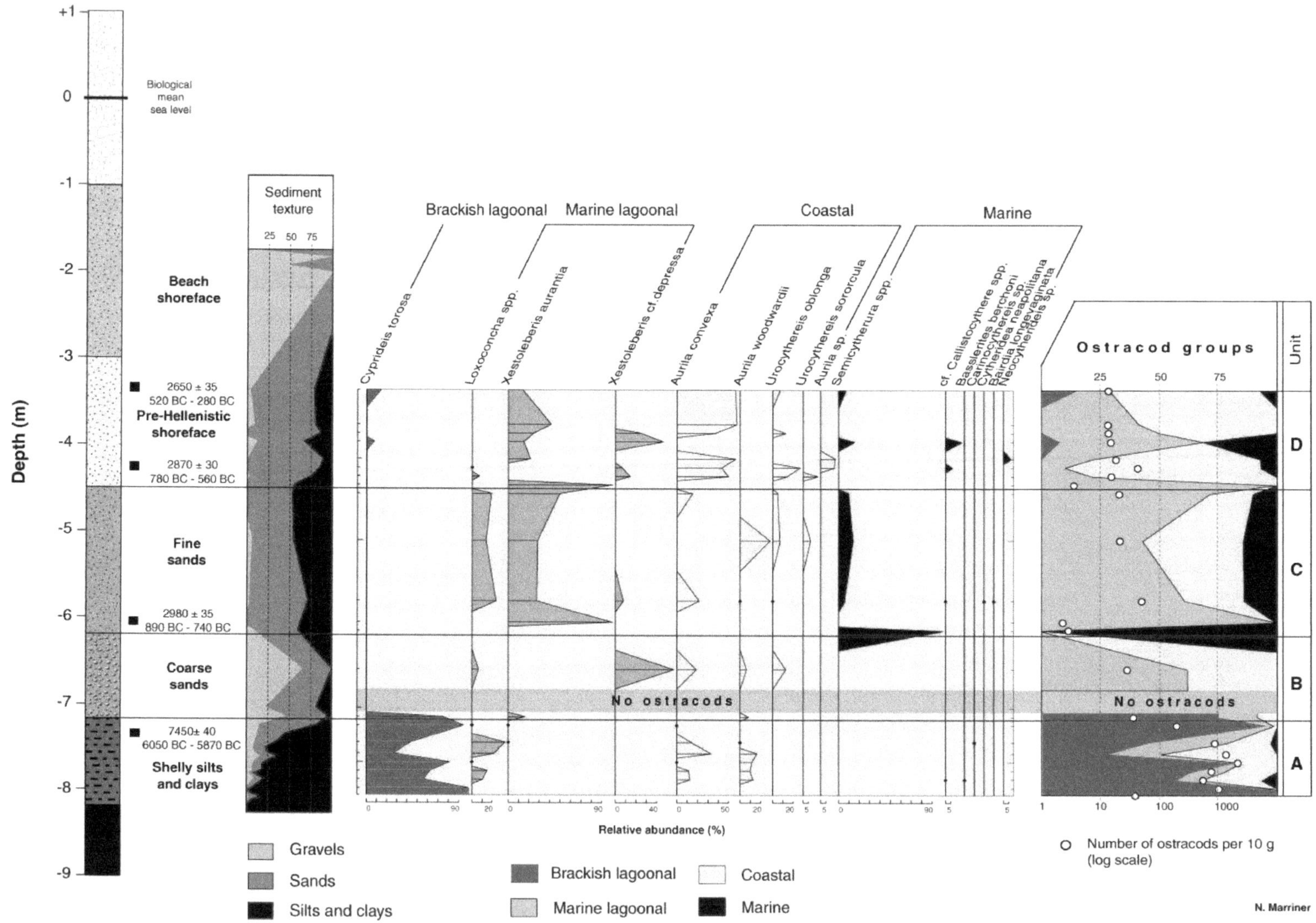

Figure 2.25: TXIII ostracods. (1) Unit A: The ostracod fauna is dominated by the brackish-lagoonal species *Cyprideis torosa*, with secondary peaks of marine lagoonal (*Loxoconcha* spp.) and coastal (*Aurila convexa*, *Aurila woodwardii*) taxa. We infer a low energy lagoon type environment. (2) Unit B: A sharp decline in *Cyprideis torosa* is countered by peaks of marine lagoonal (*Loxoconcha* spp., *Xestoleberis* spp.) and coastal (*Aurila convexa*, *Aurila woodwardii*, *Urocythereis oblonga*) ostracod species. Outer marine species such as *Semicytherura* spp., *Callistocythere* spp. and *Neocytherideis* sp. are also drifted in. High relative abundances of marine lagoonal and coastal ostracod taxa are consistent with a middle to low energy shoreface environment. (3) Unit C: The ostracod suite attests to a well-protected marine environment, dominated by *Loxoconcha* spp. and *Xestoleberis aurantia* (marine lagoonal), in addition to secondary coastal (*Aurila convexa*, *Aurila woddwardii*, *Urocythereis oblonga* and *Urocythereis sorocula*) and marine (*Semicytherura* spp.) taxa. (4) Unit D: comprises a rich suite of ostracod taxa. It is dominated by *Xestoleberis aurantia* (marine lagoonal) and *Aurila convexa* (coastal). Secondary taxa include, *Xestoleberis* cf. *depressa* (marine lagoonal), *Aurila convexa*, *Aurila woodwardii*, *Urocythereis oblonga*, *Urocythereis sororcula* (coastal) and *Semicytherura* spp. (marine). We infer a pre-Hellenistic shoreface within ~1 m of MSL.

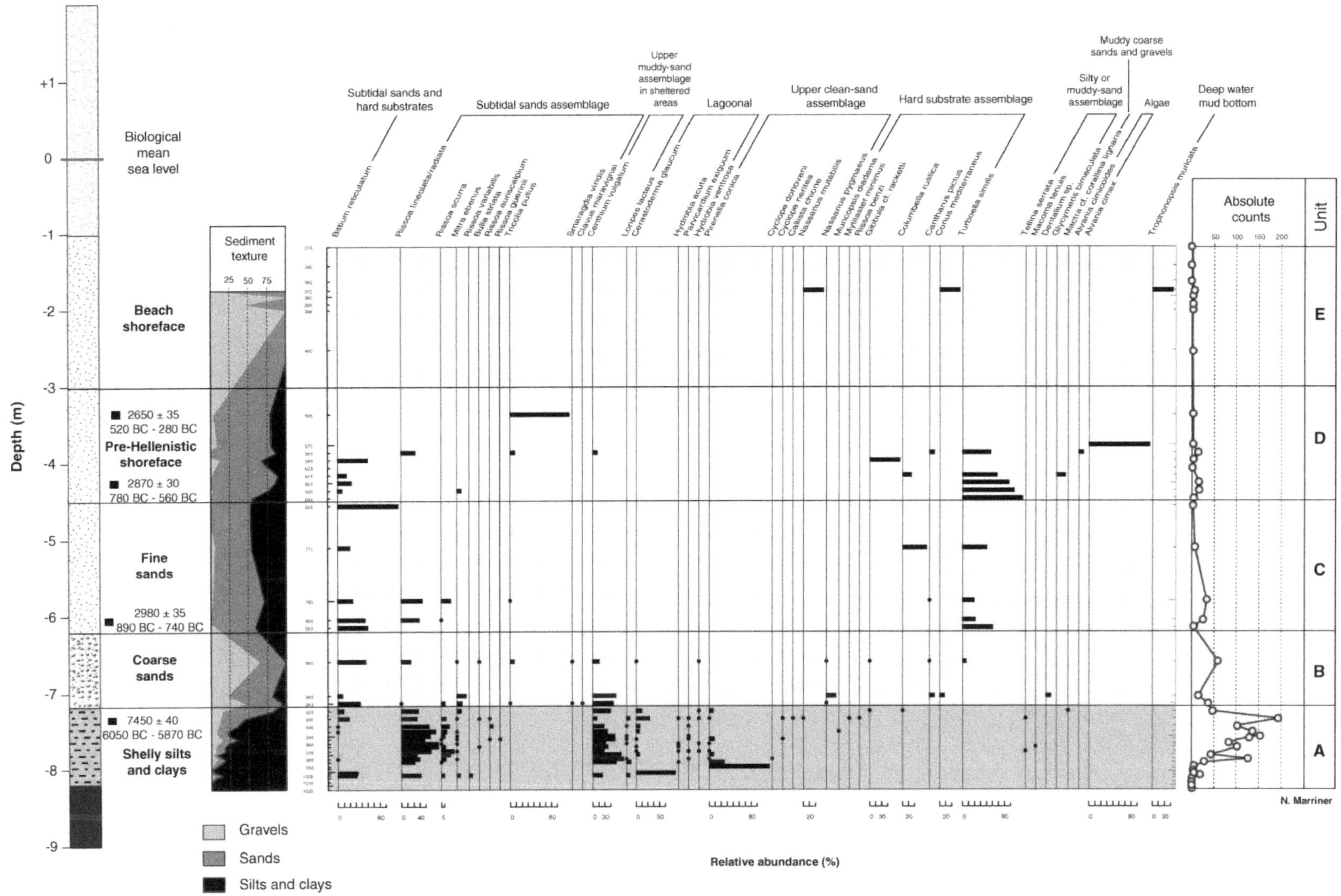

Figure 2.26: Core TXIII macrofauna. (1) Unit A: The macrofauna suite is dominated by taxa from the upper muddy-sand assemblage in sheltered areas (*Cerithium vulgatum*, *Loripes lacteus*), the subtidal sands assemblage (*Rissoa lineolata*, *Rissoa scurra*, *Mitra ebenus*) and the lagoonal assemblage (*Parvicardium exiguum*, *Hydrobia acuta*, *Cerastoderma glaucum*). We infer a protected marginal marine environment. (2) Unit B: The unit is characterised by the following macrofauna assemblages: subtidal sands (*Rissoa lineolata*, *Rissoa scurra*, *Mitra ebenus*, *Rissoa variabilis*, *Tricolia pullus*, *Smaragdia viridis*), upper muddy-sand assemblage in sheltered areas (*Cerithium vulgatum*) and the hard substrate assemblage (*Nassarius pygmaeus*, *Gibbula racketti*, *Cantharus pictus*, *Conus mediterraneus*). (3) Unit C: The subtidal sands and hard substrate taxa *Bittium reticulatum* and *Turboella similis* (hard substrate assemblage) dominate the unit. Secondary taxa include *Rissoa scurra*, *Tricolia pullus* (subtidal sands assemblage) and *Columbella rustica* (hard substrate assemblage). (4) Unit D: is dominated by molluscs from four ecological groups, the subtidal sands and hard substrates (*Bittium reticulatum*), the subtidal sands assemblage (*Rissoa lineolata*, *Tricolia pullus*), the hard substrate assemblage (*Turboella similis*) and the algae assemblage (*Alvania cimex*). We infer a pre-Hellenistic shoreface. (5) Unit E: There is a relative absence of molluscan shells in this facies. We infer a rapidly prograding shorface, not conducive to the development of a molluscan biocenosis.

2.5.3 Core TXIV: description

Core TXIV lies 120 m to the south of core site TXIII, on the opposite side of the causeway. The marine flooding surface is dated ~7800 BP by a poorly sorted sand unit which transgresses the clay substratum (**Figure 2.27**). Fine sands and a shelly gravel fraction characterise the sediment texture. Four ecological groups dominate the unit's macrofauna suite: the upper clean-sand assemblage (*Pirenella conica*), the subtidal sands assemblage (*Rissoa lineolata*, *Rissoa dolium*, *Mitra ebenus*), the upper muddy-sand assemblage in sheltered areas (*Loripes lacteus*, *Cerithium vulgatum*), lagoonal (*Parvicardium exiguum*, *Hydrobia ventrosa*) and the subtidal sands and hard subtrates assemblage (*Bittium reticulatum*, **Figure 2.28**). The ostracod fauna is dominated by the brackish-lagoonal species *Cyprideis torosa*, with the presence of marine-lagoonal taxa such as *Loxoconcha* spp. and *Xestoleberis aurantia* (**Figure 2.29**). These litho- and biostratigraphical indicators attest to a low energy lagoonal environment, protected from the open sea by the extensive breakwater ridge. We also hypothesise the existence of a series of transgressing beach and berm ridges (Otvos, 2000). Small peaks of coastal and marine taxa are concurrent with phases of breaching of these protective barriers during storms and high swell episodes.

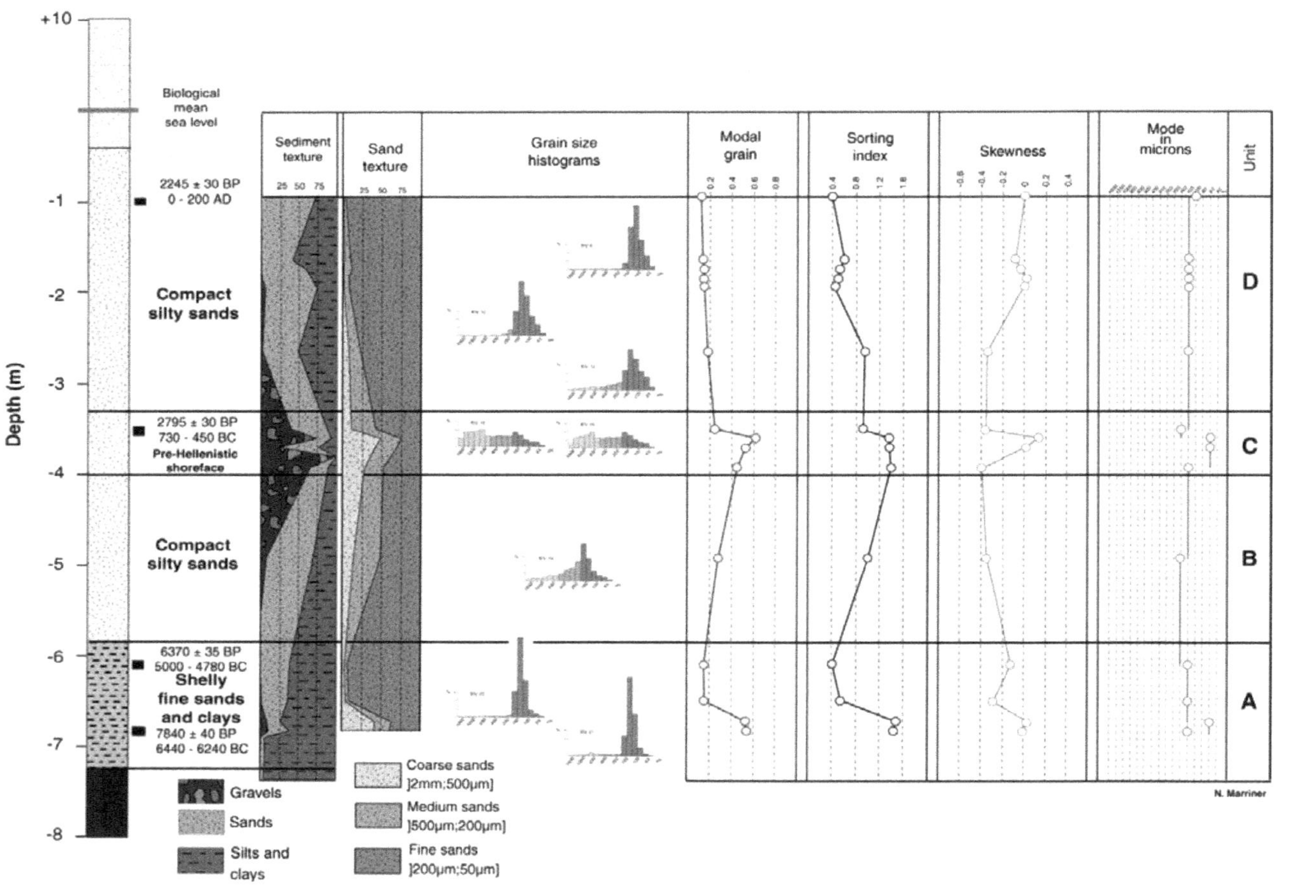

Figure 2.27: Core TXIV lithostratigraphy and grain size analyses. (1) Unit A: The marine flooding surface is dated ~7800 BP by a poorly sorted sand unit (1.45-1.49) which transgresses the clay substratum. This grades into a well sorted (0.41-0.55) fine sands unit characterised by negative skewness values of between -0.12 to -0.29. (2) Unit B: comprises an aggrading silty sand unit. A medium sorting index (1.01) and negative skewness value (-0.35) are consistent with a medium to low energy subtidal unit. (3) Unit C: is a poorly sorted gravely sand unit characterised by weakly developed histograms and contrasting positive and negative skewness values (-0.41 to 0.14), typical of a beach shoreface. (4) Unit D: is characterised by a well-sorted silty sand unit. The top of the unit is dated 2245 ± 30 BP (0-200 AD) and attests to rapid rates of salient accretion following the construction of Alexander's causeway.

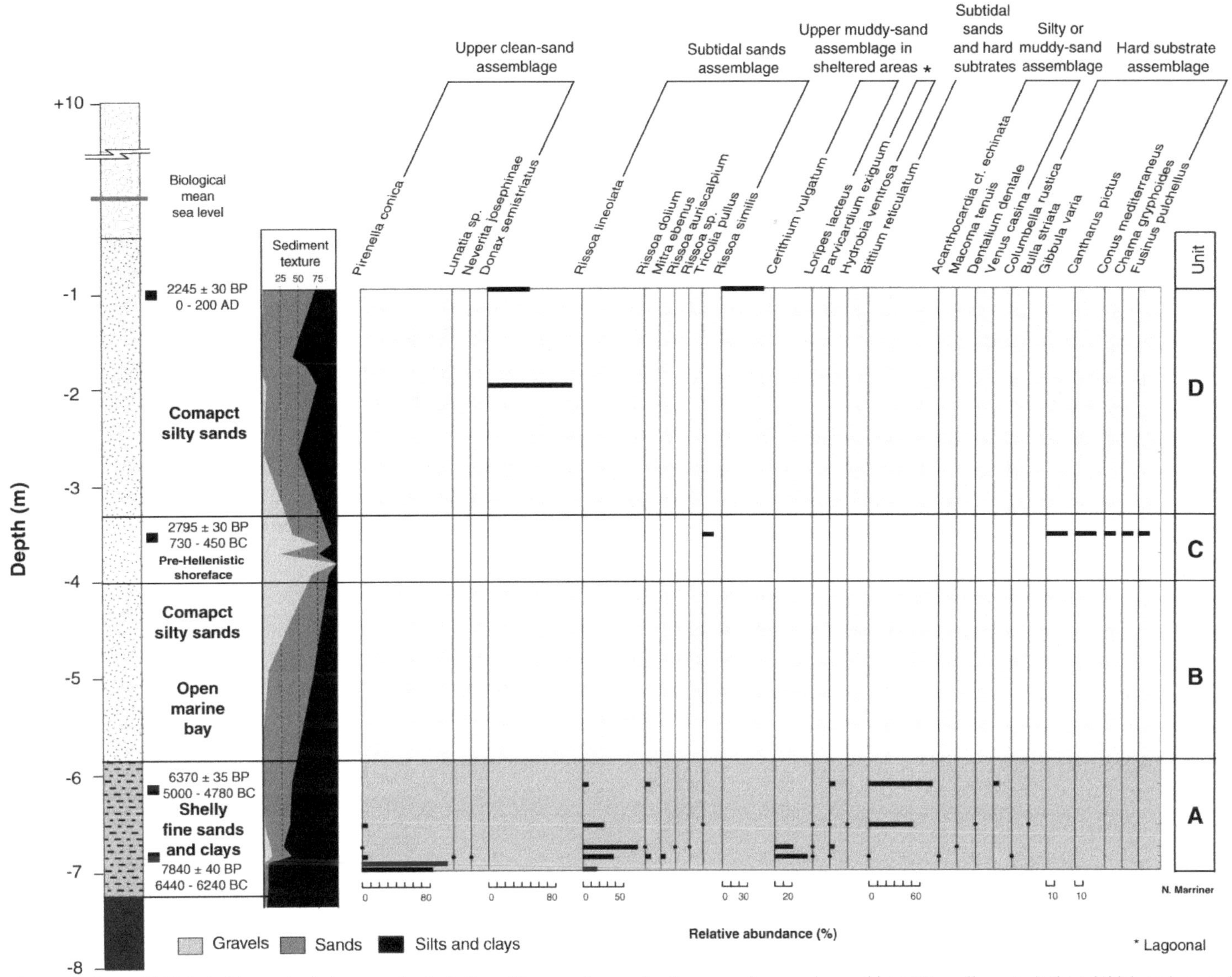

Figure 2.28: TXIV marine macrofauna. (1) Unit A: Five ecological groups dominate the unit's macrofauna suite: the upper clean-sand assemblage (*Pirenella conica*), the subtidal sands assemblage (*Rissoa lineolata*, *Rissoa dolium*, *Mitra ebenus*), the upper muddy-sand assemblage in sheltered areas (*Loripes lacteus*, *Cerithium vulgatum*), the lagoonal assemblage (*Parvicardium exiguum*, *Hydrobia ventrosa*) and the subtidal sands and hard substrates assemblage (*Bittium reticulatum*). (2) Unit B: Molluscan tests are absent from the unit. (3) Unit C: Poorly preserved hard substrate molluscan tests (*Gibbula varia*, *Fusinus pulchellus*, *Chama gryphoides*, *Cantharus pictus*, *Conus mediterraneus*) concur high energy swash zone dynamics characterised by the reworking of imported molluscan tests. The chronology is coeval with a Hellenistic beachface (2795 ± 30 BP, TXIV 14). (4) Unit D: The molluscan fauna is poor, with rare tests of *in situ Donax semistriatus* (upper clean-sand assemblage) indicating rapid rates of sedimentation that hinder the development of a biocenosis.

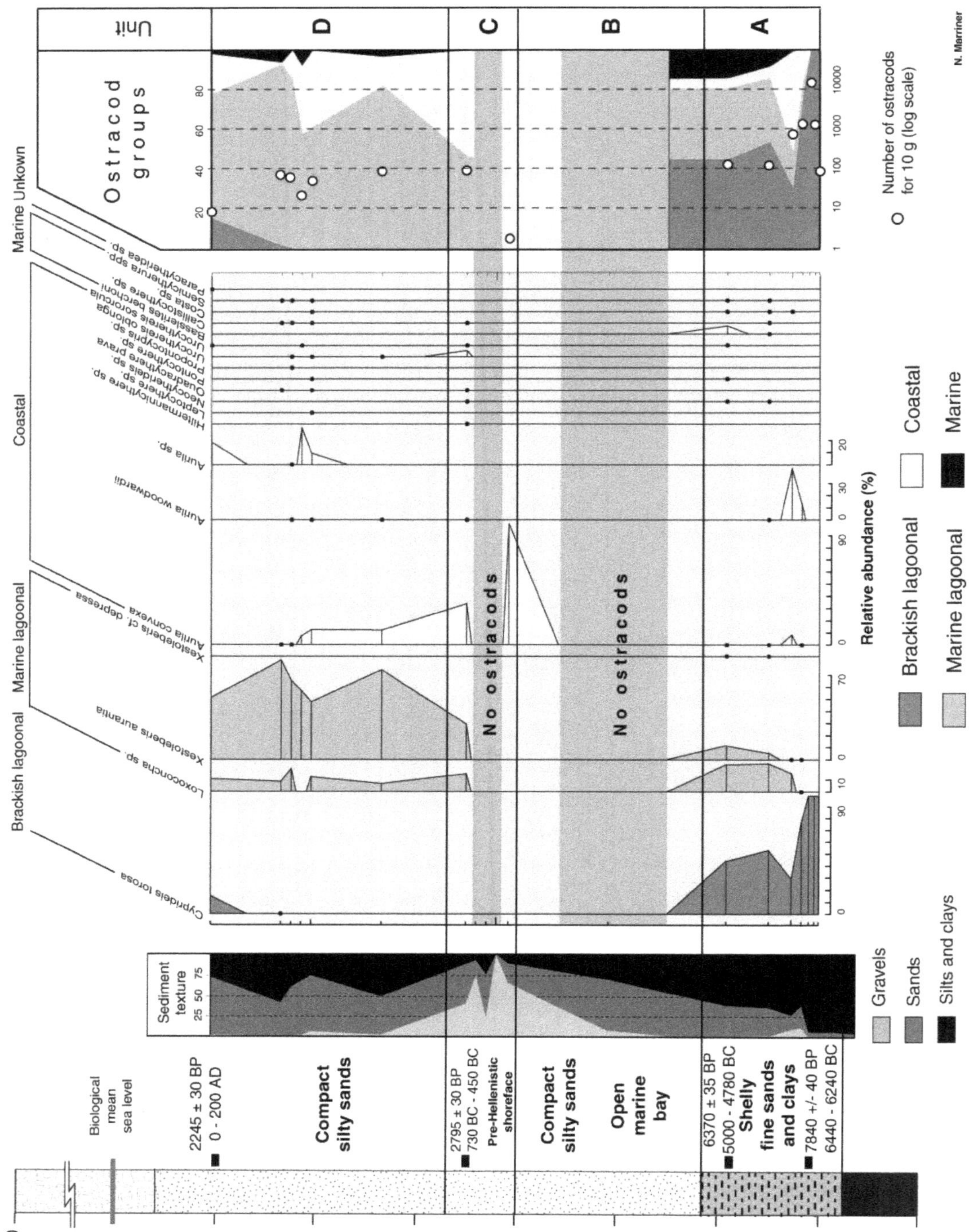

Figure 2.29: TXIV ostracods.

Unit B comprises an aggrading silty sand unit. Molluscan tests are absent from the unit and the ostracod fauna is poor, with just a few tests of the coastal species *Aurila convexa*. It is overlain by unit C, a poorly sorted gravely sand unit characterised by weakly developed histograms, typical of a beach shoreface. Poorly preserved hard substrate molluscan tests (*Gibbula varia*, *Fusinus pulchellus*, *Chama gryphoides*, *Cantharus pictus*, *Conus mediterraneus*) and the absence of an ostracod fauna concur high energy swash zone dynamics characterised by the reworking of imported molluscan tests. The chronology is coeval with a Hellenistic beach face (2795 $\pm$ 30 BP, TXIV 14).

Unit D is characterised by a well-sorted silty sand unit. The molluscan fauna is poor, with rare tests of *in situ* *Donax semistriatus* (upper clean-sand assemblage) indicating rapid rates of sedimentation that hinder the development of a biocenosis. The ostracod fauna are dominated by the marine-lagoonal species *Loxoconcha* spp. and *Xestoleberis aurantia*, with smaller peaks of coastal taxa such as *Aurila convexa* and *Aurila* sp. The top of the unit is dated 2245 ± 30 BP (0-200 AD) and attests to rapid rates of salient accretion following the construction of Alexander's causeway. This rapid salient growth significantly hindered the longshore drift. A fall in water competences around the causeway is indicated by the fine-grained nature of the unit and high relative abundances of marine-lagoonal taxa.

2.5.4 Core TXV: description

Core TXV lies 250 m to the south of the ancient causeway. 16.5 m of fill have accumulated since the Holocene marine transgression around 8000 BP. As with the other tombolo cores, the basal unit A comprises a silty-sand unit with large amounts of shell debris (**Figure 2.30**). Upper muddy-sand assemblage in sheltered areas and subtidal sands macrofauna assemblages dominate (**Figure 2.31**). Brackish lagoonal (*Cyprideis torosa*) and marine lagoonal (*Loxoconcha* spp., *Xestoleberis aurantia*) ostracod taxa characterise the microfossil fauna, data which corroborate the existence of a sheltered lagoon environment (**Figure 2.32**). In unit B, transition to a poorly sorted coarse sand fraction is consistent with retrograding berm ridges, reworking sediment stocks as the forms transgressively onlap the lagoon environment. A rise in marine-lagoonal (*Loxoconcha* spp., *Xestoleberis aurantia*) and coastal ostracod taxa (*Aurila convexa*, *Aurila woodwardii*) is to the detriment of the formerly abundant *Cyprideis torosa*.

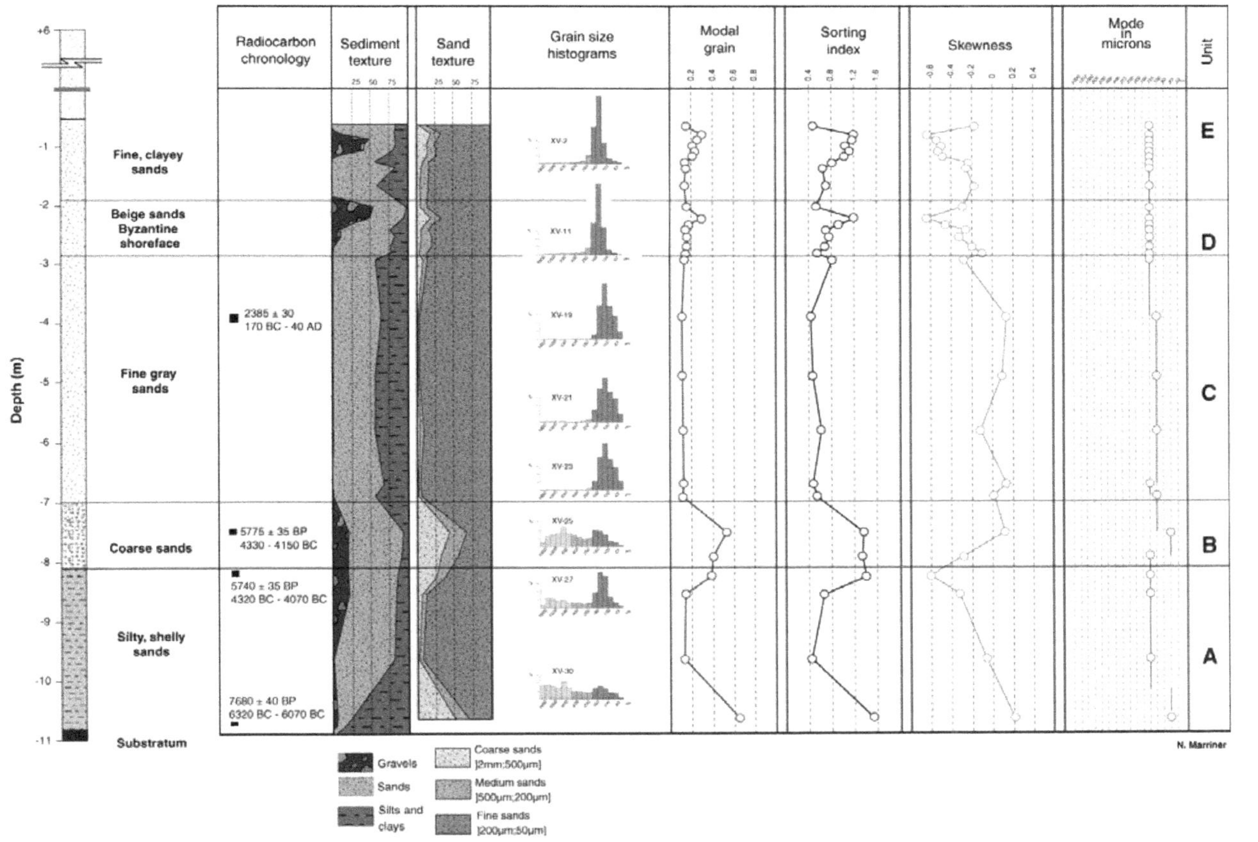

Figure 2.30: Core TXV lithostratigraphy and grain size analyses. (1) Unit A: comprises a silty-sand unit with large amounts of shell debris. We infer a low energy lagoon environment. (2) Unit B: Transition to a poorly sorted (1.34-1.32) coarse sand fraction is consistent with retrograding berm ridges, reworking sediment stocks as the forms transgressively onlap the lagoon. (3) Unit C: constitutes a well to medium sorted (0.42-0.61) fine sand unit. Fine sands dominate the unit comprising >93 % of the sand fraction. We interpret this unit as the sub-aqueous proto-tombolo phase which began accreting after ~5700 BP. (4) Unit D: comprises a coarsening-up sand unit. Negative skewness values attest to a low energy sublittoral shoreface (-0.18 to -0.45). We interpret this unit as the Byzantine shoreface. (5) Unit E: Fine-grained sand sedimentation continues into this unit, corroborating a rapidly prograding shoreface.

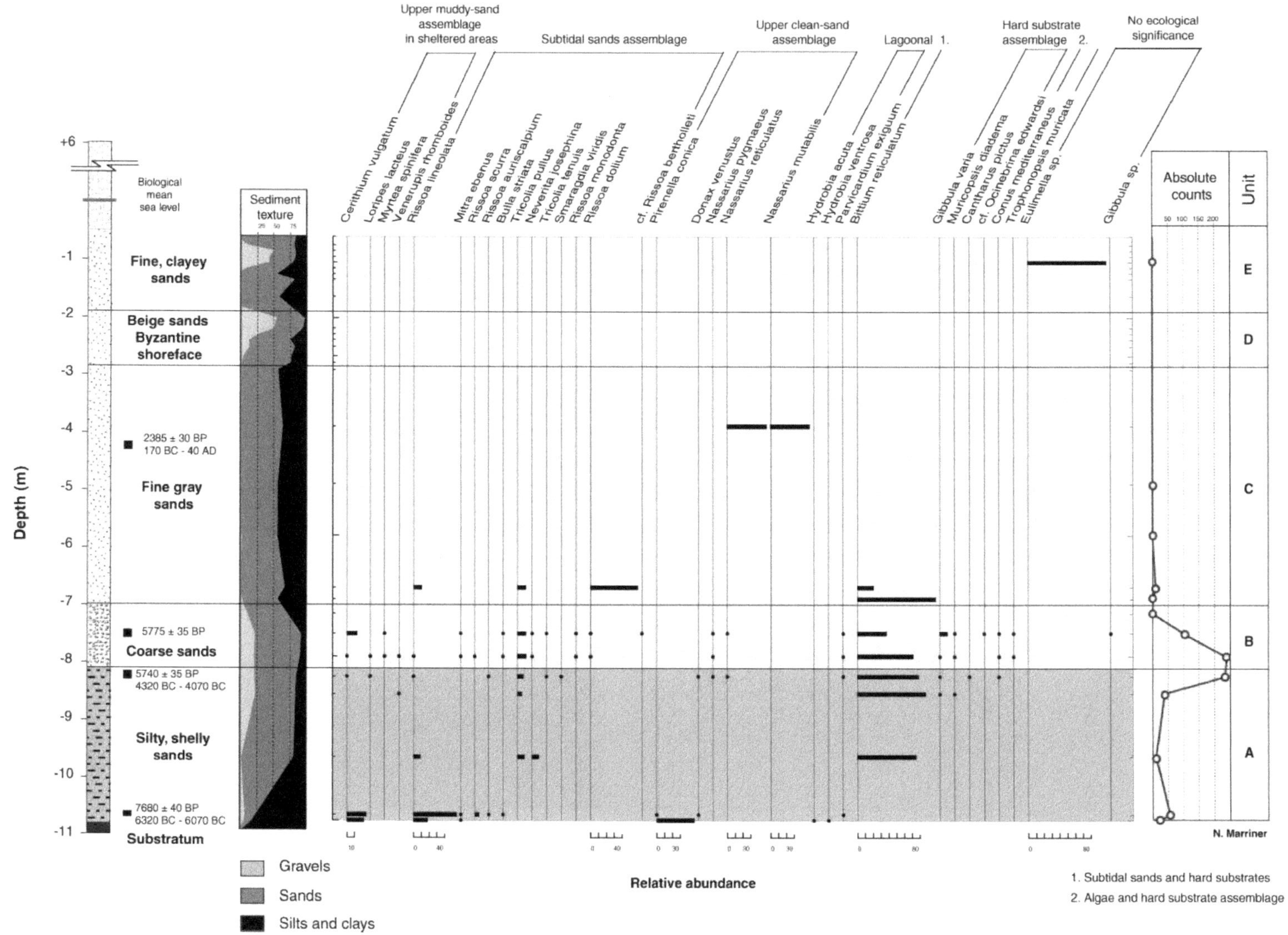

Figure 2.31: TXV marine macrofauna. (1) Unit A: Upper muddy-sand assemblage in sheltered areas (*Cerithium vulgatum*), subtidal sands (*Rissoa lineolata*, *Tricolia pullus*, *Neverita josephina*) and the subtidal sands and hard substrate assemblage (*Bittium reticulatum*) taxa dominate the unit's molluscan suites. We infer a low-energy lagoon type environment. (2) Unit B: juxtaposes a number of molluscan groups including the upper muddy-sand assemblage in sheltered areas (*Cerithium vulgatum*, *Loripes lacteus*, *Venerupis rhomboides*), the subtidal sands assemblage (*Tricolia pullus*, *Neverita josephina*, *Rissoa dolium*, *Mitra ebenus*), the upper clean-sand assemblage (*Donax venustus*, *Nassarius pygmaeus*) the subtidal sands and hard substrates assemblage (*Bittium reticulatum*) and the hard substrate assemblage (*Gibbula varia*, *Muricopsis diadema*). From these data, we infer a transgressing beach ridge characterised by the reworking of malacological stocks. (3) Unit C: comprises poor molluscan suites. Sporadic tests of *Rissoa* spp., *Tricolia pullus* (subtidal sands assemblage), *Nassarius reticulatus*, *Nassarius mutabilis* (upper clean-sand assemblage) and *Bittium reticulatum* (subtidal sands and hard substrates) are observed. (4) Unit D: is void of any molluscan shells. We interpret this unit as the Byzantine shoreface. (5) Unit E: The relative absence of shells is consistent with a rapidly prograding shoreface not conducive to the development of a molluscan biocenosis.

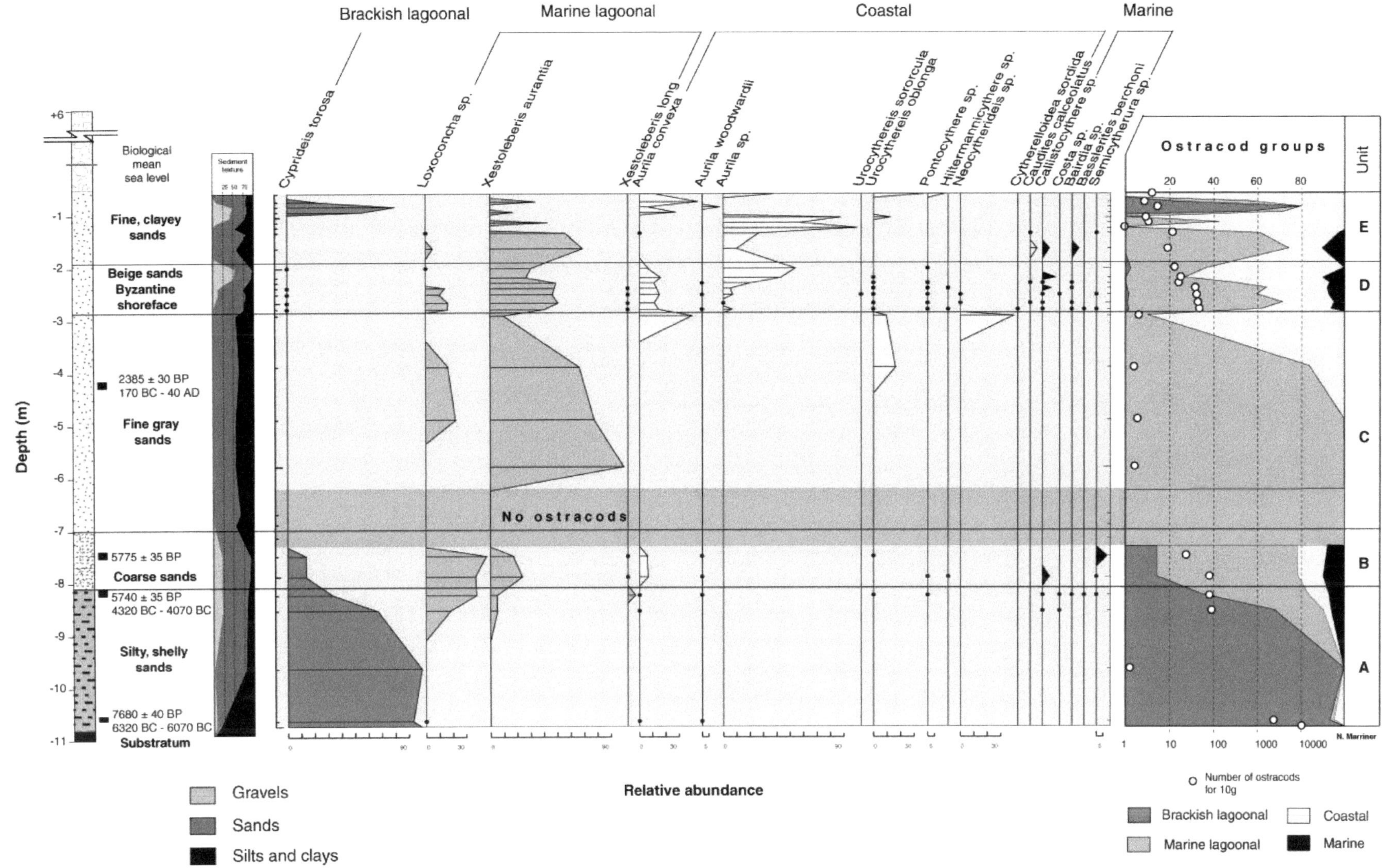

Figure 2.32: TXV ostracods. (1) Unit A: Brackish lagoonal (*Cyprideis torosa*) and marine lagoonal (*Loxoconcha* spp., *Xestoleberis aurantia*) ostracod taxa characterise the microfossil fauna. These data corroborate the existence of a sheltered lagoon environment. (2) Unit B: A rise in marine-lagoonal (*Loxoconcha* spp., *Xestoleberis aurantia*) and coastal ostracod taxa (*Aurila convexa*, *Aurila woodwardii*) is to the detriment of the formerly abundant *Cyprideis torosa*. This is consistent with the transgression of beach ridges after ~5800 BP. (3) Unit C: The unit is characterised by marine lagoonal ostracod taxa, with small peaks of coastal species (*Aurila convexa*, *Urocythereis oblonga*, *Neocytherideis* sp.). The base of the unit is void of an ostracod fauna. We interpret this unit as the early subaqueous proto-tombolo. (4) Unit D: A rise in coastal (*Aurila* spp., *Urocythereis oblonga*, *Pontocythere* sp.) ostracod taxa corroborates an accreting subaqueous salient. Marine species such as *Semicytherura* sp., *Costa* sp., *Callistocythere* spp., *Basslerites berchoni* and *Bairdia* spp. were drifted in during periods of heightened swell and storms. (5) Unit E: is dominated by marine lagoonal (*Loxoconcha* sp., *Xestoleberis aurantia*) and coastal (*Aurila convexa*, *Aurila* sp., *Urocythereis oblonga*) ostracod taxa. Towards the top of the unit (~-1 m), confined hyposaline conditions are attested to by a peak in *Cyprideis torosa*.

Unit C comprises a well-sorted, fine grey sand unit concomitant with those observed in cores TXIII and TXIV. The unit is characterised by marine lagoonal ostracod taxa, with small peaks of coastal species (*Aurila convexa*, *Urocythereis oblonga*, *Neocytherideis* sp.). The base of the unit is void of an ostracod fauna. We interpret this unit as the early subaqueous proto-tombolo phase, which began accreting after ~5700 BP. Unit D comprises a coarser beige sand, analogous with a coarsening-up sequence. A rise in coastal ostracod taxa corroborates an accreting subaqueous salient. Marine species such as *Semicytherura* sp., *Costa* sp., *Callistocythere* spp., *Basslerites berchoni* and *Bairdia* spp. were drifted in during periods of heightened swell and storms. Fine-grained sedimentary conditions persist into unit E, dominated by marine-lagoonal and coastal ostracod taxa.

2.6 Discussion and interpretations

Here we marry the chronostratigraphic and modelling data to report details on the Holocene morphodynamics and process-response of Tyre's tombolo. We used multivariate statistical analyses to better comprehend the spatial and temporal variability in sedimentary environments on our tombolo transect (**Figures 2.33 and 2.34**). High-resolution analyses shed light on three critical aspects: (1) natural accretion of Tyre's early Holocene marine bottom, leeward of the island breakwater and akin to a stratigraphic trap; (2) formation of a wave-dominated proto-tombolo after 6000 BP; and (3) the wide-reaching anthropogenic impacts of the Hellenistic causeway (**Figures 2.35 - 2.37**).

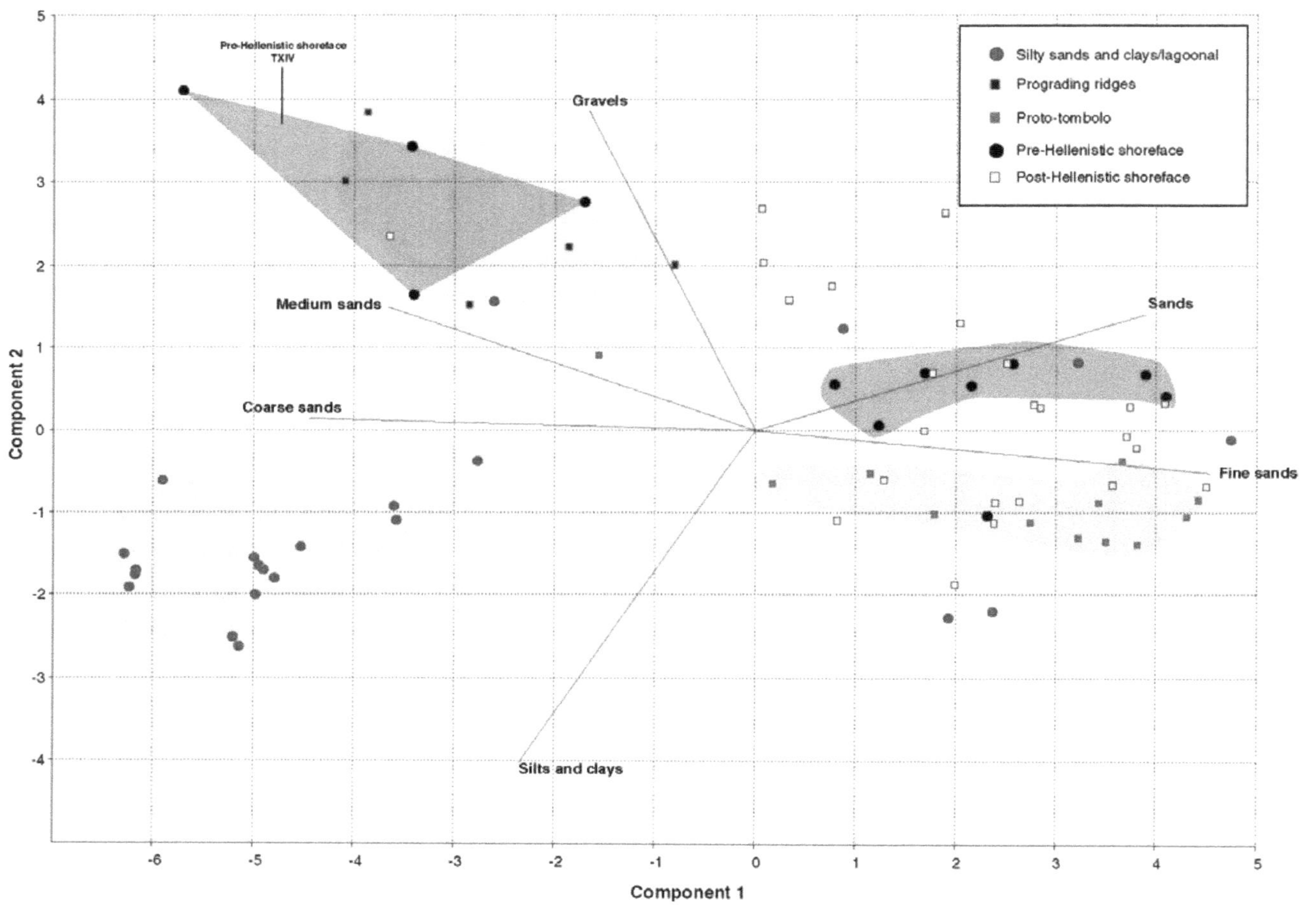

Figure 2.33: PCA scatter plot of axes 1+2 of a multivariate dataset comprising sediment texture and sand texture. 85 % of the total dataset variation is explained by components 1 and 2. Five lithostratigraphical groups have been identified on the basis of the statistical analyses. (1) A silty sands and clays unit consistent with the lagoon environment which existed behind Tyre island until 6000 BP. (2) Coarse-grained transgressive ridges which breached this environment after 6000 BP, reworking sediments from the transgressed shelf. (3) A fine-sand proto-tombolo unit. (4) A slightly coarser grained pre-Hellenistic shoreface (TXIV contrasts with TXIII and TXV by being much coarser). These two facies comprise the proto-tombolo *sensu stricto*. The units are well-constrained, with very little variance in the data consistent with a relatively ubiquitous sedimentary environment in the leeward shadow of the island breakwater. (5) The final unit comprises a post-Hellenistic shoreface. The high level of variance in this unit is attributed to human impacts and segmentation of the tombolo into two isolated bays, with a cessation of longshore currents across the form. Subaerial growth of the isthmus culminated in a cessation of the north-south sediment transport across the tombolo. Post-Hellenistic deposits in core TXV are clearly differentiated from analogous units in cores TXIV and TXIII (northern lobe), coeval with two independently evolving littoral cells.

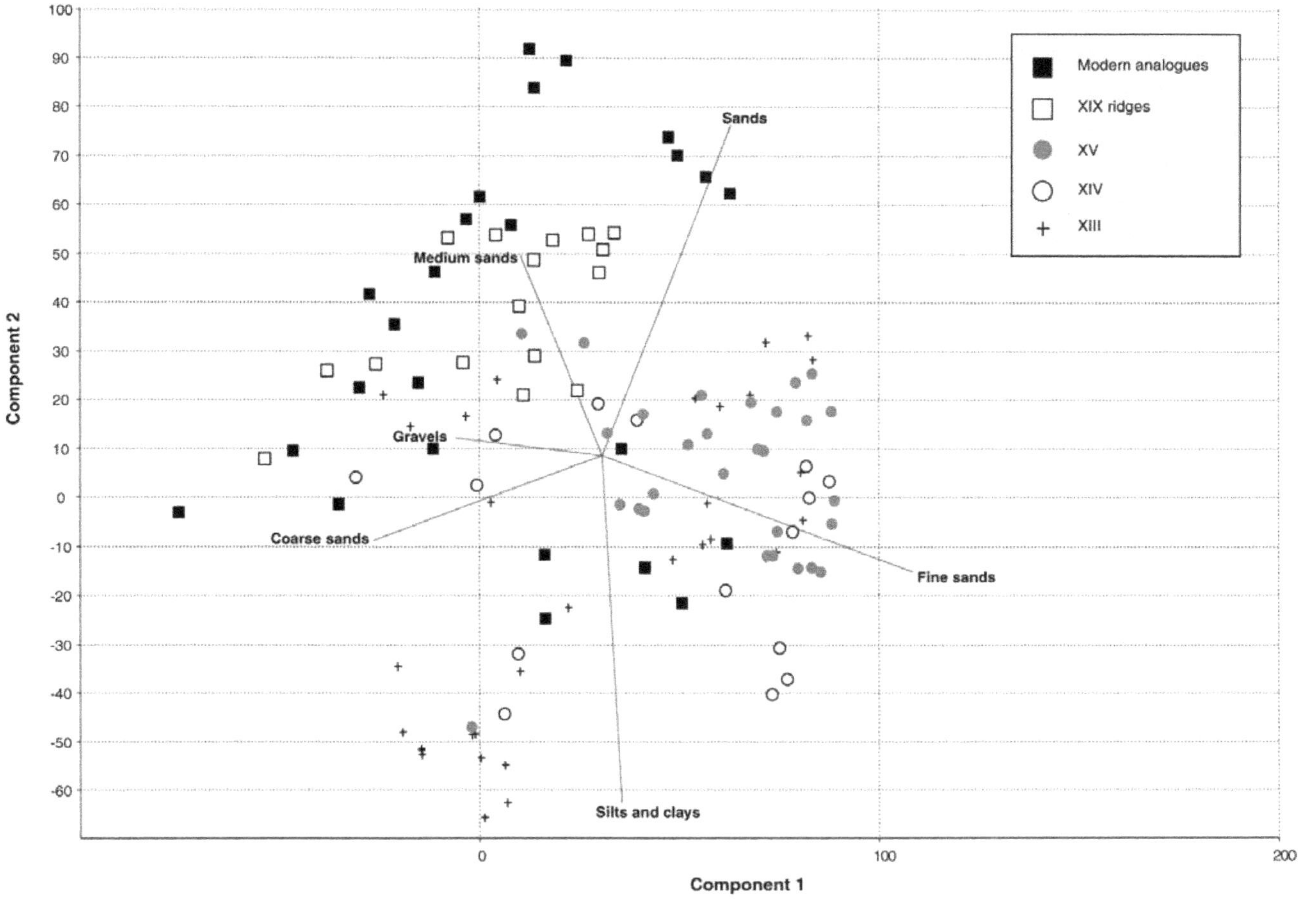

Figure 2.34: PCA scatter plot of axes 1+2 of a multivariate dataset comprising sediment texture and sand texture of tombolo, beach ridge sediments and modern sediment analogues. 76 % of the total dataset variation is explained by components 1 and 2. The analysis reveals significant discrepancies between the source (Litani) area, dominated by coarse grained sands, and sink area which is characterised by finer grained sands. While the tombolo comprises mainly fine sands, the fluvial and continental prograding ridges constitute coarser grained sediments. There is great variance in the modern analogues dataset, explained by the diversity of sedimentary environments sampled. Setting aside the basal lagoonal facies, there is relatively little variance in the proto-tombolo data. We link this to a relatively calm sedimentary environment in the shadow of the island breakwater, conducive to the formation of the salient bottom.

2.6.1 Tripartite model of tombolo morphogenesis

2.6.1.1 Origin and development - Early Holocene transgression

Our data suggest that sea-level stillstand after 6000 BP was the biggest single factor in the initial development of Tyre's tombolo. A north-south transect comprising three cores, TXIII, TXIV and TXV, elucidates the tombolo's stratigraphy (**Figure 2.38 - 2.41**). In this area, leeward of the island breakwater, the Maximum Flooding Surface (MFS) is dated ~8000 BP consistent with the transgression of deltas throughout the circum Mediterranean (Stanley and Warne, 1994). The litho- and biostratigraphical signatures of this transgression comprise low energy silts and sands rich in molluscan shells. Brackish lagoonal ostracod species (*Cyprideis torosa*), with minor peaks of marine-lagoonal (*Loxoconcha* spp.) and coastal (*Aurila convexa*, *Aurila woodwardii*) taxa characterise the unit. These fine-grained sediments and brackish and marine lagoonal faunas translate shallow, low-energy waterbodies during this period. This sedimentary unit is clearly discerned by the Principal Components Analysis (PCA), plotting between the coarse sands and silts and clays granulometric variables (**Figures 2.33 and 2.34**). This juxtaposition of opposing sedimentary conditions is explained by the shelly nature of the lagoonal silts and clays and is not coeval with high energy coastal processes. Shelter was afforded by Tyre's elongated sandstone reefs which cumulatively acted as a shore-parallel breakwater, corroborated by the numerical models.

Core TXVIII was drilled between Tell Mashuk and El Bass, the oldest portion of the continental salient *sensu stricto*. Two contrasting facies retrace the inception and progradation of the tombolo margin. The clay substratum is transgressed by fine-bedded marine sands dated ~6000 BP. Rapid beach ridge accretion is recorded with ~1 m of sediment accumulation during a 300 year period. Marine-lagoonal (*Loxoconcha* spp., *Xestoleberis aurantia*) and coastal (*Aurila convexa*, *Pontocythere* sp., *Urocythereis* sp.) ostracod taxa characterise the facies. After 5500 BP, transition from marine sands to choked lagoon sediments corroborates accretion of the salient with isolation of the back-barrier lagoon. Such rapid coastal progradation is typical of the mid-Holocene stillstand regime and high sediment supply. Given the concentration of archaeological sites in the vicinity of these lagoon deposits (Tell Mashuk, Tell Chawakir), we posit that the area served as a natural anchorage for the Bronze Age settlers of Palaeo-Tyre. Although the density of core networks does not allow the precise spatial dimensions of this lagoon to be established, local topography and geomorphological prospections in this area suggest that at its greatest extent, it could have reached the base of Tell Mashuk.

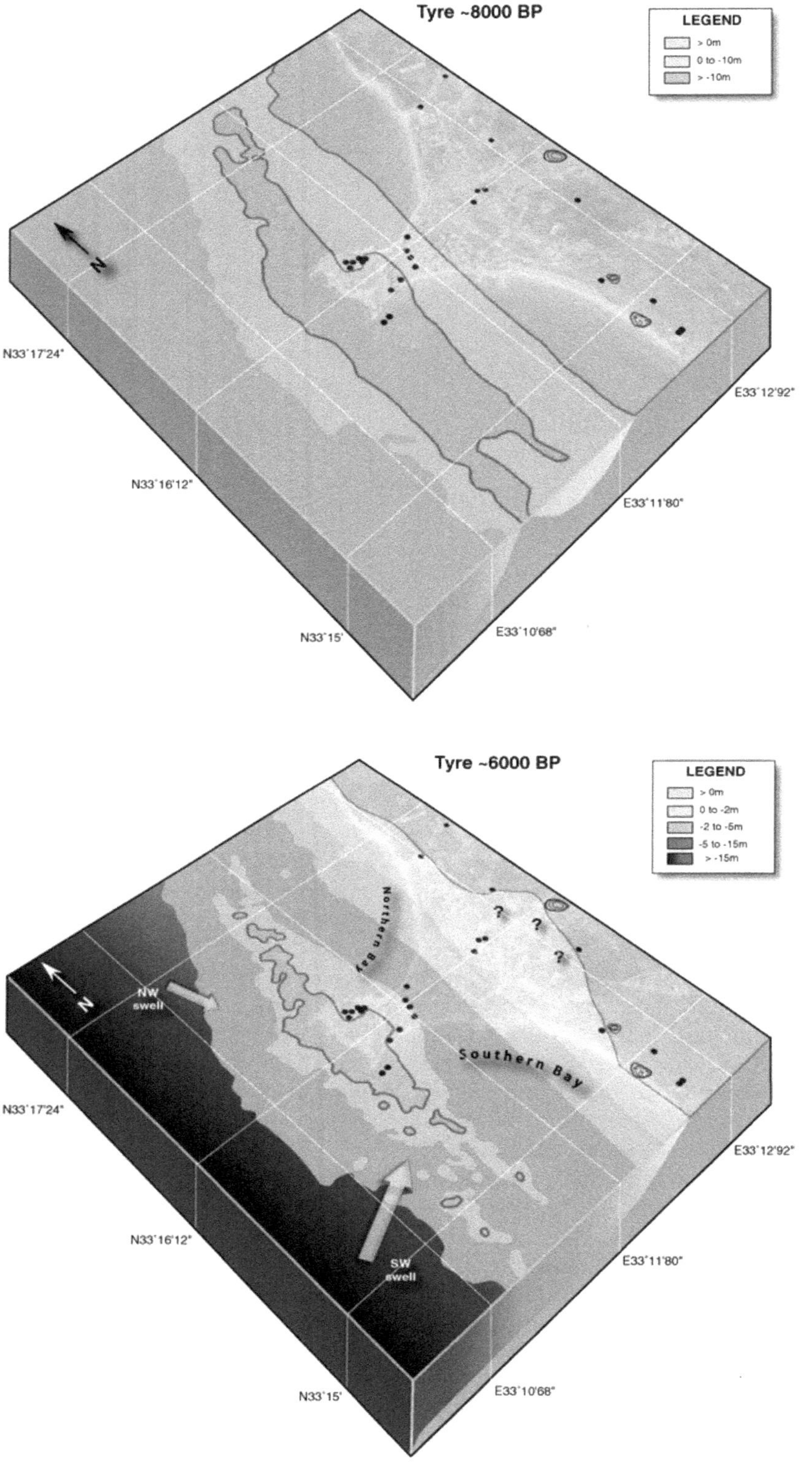

Figure 2.35: Morphodynamic evolution of Tyre's tombolo between ~8000 BP and ~6000 BP.

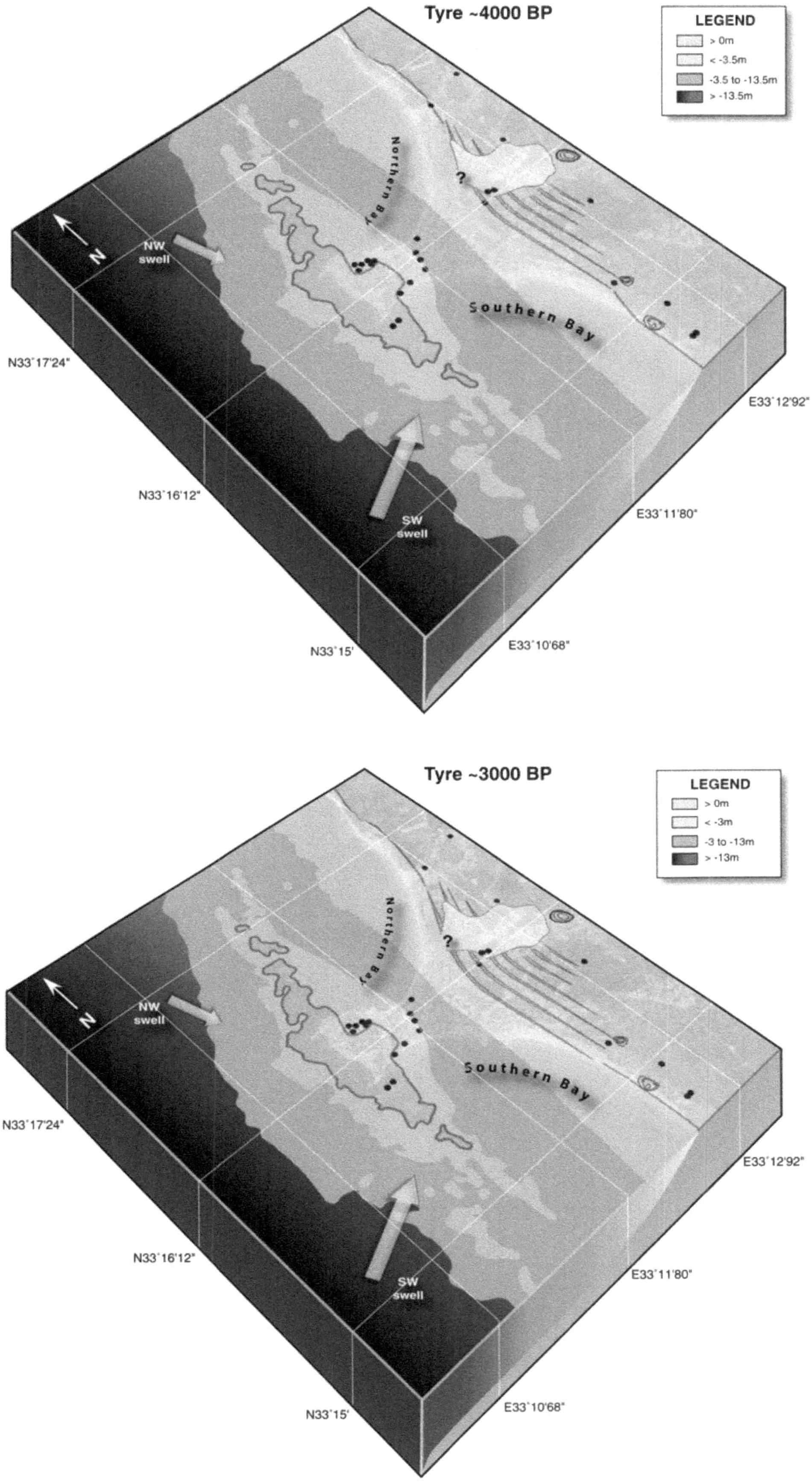

Figure 2.36: Morphodynamic evolution of Tyre's tombolo between ~4000 BP and ~3000 BP.

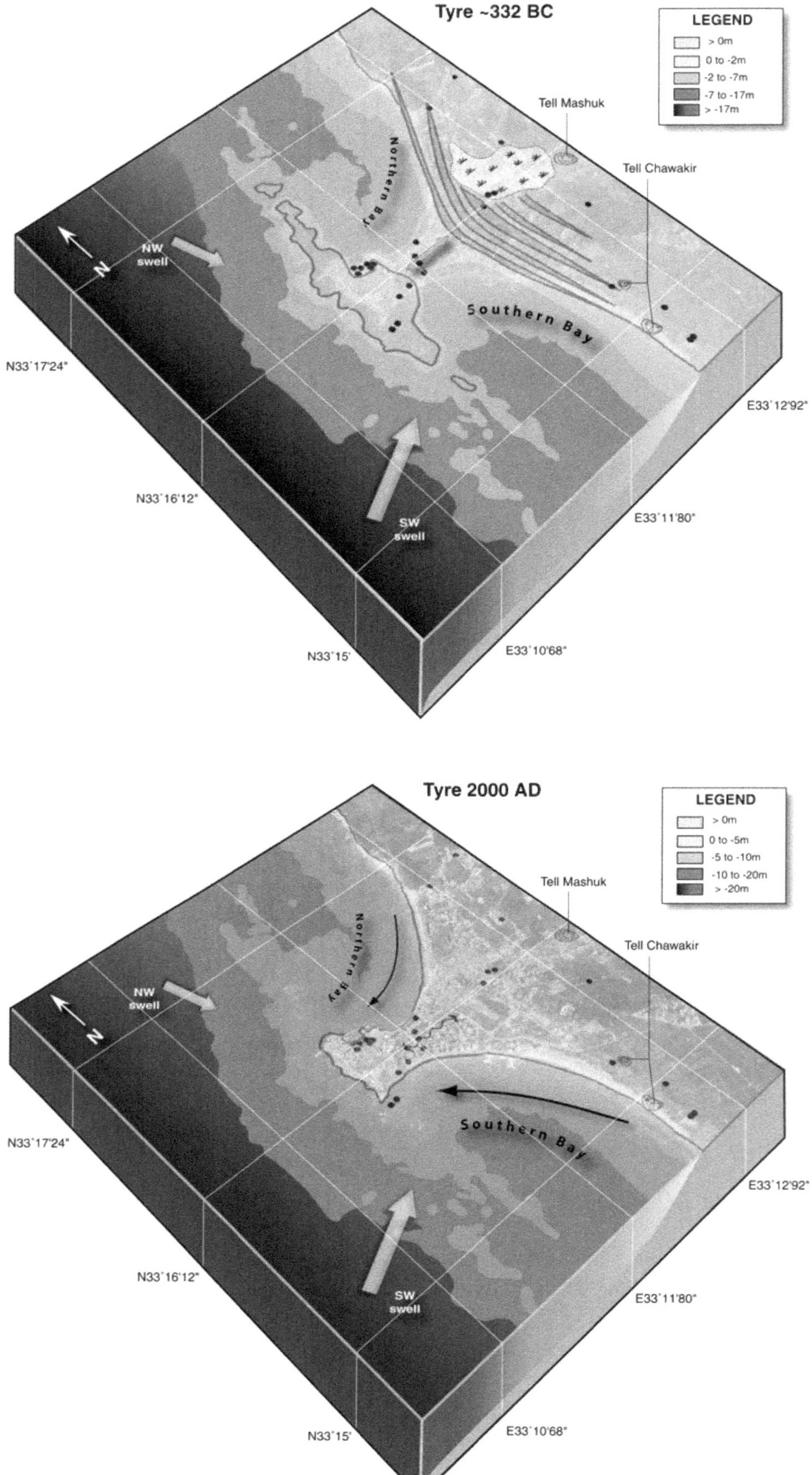

Figure 2.37: Morphodynamic evolution of Tyre's tombolo between Hellenistic times and today.

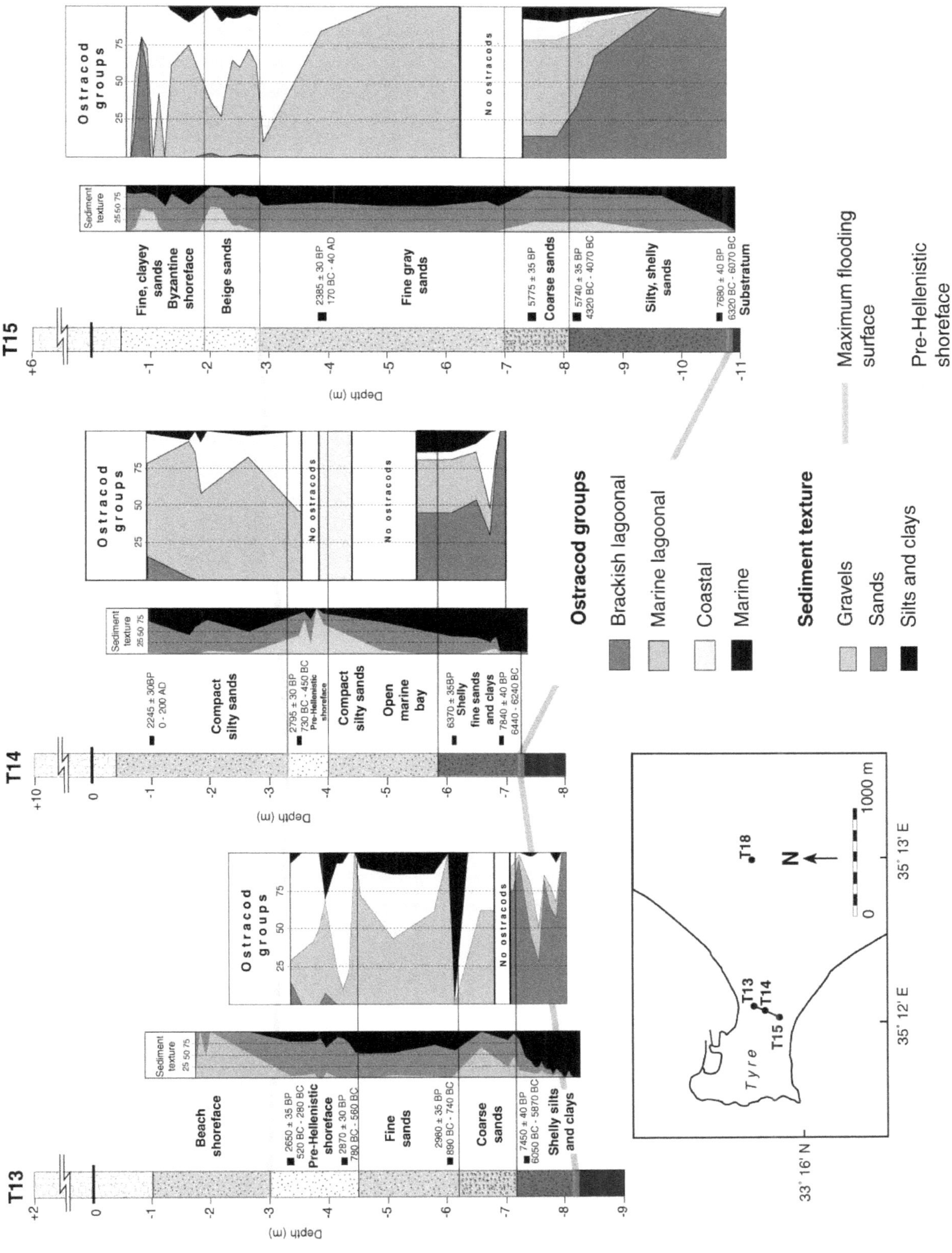

Figure 2.38: Chrono- and ostracod biostratigraphy of tombolo cores TXIII-TXV.

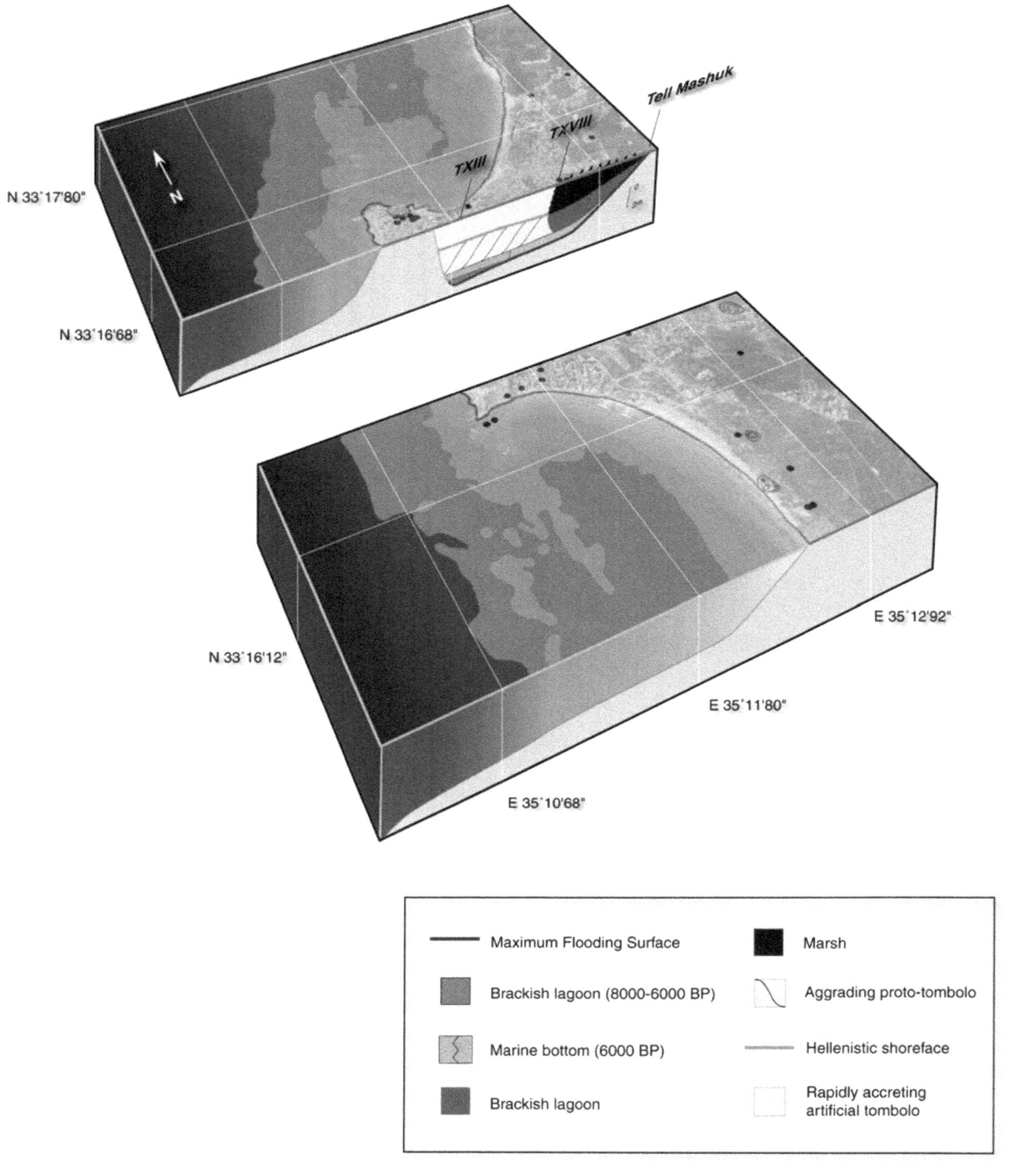

Figure 2.39: Simplified east-west transect of Tyre's tombolo stratigraphy.

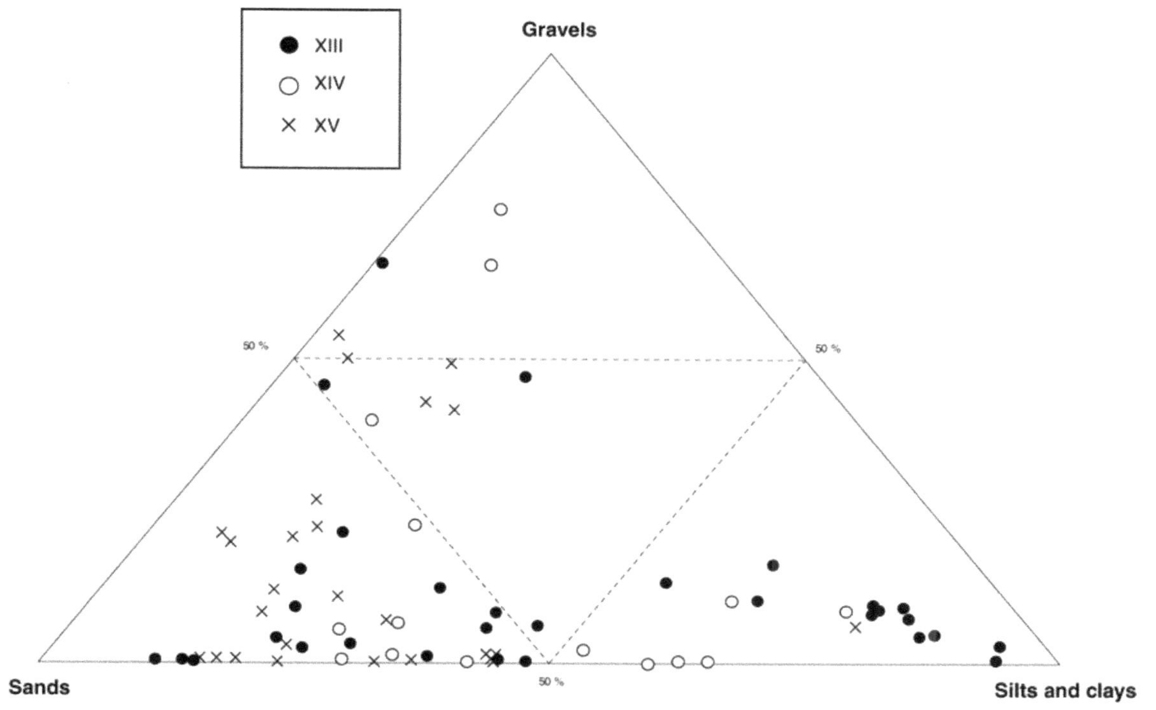

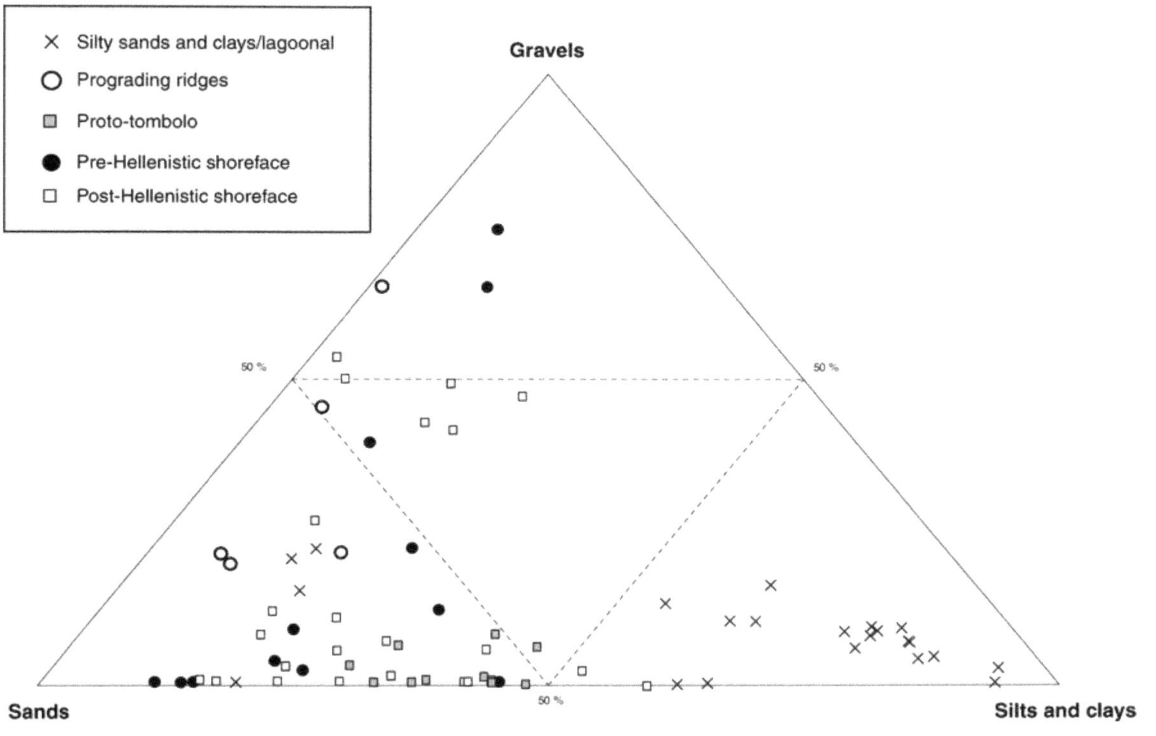

Figure 2.40: Ternary diagrams of the transect's total sediment fraction.

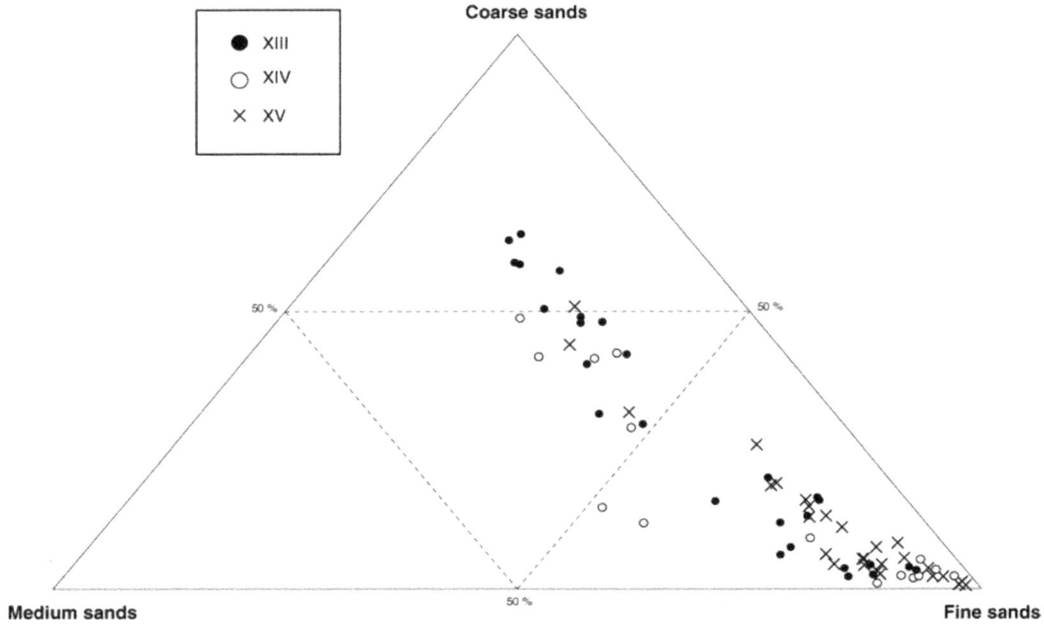

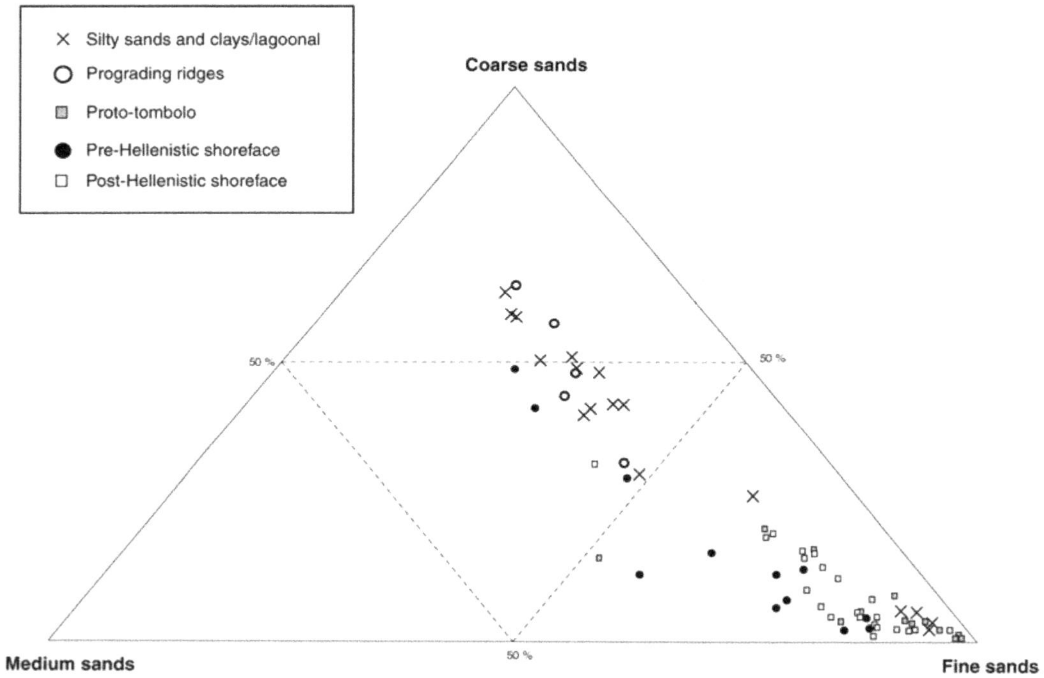

Figure 2.41: Ternary diagrams of the transect's sand fraction. Although there is overlap between the data points, there is a pattern of coarser sands to the north of the tombolo (TXIII) and slightly finer-grained material to the south (TXIV and TXV). We explain this pattern as a function of proximity to the main Holocene sediment source, the Litani river, which lies 8.5 km north of the city. The tombolo facies are much finer grained than the lagoon sands and prograding ridges. For the lagoon facies we explain this difference not as a function of energy dynamics, but rather the coarse nature of the shelly debris which constitutes this unit. The prograding ridges comprise a coarse sand consistent with the reworking of deposits during transgression of the continental shelf.

Stratigraphies north and south of this lagoon system have yielded prograding ridge topographies. Core TVII, from the base of Tell Chawakir has not yielded any lagoonal deposits; this area was transgressed before rapidly prograding to form the southern continental base of the salient (**Figure 2.42**). Between Tell Chawakir and Tell Rachidiye, the stratigraphy has revealed a series of Pleistocene paleosols (Cohen-Seffer *et al.*, 2005). No marine deposits have been elucidated in this area, pertaining to the relative stability of this portion of Tyre's maritime façade during the Holocene (**Figure 2.42**). On the northern flank of the tombolo, geological sections reveal an archetypal prograding ridge system (**Figures 2.43 and 2.44**).

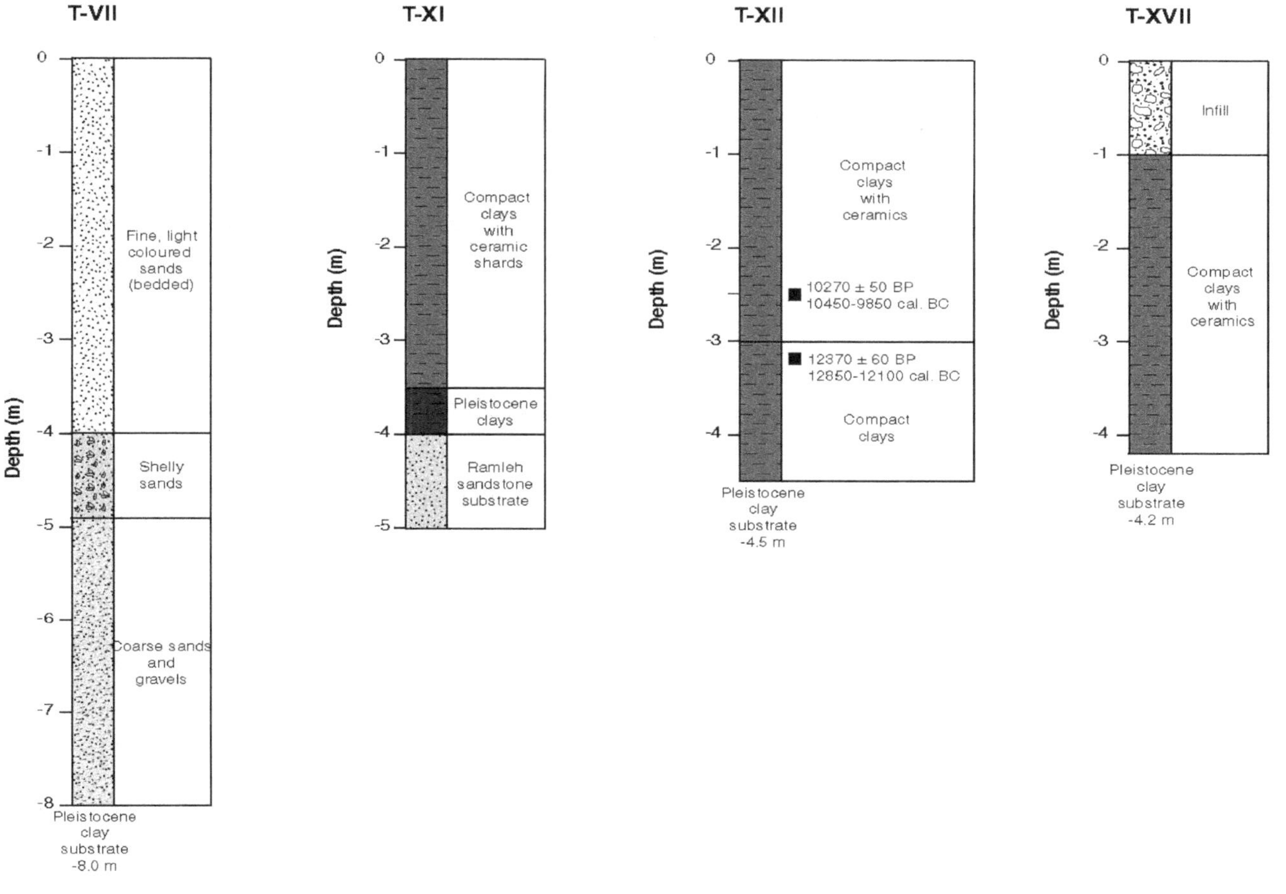

Figure 2.42: Lithostratigraphy of cores TVII, TXI, TXII and TXVII to the south of Tyre. Cores TXI, TXII and TXVII were drilled between Tell Rachidiye and Tell Chawakir. They manifest no Holocene marine facies, indicating that this portion of the Tyrian coastline has been relatively stable since the Holocene transgression. Core TVII constitutes prograding beach deposits and records the early accretion of the continental salient.

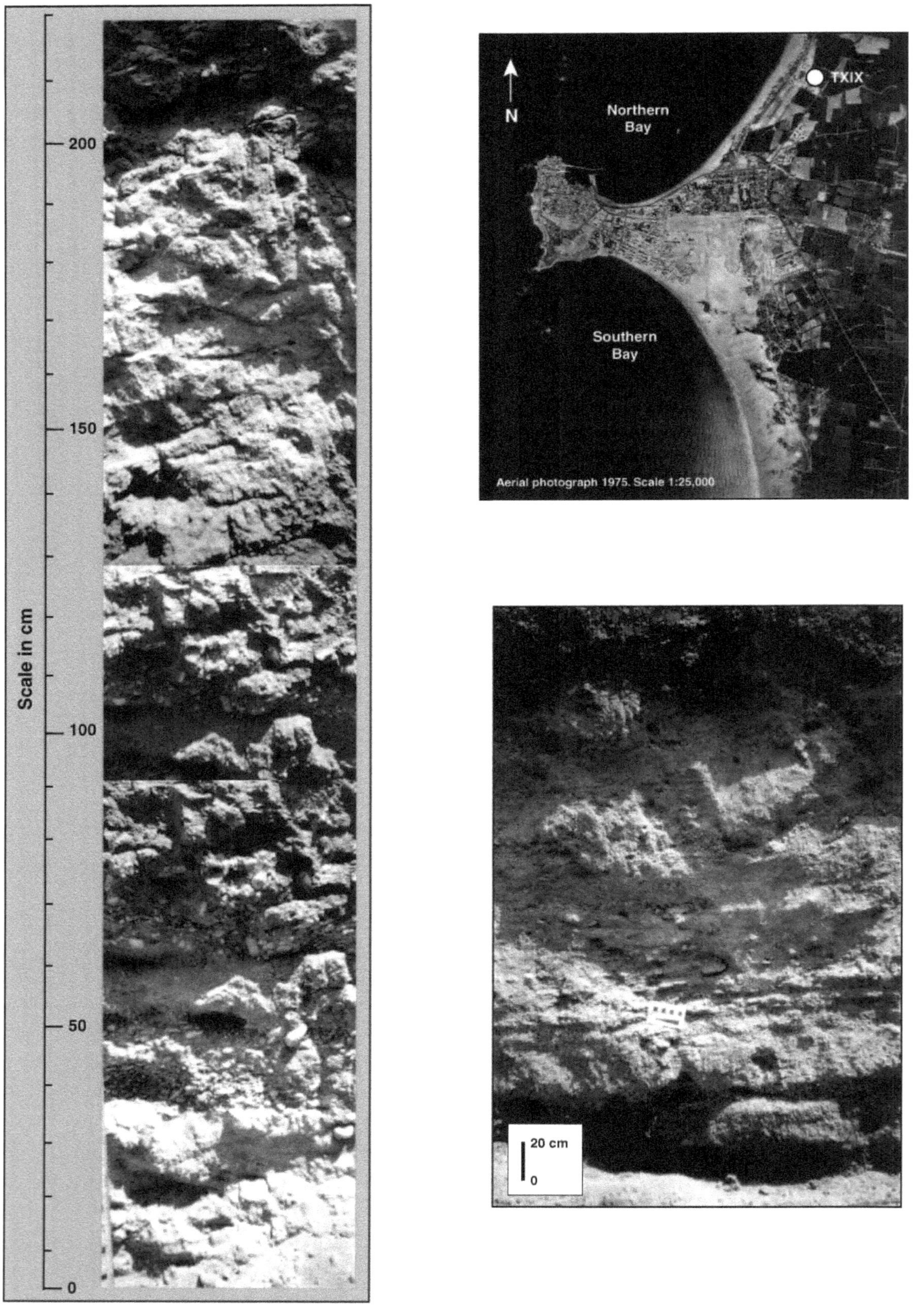

Figure 2.43: Core TXIX. Prograding ridge stratigraphy to the north of Tyre. Beach sands are intercalated with rounded gravel clasts.

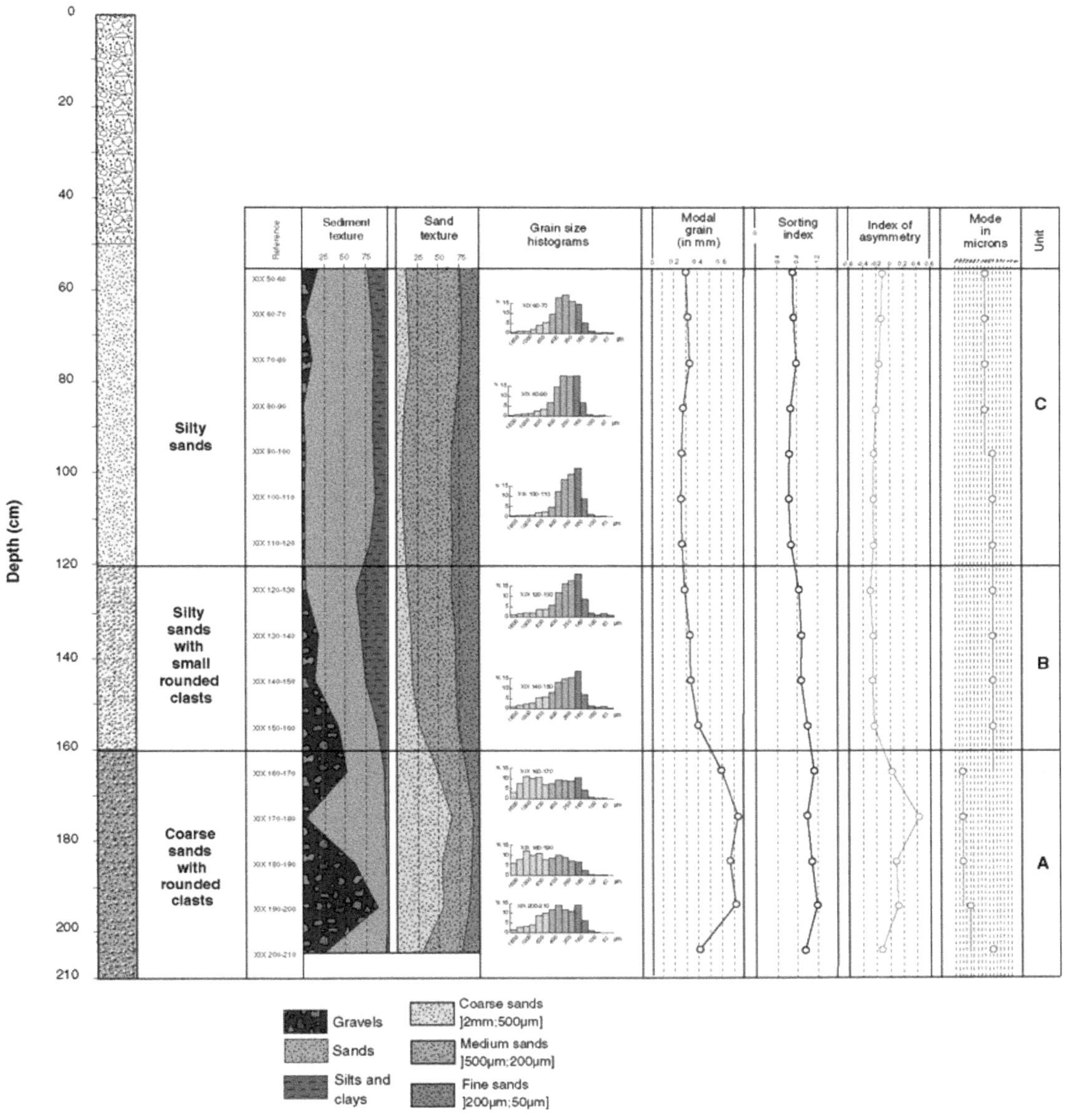

Figure 2.44: Core TXIX lithostratigraphy and grain size analyses.

2.6.1.2 Pre-Hellenistic proto-tombolo phase

Coarse sand and gravel transgressive deposits breached the leeward bay after ~6000 BP. Between 7500 BP and 6000 BP, the ridge dimensions were reduced from 6 km to 4 km. A sharp decline in *Cyprideis torosa* is countered by peaks of coastal ostracod species. Outer marine taxa, including *Semicytherura* spp., *Callistocythere* spp. and *Neocytherideis* sp., attest to an important opening-up of the environment.

The slowing of sea-level rise permitted reworking of older sediments and introduction from land of Holocene shelf sediments. Rapid rates of sedimentation, >0.3 cm/yr, following 3000 BP appear coeval with the anthropogenically forced erosion of surrounding watersheds, yielding increased sediment supply to coastal depocentres (**Figure 2.45**). Climate records from the Levantine basin indicate transition to a cool wet climate during the Late Bronze Age/Early Iron Age and the Late Roman/Byzantine periods (Bar-Matthews *et al.*, 1997; Schilman *et al.*, 2002; Enzel *et al.*, 2003; McGarry *et al.*, 2004). The expansion of agriculture and human modification of surrounding watersheds, coupled with periods of increased mountain precipitation (notably between 3500-3000 BP and 1700-1000 BP, Schilman *et al.*, 2001), would have entrained increased amounts of sediment to base-level depocentres.

Wave diffraction and a fall in water competence engendered medium to fine-grained sediment deposition on the lee of the island barriers, culminating in a natural wave-dominated proto-tombolo within 1-2 m of MSL by the time of Alexander the Great (4th century BC). High relative abundances of marine lagoonal and coastal taxa corroborate a middle energy shoreface protected by the island; during Hellenistic times Tyre island had approximate dimensions of ~2500 m long by ~750 m wide. Conditions conducive to the accretion of a proto-tomobolo are supported by our modelling of wave diffraction around Tyre island. Rapid spit growth may also have been amplified by earlier attempts to build a causeway on this underwater proto-tombolo, notably during the Babylonian siege of the city by Nabuchodonozor II (6th century BC, Katzenstein, 1997).

Analogous chrono-stratigraphy has been elucidated at Alexandria (Goiran, 2001; Goiran *et al.*, 2005) and concurs the importance of this pre-Hellenistic proto-tombolo phase, a unique sediment corpus resulting from limited accommodation space behind the island barriers. Such subtidal sedimentary systems were critical to the construction of Alexander's causeways.

2.6.1.3 Anthropogenic forcing - causeway impacts

The Hellenistic causeway entrained a complete, anthropogenically forced, metamorphosis of Tyre's coastal system accentuating many of the proto-tombolo genetic processes. After ~330 BC, the city's bays were definitively segmented into two coves; cessation of the longshore currents generated two isolated littoral cells. Stratigraphic signatures record a shift in shoreline positions characterised by rapid coastal progradation on both flanks of the tombolo. In core TXIV, for example, a sixfold incresase in sedimentation rates is observed after the construction of the causeway from 0.06 cm/yr before the Hellenistic period to 0.36 cm/yr after this time. The high level of statistical variance observed in the sedimentological data after this period translates the profound human impacts and segmentation of the tombolo into two isolated bays. Subaerial growth of the isthmus culminated in a cessation of north-south/south-north sediment transport across the tombolo. In this way, post-Hellenistic deposits in the southern bay (TXV) are clearly differentiated from northern lobe deposits (TXIII and TXIV), consistent with two independently evolving littoral cells. Rapid sub-aerial growth during the Seleucid and Roman periods can be attributed to a high sediment budget. Although we have relatively good constraints on the nature of climatic variability at these times (Schilman *et al.*, 2001; Schilman *et al.*, 2002; Enzel *et al.*, 2003; McGarry *et al.*, 2004), there is no information on the volumes of sediment being transported to base-level depocentres due to erosion in the coastal hinterland. We observe, however, that post-Hellenistic sedimentation rates increase by a factor of eight. During Roman times, this progradation accommodated urban growth (necropolis, hippodrome etc.).

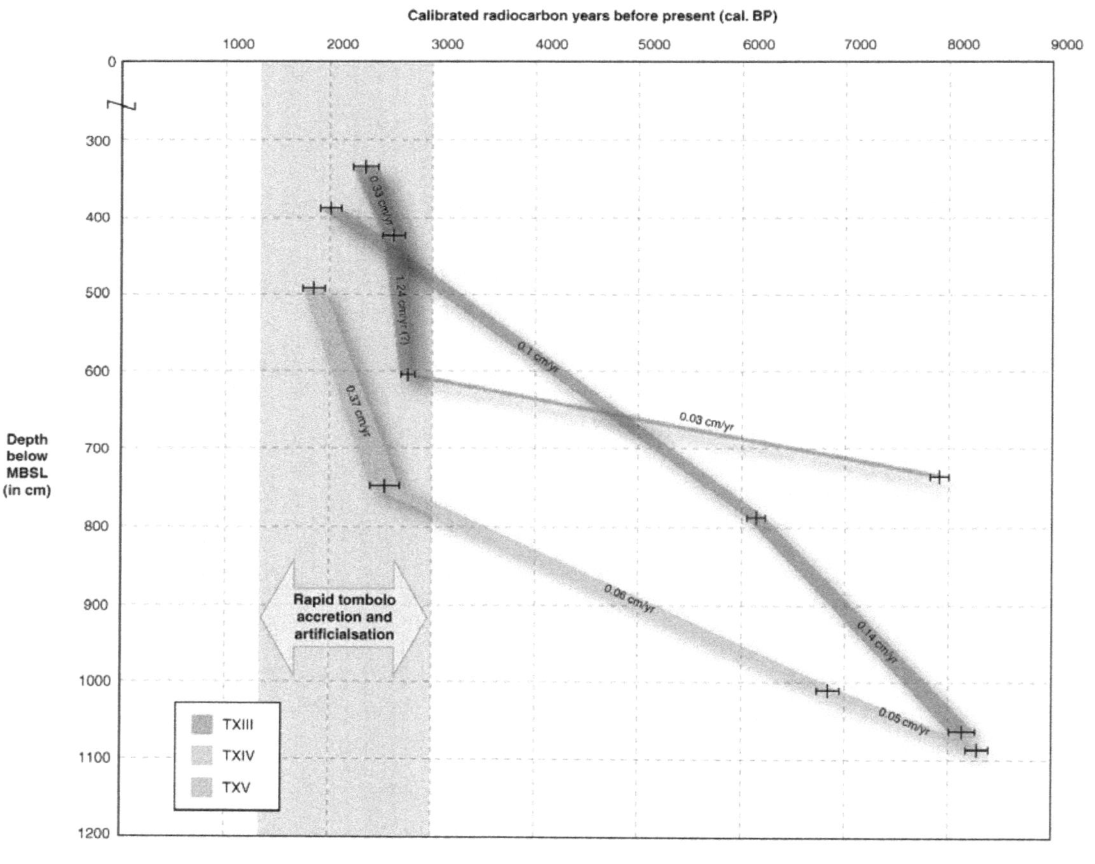

Figure 2.45: Tombolo accretion rates. Rapid rates of accretion, >0.30 cm/yr, are attested to after 3000 cal. BP. This is attributed to anthropogenically forced erosion of surrounding watersheds, yielding increased sediment supply to coastal depocentres. This entrapment was further accentuated by an artificialisation of the salient during the Hellenistic period.

Following the 3 m post-Roman tectonic collapse of the Tyre block, we record a rapid return to process-form equilibriums with infilling of the newly created accomodation space. We have very little data on the stratigraphic impacts of this event as much of the superior portion of the medieval and modern sedimentary records have been lost during recent urbanisation of the salient.

2.6.2 Tyre vs. Alexandria

One of the most celebrated tombolo examples, both in terms of its geological and archaeological scopes, is that of Alexandria ad Aegyptum studied by Goiran (2001) and Goiran *et al.* (2005). Pharos is today a heavily artificialised peninsula, 3.25 long by 1.1 km wide, linked to the continent by a 1300 m by 700 m isthmus (**Figure 2.46**). The salient separates two marine bays: a 2 x 1.5 km eastern bay, which sheltered a series of seaport complexes during antiquity, and a large western bay (10 km long by 2.5 km wide), partially protected by the drowned extension of Pharos aeolianite ridge. The presence of this offshore ridge has significantly influenced swell and wind-induced currents on its leeward side. Both bays today operate as isolated littoral cells while offshore the dominant marine currents run east to west.

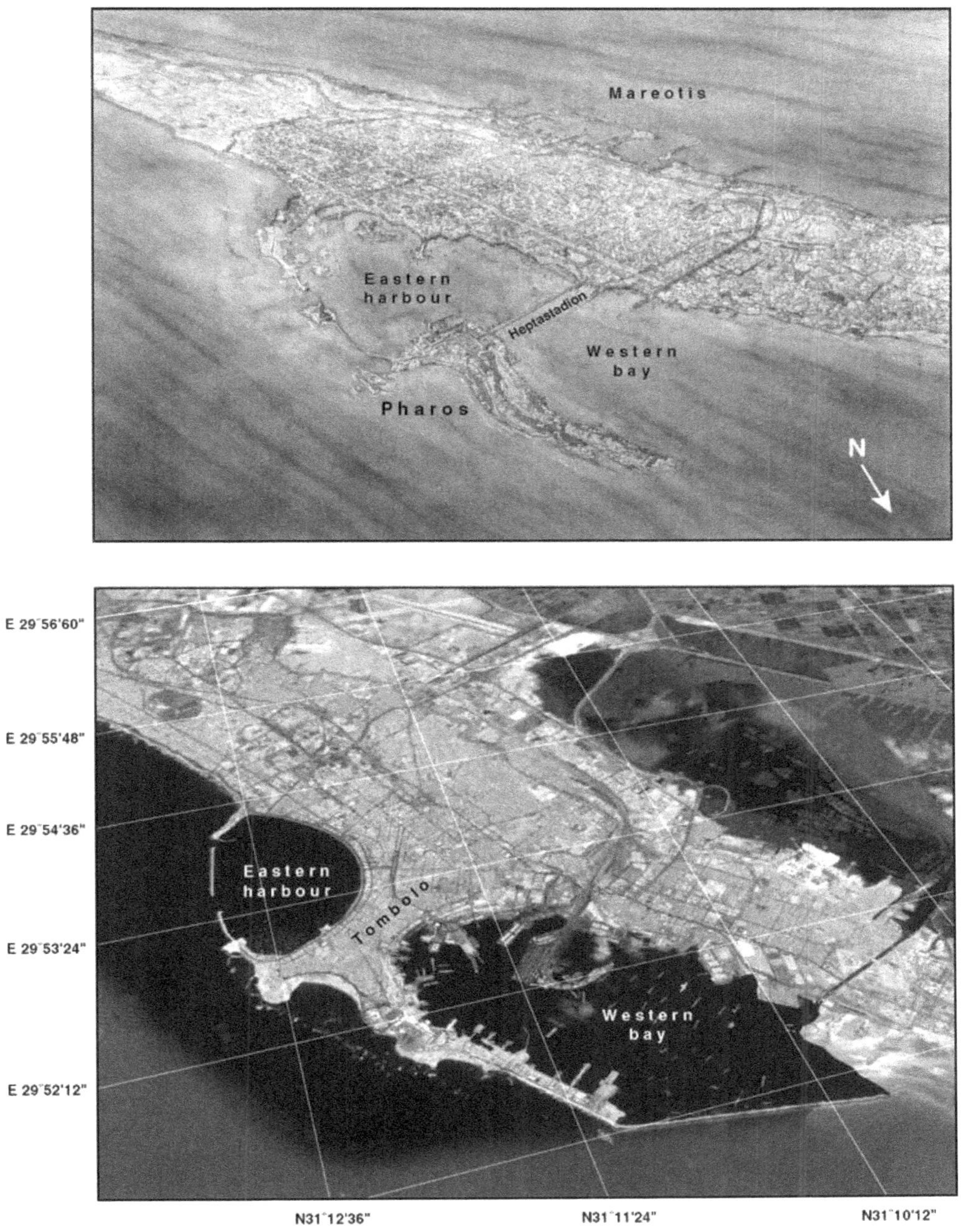

Figure 2.46: Above: Water colour reconstruction of Ptolemaic Alexandria by J.-C. Golvin (2003). Note that the reconstruction is based upon the present dimensions of Pharos island and does not account for a 5 m relative sea-level change since antiquity. Below: Alexandria and its tombolo today (image: DigitalGlobe, 2006).

There are a number of geomorphological and chronological parallels between Tyre and Alexandria rendering a direct comparison between the two sites particularly interesting: (1) the **geomorphology and topography** of both sites have been significantly influenced by the inherited Pleistocene landscapes, namely a series of partially-drowned aeolianite ridges that run parallel to the present coastline (Butzer, 1962). Great tracts of Holocene coastal sediments have accumulated between these onshore and offshore aeolianite ridges; (2) this drowned topography has given rise to **two breakwater islands** (**Figure 2.47**). The dimensions of both breakwaters have changed significantly since their marine flooding between 8000 and 7500 BP. Changes in the size of the shielding breakwaters are discernible in the Holocene sediment record. As at Tyre, ancient Alexandria comprised a long shore-parallel breakwater (Pharos ridge) whose reconstructed Hellenistic dimensions, ~5500 m long by ~1300 m wide, are approximately twice those at Tyre for the same period.; (3) both sites lie in **proximity to important fluvial systems**, the Nile and the Litani (Abd-el-Al, 1948; Said, 1993). Although the scales of these fluvial systems are not directly comparable, the two rivers were significant in dictating sediment budgets and the nature of the highstand deposits during the accretion of the tombolo (Chen *et al.*, 1992; Warne and Stanley, 1993a-b; Stanley and Warne, 1998); (4) Alexandria and Tyre were modified by **human intervention** at the same time (332 and 331 BC). Assessing the synchronicity or disynchronicity of stratigraphic impacts is interesting from a geomorphological perspective; (5) the two sites underwent significant **subsidence during the late Roman period** (~5 m at Alexandria and ~3 m at Tyre). This subsidence had two impacts: (a) from a geomorphological standpoint, it greatly diminished the dimensions of the breakwater obstacles; and (b) from a stratigraphic perspective, it created significant accommodation space for the deposition of new sediment tracts.

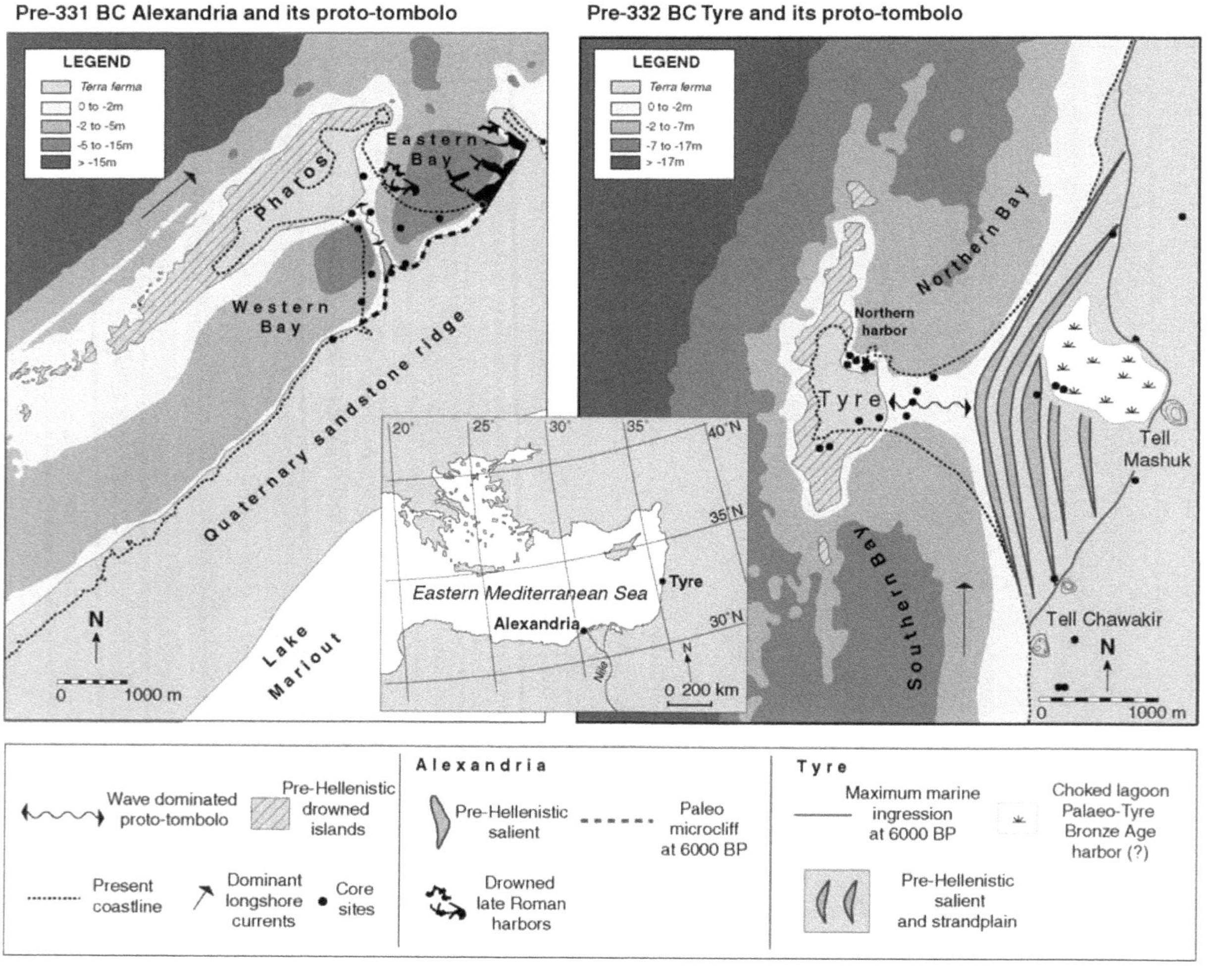

Figure 2.47: Morphodynamic evolution of Alexandria and Tyre's isthmuses since antiquity.

On the basis of a 10 m long transgressive-regressive sediment sequence, Goiran (2001) has elucidated a series of phases in the accretion of Alexandria's tombolo. The proximity of core CII to the Heptastadion renders it particularly sensitive in recording geomorphological changes and anthropogenic impacts. We note a number of stratigraphic similarities between Alexandria and Tyre.

(1) **Marine flooding:** Like Tyre, the inception of marine sedimentation is dated ~7800 BP, consistent with the flooding of deltas throughout the circum Mediterranean (Stanley and Warne, 1994). This flooding surface is onlapped by a shelly sand unit coeval with reworking of transgressed deposits.

(2) **Accretion of the proto-tombolo:** Between 5400 BP and 4200 BP, Goiran (2001) elucidates a *Cladocora* biodeposition, comprising broken coral branches enveloped in a silty clay matrix. The analogous stratigraphy at Tyre constitutes medium-grained sands. The origin of this biodeposition is explained by erosion of reefs in proximity to Alexandria. Juxtaposition of coarse and fine-grained sediment stocks evokes seasonality in currents and sediment deposition patterns; the coarse biodeposition is inferred to have been reworked and deposited during winter storms, subsequently enveloped in a silt-clay matrix during Nile summer flooding. High levels of the marine lagoonal ostracod species *Xestoloberis* spp. and *Loxoconcha* spp. evoke a medium to low energy environment protected by the partially drowned ridge system. Using topographic and bathymetric data from Goiran (2001) we have reconstructed the palaeogeography of Alexandria at 5500 BP, elucidating the importance of the subaerial ridge in protecting the site's western bay. The ridge formed a barrier over 14 km long stretching from the eastern tip of Pharos westwards towards Abu Sir (**Figure 2.48**). These litho- and biostratigraphical findings support data from Tyre, characterised by marine lagoonal fauna and a medium to low-energy sedimentology. By contrast, the lateral extension of Tyre's semi-drowned ridge (4 km) was greatly inferior to Alexandria (14 km) between 6000 BP and 4000 BP.

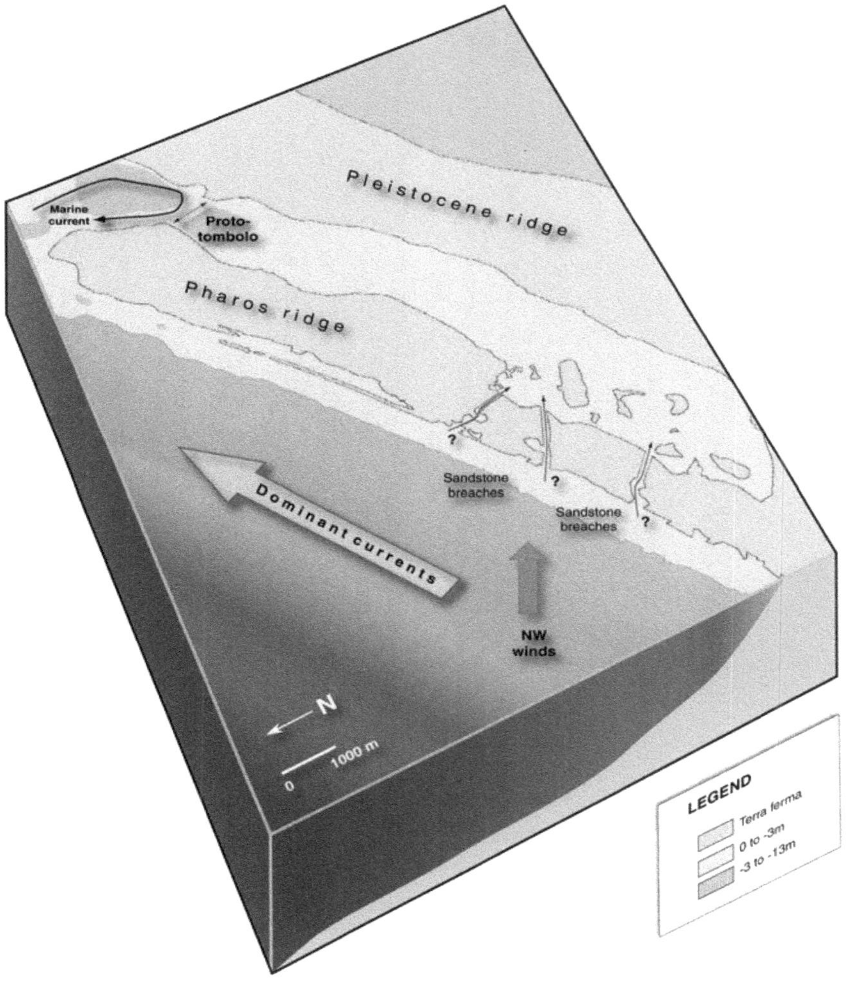

Figure 2.48: Palaeogeographical reconstruction of Alexandria's tombolo around 5500 BP (this study). The reconstruction factors in a relative sea-level change of 7 m since this time. Note the existence of an extensive 14 km breakwater ridge that shielded Alexandria's western bay. We hypothesise that a number of breaches existed in the sandstone ridge creating a low energy lagoon environment on the leeward side at this time.

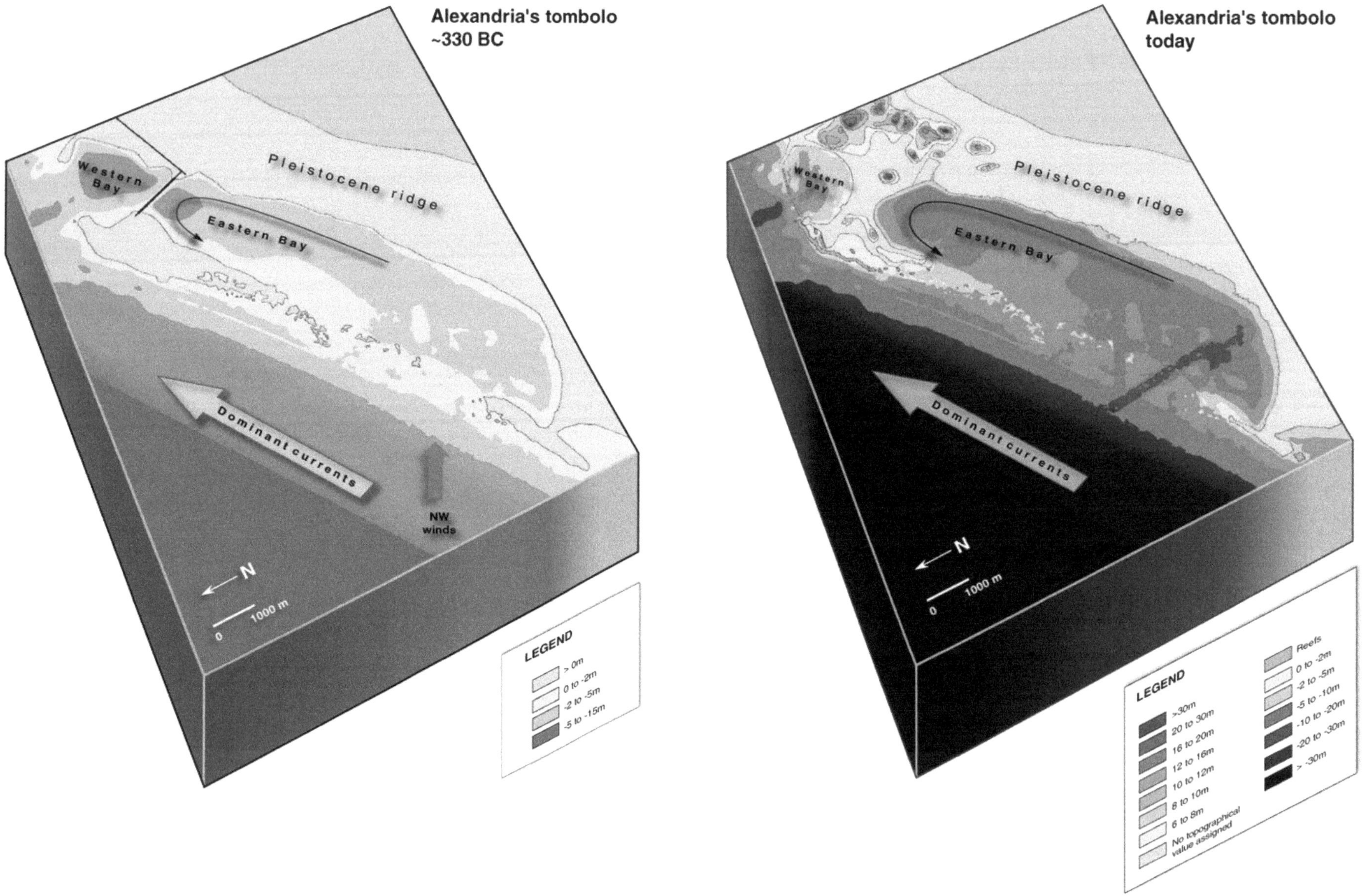

Figure 2.49: Palaeogeographical reconstruction of Alexandria's tombolo at 330BC, after data from Goiran (2001). The reconstruction factors in a 5 m relative sea-level change since the Ptolemaic period that has significantly reduced the dimensions of Pharos island.

(3) **Sediment hiatus:** A sediment hiatus is recorded between 4000 BP and 2000 BP, a time period that straddles the construction of Alexander's causeway. Goiran proposes two hypotheses to explain this absence of sediment deposition: (1) a sharp fall in sediment inputs; and/or (2) accentuation of marine currents leading to erosion and truncation of the facies. A fall in sediment supply seems improbable given the rapid and synchronous progradation of the Nile delta at this time (Warne and Stanley, 1993a-b; Stanley and Warne, 1998). Goiran has postulated that parts of the proto-salient became sub-aerial at this time, isolating areas of the spit from sediment sources. Like Tyre, this hypothesis is compliant with a proto-tombolo within 1 m of sea level by the time of Alexander the Great (**Figure 2.49**).

Goiran (2001) proposes that construction of Alexander's causeway accentuated erosion of parts of the proto-tombolo by generating more vigorous currents through the passes. Though plausible, this hypothesis is contradictory with our findings at Tyre, which manifest rapid accretion of the salient following its artificialisation. In light of this, we evoke a second alternative hypothesis, namely that significant portions of the aeolianite ridge to the west of the city were drowned and breached after 4000 BP. We hypothesise that the subsequent creation of more powerful westerly currents could be at the origin of erosion or non-deposition of parts of the proto-tombolo. After 2085 ± 45 BP (200 to 410 cal. AD), a *Serpulidae* biodeposition concurs dynamic marine currents following the construction of the Heptastadion.

(4) **Subaerial progradation phase 1**: Progradation of the tombolo is indicated by the accretion of a sand unit after 1935 ± 55 BP (360 to 610 cal. AD). The litho- and biostratigraphy concurs lateral accretion of the tombolo's eastern flank, with its passage from a marine to sub-aerial form characterised by wide sandy beaches.

(5) **Tectonic collapse**: At the end of the late Roman period, tectonic collapse accounts for the rapid submersion of great tracts of the salient. This 5 m collapse generated considerable accommodation space and led to the deposition of a low-energy plastic clays unit. Gypsum crystals are consistent with a low-energy hypersaline environment. We hypothesise that the environment was sheltered by subaerial portions of the tombolo to the west. At this time, the large eastern bay was a low-energy environment that shielded a number of anchorage complexes.

(6) **Subaerial progradation phase 2**: The coastal response of this catastrophic submersion was rapid as the salient sought to establish a new equilibrium profile in balance with the high local sediment budget. After 1635 ± 35 BP (eighth to ninth centuries AD), transition to a rounded pebble then coarse sand facies corroborates the progradation of the tombolo flank following the creation of new accommodation space. Rapid coastal deformation during the Islamic and medieval periods is concurrent with data from Tyre, and is typical of micro-tidal environments with a high sediment budget.

Largely concomitant chrono-stratigraphies at Alexandria and Tyre attest to the importance of the offshore island breakwater in creating pre-Hellenistic proto-tombolos. Goiran (2001) suggests that parts of the Alexandrian proto-tombolo were subaerial by 331 BC. At Tyre, our data indicate that the proto-tombolo lay within 1-2 m of Hellenistic sea level. These geomorphological forms, in close proximity to the water surface, considerably facilitated the construction Alexander's causeways. At both sites, the causeway impacts are quasi-identical, with irreversible segmentation of the original bays coupled with a sharp fall in water competence.

2.7 Concluding remarks

We expound a three phase stratigraphic model of tombolo morphogenesis, comprising two natural forcing factors (geomorphic and RSL change) and one anthropogenic. (1) Newly transgressed and protected shallow-marine environments are recorded at Tyre between ~8000 and 6000 BP. Natural protection was afforded by Tyre's Quaternary ridge. At this time relative sea-level changes were the main controls on salient dynamics. This transgression caused an onshore movement of gravel and sand. (2) After 6000 BP, decelerating sea-level rise, high sediment supply (especially after 3000 BP) and diffractive wave processes set up by the island obstacle engendered rapid spit growth. By Hellenistic times, we show the existence of a proto-tombolo within 1-2 m of MSL. The salient deposits can be qualified as accretionary regressive deposits. (3) After 332 BC, construction of Alexander's causeway entrained a complete anthropogenic metamorphosis of the Tyrian coastal system. Rapid coastal progradation definitively segmented the leeward bay into two discrete coves. The rate of salient growth was initially high, and it slowed down as equilibrium was approached. Although there are a number of important short-term controls on tombolo accretion, the chronological and spatial resolutions of our chronostratigraphic datasets do not allow us to account for these in our model.

Analogous morphogenetic phases for both Alexandria and Tyre's tombolos suggest that, on microtidal wave-dominated coasts, there is a great deal of predictability to tombolo geology and stratigraphy. Tyre's tombolo is an archetype of anthropogenic forcing on Mediterranean coastal systems and stratigraphy. On a Holocene timescale, the synchronous stratigraphic evolution emphasises the significance of geological heritage and the irreversibility of pluri-millennial anthropogenic impacts in shaping coastal landforms.

Chapter 3

Geoarchaeology of Tyre's ancient harbours

The exact location and evolution of Tyre's ancient harbour areas have been matters of scholarly conjecture since the sixteenth century. Ancient texts and iconographic evidence suggest that four harbour complexes existed in the Tyre area: (1) a northern seaport looking towards Sidon, Beirut and Byblos (sometimes referred to as the Sidonian harbour); (2) a southern anchorage facing Egypt; (3) a number of outer harbours, taking advantage of the exposed sandstone reefs at this time; and (4) a fourth continental complex, located around Tell Mashuk and Tell Chawakir, which served as a transport hub for the inhabitants of Palaeo-Tyre. In this chapter we combine coastal geomorphology and the multidisciplinary study of sediment archives to precisely reconstruct where, when and how Tyre's ancient harbour complexes evolved during the Holocene. We also investigate the chronostratigraphic impacts of Roman harbour dredging and its implications on how ancient harbour archives should be read.

3.1 In search of Tyre's northern harbour: geoarchaeological context

Tyre's northern harbour has been a source of archaeological speculation since the sixteenth century when a number of religious pilgrims visited the Phoenician coast *en route* to the Holy Land shrines (Villamont, 1596; Van Cotvyck, 1620; Stochove, 1650; Besson, 1660; Maundrell, 1703; Arvieux, 1735; Pococke, 1745). Although the use of this northern cove as the ancient city's main seaport has never been questioned, for many early erudites its shallow basin appeared enigmatic with Tyre's former maritime glory (Shaw, 1743). Indeed Volney (1791, for the English translation see 1792) noted that the "harbour, dug by the hand of man" was silted up to such an extent that children could walk across it without wetting their upper body.

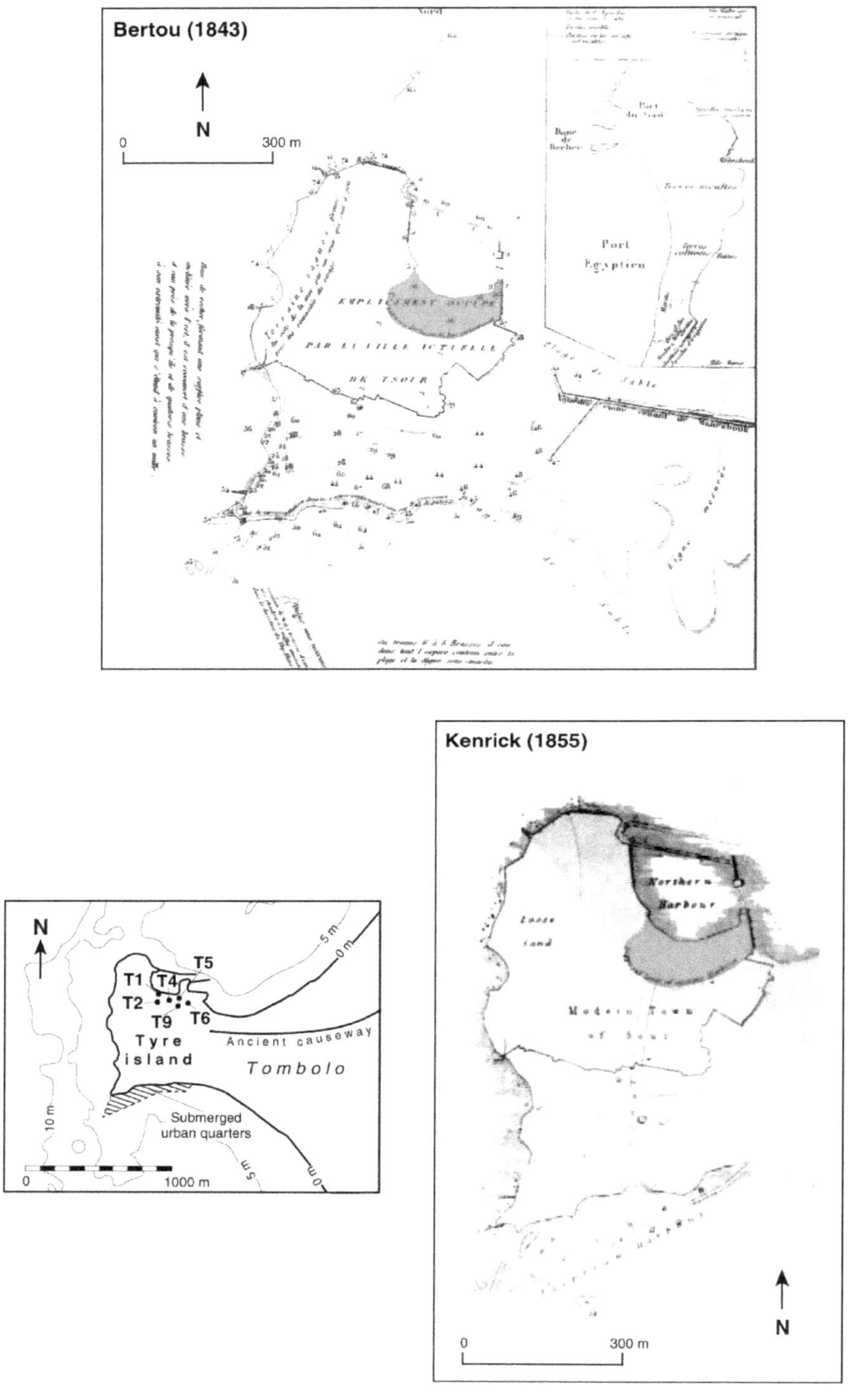

Figures 3.1: Proposed reconstructions of Tyre's silted northern harbour by Bertou (1843) and Kenrick (1855).

De Bossay (1861, 1863)

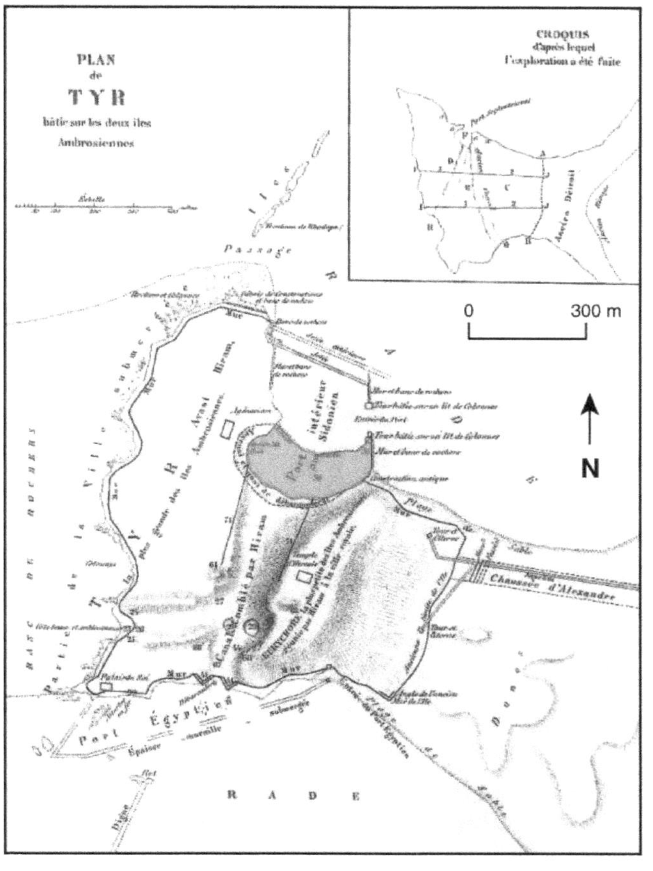

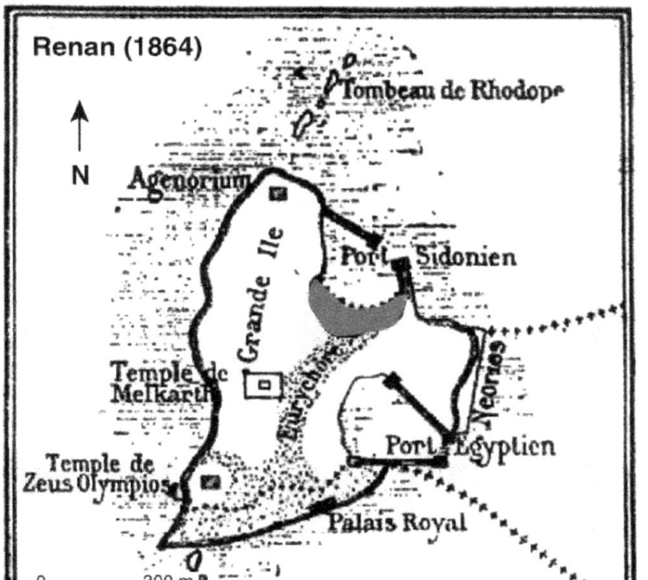

Figures 3.2: Proposed reconstructions of Tyre's silted northern harbour by De Bossay (1861, 1863) and Renan (1864).

While sixteenth to eighteenth century travellers contented themselves with landscape descriptions and nostalgic reveries of the city's former grandeur, the scientific ascendancy of the nineteenth century ushered in more robust ideologies on history and archaeological inquiry (Marriner and Morhange, 2007). It was at this time that a number of scholars intuitively drew the link between harbour silting and coastal progradation to hypothesise that great tracts of the ancient northern port lay beneath the medieval and modern centres (Bertou, 1843; Kenrick, 1855; Poulain de Bossay, 1861, 1863; Renan, 1864; Rawlinson, 1889; **see Figures 3.1 and 3.2**). This basin, it was argued, had been protected by an ancient breakwater which protruded from the eastern tip of the harbour and was known to many of the early workers (Kenrick, 1855; Renan, 1864; **Figure 3.3**). Logistic difficulties meant that it was not until the 1930s, under the auspices of Father Antoine Poidebard, that precise archaeological surveying of this mole was undertaken (Poidebard, 1939). Despite very recent diving prospections (Noureddine and Helou, 2005; Descamps, personal communication) the exact age of the structure remains ambiguous, and in reality very little is known about this basin, the city's most important transport hub (**Figure 3.4**).

Given the dearth of data in this northern area, we drilled a series of cores around the edges of the present basin (**Figure 3.3**) with an aim to accurately reconstruct (1) the harbour's ancient dimensions and topography; (2) its stages of evolution since the Holocene marine transgression; and (3) attempt to link discrete stratigraphic signatures with changes in harbour infrastructure.

3.2 Methods

The Holocene paleoenvironments have been reconstructed using a series of four stratigraphic cores. All cores have been GPS levelled and altitudinally benchmarked relative to present MSL. Interpretations are based upon high-resolution litho- and biostratigraphical studies of the sedimentary cores, as described in chapter 2. Radiocarbon and archaeological dates provide a chronological framework for the landforms and sedimentary bodies observed (**Table 3.1**).

3.3 Results and discussion

3.3.1 Where was the ancient northern harbour?

Our geoarchaeological data indicate that the heart of Tyre's ancient seaport is presently buried beneath the medieval and modern city centres (**Figure 3.5**). These centres lie upon ~8-10 m of medium to low energy coastal sediments which began accreting between 8000 to 6000 BP. The findings confirm earlier interpretations made by Kenrick (1855), Poulain de Bossay (1861, 1863) and Renan (1864) who, during the nineteenth century, were the first to suggest that silting up had significantly diminished the size of the former basin. By coupling a study of the bedrock and urban topography with the landward position of Holocene coastal strata, we have reconstructed an ancient harbour approximately twice as large as present (Marriner *et al.*, 2005; Marriner *et al.*, 2006a). Coastal progradation of 100-150 m has occurred since the Byzantine period, encroaching on the sea to landlock the heart of the ancient seaport. We discuss these stratigraphic data in more detail in part 3.3.2.

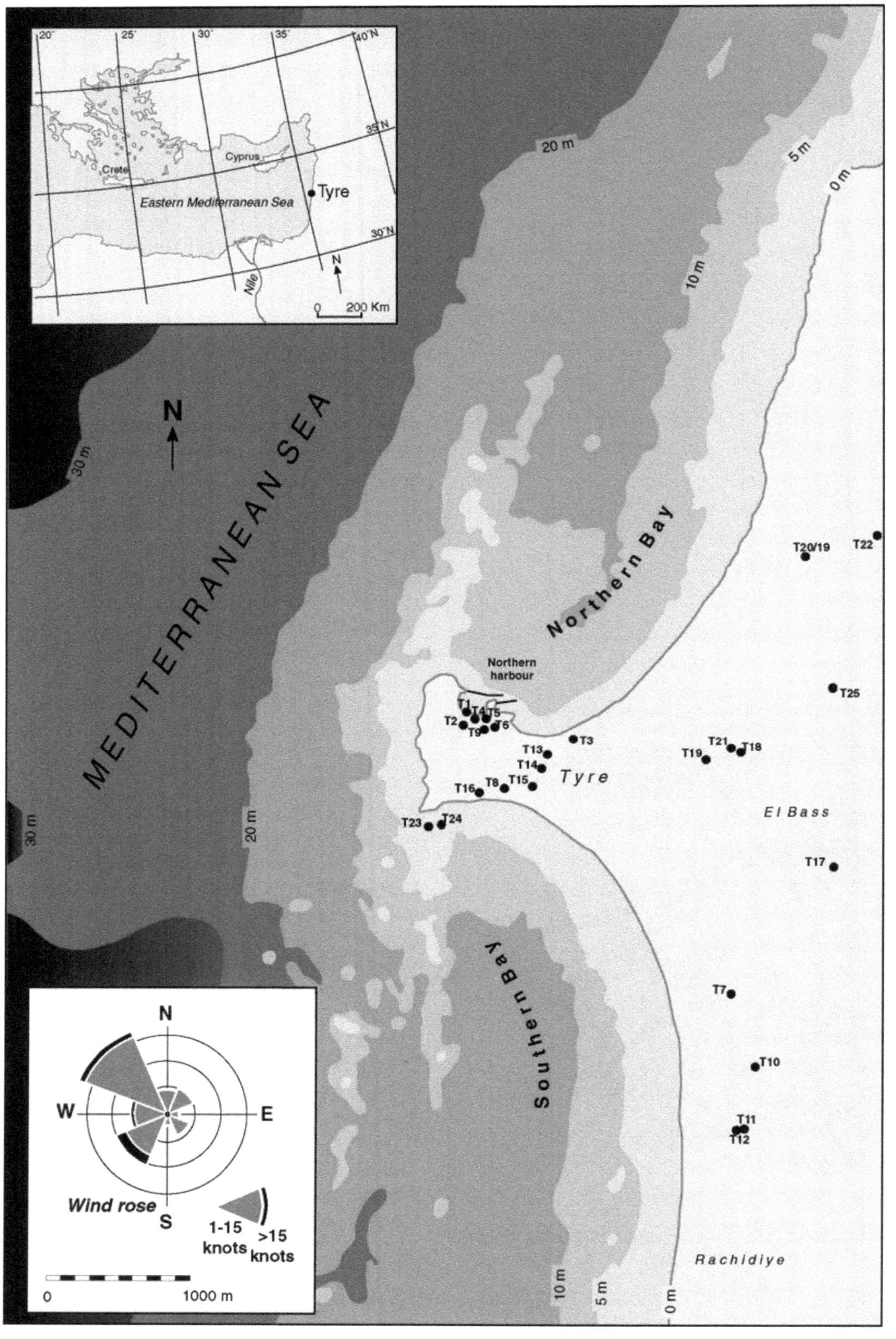

Figure 3.3: Tyre's coastal bathymetry at present and location of core sites (denoted by black dots).

Figure 3.4: Oblique aerial view of Tyre's northern harbour around 1934, taken from the east (photograph: Poidebard archives at the Université Saint Joseph). The two medieval jetties are clearly visible in the picture. The ancient Iron Age/Roman breakwater is not manifest, but trends west-east from the eastern tip of the island.

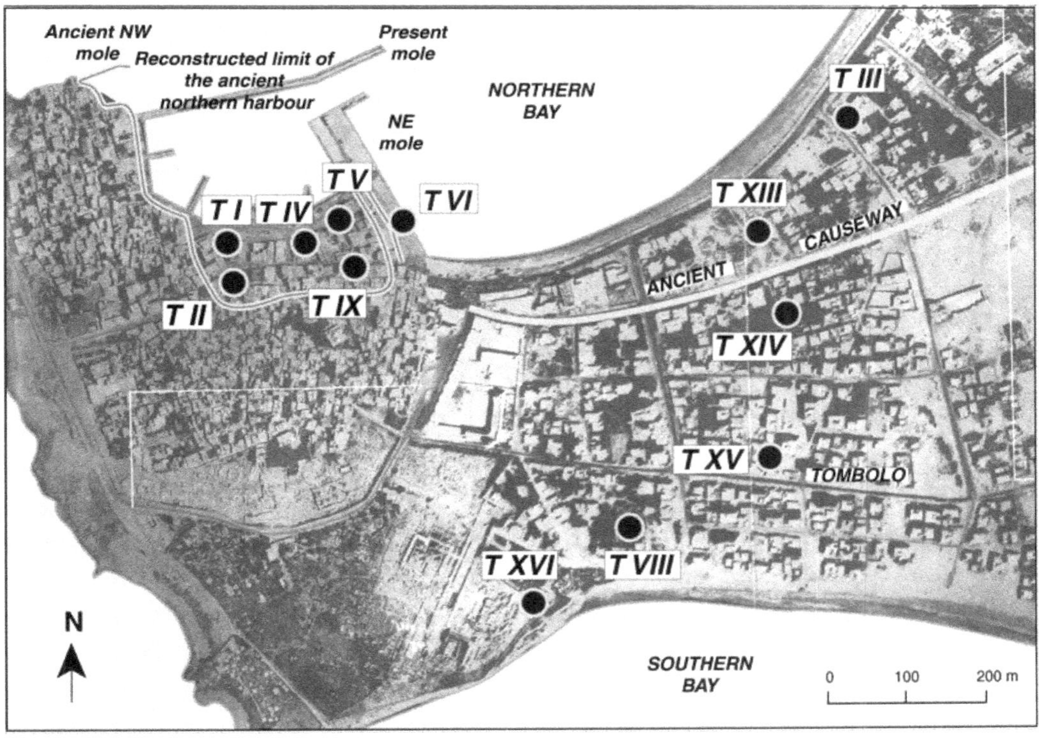

Figure 3.5: Reconstructed limits of Tyre's ancient northern harbour. Silting up since antiquity has led to a 100-150 m progradation of the cove's coastline, isolating the heart of the ancient seaport beneath the medieval and modern city centres.

Sample	Code	Sample	$^{13}C/^{12}C$ (‰)	^{14}C BP	±	Cal. BP	Cal. BC/AD
TI 4	Lyon-1469(GRA-17972)	Marine shells	1.14	1560	35	1210-1030	740 AD - 920 AD
TI 12	Lyon-1470(GRA18730)	Marine shells	-0.31	1965	35	1600-1410	350 AD - 540 AD
TI 30	Lyon-1472(GRA-18732)	Marine shells	0.08	2255	35	1940-1770	10 AD - 180 AD
TI 24	Lyon-1471(GRA-19731)	Marine shells	-1.4	2055	35	1700-1530	250 AD - 420 AD
TI 31	Lyon-1602(GRA-19345)	Marine shells	0.89	2635	45	2430-2170	480 BC - 220 BC
TI 36	Lyon-1603(GRA-19346)	Marine shells	0.94	5520	50	6000-5750	4050 BC - 3800 BC
TI 39	Lyon-1473(GRA-18733)	Marine shells	-1.67	2375	35	2100-1910	150 BC - 40 AD
TII 13	Lyon-1604(GRA-19347)	Marine shells	-1.35	1990	45	1680-1430	270 AD - 520 AD
TII 18	Lyon-1474(GRA-18735)	Marine shells	-0.72	2370	35	2090-1900	140 BC - 50 AD
TV 19	Poz-2500	Charcoal	-22.6	1485	30	1420-1300	530 AD - 650 AD
TV 41	Poz-2502	Charcoal	-27.4	1910	30	1930-1770	20 AD - 180 AD
TV 43	Poz-5768	Charcoal	-18.9	2265	30	2350-2150	400 BC - 200 BC
TV 58	Poz-5752	*Nassarius mutabilis*	4.4	2360	30	2060-1890	110 AD - 60 AD
TV 60	Poz-5769	Charcoal	-23.3	2245	35	2350-2150	400 BC - 200 BC
TV 70	Poz-2445	*Cerithium vulgatum* juv.	2.8	5730	30	6240-6030	4290 BC - 4080 BC
TV 76	Poz-2451	*Cerithium vulgatum* juv.	-1.2	6400	35	6970-6750	5020 BC - 4800 BC
TV 77	Poz-2446	*Cerithium vulgatum* juv.	-1.5	7300	40	7840-7660	5890 BC - 5710 BC
TV 81	Poz-2447	*Cerithium vulgatum* juv.	-4.1	7760	40	8330-8130	6380 BC - 6180 BC
TV 82	Poz-2448	*Cerithium vulgatum* juv.	-2.4	7780	40	8340-8150	6390 BC - 6200 BC
TV 106	Poz-2449	*Cerithium vulgatum* juv.	-5	7800	40	8350-8160	6400 BC - 6210 BC
TIX 6	Poz-5770	Olive pip	-27.2	1615	30	1570-1410	380 AD - 540 AD
TIX 20	Poz-5771	3 grape seeds	-26.9	1855	30	1870-1710	80 AD - 240 AD
TIX 25	Poz-5773	1 *Nassarius mutabilis*	-0.8	2220	35	1900-1720	50 AD - 230 AD
TIX 26	Poz-5774	1 seed	-9.8	2385	30	2710-2340	760 BC - 390 BC
TIX 35ch	Poz-5777	1 charcoal	-21.3	2215	30	2330-2140	380 BC - 190 BC
TIX 35co	Poz-5775	3 *Loripes lacteus*	4.6	2505	30	2290-2080	
TIX 43	Poz-5778	1 seed	-24.8	2055	30	2120-1920	170 BC - 30 AD
TIX 45	Poz-5779	1 charcoal	-23.1	2320	35	2440-2150	490 BC - 200 BC
TIX 49	Poz-5780	1 *Cyclope neritea*	3.5	2140	30	1810-1630	140 AD - 320 AD
TIX 52	Poz-5781	3 *Donax* sp.	2.9	2210	30	1880-1720	70 AD - 230 AD
TIX 53	Poz-7184	2 *Donax venustus*	-1.8	2300	30	1980-1830	30 BC - 120 AD
TIX 62	Poz-5783	2 articulated *L. lacteus*	4.7	5850	40	6350-6170	4400 BC - 4220 BC
TIX 63	Poz-5784	3 *L. lacteus*	4.3	5830	40	6310-6160	4360 BC - 4210 BC
TIX 76	Poz-5785	2 *P. exiguum*	-1.6	7840	50	8390-8180	6440 BC - 6230 BC
TIX 56	Poz-9897	1 *Donax semistriatus*	1.9	2380	30	2110-1910	160 BC - 40 AD
TIX 59	Poz-12140	1 *Donax semistriatus*	4.4	2385	30	2120-1910	170 BC - 40 AD
TVI 8	Poz-9893	1 *Donax semistriatus* & 1 *L. lacteus*	5.1	225	30	60 BP - -1 BP (present)	1890 AD - 1950 AD (present)
TVI 18	Poz-9894	1 *Neverita josephinae*	-0.2	2120	30	1810-1600	140 AD - 350 AD
TVI 38	Poz-9896	2 articulated *L. lacteus*	5.9	6065	30	6590-6390	4640 BC - 4440 BC

Table 3.1: Tyre radiocarbon dates.

Sample	Code	Sample	$^{13}C/^{12}C$ (‰)	^{14}C BP	±	Cal. BP	Cal. BC/AD
TXVIII 39	Poz-13966	1 *Parvicardium exiguum*	-5.6	4180	30	4380-4150	2430 BC - 2200 BC
XVIII 59	Poz-9946	1 *Neverita josephinae*	-0.8	5710	40	6240-5990	4290 BC - 4040 BC
XVIII 64	Poz-9900	1 *L. lacteus*	3.7	5980	40	6490-6290	4540 BC - 4340 BC
XIII 9	Poz-9943	1 *Tricolia pullus*	0.9	2650	35	2470-2230	520 BC - 280 BC
XIII 15	Poz-9898	4 *Rissoa* sp.	2	2870	30	2730-2510	780 BC - 560 BC
XIII 21	Poz-9944	6 *Rissoa* sp.	-1.9	2980	35	2840-2690	890 BC - 740 BC
XIII 25	Poz-9945	2 *L. lacteus*	4.7	7450	40	8000-7820	6050 BC - 5870 BC
TXIII 39	Poz-13916	1 *Parvicardium exiguum*		7210	40	7780-7580	5830 BC - 5630 BC
TVIII 25	Poz-12135	*Columbella rustica*	0.4	2510	30	2300-2080	350 BC - 130 BC
TVIII 51	Poz-12139	1 *Gastrana fragilis*	4.5	6300	40	6880-6650	4930 BC - 4700 BC
TXV 19	Poz-12154	1 *Nassarius reticulatus*	-0.4	2385	30	2120-1910	170 BC - 40 AD
TXV 25	Poz-13965	2 *Mitra ebenus*	4	5775	35	6280-6100	4330 BC - 4150 BC
TXV 26	Poz-12133	3 *L. lacteus*	4.4	5740	35	6270-6020	4320 BC - 4070 BC
TXV 30	Poz-12134	1 *Cerithium vulgatum*, 2 *P. exiguum*	4.3	7680	40	8270-8020	6320 BC - 6070 BC
TXIV 6	Poz-13964	*Donax semistriatus*	-0.3	2245	30	1950-1750	0 AD - 200 AD
TXIV 14	Poz-12137	1 *Tricolia pullus*	3.3	2795	30	2680-2400	730 BC - 450 BC
TXIV 20	Poz-13915	3 *Bittium* spp., 1 *Rissoa* sp., 1 *Venus casina*	-0.4	6370	35	6950-6730	5000 BC - 4780 BC
TXIV 23	Poz-12136	4 *L. lacteus*	3.3	7840	40	8390-8190	6440 BC - 6240 BC

Table 3.1 continued: Tyre radiocarbon dates.

Core TVI, drilled just east of the harbour's south-north trending mole, is void of diagnostic harbour facies affirming that the hub was not more extensive in this direction (**Figures 3.6, 3.7 and 3.8**).

Description: Unit B is characterised by a medium to coarse grained shelly sands deposit, dominated by molluscan taxa indicative of a low energy beach environment, including *Bittium reticulatum* (subtidal sands and hard substrate assemblage), *Tricolia tenuis*, *Rissoa dolium*, *Rissoa lineolata* (subtidal sands assemblage), *Nassarius louisi*, *Nassarius pygmaeus*, *Nassarius mutabilis* and *Donax semistriatus* (upper clean-sand assemblage). In unit A, a net transition to fine-grained sands (~90 %) is attested to after ~1700 cal. BP. Modal grain values of between 0.2 and 0.4 µm and medium sorting indices concur a low energy beach environment. These sedimentological data are consistent with a fall in water competence linked to the construction or reinforcement of the south-eastern harbour mole. The molluscan fauna is dominated by sand-loving taxa including the subtidal sands assemblage (*Rissoa lineolata*, *Tricolia tenuis*, *Smaragdia viridis*, *Bulla striata*), the upper clean-sand assemblage (*Nassarius pygmaeus*, *Donax semistriatus*) and the upper muddy-sand assemblage in sheltered areas (*Cerithium vulgatum*).

The top of unit B is dated to modern times. Both units B and A comprise marine lagoonal (*Loxoconcha* spp. and *Xestoleberis aurantia*) and coastal (*Aurila convexa*) ostracod species.

Interpretation: *A priori* the eastern mole has been continuously adapted and reinforced from antiquity through to present, as is commonplace with ancient seaport infrastructure.

The submerged northern breakwater, which protrudes seawards from the extreme eastern tip of the island, closes the basin. This structure lies ~40 m north of the present mole. First mentioned in the literature by Bertou (1843), it has manifestly been known to local fishermen and sailors for much longer. The structure was not precisely surveyed until the work of Poidebard during the 1930s (Poidebard, 1939). Since then, two surveys of note have been undertaken by Noureddine and Helou (2005; **Figure 3.9**) and Descamps *et al.* (Descamps, personal communication). These studies have elucidated an 80 m long breakwater, with a constant width of 12.7 m. The mole comprises two parallel walls made-up of 40/50 x 50/60 x 190 cm ramleh headers, and whose surfaces presently lie ~2 m below MSL (Descamps, personal communication). Excavations by Descamps *et al.* have elucidated five assises constituting a 3.1 m high wall. On the basis of ceramics and engravings on the stone blocks, Descamps has attributed the mole to the Graeco-Roman period.

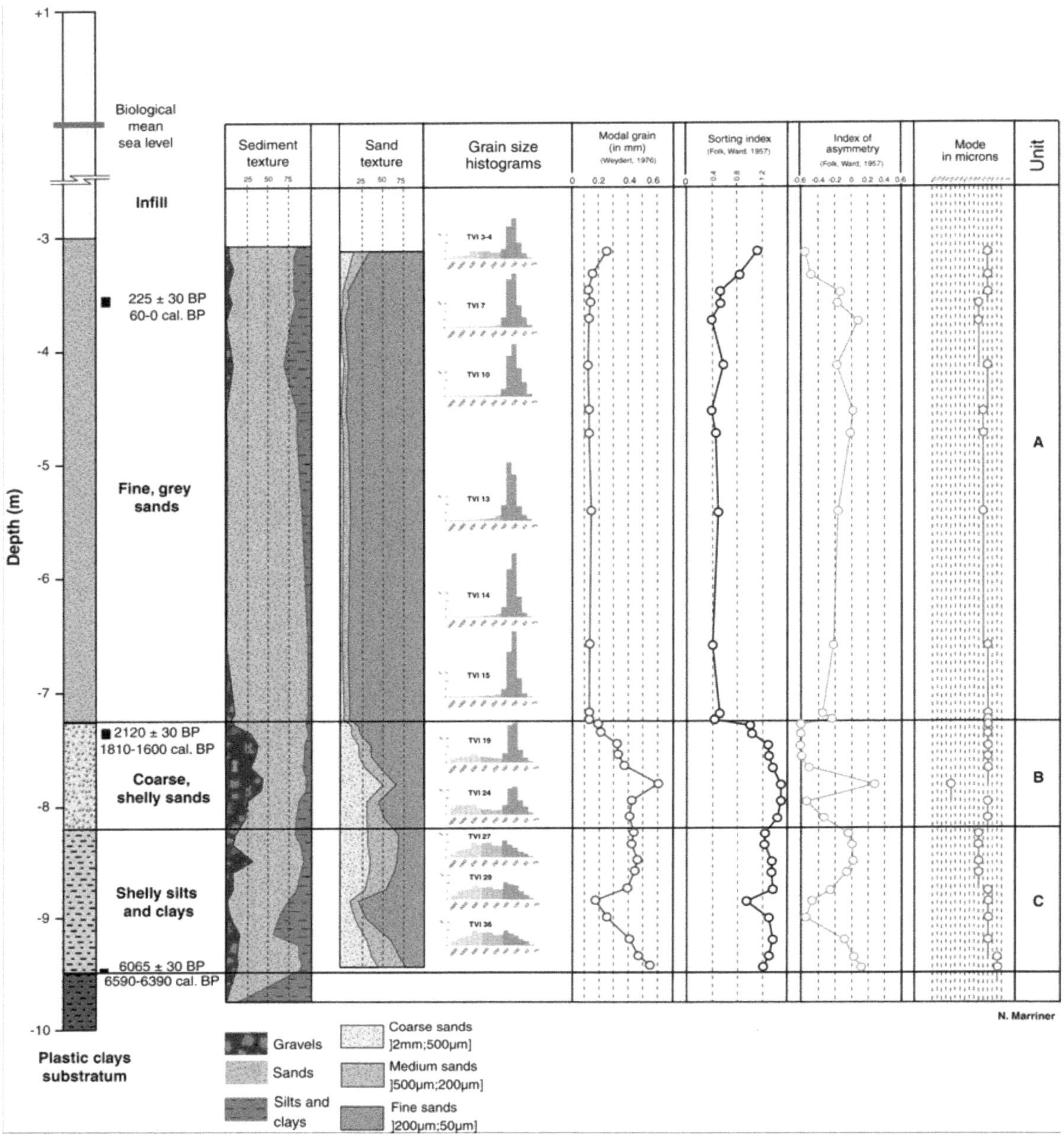

Figure 3.6: Grain size analyses of core TVI. (1) Unit C comprises a low energy silts deposit rich in molluscan shells. The transgressive contact is dated ~6400 cal. BP. Bimodal grain size histograms and high levels of fine sands (25-80 % of the sands fraction) are consistent with a reworking of the underlying clay substratum. (2) Unit B is characterised by a medium to coarse grained shelly sands deposit, concomitant with the pocket beach deposits observed in cores TI, TV and TIX. The top of the facies is dated ~1700 cal. BP. (3) After this date, a net transition to fine-grained sands is observed in unit A, comprising ~90 % fine sands. Modal grain values of between 0.2 and 0.4 µm and medium sorting indices concur a low energy beach environment. These sedimentological data appear consistent with a fall in water competence linked to the construction/reinforcement of the south-eastern harbour mole.

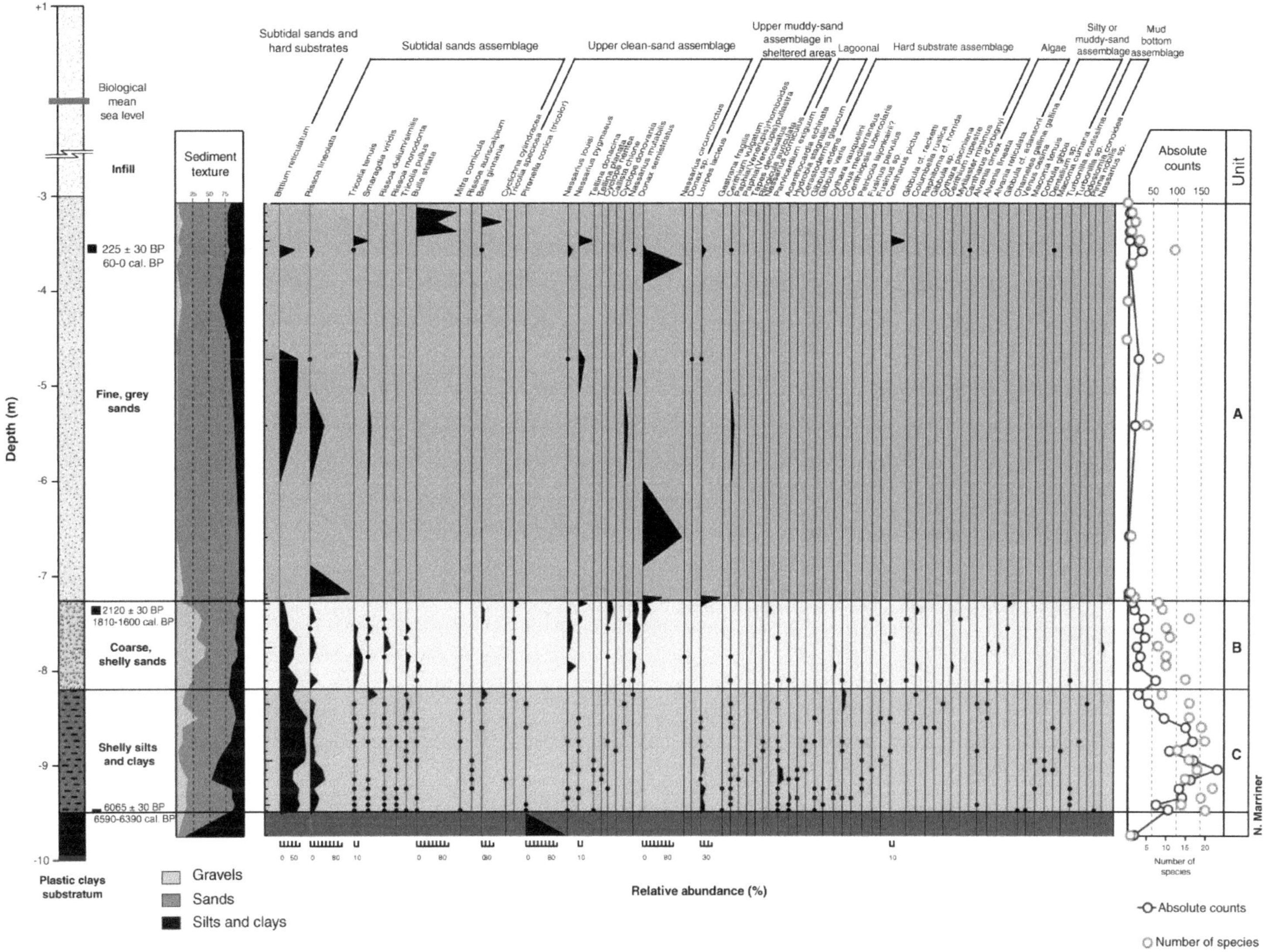

Figure 3.7: Macrofauna of core TVI. (1) Unit C comprises a rich and diversified molluscan suite dominated by two species *Bittium reticulatum* and *Rissoa lineolata*. *Rissoa lineolata* prefers subtidal sands rich in organic matter and is consistent with the confined sedimentary environment and a reworking of the organic rich clay substratum. A rich variety of other molluscan taxa is consistent with the development of a lagoonal/well-protected coastal biocenosis. Dominant secondary taxa include *Loripes lacteus*, *Cerithium vulgatum* (upper muddy-sand in sheltered areas assemblage) and *Parvicardium exiguum* (lagoonal assemblage). (2) Both *Bittium reticulatum* and *Rissoa lineolata* persist into unit B. Other species present in significant numbers include *Tricolia tenuis*, *Rissoa dolium* (subtidal sands assemblage), *Nassarius louisi*, *Nassarius pygmaeus*, *Nassarius mutabilis* and *Donax semistriatus* (upper clean-sand assemblage). These taxa concur a low to medium energy beach environment. (3) Unit A comprises species from diverse ecological groups including the subtidal sands assemblage (*Rissoa lineolata*, *Tricolia tenuis*, *Smaragdia viridis*, *Bulla striata*), the upper clean-sand assemblage (*Nassarius pygmaeus*, *Donax semistriatus*) and the upper muddy-sand assemblage in sheltered areas (*Cerithium vulgatum*). Again, these groups are concomitant with a well-protected beach environment.

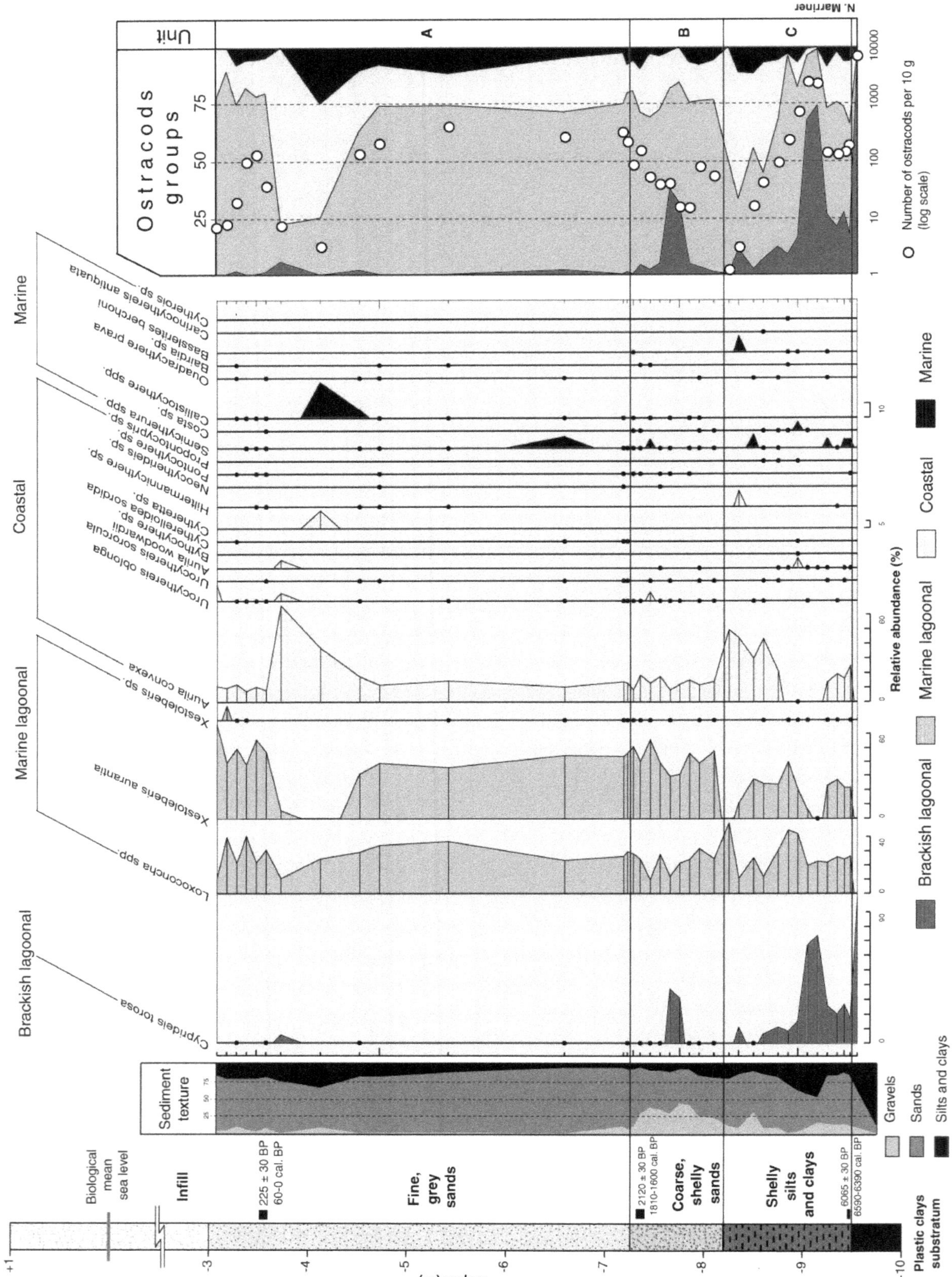

Figure 3.8: Ostracods data for core TVI.

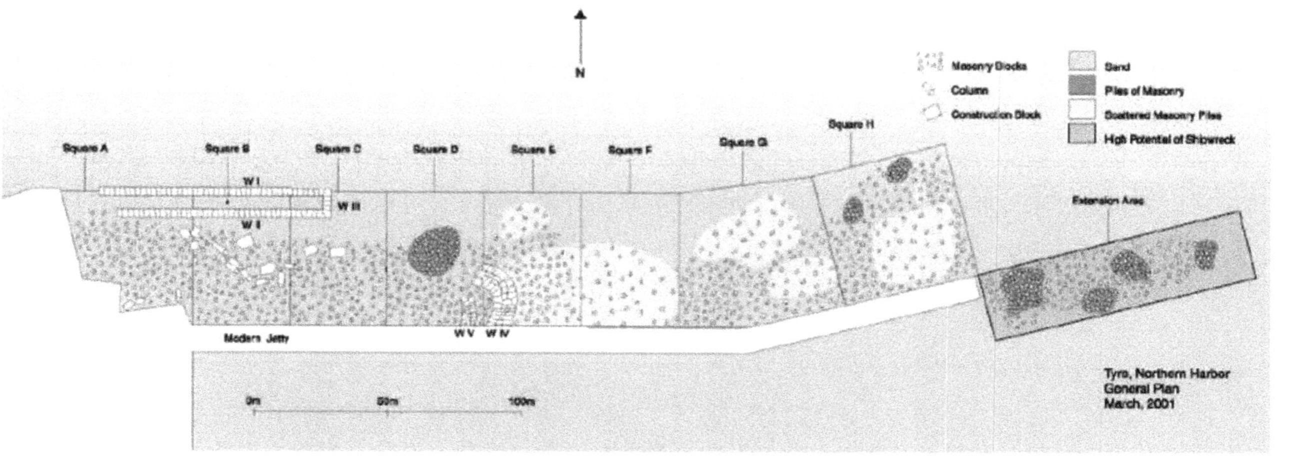

Figure 3.9: Northern harbour mole in Noureddine and Helou (2005).

3.3.2 When and how did the northern seaport evolve?

Coastal stratigraphy and high-resolution laboratory analyses allow us to elucidate six phases in the evolution of this semi-sheltered cove. Here we report its palaeogeographical history since the flooding of the cove during the Holocene marine transgression, and attempt to interpret the stratigraphic data with reference to the archaeological record. We demonstrate that harbour history can be clearly chronicled by diagnostic litho- and biostratigraphies. These depocentres are also appropriate for the analysis of numerous archaeological problems and cultural processes, providing a diversity of research possibilities (Goiran and Morhange, 2003).

3.3.2.1 Maximum Flooding Surface and sheltered lagoon environment (~8000 to ~6000 years BP)

Description: The Holocene Maximum Flooding Surface is dated ~8000 BP, overlain by shelly silt and clay deposits that accreted in the cove until ~6000 BP (**Figures 3.10 and 3.18**). The sedimentology of this basal layer is consistent with a reworking of the underlying clay substratum, giving rise to a lithodependent unit. The facies is dominated by upper muddy-sand (*Loripes lacteus*) and lagoonal molluscan taxa (*Parvicardium exiguum*; **Figures 3.11, 3.16 and 3.19**). High species diversity and the presence of both juvenile and adult tests are consistent with a rich biocenosis. The brackish lagoonal ostracod *Cyprideis torosa* attains relative abundance figures of >80 % with intercalated peaks of *Aurila woodwardii*, *Aurila convexa* and outer marine species (**Figures 3.15, 3.17 and 3.20**). Foraminifera are dominated by the brackish water species *Ammonia beccarii*.

Interpretation: The dominant litho- and biostratigraphical proxies attest to a shallow, low energy marine embayment newly transgressed by the post-glacial sea-level rise. With relative sea level approximately 7 ± 1 m below present, the coastal bathymetry shows the existence of a north-south trending subaerial ridge ~6 times longer than the present 1 km island (**Figure 3.21, Inset 1**). This geomorphological disposition led to the creation of a very low energy sediment depocentre, sheltered from the onshore south-westerly winds and fetch by this extensive drowned ridge. A series of proximal beach ridges are inferred to have closed the marine embayment, which, the biostratigraphic data demonstrate, was breached during periods of storm and high swell.

3.3.2.2 Pocket beach unit and Bronze Age proto-harbour

Description: After the onset of relative sea-level stability around 6000 BP, the northern coast of Tyre remained naturally protected by the aeolianite reef system. **Figure 3.21 (Inset 2)** depicts a tentative coastal reconstruction of Tyre island ~6000 BP, with a projected relative sea-level scenario at 5 m below present.

The molluscan fossils are consistent with a semi-sheltered environment in which both *in situ* taxa and reworked *extra situ* tests are represented. The dominant taxa include species from the hard substrate assemblage (*Cerithium rupestre*, *Fusinus pulchellus*, *Gibbula varia*) and *in situ* individuals from the upper-muddy sand in sheltered areas assemblage (*Nassarius corniculus*, *Cerastoderma glaucum*). An increase in coastal ostracod species, notably *Aurila convexa*, is to the detriment of *Cyprideis torosa*. Marine lagoonal taxa *Loxoconcha rhomboidea* and *Xestoleberis aurantia* attest to a relatively confined littoral environment. This unit corresponds to a semi-sheltered shoreface, which began accreting at the end of the mid-Holocene marine transgression. Presence of numerous marine taxa including *Semicytherura* sp., *Bairdia* sp., *Cistacythereis* sp., *Jugosocythereis* sp., *Loculicytheretta* sp., and *Callistocythere* spp. are drifted in and corroborate communication with the open sea.

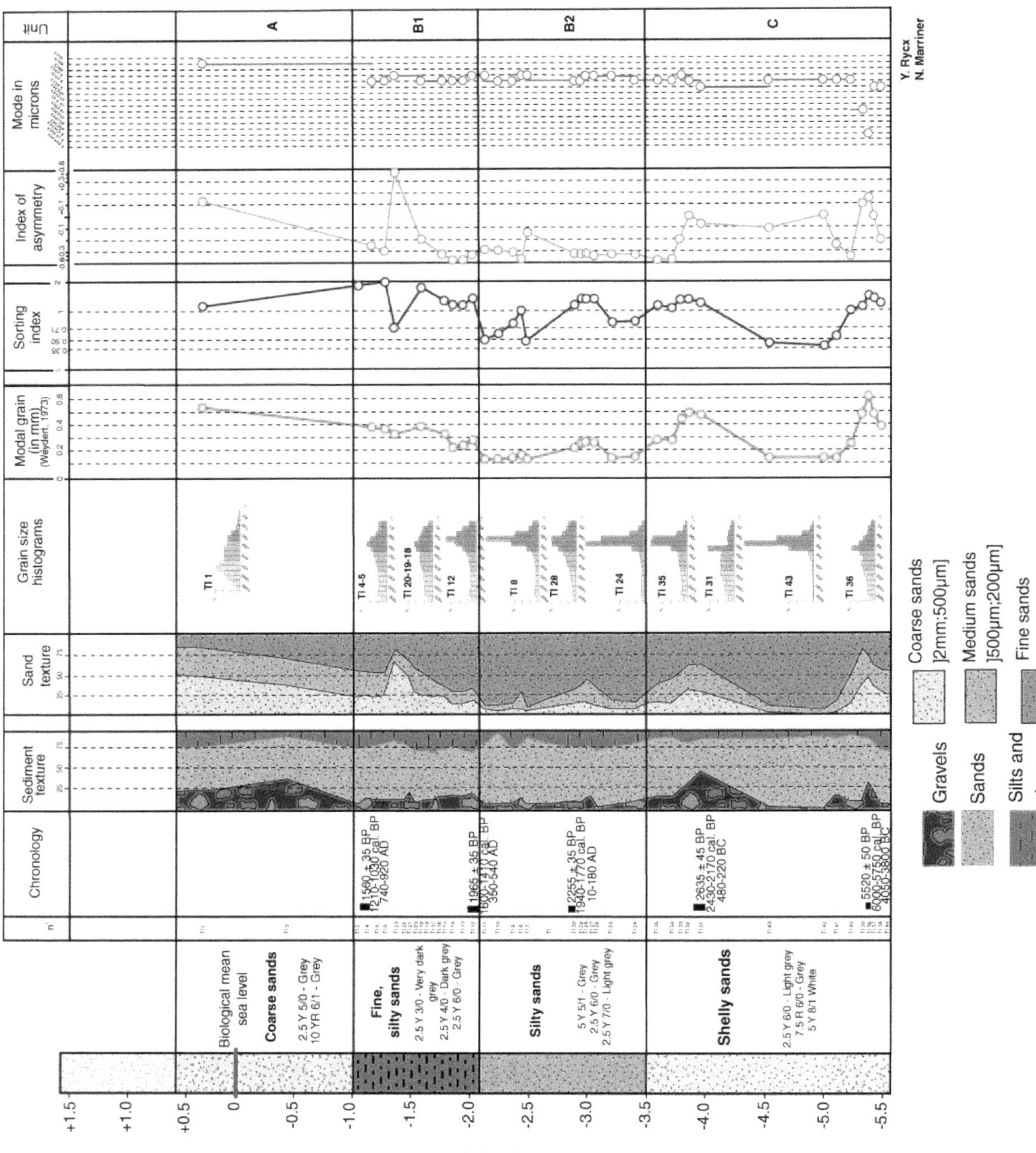

Figure 3.10: Grain size analyses of core TI.

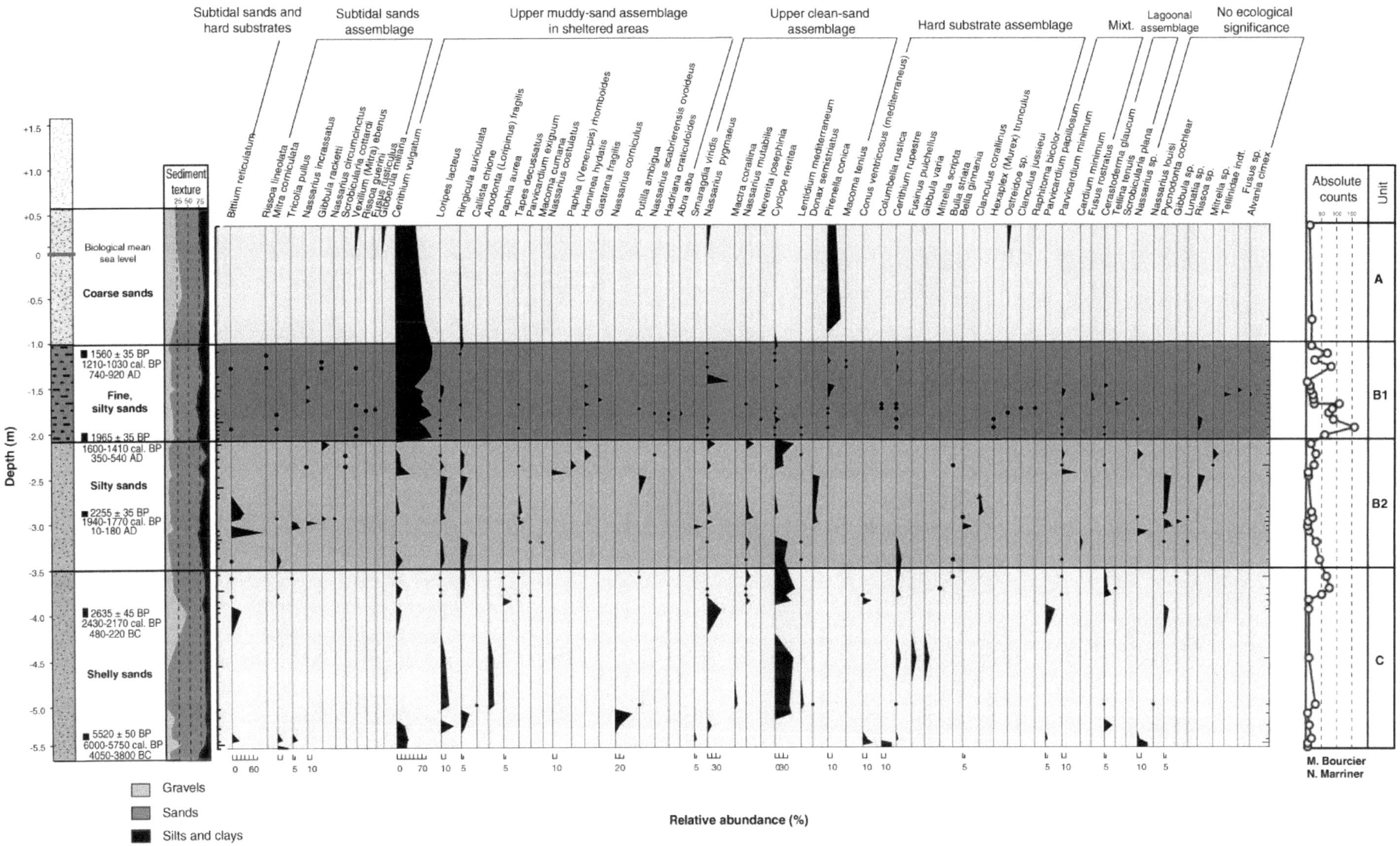

Figure 3.11: Macrofauna data of core TI. The diagram gives relative abundances and the absolute counts for each level. We identify three main units. Autochthonous and allochthonous taxa were identified on the basis of their taphonomic condition and their consistency with the lithofacies. (1) Unit C is dominated by taxa from the upper-clean sand assemblage (*Cyclope neritea*) and the upper muddy sand assemblage in sheltered areas (*Cerithium vulgatum*, *Loripes lacteus*). *Extra situ* taxa from the hard substrate assemblage indicate communication with the open sea. (2) Unit B2 is marked by taxa from the upper muddy sand assemblage in sheltered areas and the upper clean-sand assemblage. (3) Unit B1 is dominated by *Cerithium vulgatum* (upper muddy-sand assemblage in sheltered areas). This species attains relative abundance figures >70 %. *Cerithium vulgatum* is also adapted to hard substrates and we interpret its rapid expansion as being concomitant with nearby stone harbourworks. (4) Unit A is characterised by just a few dominant species including *Cerithium vulgatum* and *Pirenella conica*. We interpret this unit as an exposed beach, in proximity to the swash zone.

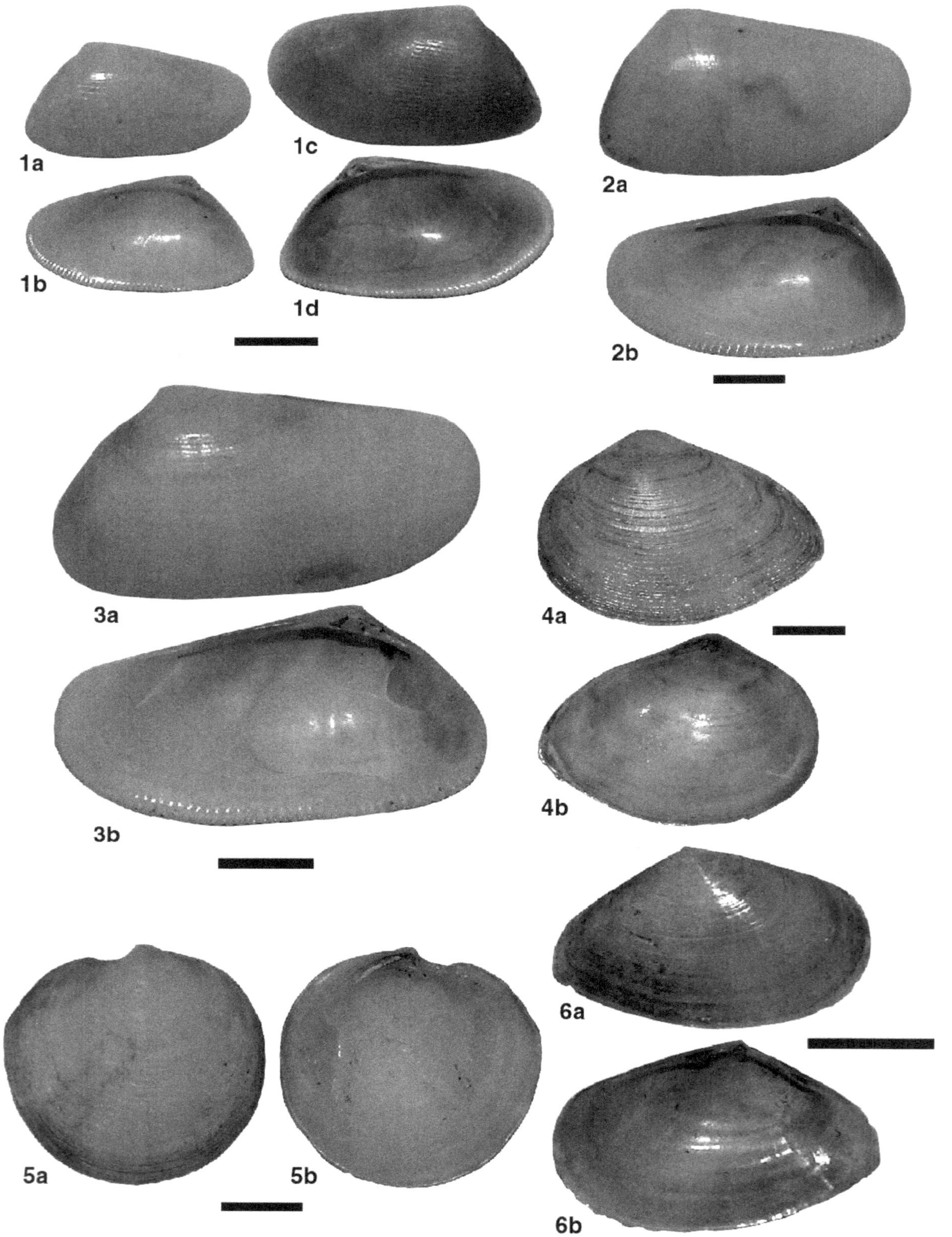

Figure 3.12: Scale bar: 5 mm. **1a-d:** *Donax semistriatus* (upper clean-sand assemblage); **2a-b:** *Donax trunculus* (upper clean-sand assemblage); **3a-b:** *Donax venustus* (upper clean-sand assemblage); **4a-b:** *Gastrana fragilis* (upper muddy-sand assemblage in sheltered areas); **5a-b:** *Loripes lacteus* (upper muddy-sand assemblage in sheltered areas); **6a-b:** *Tellina donacina* (upper clean-sand assemblage). Photographs: N. Marriner.

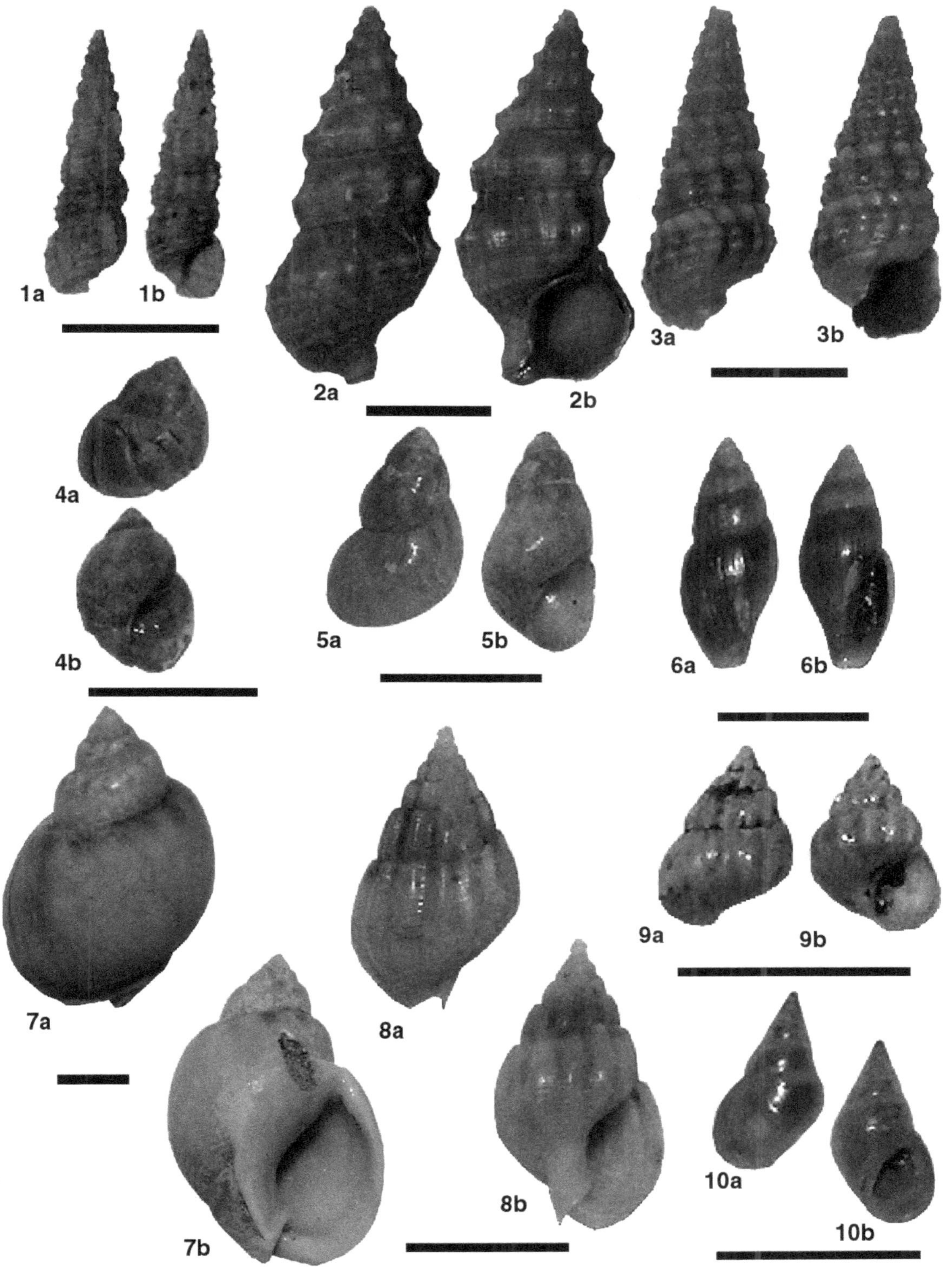

Figure 3.13: Scale bar: 5 mm. **1a-b:** *Bittium reticulatum* (hard substrate and subtidal sands assemblage); **2a-b:** *Cerithium vulgatum* (upper muddy-sand assemblage in sheltered areas); **3a-b:** *Pirenella conica* (subtidal sands assemblage); **4a-b:** *Tricolia pullus* (subtidal sands assemblage); **5a-b:** *Tricolia tenuis* (subtidal sands assemblage); **6a-b:** *Mitra ebenus* (subtidal sands assemblage); **7a-b:** *Nassarius mutabilis* (upper clean-sand assemblage); **8a-b:** *Nassarius pygmaeus* (upper clean-sand assemblage); **9a-b:** *Rissoa lineolata* (subtidal sands assemblage); **10a-b:** *Rissoa monodonta* (subtidal sands assemblage). Photographs: N. Marriner.

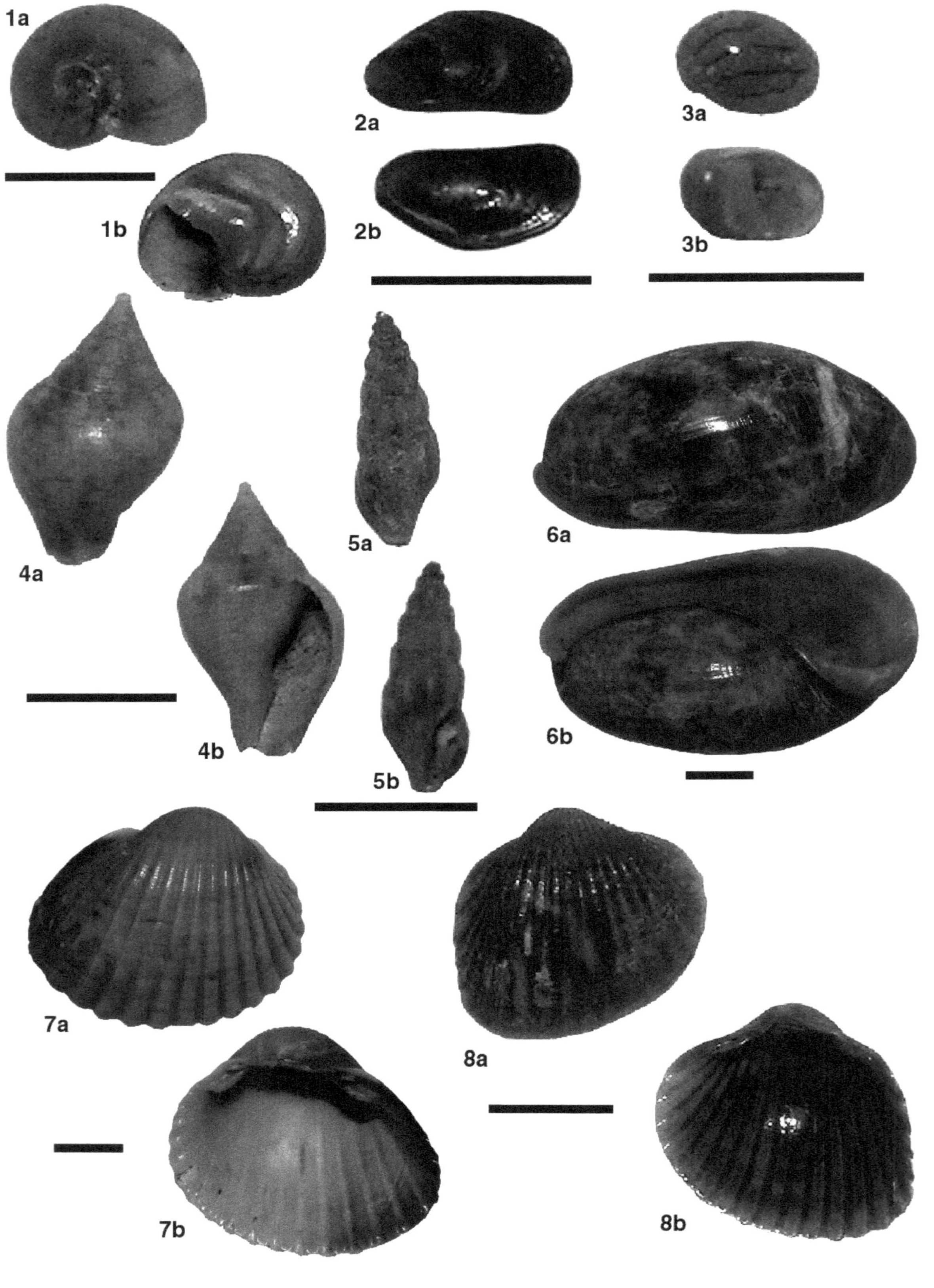

Figure 3.14: Scale bar: 5 mm. **1a-b:** *Cyclope neritea* (upper clean-sand assemblage); **2a-b:** *Mytilaster minimus* (hard substrate assemblage); **3a-b:** *Smaragdia viridis* (subtidal sands assemblage); **4a-b:** *Columbella rustica* (hard substrate assemblage); **5a-b:** *Bela ginnania* (subtidal sands assemblage); **6a-b:** *Bulla striata* (subtidal sands assemblage); **7a-b:** *Cerastoderma glaucum* (lagoonal); **8a-b:** *Parvicardium exiguum* (lagoonal). Photographs: N. Marriner.

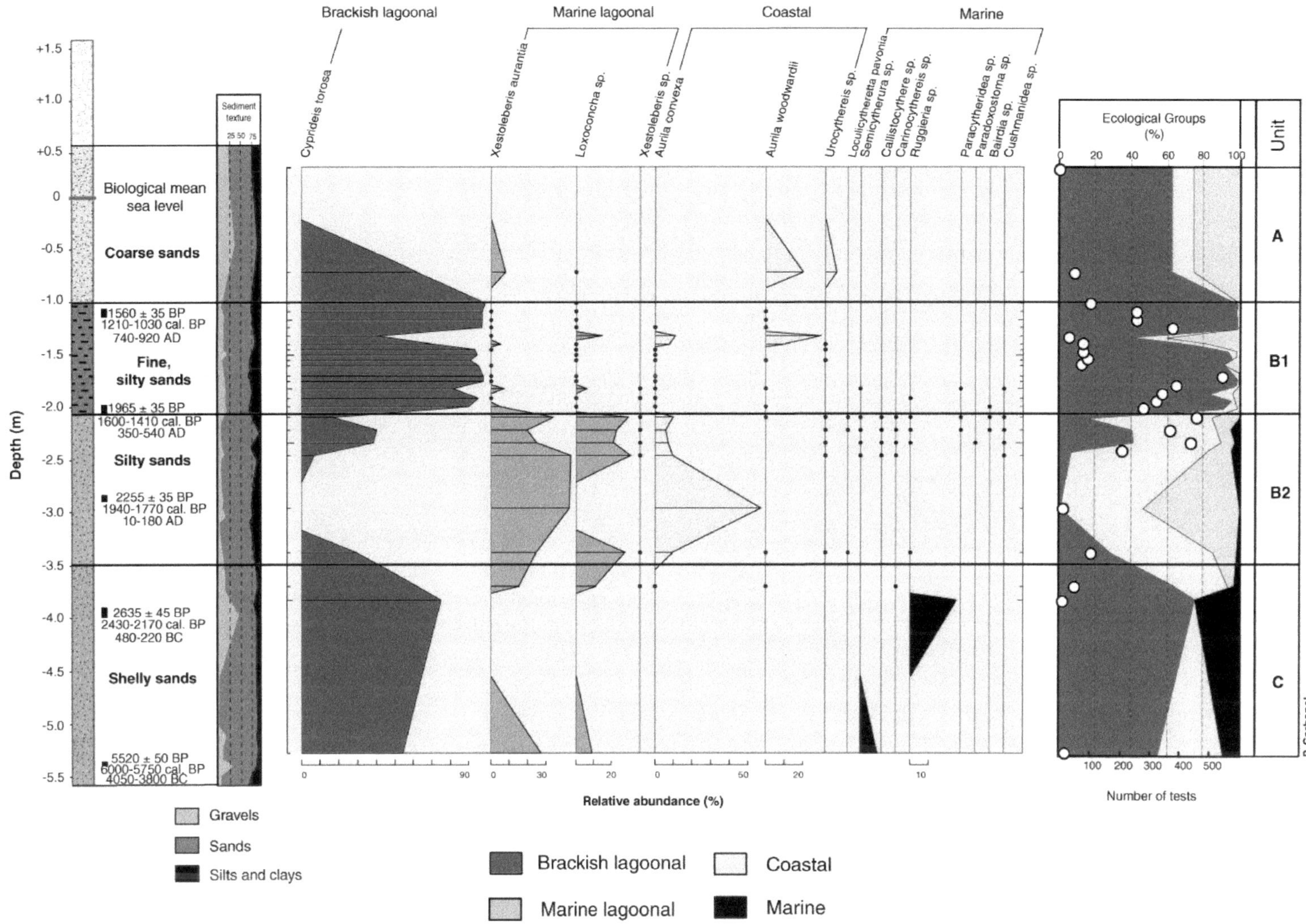

Figure 3.15: Ostracod data of core TI. (1) Unit C is characterized by a poor fauna (<50 individuals) and is dominated by the opportunist brackish lagoonal species *Cyprideis torosa*, which attains relative abundance figures in excess of 50 %. Secondary taxa include *Xestoleberis* sp. and *Loxoconcha* sp. (marine lagoonal), and *Semicytherura* sp. (marine).We interpret this unit as a semi-exposed pocket beach. (2) Unit B2 is marked by a strong increase in the faunal density (200-500 individuals), with a rise in marine lagoonal (*Xestoleberis* sp. and *Loxoconcha* sp.) and coastal taxa (*Aurila convexa*) to the detriment of *Cyprideis torosa*. Both *Loxoconcha* sp. and *Xestoleberis aurantia* are marine lagoonal species with oligo mesohaline salinity ranges, preferring muddy to sandy substrates with plants (Mazzini *et al.*, 1999). A few transported valves of *Cushmanidea* sp., *Bairdia* sp, *Paracytheridea* sp., *Callistocythere* sp., *Semicytherura* sp., *Urocythereis* sp., and *Aurila woodwardii* are recognised by their low frequency and underline continued exposure of the environment to offshore marine dynamics. This unit corresponds to a semi-protected harbour environment, sheltered from the dominant swell. (3) In unit B1 a renewed sharp rise in *Cyprideis torosa* (relative abundance levels >80 %), is consistent with a sheltering of the harbour basin which acts like a restricted leaky lagoon. Small peaks of marine lagoonal and coastal taxa are consistent with high-energy storm events. (4) In unit A, exposure of the harbour environment is marked by a decline in the faunal density, with a fall in *Cyprideis torosa* and a rise in coastal taxa (*Aurila woodwardii* and *Urocythereis* sp.). This evolution corresponds to the semi-abandonment of the harbour during the early Islamic period.

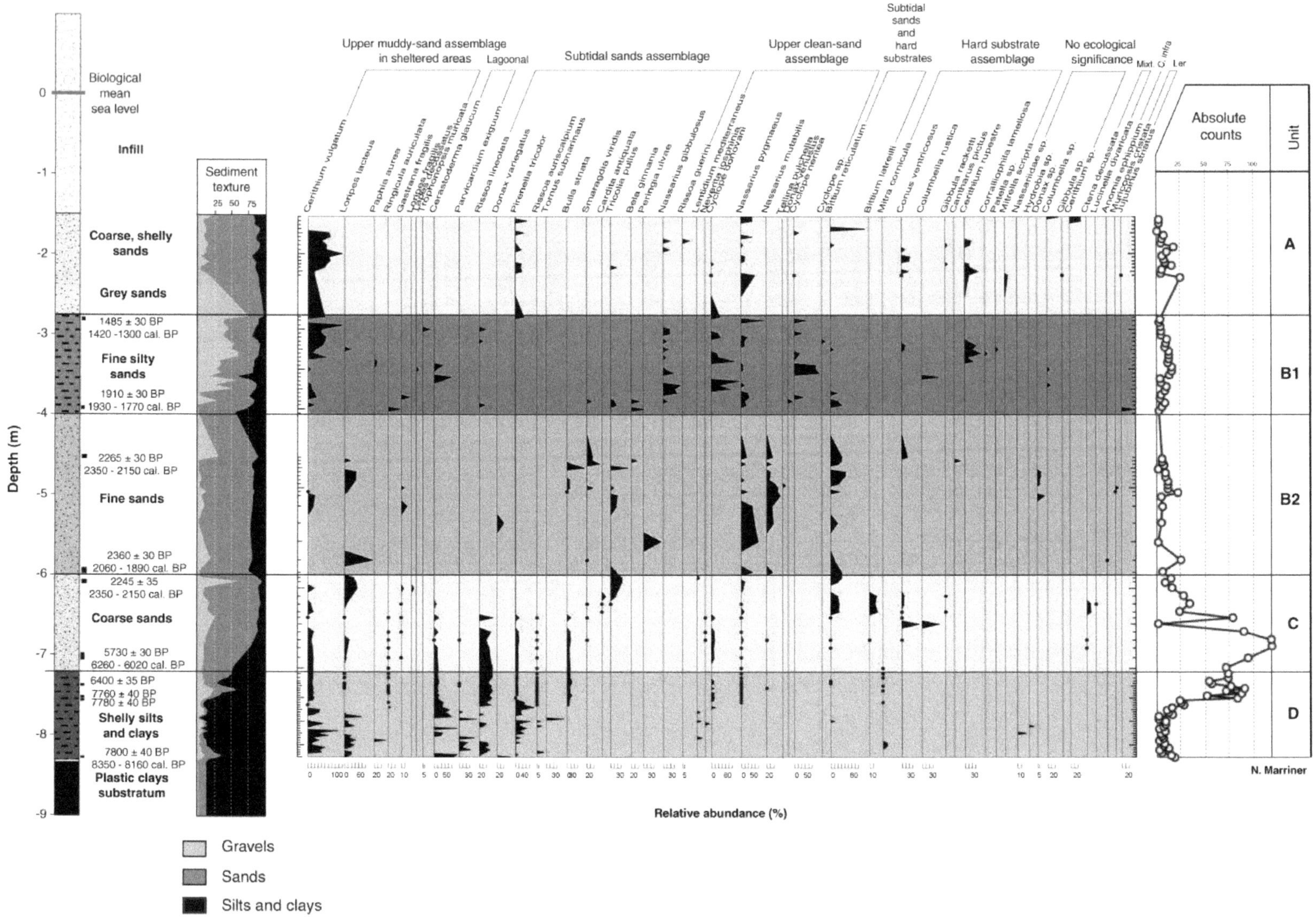

Figure 3.16: Macrofauna core TV. (1) Unit D is characterised by taxa from the upper muddy-sand assemblage in sheltered areas (*Cerithium vulgatum* and *Loripes lacteus*), lagoonal assemblage (*Cerastoderma glaucum* and *Parvicardium exiguum*) and subtidal sands assemblage (*Pirenella tricolor*). These data are consistent with a shallow marine environment protected by the extensive Quaternary ridge complex. (2) In unit C, taxa from the lower facies persist, with a rise in hard substrate assemblage species (*Bittium reticulatum, Bittium latreilli, Conus ventricosus* and *Columbella rustica*) indicating increased exposure to offshore marine dynamics. (3) In unit B2, a rise in upper clean-sand assemblage taxa, notably *Nassarius pygmaeus* and *Nassarius mutabilis*, is consistent with a fine-grained sand bottom well-protected from offshore dynamics. Other important species include *Bulla striata*, *Smaragdia viridis* and *Tricolia pullus* (subtidal sands assemblage) and *Loripes lacteus*, characteristic of a protected harbour. (3) Unit B1 is consistent with a well-sheltered harbour environment. The unit is dominated by species from the upper muddy-sand assemblage in sheltered areas (*Cerithium vulgatum*) and upper clean-sand assemblage (*Cyclope donovani*, *Nassarius pygmaeus* and *Cyclope neritea*). (4) Unit A is dominated by *Cerithium vulgatum*. Low percentages of other taxa suggest a reopening of the environment to offshore marine dynamics, such as storms.

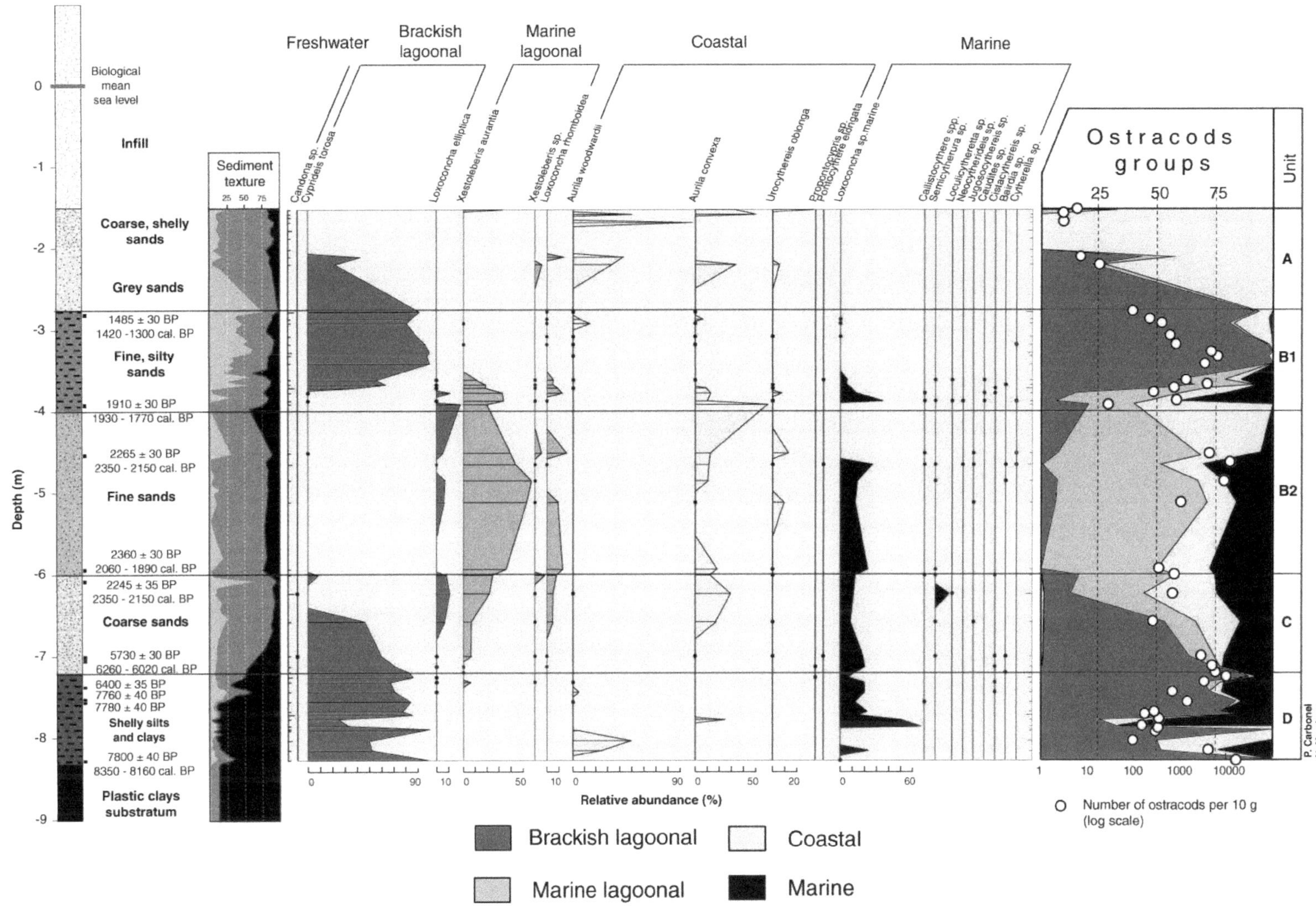

Figure 3.17: Ostracod data of core TV. (1) Unit D, relatively poor in ostracods, is dominated by *Cyprideis torosa*, which attains relative abundance of >80 %. It indicates a sheltered, brackish lagoonal environment. Peaks of *Aurila woodwardii* and *Aurila convexa* are consistent with higher energy episodes, such as storms and high swell. Presence of the marine species *Loxoxoncha* (type marine) attests to interaction with the open sea. (2) In unit C, the increase in coastal species, notably *Aurila convexa*, is to the detriment of *Cyprideis torosa*. Presence of the marine lagoonal taxa *Loxoconcha rhomboidea* and *Xestoleberis aurantia* do however indicate a relatively sheltered marine context. This unit corresponds to a semi-sheltered shoreface, resulting from the immersion of the Quaternary ridge at the end of the Holocene marine transgression. Presence of numerous marine taxa including *Semicytherura* sp., *Bairdia* sp., *Cistacythereis* sp., *Jugosocythereis* sp., *Loculicytheretta* sp., and *Callistocythere* spp. are drifted in and corroborate communication with the open sea. (3) Unit B2 is dominated by marine lagoonal taxa such as *Xestoleberis* spp. and *Loxoconcha rhomboidea*. Presence of the brackish lagoonal species *Loxoconcha elliptica* is consistent with a sheltered harbour in connection with the open sea. Coastal taxa such as *Urocythereis oblonga* and *Aurila convexa* are also present in significant numbers. (4) Unit B1 is marked by a sharp rise in *Cyprideis torosa* (in number and ratio) to the detriment of marine lagoonal, coastal and marine taxa. We interpret this as translating a sudden artificial sheltering of the harbour environment, brought about by increased protective infrastructures.

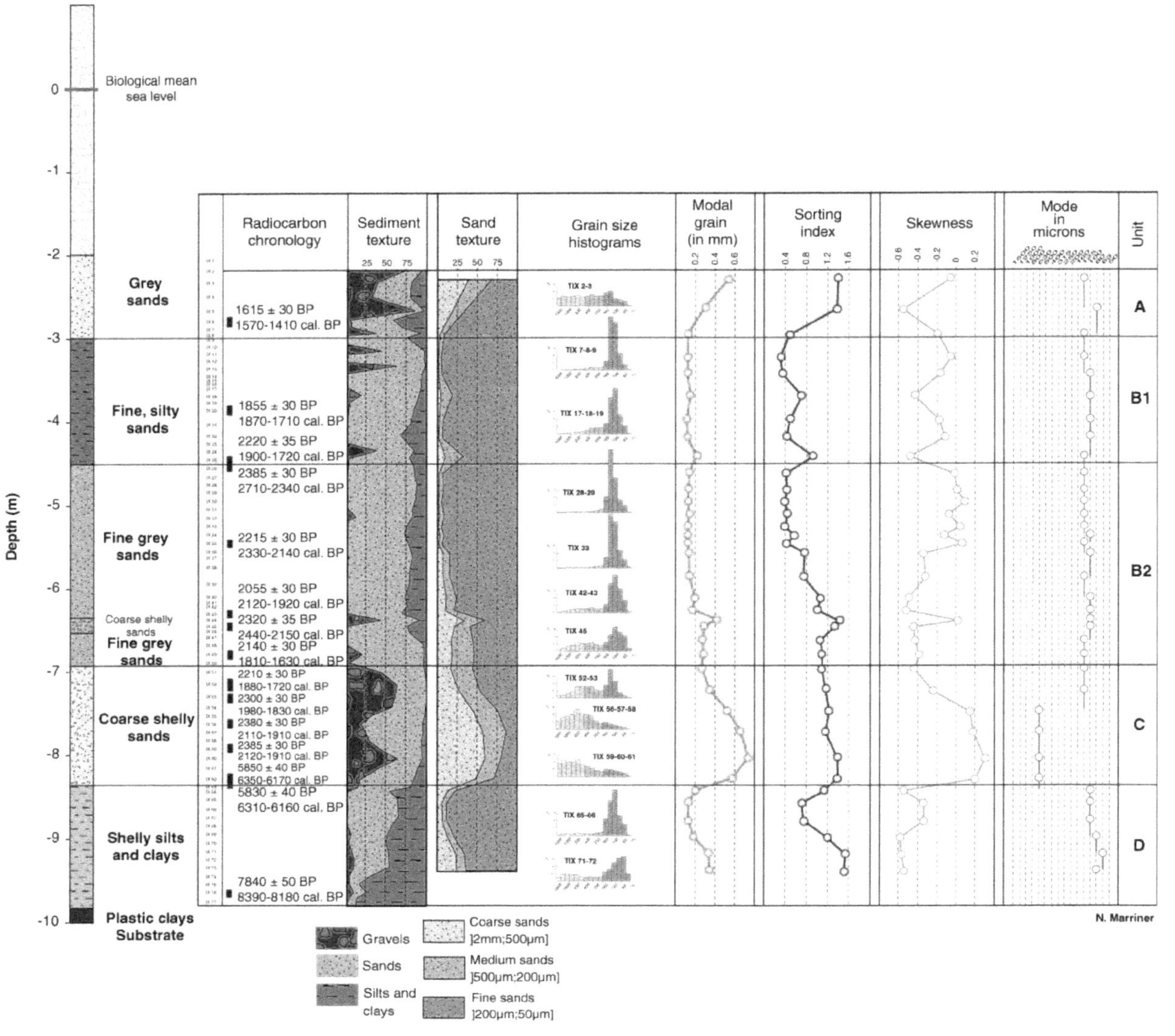

Figure 3.18: Grain size analyses core TIX. (1) Unit D is characterised by a high proportion of silts and clays decreasing from 95 % at the base of the unit to 20 % at the top. These characteristics are consistent with a reworking of the underlying Pleistocene substratum. Low energy conditions are indicated by grain size histograms well-developed in the finer end of the scale. The modal grain size ranges between 0.34 mm and 0.21 mm. (2) In unit C, higher energy conditions relative to unit D are indicated by a shift in sediment texture: gravels comprise 65-25 %, sands 28-75 %, and silts and clays 1-16 %. Sorting indices of between 1.4 and 1.1 indicate a poorly sorted sediment. (3) In unit B2, a fall in water competence is indicated by a sharp rise in the fine sands fraction (between 76-95 % of the total sand fraction), with well-developed unimodal histograms. The base of the unit (IX44 and IX45) is marked by higher energy deposits which we interpret as either (i) a storm event intercalated in the fine sediments, or (ii) harbour dredging with reworking of the bottom sediments. (4) Low energy conditions persist into unit B1, with modal grain sizes of between 0.13-0.21 mm. Sorting indices of 0.31-0.9 indicate a good to mediocre sorted sand fraction. (5) Unit A is marked by amodal grain size distributions of the sand fraction and indicates an open coastal environment.

Overall, the stratigraphy from Tyre's northern harbour demonstrates the peculiarity of this basin. There are essentially two major allochthonous sediment provenances: (1) the fine sediments derive from the erosion of Pleistocene outcrops; and (2) the sands derive from the Litani river system to the north of the city. By comparison with other ancient harbours, Tyrian sediments are coarser and we do not observe the deposition of a classic plastic harbour clay (i.e. Alexandria, Sidon, Marseilles). This is possibly due to a large harbour inlet.

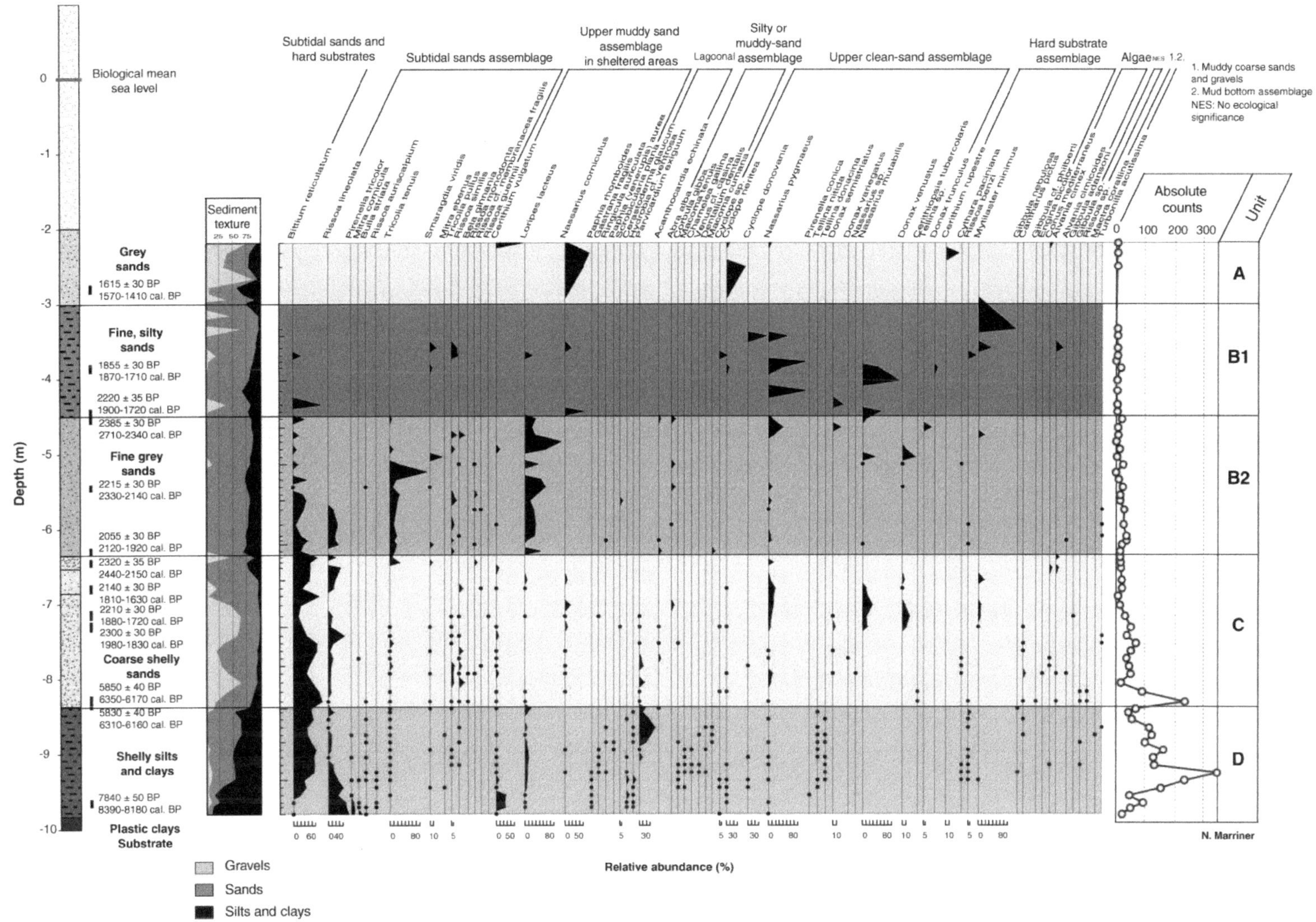

Figure 3.19: Macrofauna data of core TIX. (1) Unit D is marked by a large number of different species. Dominant taxa derive from the subtidal sands assemblage (*Rissoa lineolata, Tricolia tenius, Rissoa auriscalpium, Pirenella tricolor, Bulla striata*), the upper muddy sand assemblage in sheltered areas (*Loripes lacteus, Cerithium vulgatum*) and the lagoonal assemblage (*Parvicardium exiguum, Hydrobia ventrosa, Cerastoderma glaucum*). Significant numbers of *Bittium reticulatum* (up to 60 % relative abundance) are consistent with proximity to the sandstone reefs. Average specimens are between 6-8 mm long and are easily transported by bottom currents. (2) Unit C is dominated by the subtidal sands assemblage (*Tricolia tenuis*, *Tricolia pullus*, *Smaragdia viridis*, *Rissoa lineolata*) and the upper clean sand assemblage (*Nassarius pygmaeus, Nassarius mutabilis, Donax venustus*). High relative abundance of *Bittium reticulatum* continues into this unit. (3) In unit B2, transition to lower energy artificial harbour conditions is manifested by a sharp rise in *Loripes lacteus* (upper muddy sand assemblage in sheltered areas) and *Tricolia tenuis* (subtidal sands assemblage). Artificial protection is corroborated by a fall in species diversity relative to units D and C. (4) Unit B1 is characterised by few species from the upper clean sand assemblage (*Nassarius pygmaeus, Nassarius mutabilis*), with a very low number of taxa attesting to a confined harbour environment. (5) Unit A is characterised by the upper muddy sand assemblage in sheltered areas (*Cerithium vulgatum*, *Nassarius corniculus*), concomitant with a reopening of the harbour.

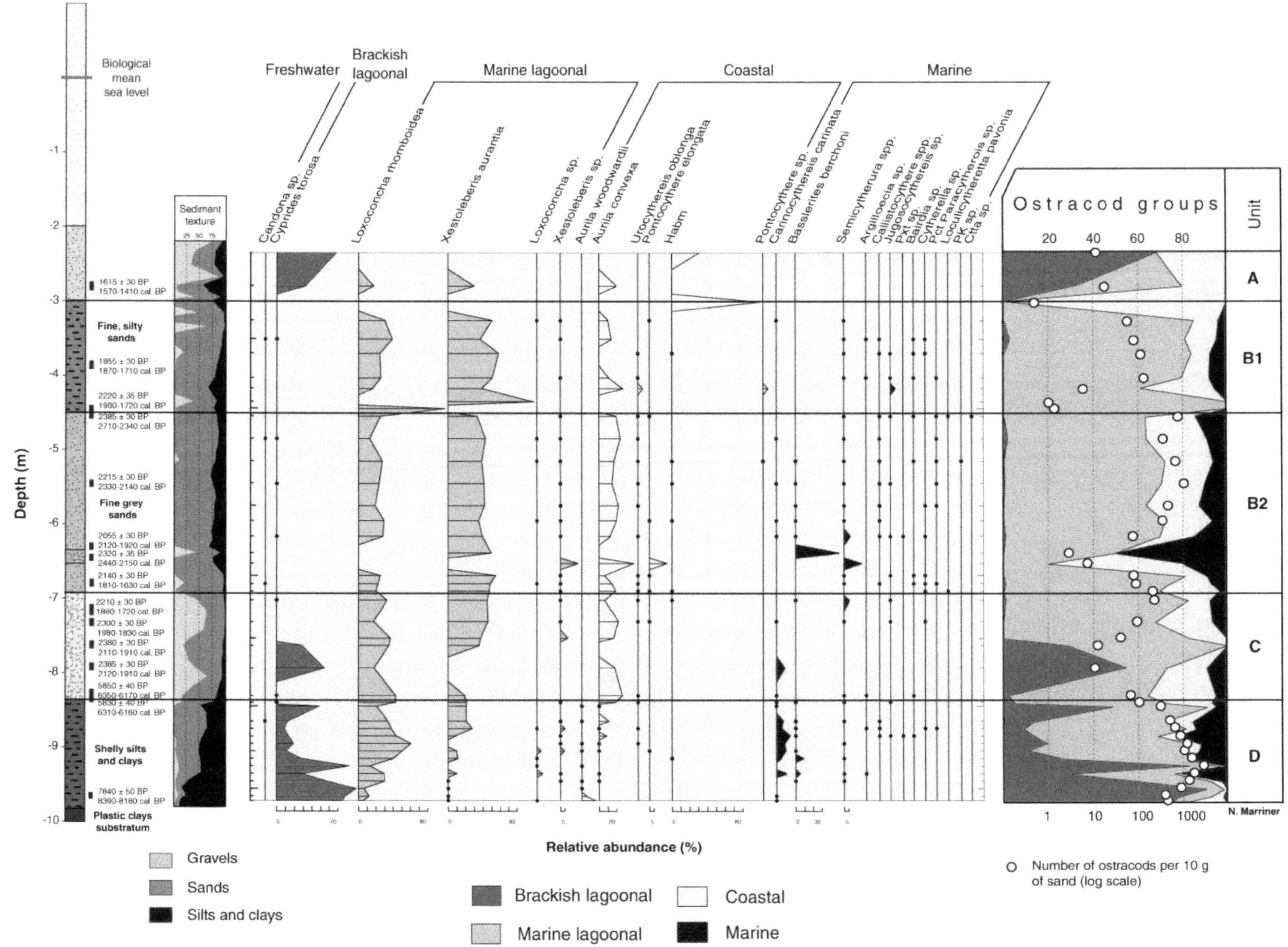

Figure 3.20: Ostracods data of core TIX. (1) The base of unit D has a rich faunal density of >1000 tests per 10 g of sand, and is dominated by the taxon *Cyprideis torosa* consistent with a shallow brackish water environment. Rising sea level led to the exposure of this brackish environment to increased marine influence as indicated by the marine lagoonal species *Loxoconcha rhomboidea* and *Xestoleberis aurantia*. Coastal and marine valves are also drifted in. (2) Marine lagoonal taxa continue to dominate unit C. Higher energy dynamics are indicated by the presence of the coastal taxa *Aurila convexa*. There is a sharp fall in faunal densities to between 10 and 100 tests per 10 g of sand. (3/4) There is very little biostratigraphic variation between units B2 and B1. The marine lagoonal taxa *Loxoconcha rhomboidea* and *Xestoleberis aurantia* persist into this unit. *Aurila convexa* is also present in significant numbers. The unit is intercalated with peaks of coastal (*Pontocythere elongata*) and marine (*Basslerites berchoni*, *Semicytherura* spp.) species. (5) Unit A juxtaposes taxa from the three ecological groups, brackish lagoonal (*Cyprideis torosa*), marine lagoonal (*Loxoconcha rhomboidea*, *Xestoleberis aurantia*) and coastal assemblages (*Aurila convexa*).

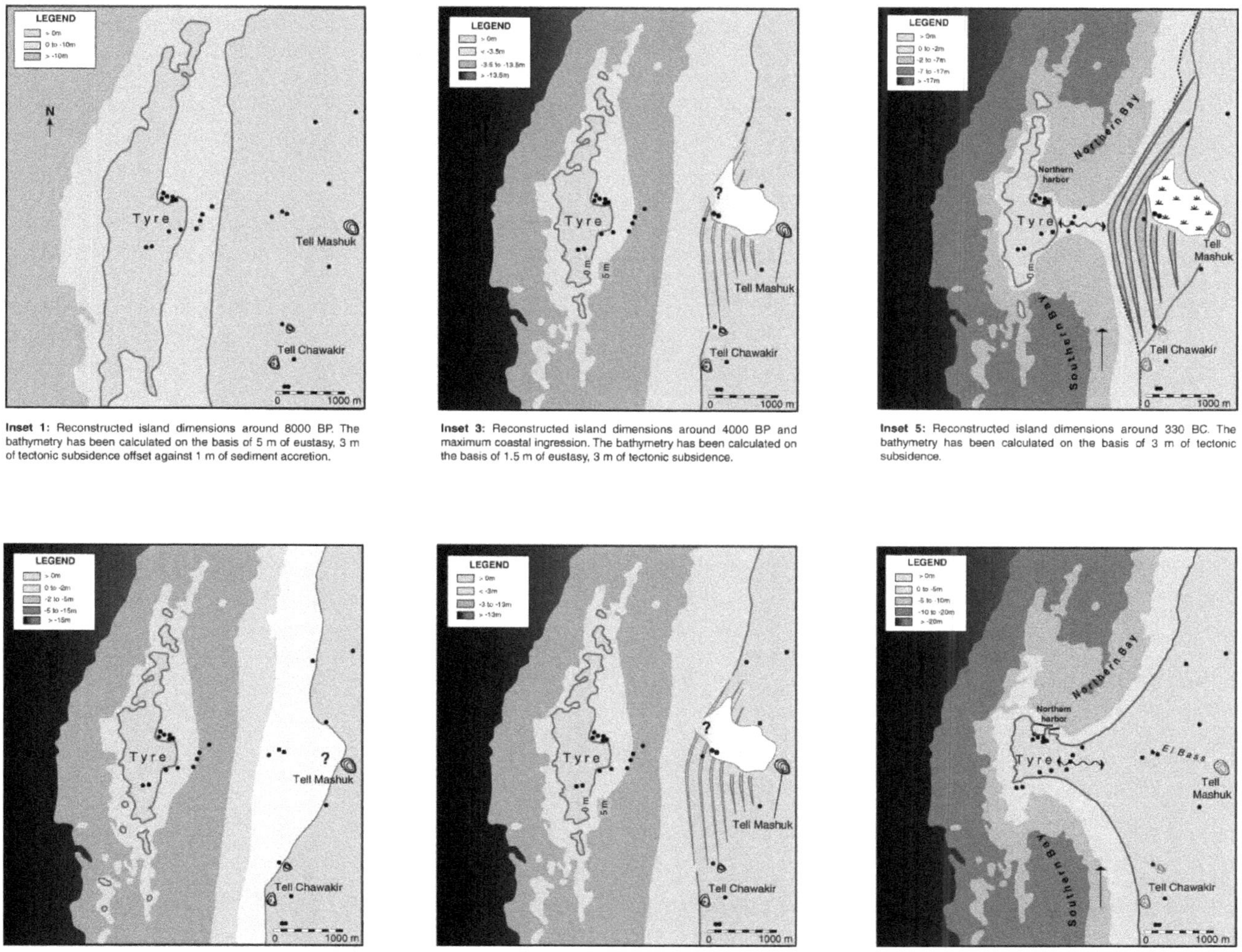

Inset 1: Reconstructed island dimensions around 8000 BP. The bathymetry has been calculated on the basis of 5 m of eustasy, 3 m of tectonic subsidence offset against 1 m of sediment accretion.

Inset 3: Reconstructed island dimensions around 4000 BP and maximum coastal ingression. The bathymetry has been calculated on the basis of 1.5 m of eustasy, 3 m of tectonic subsidence.

Inset 5: Reconstructed island dimensions around 330 BC. The bathymetry has been calculated on the basis of 3 m of tectonic subsidence.

Inset 2: Reconstructed island dimensions around 6000 BP and maximum coastal ingression. The bathymetry has been calculated on the basis of 2 m of eustasy and 3 m of tectonic subsidence.

Inset 4: Reconstructed island dimensions around 3000 BP and maximum coastal ingression. The bathymetry has been calculated on the basis of 1 m of eustasy, 3 m of tectonic subsidence.

Inset 6: Present day coastal morphology and bathymetry

Figure 3.21: Tentative coastal reconstruction of Tyre's sandstone ridge since 8000 BP based on relative sea-level variations at the site.

Interpretation: The biosedimentological proxies asseverate a semi-open marine cove conducive to early settlement of the island by human societies. Unequivocally, the geomorphology of the Tyrian coastline – an easily defendable offshore island shielding a number of natural low-energy anchorages – is significant in explaining the foundation of Tyre during the third millennium BC. At this time, the northern cove was used as a proto-harbour, where seafarers would have the choice of beaching their ships on the sandy shore, or leaving them at anchor in the bay, depending on the daily or seasonal weather and the direction of the wind. The use of lighters, small ferry boats used to load and unload larger vessels, was also widespread at this time (Marcus, 2002a). The oldest documentary evidence we have for the beaching of seafaring vessels during the Bronze Age derives from Akrotiri, on the island of Thera (Gillmer, 1975, 1978, 1985; Giescke, 1983). The so-called Flotilla Fresco, dated ~1550 BC, depicts a maritime scene of small harbours in which boats are moored or pulled up onto the beach (Marinatos, 1971, 1974, 1976; Casson, 1975, 1978a; Doumas, 1992; Televantou, 1990; Wachsmann, 1998; **Figure 3.22**). On the basis of an architectural study of the larger processional boats in the frieze, Cervin has proposed a hypothetical reconstruction of a ship being beached using lines attached to the stern device (Cervin, 1977, 1978; Figure **3.23**). It is interesting to note that boat beaching is still used to this day by Mediterranean fisherman with light, shallow draught vessels (**Figure 3.24**).

During the Middle to Late Bronze Age, dynamic interaction throughout the eastern Mediterranean brought about a period of internationalism characterised by developed trade routes which interlaced the Levant, Egypt and the Aegean (Holmes, 1975; Hankey, 1993; Knapp, 1993; Dickinson, 1994; Ilan, 1995; Marcus, 2002b; Kristiansen and Larsson, 2005). For the societies ringing the eastern basin of the Mediterranean, contact was established primarily by the sea (Wachsmann, 1998). The movement of ships around the eastern

Mediterranean is implied in the distribution of trade goods along its shores, and demonstrated by the discovery of numerous Bronze Age shipwrecks (Bass 1987, 1991). At this time, Tyre and other coastal settlements along the Phoenician coast, started to consolidate their positions as major emporia for trade with Cyprus, Crete, and Mycenaean Greece (Barnett, 1956; Doumet-Serhal, 2004a). Although it is widely believed that this expanding Mediterranean trade prompted coastal populations into modifying their natural anchorages, the exact timing and technology used to achieve this is open to conjecture (Raban, 1995).

For many scholars, the ancient Near East is the cradle of coastal harbour development, logically adapted from the deltaic contexts of Mesopotamia and Egypt (Fabre, 2004/2005). From Egypt alone, we have iconographic depictions of ships and fluvial quaysides in many tomb paintings (Edgerton, 1922-1923; Haldane, 1990; Wachsmann, 1998). A great number of tombs have also yielded model boats of various descriptions, which, along with their crews, are poised to transport and provision the deceased in the afterlife (El Baz, 1988). The Cheops ship, for example, is so technically advanced that development over many thousands of years must be assumed (Casson, 1995). Indeed, primitive river transportation probably existed on the Nile by Palaeolithic times (Hornell, 1970; Fabre, 2004/2005).

Although opinions vary on whether Egypt was a major sea-going culture (Säve-Söderbergh, 1946; Nibbi, 1979; Wachsmann, 1998; Fabre, 2004/2005), proximity to the two major Egyptian and Mesopotamian riverine cultures appears consistent with rapid diffusion and adaptation of this *savoir faire* to the Levantine seaboard; by contrast, contemporaneous Aegean harbour infrastructure was much less evolved at this time (Shaw, 1990; Raban, 1991). Although fluvial construction techniques had to be adapted to the difficult maritime context, the quasi-absence of tides in the Mediterranean was a significant advantage in the construction of permanent harbour infrastructure. Active research along the Levantine seaboard, notably in Israel, has yielded valuable new information about shipping and harbours (Raban, 1984, 1985a-c, 1987a-b, 1991, 1997a-b; Raban and Holum, 1996; Marcus, 2002a-b; Galili *et al.*, 2002). Although there is general agreement about the attribution of numerous moles and quays to the Romans (Raban and Holum, 1996) or even the Phoenicians (Haggai, 2006), very few can be unequivocally constrained to the Bronze Age due to the difficulties in dating rock cut structures. Broadly speaking two groups of archaeological evidence, indirect and direct, can be used in support of artificial port infrastructure at this time (**Figure 3.25**).

(1) Indirect evidence: Carayon (PhD in progress) has compiled three lines of evidence in support of artificial quays during the Bronze Age. As the first two examples demonstrate, the epigraphic record of Egyptian hieroglyphs and Mesopotamian cuneiform tablets has been particularly rich: (i) **The poem of Pentaour** was engraved by Ramses II at Louxor, Karnak and Abydos to commemorate Egyptian victory over the Hittites at Qdesh in 1293 BC. The hieroglyphic poem depicts sea boats with three masts. If one assumes that the central mast was used to harness the wind and propel the vessel, and the second to steer the rudder, the role of the third mast is unclear. Basch (1987) argues that this final mast was used in tandem with its central counterpart, by way of ropes and pulleys, to load and unload goods onto the boat. Such a lateral loading and unloading system assumes the presence of a harbour quay (Casson, 1976).

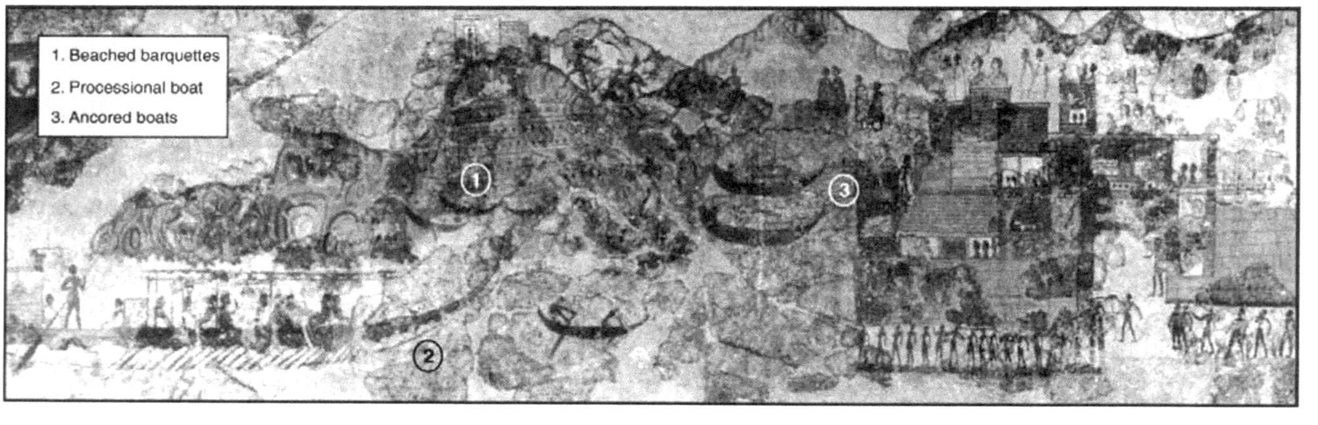

Figure 3.22: The waterborne procession in the south miniature frieze at Thera (from Wachsmann, 1998). The smallest barquettes have been drawn from the water (1), while the larger vessels are anchored in the pocket cove to the left of the main settlement (3).

(ii) **An epistolary document (RS 17.133)** found during excavations at the Royal Palace of Ugarit mentions an Ugarit boat sunken in a non-identified harbour. The text explains that the vessel was damaged against port quays, and would appear to lend credence to artificial harbourworks in northern Syria during the Late Bronze Age. (iii) The existence of **stone anchors**, first identified by Frost (1969a-b) at Byblos and Ugarit, is a more speculative candidate for artificial quays. At Ugarit, the heaviest example weighs 600 kg and calculations by Saadé (1995) suggest that a 500 kg anchor corresponds to a 20 m long boat. Beaching of such large vessels would not have been possible and assumes the existence of either artificial docks or lighter vessels that ferried goods to and from the shoreline. Such anchors are widespread at Bronze Age Levantine sites.

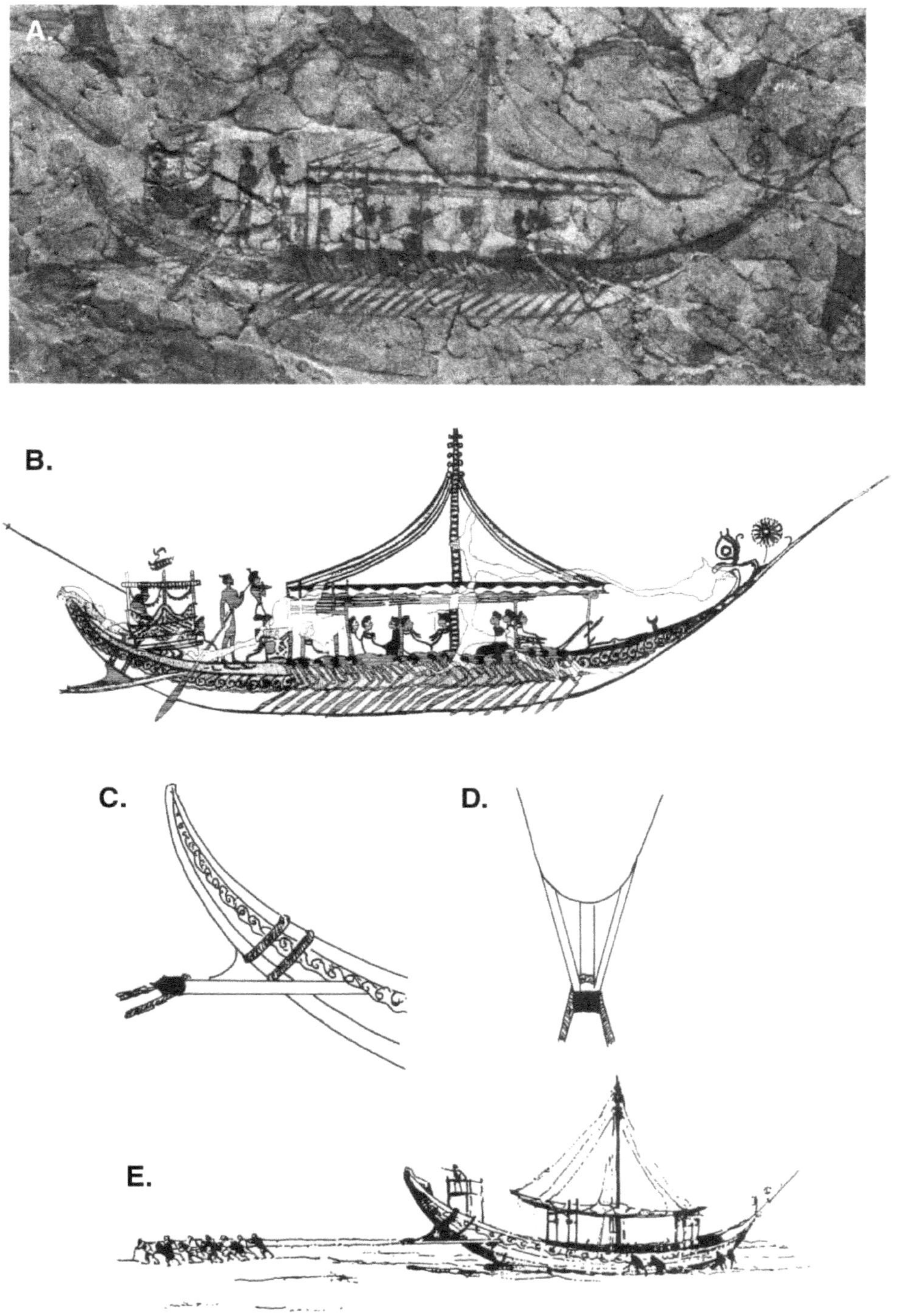

Figure 3.23: Cervin's reconstruction of the stern device used to haul the processional vessel from the water. (A) The processional boat depicted in the south Miniature Frieze, Room 5, "West House" at Thera (photo in Wachsmann, 1998). (B) Line drawing of the vessel in A (in Marinatos, 1974). (C) The stern device and knee support. (D) Plan view of the stern. (E) Hypothetical reconstruction of the vessel being beached (C-E from Cervin, 1977).

Figure 3.24: Beaching of boats and the use of lighter vessels in Tyre's northern harbour during the1830s. There are a number of things to note in this lithograph: (1) the fishermen nearest the shore are only knee deep in water, attesting to the shallow, silted up nature of the harbour; (2) the fishermen in the foreground are standing upon the prograded beach unit elucidated in our harbour stratigraphy; (3) finally, note the total absence of harbourworks in this beach area (from D. Roberts: Port of Tyre dated April 27th 1839 in Roberts, 2000).

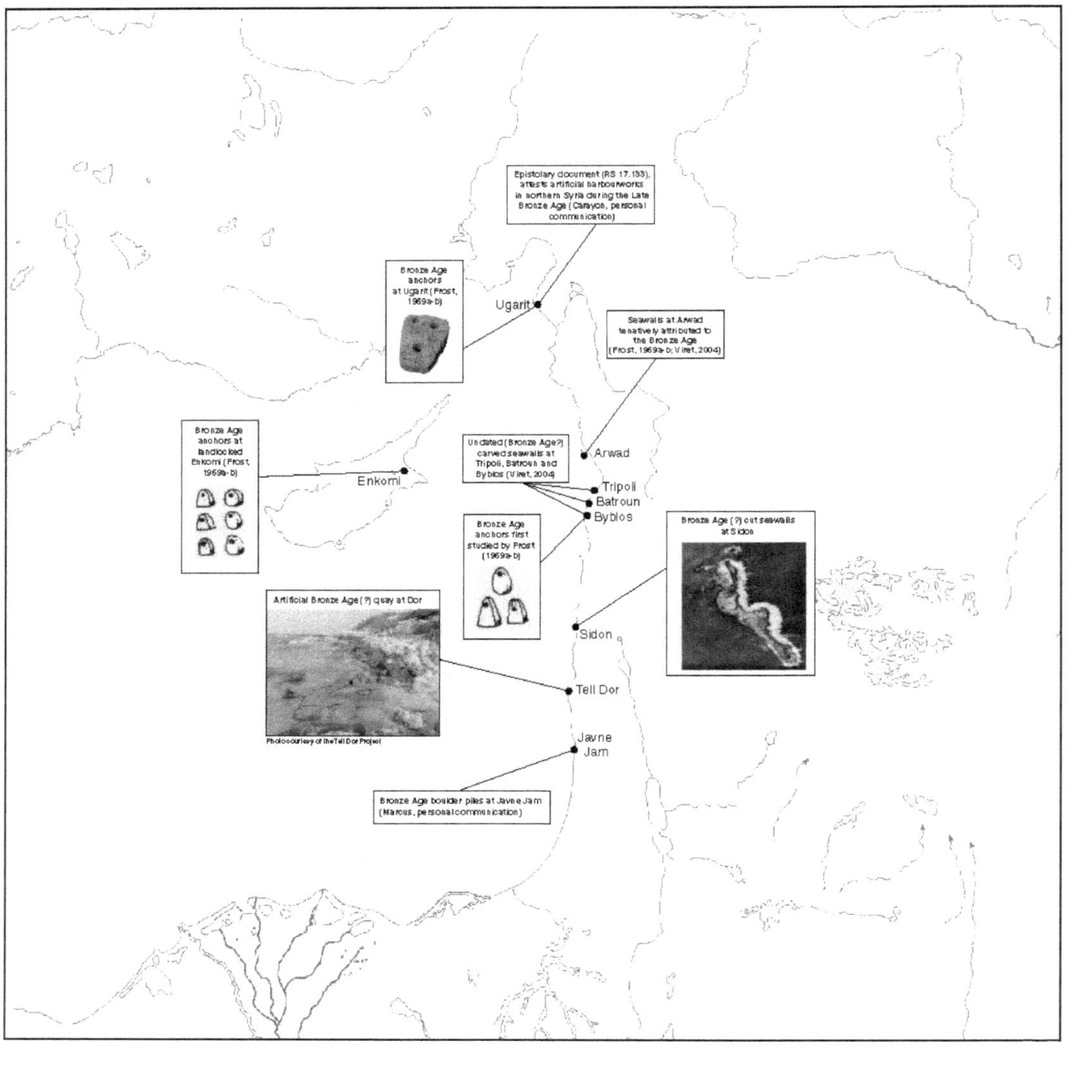

Figure 3.25: Tentative compilation of evidence for artificial Bronze Age harbourworks on the Levantine seaboard.

(2) Direct evidence. (i) For many, the most likely candidate for a constructed Bronze Age facility is the quay found at **Dor** in Israel (Kingsley, 1996). This quay is established along a coastal lagoon to the south of the settlement (**Figure 3.26**). Raban has dated the 35 m by 11-12 m platform to the thirteenth/twelfth centuries BC, on the basis of ceramics at the foot of the structure (Raban, 1984, 1985a, 1987b). The masonary techniques employed are typical of early Canaanite and Phoenician societies.

(ii) At **Yavne Yam**, Israel, recent work by Marcus (personal communication) has brought to light boulder piles on a submerged ridge, inferred to have been used to improve the quality of the ancient anchorage. (iii) On the northern Levantine coastline, Frost (1964, 1966) has also attributed the early harbour infrastructure at **Arrados** (Arwad, Syria) to the Bronze Age. Like Tyre, Arrados was an insular city founded upon a partially drowned sandstone ridge, 2.5 km from the present coastline. Extensive seawalls also are present on the island (Viret, 2005). The leeward side of the island forms a natural bay separated into two basins by a semi-artificial jetty. Renan (1864) and Frost (1966) have hypothesised that the southern basin was the larger of the two, silting up with coastal sediments since antiquity; (iv) At **Sidon**, modification of the sandstone ridge is constrained to the Middle to Late Bronze Age (~1700-1450 cal. BC) on the basis of sedimentological evidence (see chapter 4); (v) Other examples of carved seawalls are known from Tripoli, Batroun and Byblos but these have no clear chronological control and cannot be unequivocally attributed to the Bronze Age (Viret, 2004).

At Tyre, linking the coastal stratigraphy to early artificial harbourworks is difficult for two reasons: (1) the dearth of Bronze Age archaeological finds in and around the basin; and (2) the relative absence of Middle to Late Bronze Age harbour sediments, due to Roman dredging practices. We expand upon this point later in the chapter.

Figure 3.26: Top left: Speculated artificial Bronze Age quay at Dor (photograph: C. Morhange). Raban has dated the 35 m by 11-12 m platform to the thirteenth/twelfth centuries BC, on the basis of ceramics at the foot of the structure (Raban, 1984, 1985a, 1987b). Bottom left: Undated slipway at Dor (photograph: C. Morhange).

3.3.2.3 Phoenician and Persian periods: archiveless harbour

Description: Paradoxically, Graeco-Roman dredging means that very little stratigraphic evidence exists for Tyre's Iron Age harbours. Fine-grained clay deposits found to the south of the city, in the drowned southern quarters, have been dated to the Persian period (see below). On the basis of granulometric analyses, we hypothesise that these units are dredged Iron Age sediments from the northern harbour deposited in this area for use in the ceramics and construction industries. The sediments are fine-grained comprising 4-10 % gravels, 10-30 % sands and 60-90 % silts and clays, while the ostracod fauna is dominated by marine lagoonal taxa (*Loxoconcha* spp. and *Xestoleberis* spp.). Although great tracts of Iron Age strata are missing from the northern harbour, these litho- and biostratigraphic data attest to a well-protected harbour during the Persian period. This is corroborated by findings from Sidon's northern harbour (Marriner *et al.*, 2006b).

Interpretation: During the Late Bronze Age, destruction of the Minoan civilisation around 1400 BC and decline of the Egyptian empire left a power vacuum in the eastern Mediterranean's mercantile network (Hallager, 1988). It was at this time that Tyre consolidated its position as an important trade centre to emerge during the Iron Age as a commercial dominion with hegemony over the region's sea lanes (Aubet, 2001). The city drew its wealth from the coastal hinterland, namely metal, ivory, glass and cedar, commodities which it subsequently traded throughout the eastern Mediterranean. It also became an important transit seaport for Egyptian and Mesopotamian goods *en route* to other Mediterranean destinations. At the turn of the first millennium BC, Tyre had surpassed Byblos – Phoenicia's principal Bronze Age seaport – to become the Levant's most eminent trade emporia (Katzenstein, 1997; Aubet, 2001). This city of merchants had large marketplaces, the most prominent of which was centred on its northern harbour (Lehmann-Harteleben, 1923). Improvements in water storage also meant that the island could support much larger populations, and it has been speculated that during its golden age Tyre was the most populated city in the Levant, ahead of Jerusalem (Katzenstein, 1997).

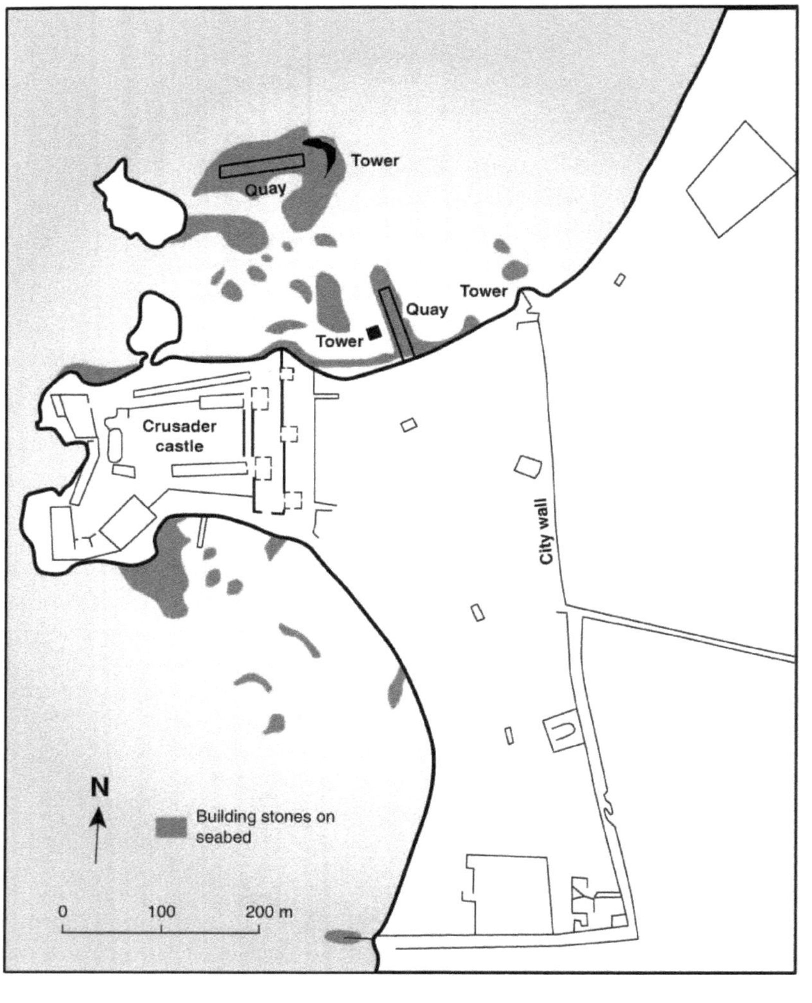

Figure 3.27: Athlit's ancient harbour (after Haggai, 2006). Like Tyre, Athlit's harbour lies north of a natural promontory and has yielded some of the best evidence for early Iron Age harbourworks on the Levatine coast. The Phoenician basin is separated into two areas (1) a southern quay and N-S pier; and (2) an eastern quay and northern mole.

Significant advances in naval technology have been attributed to this period (Bass, 1974; Casson, 1994; Kemp, 2001). Most importantly, the use of iron in shipbuilding, from the thirteenth century BC onwards, served as a major impetus to the construction of larger and stronger vessels. It is indeed the Tyrians who are attributed with the invention of the cargo-ship, also referred to as "ships of Tarshish" in the Bible (White, 2002). These new vessels, capable of sailing great distances, permitted trading outposts to be founded at various points throughout the Mediterranean basin (Aubet, 2001). Unmistakably, the development of maritime trade and rising population levels necessitated a complex port infrastructure containing installations for the overhauling of vessels, dockyards, quays and sheds in which boats could be kept. Direct evidence of this infrastructure is at present sparse and insights into Tyre's Phoenician and later Persian ports are marred by the relative absence of Iron Age sediments.

One must look to other Levantine harbours in attempt to better understand the construction methods and infrastructure used at this time. Archetypal Phoenician moles and quays are known from Athlit (Israel) and Tabbat al-Hammam (Syria). (1) The harbour at Athlit is the best preserved Phoenician seaport on the Levantine seaboard, and its remains have solicited the attentions of archaeologists looking to understand Iron Age building techniques and harbour engineering for a number of decades (Raban, 1985b, 1997a-b; Haggai, PhD in progress). Athlit's harbour lies north of a natural promontory; the Phoenician basin is separated into two areas (**Figure 3.27**): (i) **A southern quay and N-S pier**. This southern quay comprises narrow headers, 1.2 x 0.5 x 0.5 m, and extends eastward along the shoreline for a length of 38 m. A 10 m wide pier lies perpendicular to this quay and runs northwards 100 m into the sea. The structure was built of two walls of headers, with a mixture of ashlar and rubble fill. (ii) **An eastern quay and northern mole.** A 43 m long quay has been constructed on a partially drowned sandstone reef and is protected by an eastward trending mole that runs from the northern extremity of the quay. This structure is identical in construction to its counterpart on the southern shore. Haggai (2006) has recently dated wooden fragments contained within these two moles, constraining the inception of the structures to the ninth to eighth centuries BC. The separation of the harbour into two mooring areas enabled a distinction between the "home quay", for Phoenician ships to anchor, and the "free quay" or emporium for foreign ships. Cargo would have been transhipped in lighters to harbours such as Acre, Tyre and Sidon. (2) These chronological findings support similar evidence from Tabbat al-Hammam, where a 130 m long mole has been dated to the same period (Braidwood, 1940). The mole comprises a sheltered face made of ashlar headers that almost certainly served as a harbour quay. This artificially cut quay is protected by a more elevated backing wall that faces seawards.

At Sidon, the sandstone reef which encloses the ancient northern harbour appears to have been cut to create semi-artificial docking stations on the rock itself. Such docking facilities are still clearly visible on Zire island. The quality of this outer anchorage was improved during the Persian period, when two jetties were erected to protect a series of quay dockings. At present, however, it is unclear at what date the semi-artificial quays were fashioned (Carayon, 2003; Carayon and Viret, 2004).

Our palaeogeographical reconstructions for Iron Age Tyre evidence a ~3000 m long island (**Figure 3.23, Insets 4 and 5**). Archaeological finds such as amphorae and stone anchors corroborate the use of these reefs as outer harbours, functioning in tandem with the city's artificial anchorage(s) (El Amouri *et al.*, 2005).

3.3.2.4 Graeco-Roman harbour

Description: Transition from a medium grained sand unit to fine grained sands and silts is the most conspicuous geological evidence we have found for an artificially closed harbour at Tyre. This facies comprises a series of radiocarbon dates that cluster between ~2400 BP and ~2000 BP, consistent with the Graeco-Roman period. The sediment texture comprises 2-22 % silts and clays, 59-92 % sands and 2-30 % gravels. Unimodal histograms well developed in the fine sands, allied with skewness values of –0.16 to –0.52, attest to a low energy depositional environment (**Figures 3.10 and 3.18**).

The molluscan faunas comprise taxa from diverse biocenoses (**Figures 3.11, 3.16 and 3.19**). Despite the increasingly sheltered nature of the environment, significant numbers of *extra situ* species continue to be represented. These are indicative of either (1) the periodic incursion of strong marine currents; or (2) cultural inputs from boats and ships (shells caught in fishing nets etc.). *In situ* taxa are dominated by species from the upper clean-sand assemblage (*Cyclope neritea*, *Smaragdia viridis*, *Nassarius pygmaeus*, *Nassarius mutabilis*) and the upper muddy-sand assemblage (*Macoma cumana*, *Haminea hydatis*, *Loripes lacteus*).

Four diagnostic taxa dominate the ostracod fauna of this harbour facies: *Loxoconcha elliptica*, *Loxoconcha* spp., *Xestoleberis aurantia* and *Aurila convexa*. *Loxoconcha* spp. and *Xestoleberis aurantia* have oligo-mesohaline salinity ranges and prefer muddy to sandy substrates; they attest to a marine lagoonal environment (**Figures 3.15, 3.17 and 3.20**). In TV, presence of the brackish water species *Loxoconcha elliptica* is consistent with a sheltered harbour. Continued exposure to outer marine dynamics is manifested by low percentages of coastal and marine species such as *Aurila convexa*, *Aurila woodwardii*, *Cushmanidea* sp. and *Urocythereis oblonga*.

Interpretation: The northern harbour's geological record clearly translates a number of technological advances in port engineering, which can be attributed to the Romans

(Oleson, 1988). The most important of these breakthroughs was their mastery of concrete (Fitchen, 1988; Fletcher, 1996; Brandon, 1996, 1999; Garrison, 1998; Oleson *et al.*, 2004a-b). The Romans had two distinct types of concrete mortar:
(1) The first was made with simple lime and river sand, mixed at a ratio of three parts sand to one part lime, and was widely used in terrestrial construction (Ginouves, 1998; Lancaster, 2005). Its use in seaport contexts was more problematic as it could not harden underwater, which implied building structures on dry land or in a previously drained area (using cofferdams, compacted clays etc.).
(2) The other type used pozzolan instead of river sand and was mixed at a ratio of two parts pozzolan to one part lime (Humphrey *et al.*, 1998; Garrison, 1998). Pozzolan, or pozzolanic ash, is a siliceous and aluminous material which reacts with calcium hydroxide in the presence of water to form compounds possessing cementitious properties at ambient temperature (Mehta, 1991; Malhotra and Mehta, 1996). Its discovery was a revolution in harbour engineering, as it could set underwater. It derives its name from the Bay of Pozzuoli, where the ash is found in abundance (Fischer *et al.*, 1997); indeed, recent research suggests that the Pozzuoli vicinity was the primary source for Mediterranean pozzolana during Roman times (Oleson *et al.*, 2004a).

In his second book on ancient architecture, Vitruvius dedicates a whole chapter to the construction possibilities of this particular concrete (Morgan, 2000). He notes that pozzolan, "when mixed with lime and rubble, not only lends strength to buildings of other kinds, but even when piers of it are constructed in the sea, they set hard under water." Work at Caesarea has brought to light a prefabrication method which greatly facilitated the installation of harbour breakwaters and quay infrastructures (Brandon, 1996, 1997). The method involved building a double-walled wooden enclosure in the dry. This bottomless caisson was then floated to the desired location and the double walls were filled with hydraulic mortar, sinking the caisson on to a previously prepared base. It was then simply a matter of filling the caisson with pozzolana mortar to obtain a gigantic monolith. At Tyre, there is a paucity of data pertaining to the nature of harbourworks in antiquity. However, unlike Caesarea's outer harbour, which is exposed to the southern Levant's high-energy meteo-marine regime, Tyre was artificially protected by the partially drowned sandstone ridge. Large scale seaport enterprises were thus not required and we speculate that preceding structures were reinforced and added to at this time.

Whereas earlier Bronze Age and Iron Age societies exploited natural roadsteads to establish anchorages, the Romans' mastery of hydraulic concrete engendered significant new construction possibilities in coastal areas (Brandon, 1999; Oleson *et al.*, 2004a). It appears that by ~200 BC, they were using pozzolan concrete to line the harbourworks at Pozzuoli (Puteoli), which already indicates a striking degree of sophistication. The use of hydraulic concrete in Roman port construction reached its zenith during the first and second centuries AD as bear witness the totally artificial harbours at Caesarea, Cosa and Portus (McCann *et al.*, 1987; Oleson and Branton, 1992; Raban and Holum, 1996; Roller, 1998; McCann, 2002; Gazda, 2001; Oleson *et al.*, 2004b). Throughout the circum Mediterranean, geoarchaeological studies of Roman harbours indicate a diagnostic silt or plastic clay unit that is consistent with the significant technological advances of the period (Hesnard, 1995, 2004a-b; Morhange, 2001; Goiran, 2001; Goiran and Morhange, 2003; Giampaola *et al.*, 2004; Marriner and Morhange, 2006a-b; Marriner and Morhange, 2007).

At Tyre, there is presently very little archaeological information pertaining to the nature of the harbourworks in the northern cove. The main west-east trending breakwater shows evidence of Roman courses (Descamps, personal communication), and it seems probable that existing Iron Age port infrastructure would have been reinforced and added to at this time. The cluster of radiocarbon dates centred on the first century BC and second century AD, suggests a major overhaul of harbour infrastructure when Tyre fell under Roman control in 64 BC.

3.3.2.5 Byzantine harbour

Description: The late Roman and Byzantine periods are marked by transition to a fine grained silty sand facies. This second pronounced phase of harbour confinement is consistent with highly sophisticated port infrastructure creating a brackish lagoon type environment isolated from the sea. We observe a sharp decline in molluscan and ostracod species diversities. *In situ* molluscan taxa include individuals from the upper clean-sand assemblage (e.g. *Pirenella conica*, *Cyclope neritea*), the lagoonal assemblage, and the upper muddy-sand assemblage in sheltered areas (*Cerithium vulgatum*). The brackish water ostracod *Cyprideis torosa* attains relative abundance levels >90 %. Relative absence of outer marine species is a result of confinement from the open sea.

Interpretation: These litho- and biostratigraphical data all point to an infrastructural apogee during the Byzantine period, culminating in a protected hyposaline basin (Kjerfve and Magill, 1989). The Byzantines inherited from the Romans' rich legacy of engineering *savoir faire*. Many of the techniques were improved upon and consolidated during this period (Hohlfelder, 1997). Indeed, never again did such a well-protected basin exist at Tyre (Borrut, 1999-2000, 2001).

3.3.2.6 Semi-abandonment phase: sixth to eighth centuries AD

Description: The transition to unit A postdates the Byzantine period and is constrained to between the sixth to eighth centuries AD. The unit comprises a grey, shelly sand unit rich in gravels (3 % to 31 %) and coarse sands (58 % to 83 %), against a mere 9 % to 18 % for the silts and clays (**Figures 3.10 and 3.18**).

Cerithium vulgatum and *Pirenella conica* dominate the molluscan suites, with numerous secondary species from diverse biocenoses (*Ringicula auriculata*, *Nassarius pygmaeus*, *Gibberula miliaria*; **Figures 3.11, 3.16 and 3.19**). The increase in coastal ostracod taxa, such as *Urocythereis* sp. and *Aurila woodwardii*, is to the detriment of the formerly abundant lagoonal and marine lagoonal taxa of the Byzantine harbour (**Figures 3.15, 3.17 and 3.20**). This translates a re-exposure of the environment to the influence of the marine swell and currents. The tests of many of these individuals have been broken by wave action, consistent with a rise in energy dynamics due to the demise of the port. This semi-abandonment phase led to rapid coastal progradation diminishing the size of the basin by around 40 %. Seaward dislocation of the land accommodated urban growth during the medieval period, at which time it became Tyre's main market centre.

Interpretation: Presence of this coarse sand unit is a classic feature of semi-abandoned ports not only in the Levant, but throughout the Mediterranean (Marriner and Morhange, 2007). Explaining its presence at Tyre is complex and appears to be linked to a number of different phenomena. (1) **Cultural:** Byzantine control of Tyre lasted until the seventh century AD at which time a series of political crises (internal quarrels, Arab onslaught) forced the empire to shrink back to its Anatolian core (Norwich, 1993; Treadgold, 2000). By 650 AD Arab forces, unified under the banners of Islam and anxious to expand their faith, had conquered all of the Levant (up to Syria), Persia, and Egypt (Bonner, 2005). This epoch was marked by significant permutations in the eastern Mediterranean's trade network which directly impacted upon the area's seaports. (2) **Natural catastrophes:** the fourth to eleventh centuries AD are well-documented as being a period of tectonic and tsunamogenic instability on the Levantine coast (Guidoboni *et al.*, 1994). We discuss these factors in more detail in chapter 6. In the rest of this chapter, we look to elucidate new evidence for Tyre's other harbour complexes, notably the hypothesised southern basin, the city's outer harbours and the series of continental anchorages that functioned in tandem with the Tells Mashuk, Chawakir and Rachidiye.

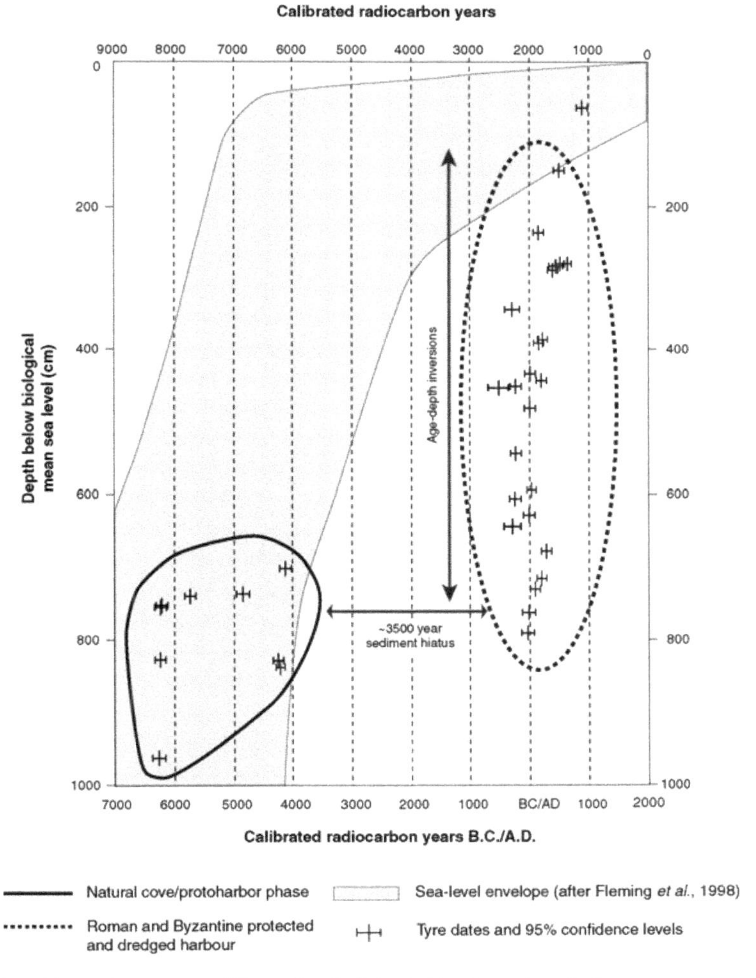

Figure 3.28: Chronostratigraphic evidence for Roman and Byzantine dredging in Tyre's northern harbour.

3.3.2.7 Roman and Byzantine dredging

Until recently, ancient harbours were considered to be quasi-continuous archives of human-environment interactions spanning the late Neolithic to present (Morhange, 2001; Goiran and Morhange, 2003). During the course of our project, this premise has come under scrutiny and indeed mounting evidence suggests that scholars must now nuance how they interpret these records. At Tyre and Sidon, persistent dating discrepancies have been evidenced throughout the fine-grained harbour units. Initially, these were problematic to interpret and evoked an unquantified reservoir effect for the coasts of southern Lebanon. In the absence of pre-1930 molluscan shells to more precisely elucidate this local reservoir age, we performed a series of radiocarbon dates on charcoal (2215 ± 30 BP; Poz-5777) and *in situ Loripes lacteus* molluscan shells (2505 ± 30 BP; Poz-5775) from the same stratigraphic layer, TIX 35. The offset confirms findings from elsewhere in the Mediterranean (Reimer and McCormac, 2002), and indicates that radiocarbon anomalies are not responsible for the empirical chronostratigraphic patterns. These data, coupled with mounting evidence from other ancient harbours (Marseilles and Naples), suggest the Romans significantly overhauled their seaports at the turn of the Christian era, notably removing great tracts of Bronze Age and Iron Age sediments (Hesnard, 2004a; Giampaola *et al.*, 2004; Giampaola and Carsana, 2005). This has created a stratigraphic paradox of archiveless Phoenician harbours. In this section we elucidate these findings.

Roman and Byzantine societies had a significant impact upon sedimentation patterns within Tyre's northern basin. Whilst in natural coastal systems, sediment accumulation rates and sediment supply converged towards a state of equilibrium, the influence of man in this low energy environment brought about a number of 'anthropogenic disequilibria'. Rapid rates of sedimentation, around 10 times greater than nearby natural coastlines, are recorded at this time. For example, for the period 6000 to 4000 BC, sedimentation rates of 0.5-1 mm/yr are contrasted against 10 mm/yr for the period 500 BC to 500 AD. The basin acted like a sediment sink, or depocentre, accumulating thick sequences of fine-grained sediments. Silting up rates of between 10 to 20 mm/yr have also been observed at Alexandria (Goiran, 2001) and Marseilles (Morhange *et al.*, 2003a). By Roman times the problem had grown so acute that it threatened the viability of Tyre's northern harbour and necessitated a clear management response. Although desilting infrastructure, such as sluice gates and channels, could have partially attenuated the problem in the long term these were relatively ineffective (Blackman, 1982a-b). Repeated dredging remained, therefore, the only means of creating artificial accommodation space and ensuring long-term harbour viability.

At Tyre there are two lines of evidence for these dredging practices (**Figure 3.28**): (1) **Numerous age-depth inversions** are observed in the ancient harbour facies. Radiocarbon dates from within this unit consistently cluster between ~500 BC to ~1000 AD. (2) **A pronounced sediment hiatus** between ~4000 BC to ~500 BC is not consistent with a natural base-level sediment sink. These data support removal of Middle Bronze Age to Persian period sediment strata, with deliberate overdeepening of the harbour bottom by Graeco-Roman and Byzantine societies.

Until recently, direct evidence for ancient dredging had been difficult to corroborate in the field, and consequently neglected in the literature (Hesnard, 2004a-b). These findings from Tyre support data from Marseilles and more recently Naples. Mounting evidence from other ancient Mediterranean seaports also supports similar practices.

3.3.2.7.1 Dredging evidence at Marseilles

Archaeological excavations at Marseilles have uncovered around 8000 m² of the city's buried port. Litho- and biostratigraphical studies elucidate a long history of human impacts stretching back to the late Neolithic period (Morhange, 1994; Morhange *et al.*, 2003a). Rapid shoreline progradation is recorded following the foundation of the Greek colony in 600 BC. The eventual demise and fall of the Phocean city, some 500 years later, is translated by wide-reaching changes in the spatial organisation of the harbour area during the first century BC. Although dredging phases are recorded from the third century BC onwards, the most extensive enterprises were undertaken during the first century AD, at which time huge tracts of Greek sediment were extracted down to a hard layer of oyster shells (Hesnard, 2004a). Notwithstanding the creation of artificial accommodation space, this was rapidly infilled and necessitated regular intervention. Repeated phases are evidenced up until late Roman times.

Figure 3.29: Jules Verne 3, a Roman dredging boat unearthed in Marseilles' ancient harbour. The vessel dates from the 1st to 2nd centuries AD. The central dredging well measures 255 cm by 50 cm. Photograph: C. Morhange.

3.3.2.7.2 Dredging evidence at Naples

At Naples, recent excavations at the Piazza Municipo show the absence of pre-fourth century BC layers due to extensive dredging between the fourth to second centuries BC (Giampaola *et al.*, 2004; Giampaola and Carsana, 2005). Unprecedented scouring marks, 165 to 180 cm wide and 30 to 50 cm deep, attest to powerful dredging technology which cut into the volcanic tufa substratum.

Dateable archaeological artefacts contained within the deposits allow for a very detailed time series of sediment fluxes to be established, with much greater temporal resolution than traditional radiometric methods. Investigated stratigraphic sections were dated to the third century BC and the beginning of the sixth century AD. Calculated fluxes are concurrent with centennial variability throughout this period. Rapid settling velocities of 17 to 20 mm/yr are recorded during the second century BC and the first and fifth centuries AD. Low sedimentation fluxes of 0 to 5 mm/yr are evidenced during the first century BC, and the late second and early fifth centuries AD. The most rapid rates are consistent with data from Archaic Marseilles (20 mm/yr; Morhange, 1994), Roman Alexandria (15 mm/yr; Goiran, 2001) and Roman and Byzantine Tyre (10 mm/yr; Marriner and Morhange, 2006a).

Figure 3.30: Example of a scouring talus at Marseilles. The highly cohesive nature of the harbour muds (>90 % silts) means these have been well-preserved in the stratigraphic record. Photograph: C. Morhange.

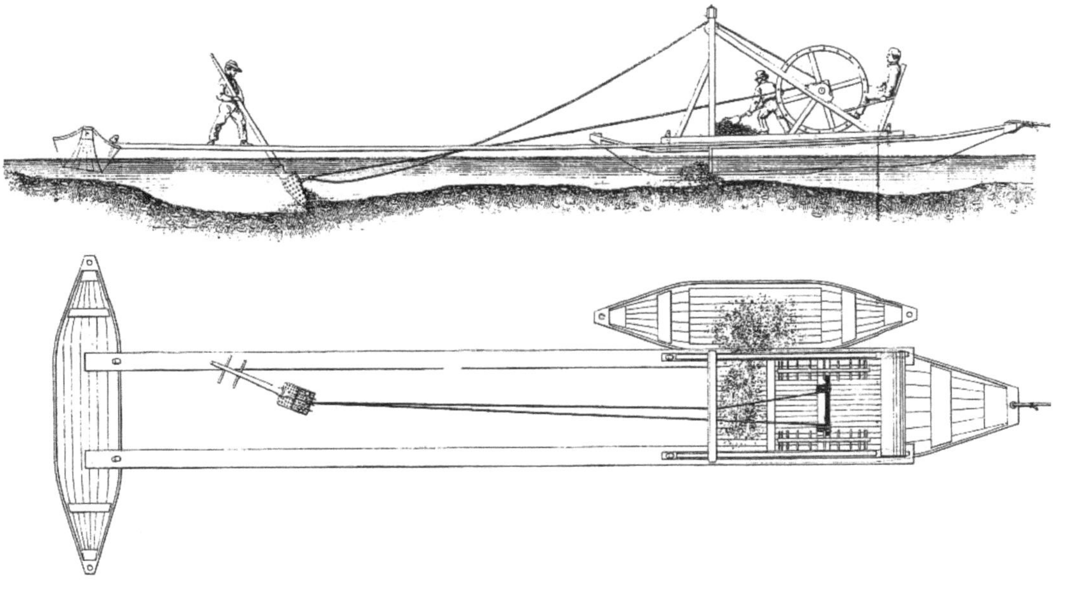

Figure 3.31: Eighteenth and nineteenth century dredging equipment used in European fluvial contexts. The vessel immediately adjacent to the 'pull boat' is equipped with a removable bottom, to facilitate the dumping of dredged material at a distal location. Engraving in Billaud and Marguet (2006).

Three possible explanations can be evoked for the contrasting detritic fluxes: (1) hypothetic changes in climatic conditions. Palaeo-climate records from around the circum Mediterranean demonstrate that the region has been subjected to a series of notable climatic fluctuations during the past 6000 years (Issar, 2003; Devillers, 2005; Duplessy *et al.*, 2005; Vella *et al.*, 2005; Robinson *et al.*, 2006; Rosen, 2006; Migowski *et al.*, in press). Climate cooling and increases in storm flooding can, for example, erode and mobilise sediment stocks (slope failure, mud flows) for deposition in base-level depocentres (Thomas and van Schalkwyk, 1993; Dubar and Anthony, 1995; Vandaele and Poesen, 1995; Nogueras *et al.*, 2000; Cerdan *et al.*, 2002; Wittenberg *et al.*, in press); (2) human agency and changes in land use patterns, both regionally in the watershed and locally in the urban settlement. The semi-desert environments of the Mediterranean region are most sensitive to climatic changes and human land exploitation (Schulte, 2002). Deforestation of river catchments from the late Neolithic onwards rendered clastic sediment stocks more susceptible to erosion (Chester and James, 1991; Arnaud-Fassetta *et al.*, 2000; Marchetti, 2002; Knox, 2006). In the Mediterranean, this process is accentuated by the region's high-intensity winter rainfall. Erosion of adobe infrastructure in coastal settlements can also yield significant amounts of sediment for deposition in the harbour basin (Rosen, 1986); or (3) dredging activity. This can be both a cause and response to changes in the empirical sediment fluxes.

3.3.2.7.3 How? Archaeological insights into dredging technology

The question of how the Romans dredged their harbours is an interesting one. Unequivocally, the mounting geoarchaeological data suggests that Roman dredging was a well-organized management technique, not as crude as previously speculated. Bed shear stress in cohesive harbour clays is considerable, and powerful vessels are inferred from the depth of scour marks and the volume of sediment removed. Three dredging boats, named Jules Verne 3, 4 and 5, have been unearthed and studied at Marseilles (Pomey, 1995). These were abandoned at the bottom of the harbour during the first and second centuries AD. All three vessels are characterised by an open central well that is inferred to have accommodated the dredging arm. For Jules Verne 3 (**Figure 3.29**), the boat's reconstructed length is ~16 m and the central well measures 255 cm long by 50 cm wide. Although the exact nature and mechanics of the dredging arms are not known, dredging taluses 50 to 60 cm wide and 30 to 50 cm deep have been fossilised in the stratigraphic record at Marseilles (**Figure 3.30**). No such dredging vessels have been unearthed at Naples, although analogous taluses and consistent chronostratigraphic inversions are widespread throughout the basin (**see Figure 1.33**). **Figure 3.31** depicts eighteenth and nineteenth century dredging equipment used in fluvial contexts. The nature of the fossilised dredging scours at Marseilles and Naples leads us to hypothesise that similar methods were employed during antiquity. It seems likely that the technique had evolved very little since this time.

3.4 In search of Tyre's southern harbour

Although the question of a southern seaport was not properly evoked until the nineteenth century (Kenrick, 1855; Poulain De Bossay, 1861; Guérin, 1880), drowned archaeological remains along the city's southern coastal fringe had long attracted the curiosity of early pilgrims and travellers. As early as the twelfth century AD, Benjamin of Tudel, a Jewish traveller in Phoenicia, described the towers, palaces, squares and streets he observed drowned on the southern coast of the city. Later Maundrell (1703), briefly touches on the subject when he observes that the northern and southern "bays are, in part, defended from the ocean, each by a long ridge, resembling a mole, stretching directly out, on both sides, from the head of the island; but these ridges, whether they were walls or rocks, whether the work of art or nature, I was too far distant to discern." The idea of building a double harbour is motivated by the fact that there are two main wind and offshore wave directions at Tyre. Indeed, dual seaports are a recurrent theme in Iron Age and later period maritime façades (Lehmann-Harteleben, 1923).

Figure 3.32: Ninth century BC depiction of Tyre from the Assyrian site of Balawat in Iraq. This particular representation of the Phoenician island raises a number of interesting palaeogeographical questions: (1) there are two gates in the offshore bastion. Are these a symbol of Tyre's twin harbours?; (2) secondly, the sea is clearly rendered between the offshore fortress and the mainland. However, today Tyre comprises a peninsula, connected to the mainland by a sandy isthmus; (3) finally, the beaching of the vessels evokes Bronze Age/Iron Age anchorage havens around Palaeo-Tyrus on the mainland (Tell Mashuk and Tell Rachidiye).

3.4.1 What is the epigraphic and iconographic evidence for two harbours at Tyre?

The development of inter-state trade during the Early Iron Age almost certainly demanded an expansion of the city's docking capacities (Katzenstein, 1997). Various lines of indirect evidence exist in support of two basins on the island. (1) The first, and perhaps most interesting from a palaeogeographical standpoint, derives from an Assyrian iconographic depiction of Tyre at Balawat (near Mosul in modern Iraq [**Figure 3.32**]). The upper register of this band shows Shalmaneser in the first full year of his reign (858 BC), receiving tribute from the Tyrians. The Phoenicians, distinguished by their pointed caps, ferry goods across from their island fortress to King Shalamaneser (?) who awaits on the mainland with a bow and arrow as symbols of the conqueror. The offshore bastion is represented by tall turreted walls with two gates leading into the city. Some scholars have interpreted these separate entrances as a symbol of the city's two ports (Barnett, 1969; Katzenstein, 1997). (2) The idea of double harbours on Tyre is further corroborated by an Old Testament description of Tyre as "throned above your harbours" (Ezek. 27: 3). (3) Finally, Arrian, writing at the time of Alexander the Great, noted that Tyre had two harbours, one a natural bay and the other artificial (Arrian II, 20, 10: "towards Sidon" and "towards Egypt" in Katzenstein, 1997 pg. 11).

3.4.2 Poidebard's southern harbour

By the turn of the twentieth century, two schools of thought were juxtaposed regarding the interpretation of these southern archaeological remains: (1) an 'ancient harbour school', with proponents such as Kenrick and De Bossay; and (2) a 'drowned city school', led notably by Renan. Significantly, technological limitations meant that neither group had strong archaeological evidence in which to ground their hypotheses.

Fascinated by the conjectures of his predecessors, Father Antoine Poidebard coupled aerial photography and diving surveys to investigate the underwater archaeology of Tyre's southern coastal fringe (Poidebard, 1939; **Figure 3.33**). His interest in Tyre began in 1934 when, as an officer of the French air force, he began aerially photographing the Phoenician coast. From this exceptional vantage point, Poidebard came to fully comprehend the rich archaeological potential of the city and saw scope to extrapolate the techniques he had developed working in the Syrian desert.

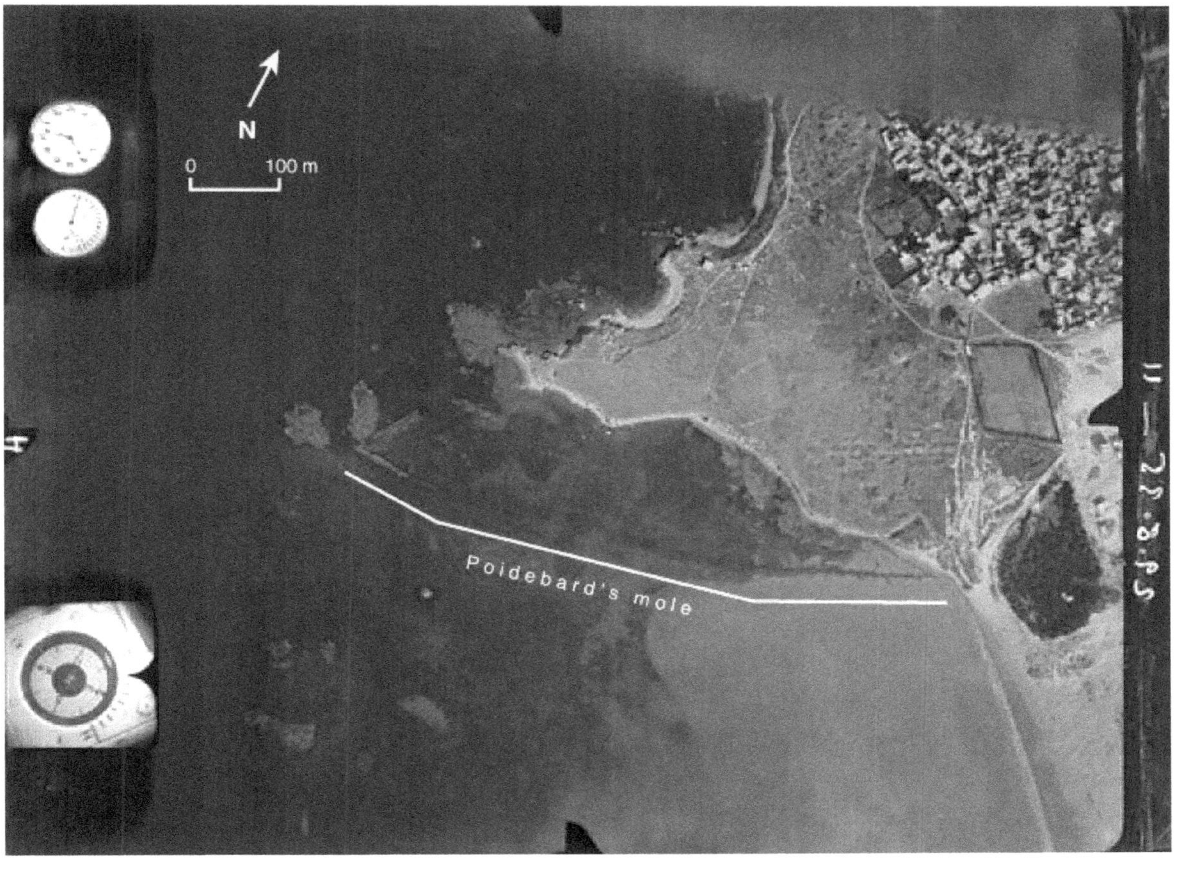

Figure 3.33: Aerial photograph of Tyre's southern basin and polder walls in the 1930s. Since the work of Poidebard it has been widely accepted that this southern basin was an ancient anchorage. Our new geoarchaeological data refute this hypothesis. Photograph: Poidebard (1939).

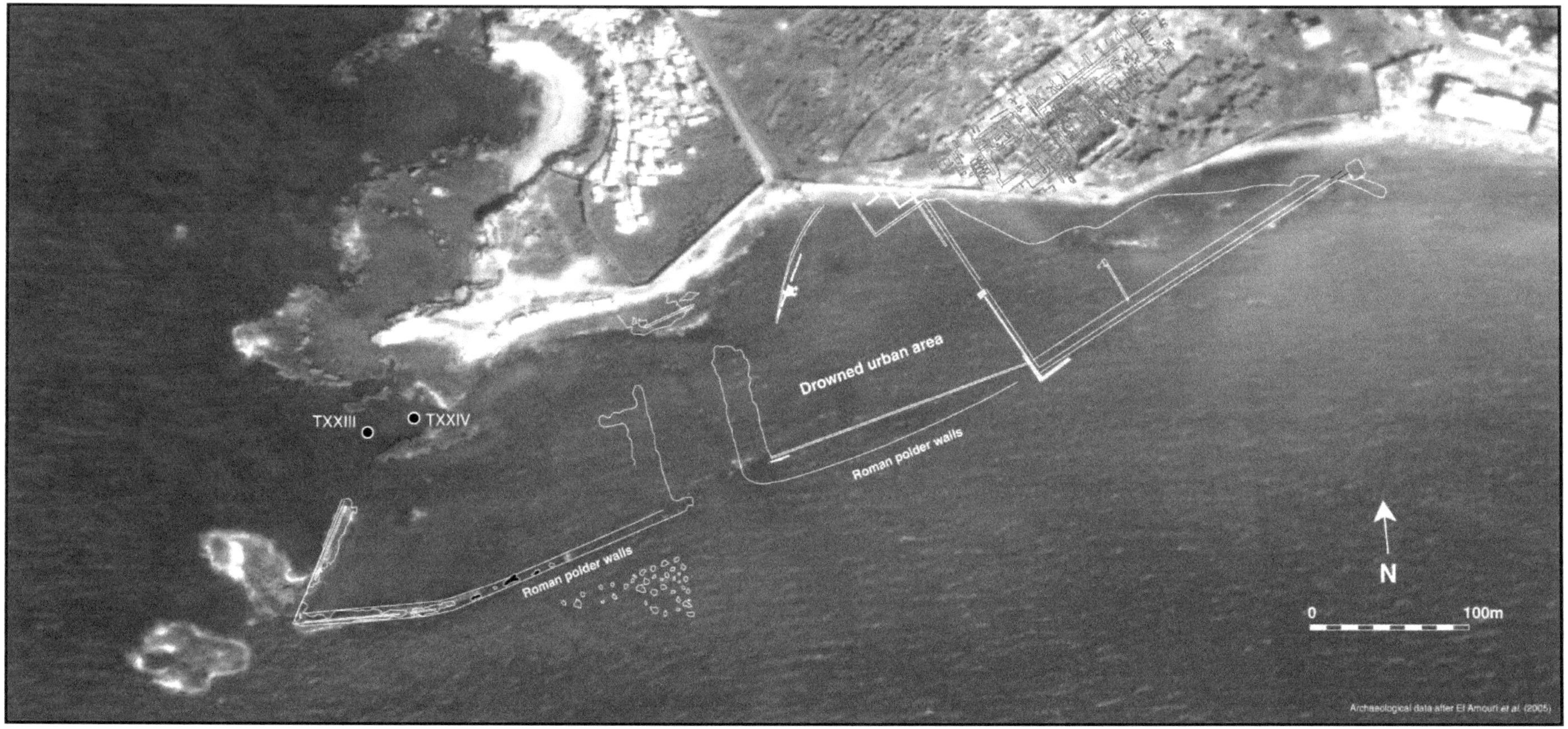

Figure 3.34: Aerial photograph of Tyre's southern harbour and outlay of its drowned polder walls (after Poidebard, 1939 and El Amouri *et al.*, 2005). The drowned southern basin has yielded extensive archaeological material pertaining to a submerged urban quarter still active during the late Roman period. Base image: DigitalGlobe.

Figure 3.35: Drowned urban structures dating from the Hellenistic and Roman periods, presently drowned in Tyre's southern basin. Photographs courtesy of M. El Amouri.

This southern area comprises an east-west trending basin around 700 m long by 200 m wide and runs parallel to the present island coastline (**Figure 3.34**). The marine bottom lies ~3-4 m below present mean sea level. Today, the area is sometimes used as a fair weather anchorage by local fishermen, but is at present not sufficiently protected to be considered a harbour basin *sensu stricto*. Indeed, it seems almost paradoxical that ancient societies should have built such a large harbour basin on the most exposed sector of the island.

After extensive research spanning 1934 to 1936, Poidebard eventually diagnosed what he believed to be an artificial harbour basin enclosed by an imposing 670 m long mole. For many scholars Tyre's southern seaport had been rediscovered and, despite questions raised by Honor Frost during the 1960s and 1970s (Frost, 1973), archaeologists and historians continue to promulgate Poidebard's (mis)interpretations to this day (Bikai, 1979; Bikai and Bikai 1987; Jidejian, 1996; Katzenstein, 1997; Kassis, 2005). Building on Frost's partial reinterpretation of Poidebard's southern harbour, new underwater research undertaken in 2002 has confirmed that this locale is in fact a drowned quarter of the ancient city (El Amouri *et al.*, 2005). Poidebard's supposed Iron Age mole has yielded late Roman ceramics, implying that it is almost certainly not of Phoenician age; El Amouri *et al.* (2005) have hypothesised this mole to be a polder wall. The presence of urban structures, walls and drowned quarries within the basin further call into question its use as a seaport during antiquity (**Figure 3.35**).

3.4.3 Reinterpretation of Tyre's supposed southern harbour

To independently test these archaeological data, we drilled two short offshore cores (TXXIII and TXIV; **Figures 3.36 to 3.41**) in Poidebard's supposed southern basin looking to answer a number of questions: (1) is there any sedimentological evidence for ancient harbour sediments in this area (be they Iron Age, Roman or Byzantine)?; and (2) if this basin was indeed an urban quarter, would it be possible to accurately date the collapse of the island?

3.4.3.1 Polder surface (unit C)

A very thin veneer (50 cm) of soft sediments was elucidated in this southern area (**Figures 3.36 and 3.39**). Unit C of cores TXXIII and TXXIV comprises coarse grained sands and poor sorting indices of between 1.2 and 1.6. Five taxa dominate the ostracod suites, *Loxoconcha rhomboidea*, *Xestoleberis aurantia* (marine lagoonal), *Aurila convexa*, *Aurila woodwardii* and *Urocythereis oblonga* (coastal). Faunal densities are low, between 20 to 40 tests per 10 g of sand (**Figures 3.37 and 3.40**). Diverse ecological groups are represented by the molluscan faunas, including the subtidal sands assemblage (*Tricolia pullus*, *Rissoa dolium*, *Mitra cornicula*), the hard substrate assemblage and the upper clean-sand assemblage (**Figures 3.38 and 3.41**). These litho- and biostratigraphical data appear consistent with polder infill, whereby nearby coastal sediments were imported to raise the general level of the land. Tracts of the island have been infilled since Phoenician times, to increase the habitable area of the ancient city (Katzenstein, 1997).

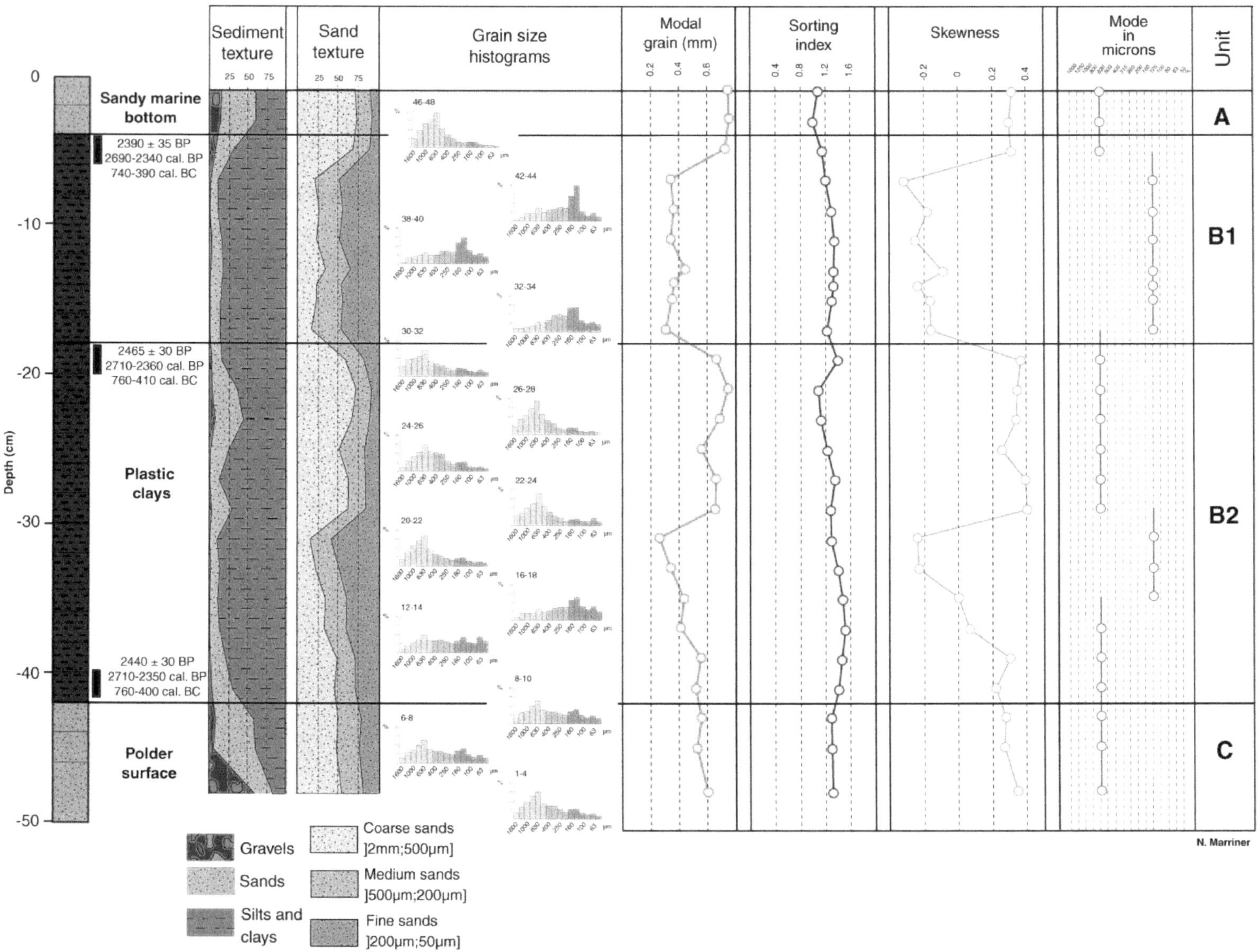

Figure 3.36: Grain size analyses of core TXXIV. (1) Unit C: comprises a coarse grain sand unit with modal grain values of ~0.6 mm. Poorly developed grain size histograms are diagnostic of a polder surface, with a reworking of coastal sediments by human societies. (2) Unit B2: is a plastic clays unit comprising >75% silts and clays. Poor sorting indices (~1.2-1.6) and the content of the gravel fractions, including seeds, charcoal and wood fragments, are very similar to the ancient harbours units observed in the northern harbour. (3) Unit B1: is a plastic clays unit similar to B2. It is, however, richer in fine sands (~40-50 %) with negative skewness values of ~-0.1 to -0.3. Both these units have been dated to around 2400 BP inferring that they were deposited contemporaneously. We hypothesize that these are dredged Iron Age harbour deposits from the northern basin, deposited in this area for use in the ceramics/construction industries. (4) Unit A: constitutes a coarse grain sand unit consistent with the present marine bottom.

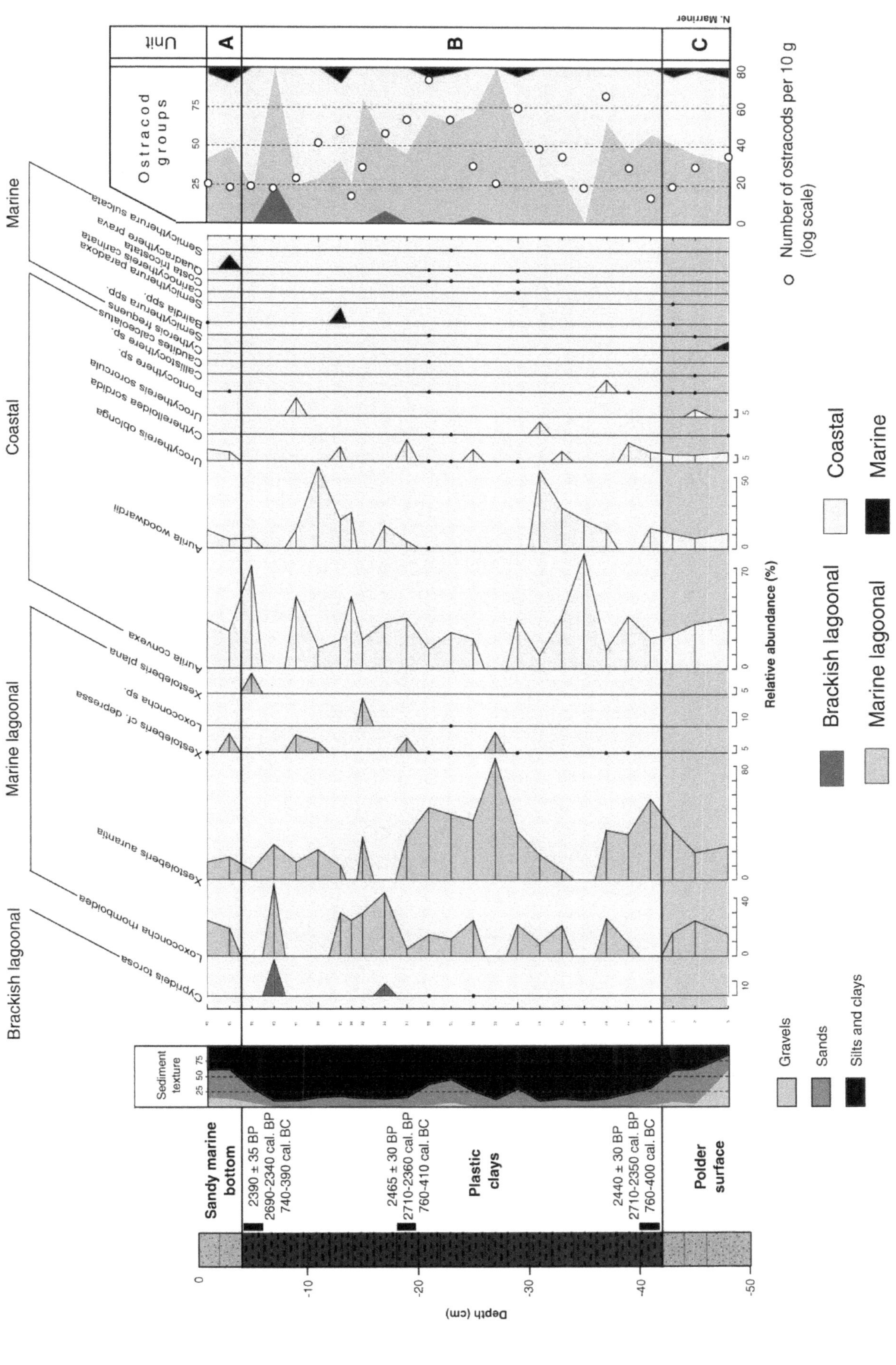

Figure 3.37: Ostracods for core TXXIV. Core TXXIV manifests very little biostratigraphic variation in its ostracod suites. The stratigraphy is dominated by *Loxoconcha rhomboidea*, *Xestoleberis aurantia* (marine lagoonal), *Aurila convexa* and *Aurila woodwardii* (coastal). The faunal density is poor, between ~20 and 80 tests for 10 g of sand.

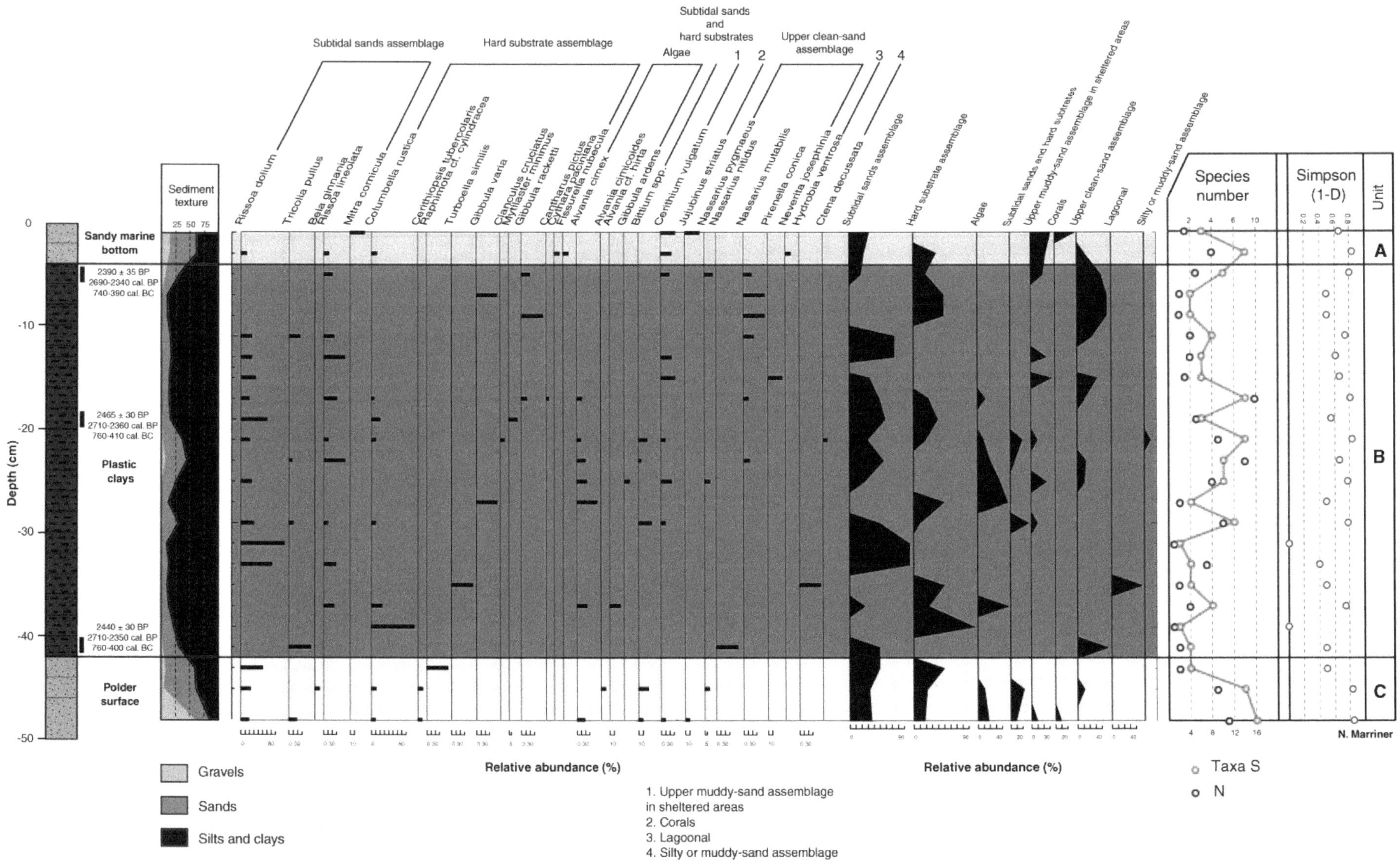

Figure 3.38: Macrofauna core TXXIV. There is very little biostratigraphic variation in the molluscan data. The dominant taxa include *Rissoa dolium*, *Rissoa lineolata* (subtidal sands assemblage), *Alvania* spp. (algal assemblage) and *Nassarius* spp. (upper clean-sand assemblage).

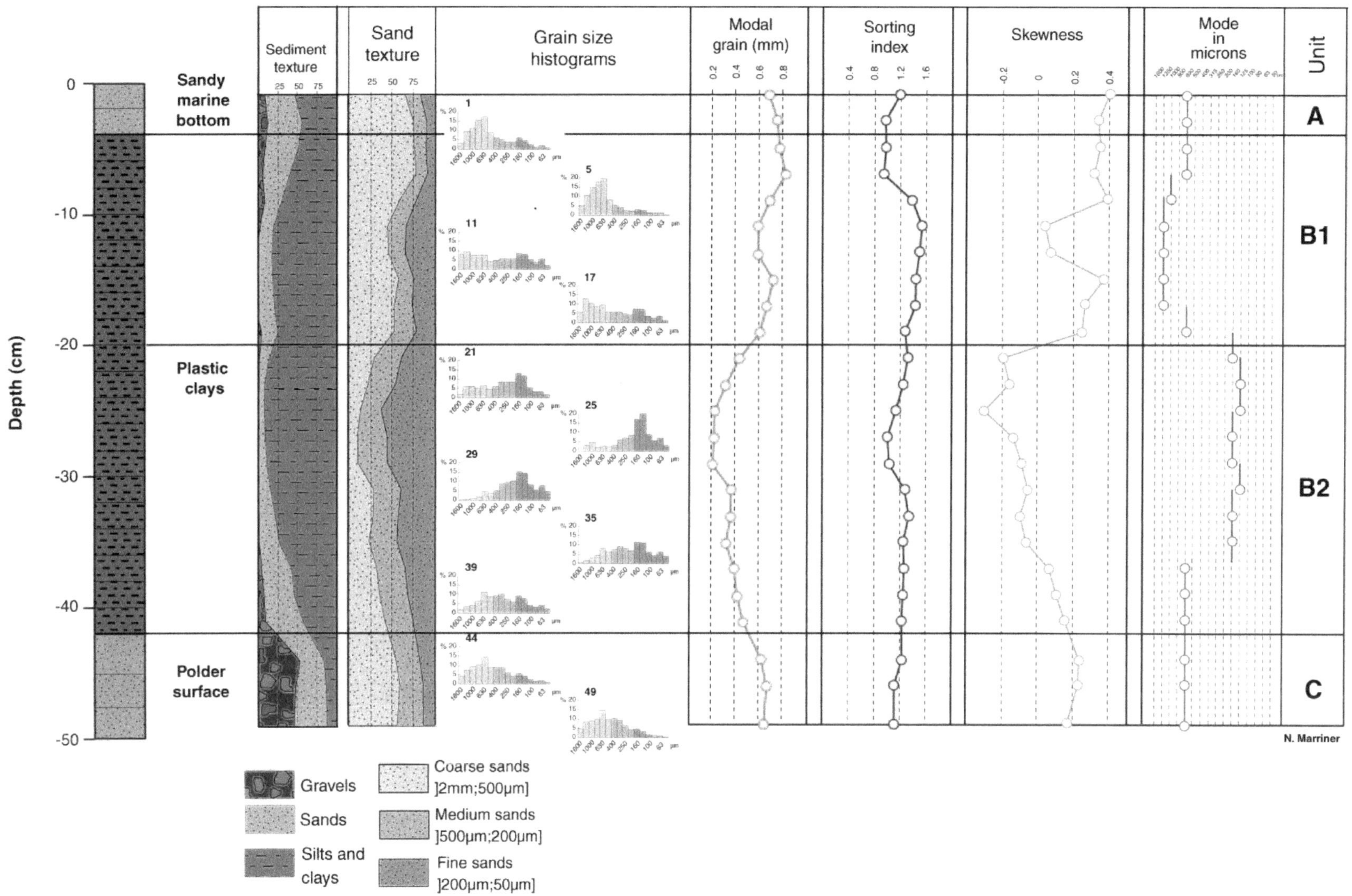

Figure 3.39: Grain size analyses of core TXXIII. (1) Unit C: is a poorly sorted gravely sand unit analogous to the polder surface observed in unit C of core TXXIV. (2) Unit B2: comprises a plastic clays unit. Poorly developed histograms and poor sorting indices of ~1.1-1.2 evoke a protected harbour environment. (3) Unit B1: is analogous to B2, but is dominated by the coarse sand fraction at 50-75 %. We interpret these units as dredged harbour deposits. (4) Unit A: A fall in the silts and clays fraction to <50 % is consistent with the present marine bottom.

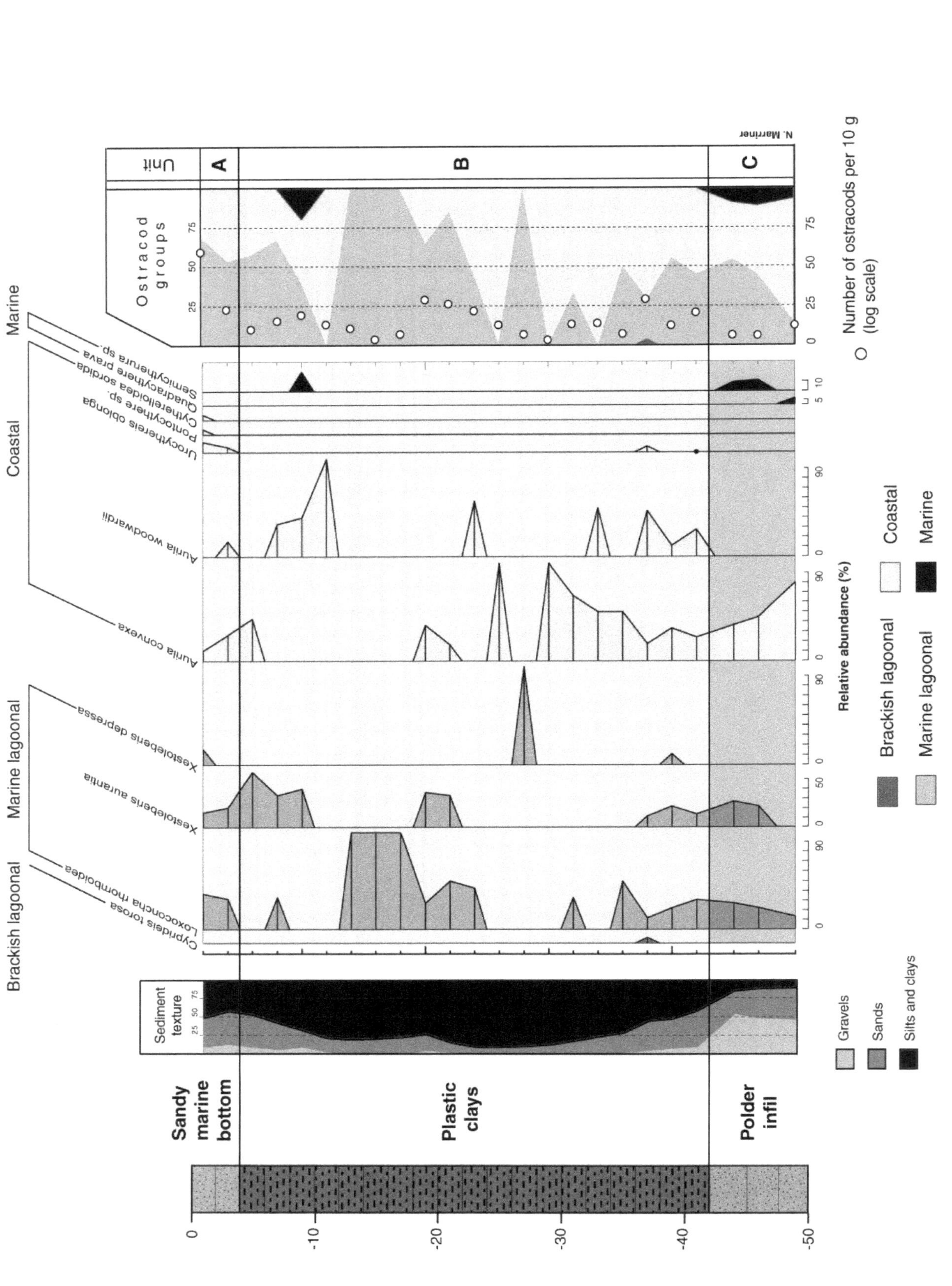

Figure 3.40: Ostracod data for core TXXIII. There is no overriding biostratigraphic pattern in the ostracod data. *Loxoconcha rhomboidea*, *Xestoleberis* spp. (marine lagoonal) and *Aurila* spp. (coastal) are diagnostic of a semi-protected near-shore environment.

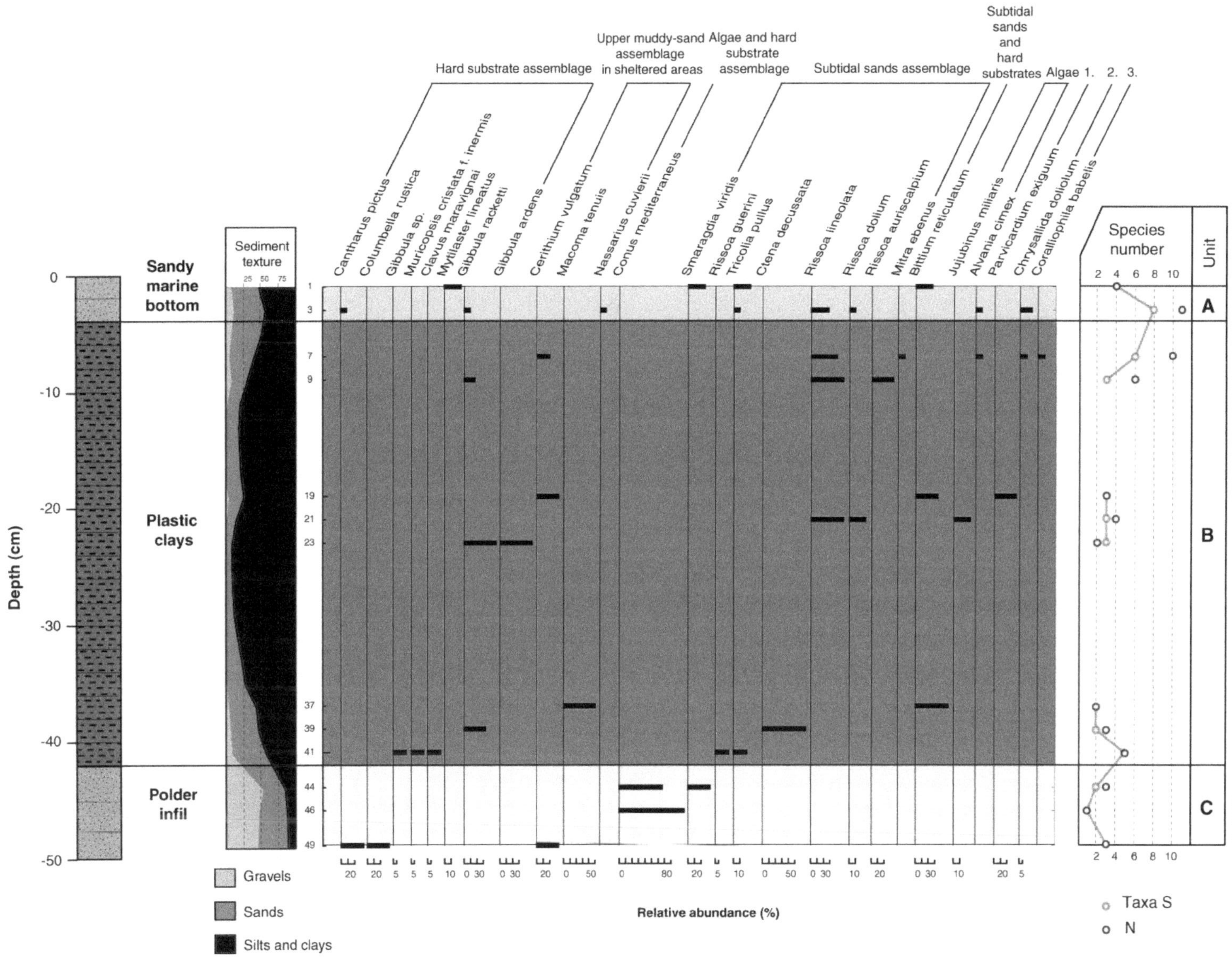

Figure 3.41: Macrofauna core TXXIII. (1) Unit C: constitutes taxa from four ecological groups, including *Cantharus pictus*, *Columbella rustica* (hard substrate assemblage), *Cerithium vulgatum* (upper muddy-sand assemblage in sheltered areas), *Conus mediterraneus* (algae and hard substrate assemblage) and *Smaragdia viridis* (subtidal sands assemblage). Poor faunal densities and the juxtaposition of ecological groups evoke artificially deposited polder sediments. (2) Unit B: comprises taxa typical of a sheltered marine environment, notably, *Cerithium vulgatum* (upper muddy-sand assemblage in sheltered areas), *Tricolia pullus*, *Rissoa* spp., *Mitra ebenus* (subtidal sands assemblage) and *Bittium reticulatum* (subtidal sands and hard substrate assemblage). These sediments evoke dredged harbour deposits. (3) Unit C: has a poor faunal density and juxtaposes species from diverse ecological environments (hard substrate assemblage, subtidal sands assemblage), typical of a medium energy marine bottom.

3.4.3.2 Dredged Iron Age harbour deposits (unit B)?

The coarse grained sand deposits of unit C are overlain by low energy fine sands and silts. A sharp rise is observed in the silts and clays fraction to between 80-90 % of the total sediment texture. Poorly developed histograms allied with the juxtaposition of coarse sands and fine-grained silts are typical of an ancient harbour deposit (Morhange, 2001). The gravels fraction comprises numerous rootlets, charcoal, wood fragments and seeds analogous to deposits found in Tyre's northern harbour during the Graeco-Roman and Byzantine periods. There is very little biostratigraphic variation in the ostracod faunas, and the unit is dominated by marine lagoonal and coastal taxa consistent with a semi-protected environment. The subtidal sands and hard substrate assemblages characterise the molluscs. Three analogous dates from core TXXIV constrain the chronology of this unit to the Persian period (750 to 400 cal. years BC). The absence of any notable bio- and chronostratigraphic variations suggests that the unit was deposited contemporaneously.

We advance two hypotheses to explain the origin of these fine-grained deposits.

(1) **Tectonic subsidence** of the coastal ridge drowned this southern urban quarter during antiquity leading to the formation of a semi-protected enclave sheltered by the polder walls. New archaeological and stratigraphical data support tectonic collapse of ~3 m during antiquity. Two types of evidence exist. (i) Archaeological: Tyre's northern Roman mole is currently 2.5 m below present sea level, translating a subsidence of ~3-3.5 m (Noureddine and Helou, 2005; Descamps, personal communication). On the southern shore, walls and drowned quarries at 2.5 m below MSL have also been elucidated (Frost, 1973; El Amouri *et al.*, 2005). (ii) Stratigraphic: Similar subsidence is translated in the city's coastal stratigraphy, notably a ~3 m offset in sediment accommodation space and radiocarbon dated strata between Tyre and Sidon. The archaeology suggests that this area of the ancient city was still fully functional during the Hellenistic and Roman periods (El Amouri *et al.*, 2005), therefore contradicting evidence for island collapse during the Persian period. Our palaeogeographical reconstructions attest to a 2800 m (N-S) by 800 m (E-W) Hellenistic/Roman island compared to 1000 m (N-S) by 700 m (E-W) after the late Roman collapse (**Figure 3.21**). In light of these chronological discrepancies, how can one explain the presence of marine harbour deposits on dry land?

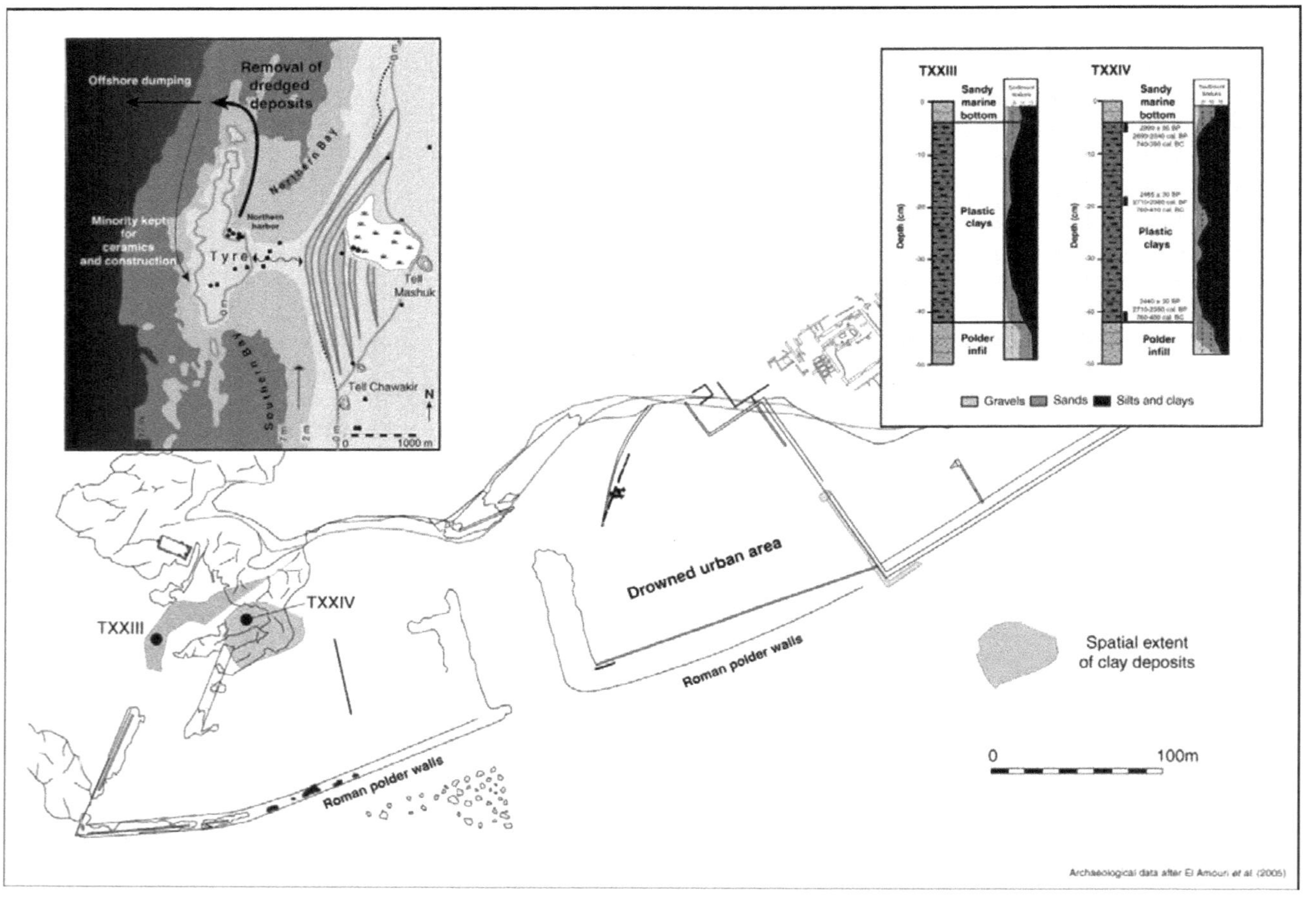

Figure 3.42: Spatial extent of dredged clay deposits in Tyre's southern drowned basin (archaeological data from El Amouri *et al.*, 2005).

(2) This question leads us on to a second, more plausible hypothesis. We speculate that these deposits comprise **dredged Iron Age material** from the northern harbour, deposited in this area for use in the ceramic and construction industries (**Figure 3.42**). This hypothesis is further corroborated by the relatively localised distribution of the clay deposits and their proximity to the Greek/Roman shoreline of the period (**Figure 3.21**). Semblable examples of clay mounds are known from a number of sites on the Phoenician coast, and the technique is still used by artisanal industries on the Nile delta (Leclant, 2005). It appears likely, therefore, that some of the dredging vessels deposited small sediment loads from the northern harbour in this area. El Amouri *et al.* (2005) have noted that these clay mounds are spatially localised, constituting <5 % of the total basin area. Evidence from Marseilles suggests that *marie-salopes* were important in transporting dredged material (Pomey and Rieth, 2005). *Marie-salopes* are designed to accommodate dredged sands and silts extracted by dredging. Once the vessel had been filled, it would either offload its deposit on land, or dump the material offshore via a specially conceived trap in the hull of the boat (Pomey, 1995; Long, personal communication).

3.4.3.3 Present marine bottom (unit A)

Coarse grained sands characterise unit A of cores TXXIII and TXXIV. This thin cap of sediments is consistent with the present marine bottom. It has not been possible to accurately date the submergence of this southern area on purely radiometric grounds. The archaeology, on the other hand, constrains this tectonic collapse to the late Roman period. As today, it is plausible that the drowned basin could have been used as a fair weather harbour by small vessels during the Byzantine and later periods. There is no evidence to suggest, however, any elaborate seaport infrastructure and Poidebard's initial interpretations can now be irrefutably rejected.

3.5 Where was Tyre's southern harbour?

Unlike many of his contemporaries, Renan (1864) always questioned the validity of a southern harbour on the exposed fringe of the island. Based on his knowledge and observations of the site, he conjectured that the most plausible location for a second anchorage lay on the southeastern side of the island, in an area presently covered by thick tracts of coastal sediments. We employed two techniques to test the validity of this hypothesis (1) numerical models of swell and wind-induced currents around Tyre island; and (2) coastal stratigraphy.

3.5.1 Numerical current models: what can they tell us about Tyrian anchorages in antiquity?

We ran a series of swell and wind-induced current models working in collaboration with B. Millet and S. Meulé from the Université Aix-Marseille. The main corpus of these results are presented and discussed in relation to the Holocene morphogenesis of Tyre's tombolo (chapter 2), however the data are also interesting in understanding the distribution of anchorage havens around Tyre island and the subsequent choices made by human societies.

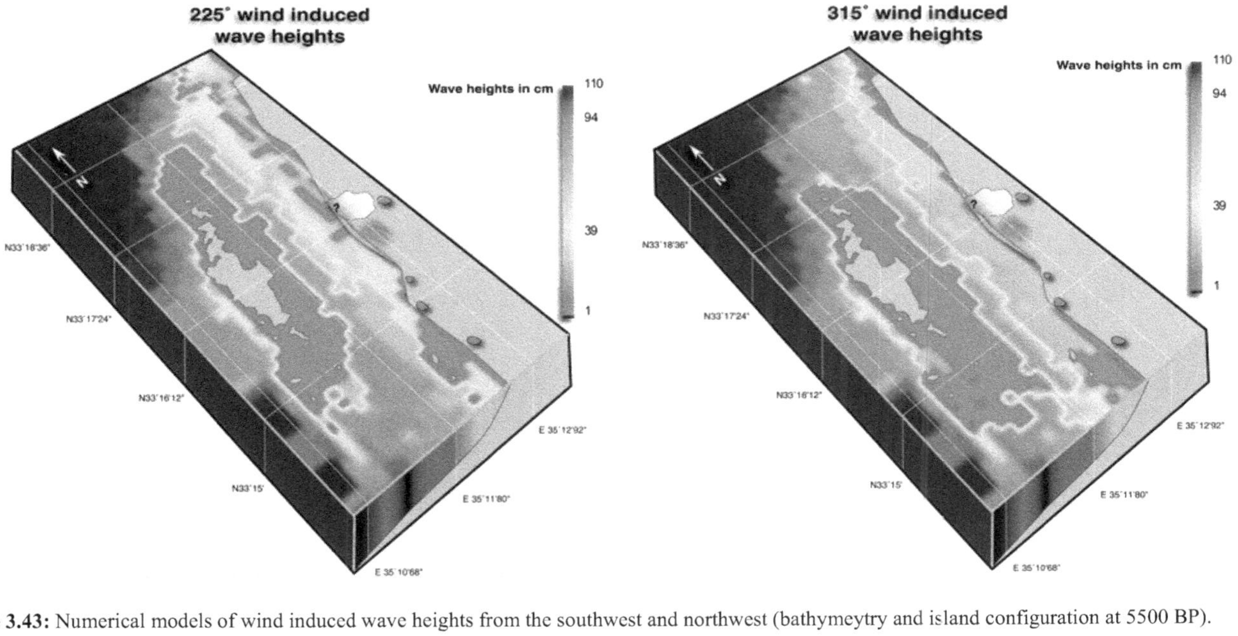

Figure 3.43: Numerical models of wind induced wave heights from the southwest and northwest (bathymeytry and island configuration at 5500 BP).

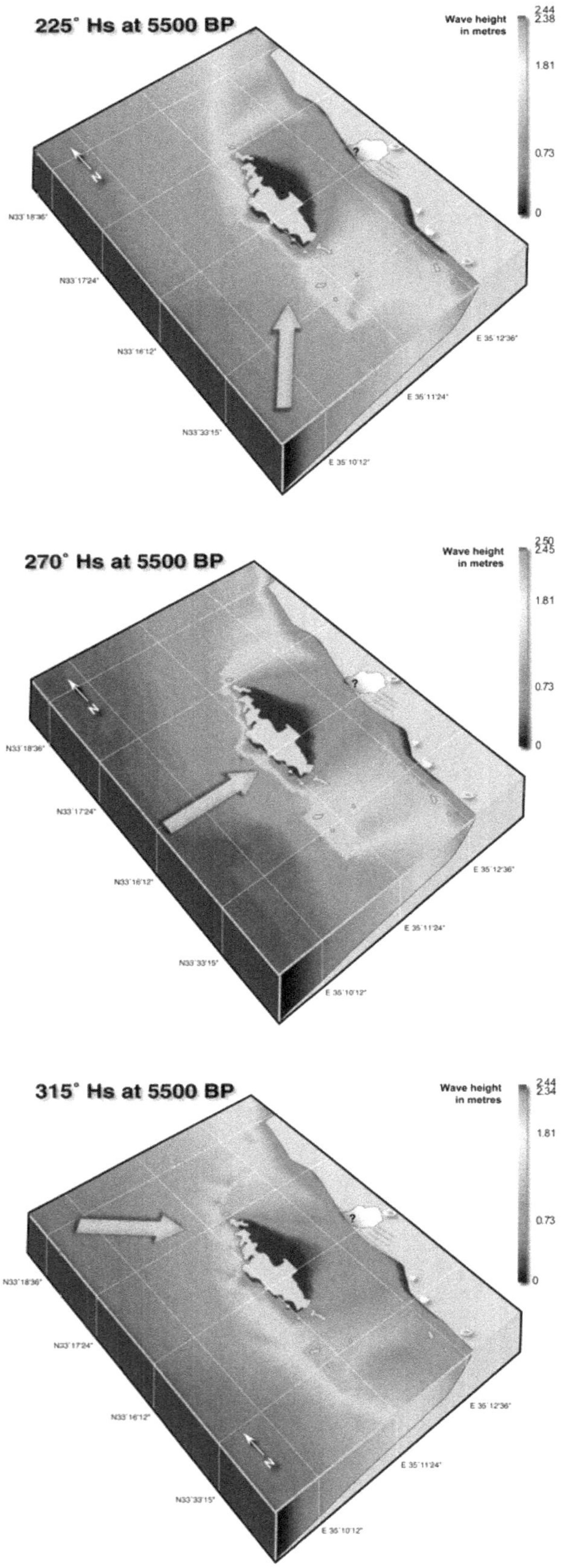

Figure 3.44: Numerical models of swell induced wave heights at Tyre (bathymetry and island configuration at 5500 BP). Note the breakwater effect of the island bastion that generates a low energy shadow on the leeward side, conducive to the establishment of a number of anchorage havens.

Two separate models were run (1) the first to elucidate the effect of **wind-induced currents** in forcing coastal processes around the breakwater (5500 BP, **Figure 3.43**) and (2) a **swell propogation** model to better comprehend the diffraction and refraction of swell energy around the breakwater island (5500 BP and present, **Figure 3.44**). Runs were performed for the dominant wind and swell directions from the southwest, west and northwest. For the historic period and antiquity, both models show the existence of a significant wave shadow on the lee of Tyre island generated by the offshore breakwater and reefs system. The incoming swell, which is initially quasi-unidirectional (225°, 270° or 315°), is spread by the lateral movement of energy into the shadowed area behind Tyre island. Modelled scenarios show that the majority of the swell incident energy was blocked by the breakwater obstacle. Wave energy passes the breakwater obstacle and spreads laterally into the shadow region where the main anchorages are to be found. The total energy is distributed over a broader area reducing the swell height leeward of the island from >1.80 m west of the island to <0.75 m on the eastern shadow side.

During antiquity, this large 3 km by 1.5 km wave shadow would have favoured a number of secondary anchorages on Tyre. One of the lowest energy zones lies in the area identified by Renan as sheltering the southern harbour (see **Figure 3.2**); this area also lies in close proximity to the city centre. The numerical models show that the reefs extending to the north and south of the city could have been used as offshore anchorages in the same way as Zire island at Sidon. This is corroborated by extensive anchorage finds from around these outer reef systems (El Amouri *et al.*, 2005). Exploitation of such an outer harbour system would have greatly increased the docking capacity of the city.

3.5.2 What can coastal stratigraphy tell us about Tyre's southern harbour?

A number of cores were also drilled on the southeastern fringe, in an area hypothesised by Renan (1864) to have sheltered Tyre's southern harbour. This area attests to a complex stratigraphic history characterised by the accretion of natural coastal sediments juxtaposed against a series of infill phases, to raise the surface of the ancient city during the Hellenistic and Roman periods.

3.5.2.1 Renan's southeastern harbour

3.5.2.1.1 TVIII marine transgression

The marine transgression in core TVIII is dated 6300 ± 40 BP (6880-6650 cal. BP). The unit comprises fine-grained sands (**Figure 4.45**) with molluscan taxa from the subtidal sands and hard substrates (*Bittium reticulatum*), upper muddy-sand assemblage in sheltered areas (*Cerithium vulgatum*) and the subtidal sands assemblage (*Rissoa lineolata*, *Rissoa dolium* [**Figure 4.46**]). A high faunal density population (~1000 tests per 10 g sand) of *Cyprideis torosa* dominates the ostracod fauna (**Figure 4.47**).

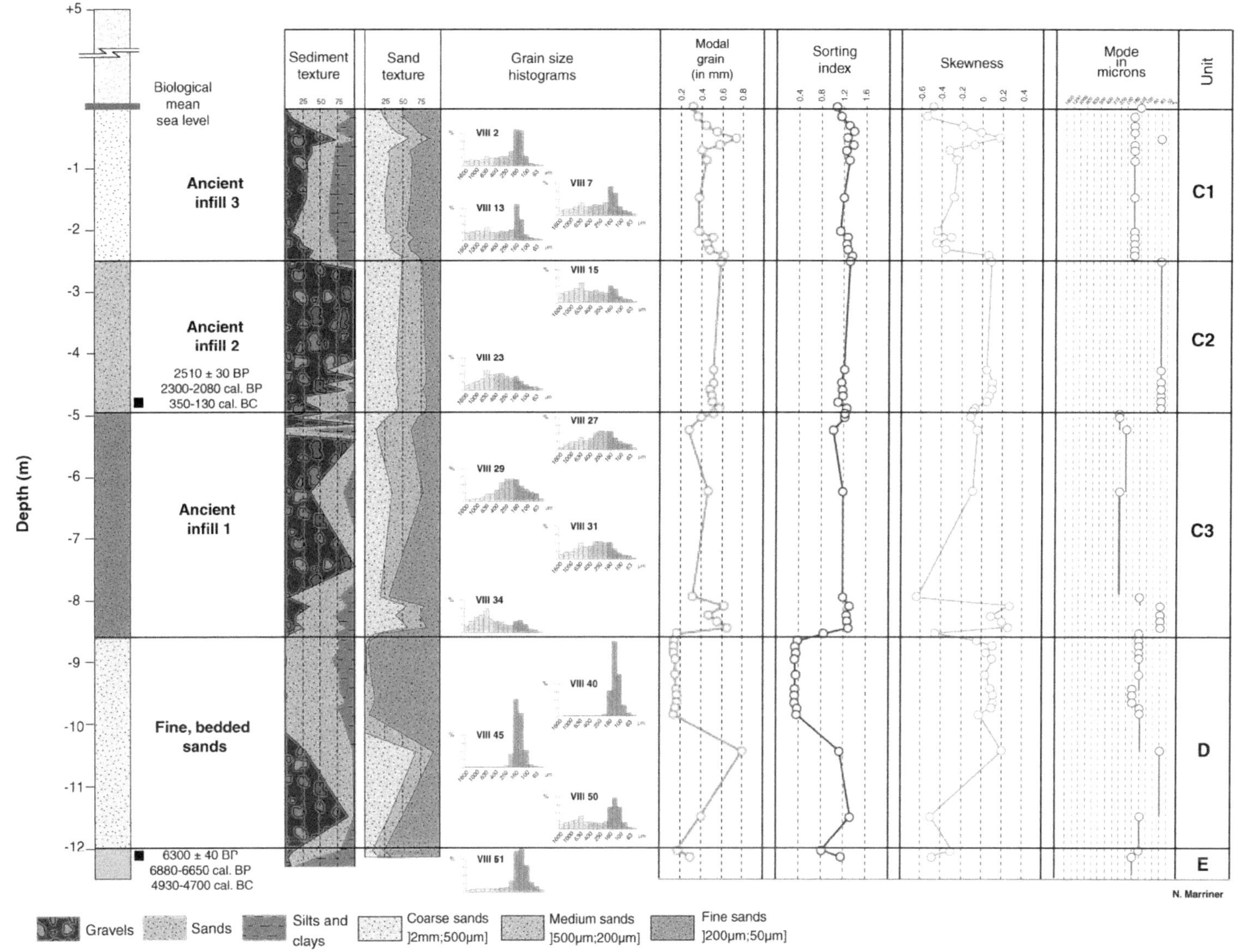

Figure 3.45: Grain size analyses core TVIII. (1) Unit E: is a plastic clays unit rich in shelly debris. We infer a sheltered lagoon type environment behind the island breakwater. (2) Unit D: After ~6300 BP transition to a bedded sands unit evokes full marine conditions. The base of the unit comprises poor sorting indices of ~1.2 and an important gravel component. This evokes the reworking of coastal deposits during the transgression of the continental shelf. Transition to a well-sorted fine sand unit (~0.2) attests to a sheltered beach environment in the wave shadow of Tyre island. (3) Units C: are characterised by poorly defined histograms and poor sorting indices of ~1.2. The juxtaposition of coarse gravels and sands with finer grain material is diagnostic of artificial polder deposits. Unit C2 has been dated to 2500 BP, or the Hellenistic period. The reworked nature of these deposits means obtaining a well-constrained chronology has been problematic.

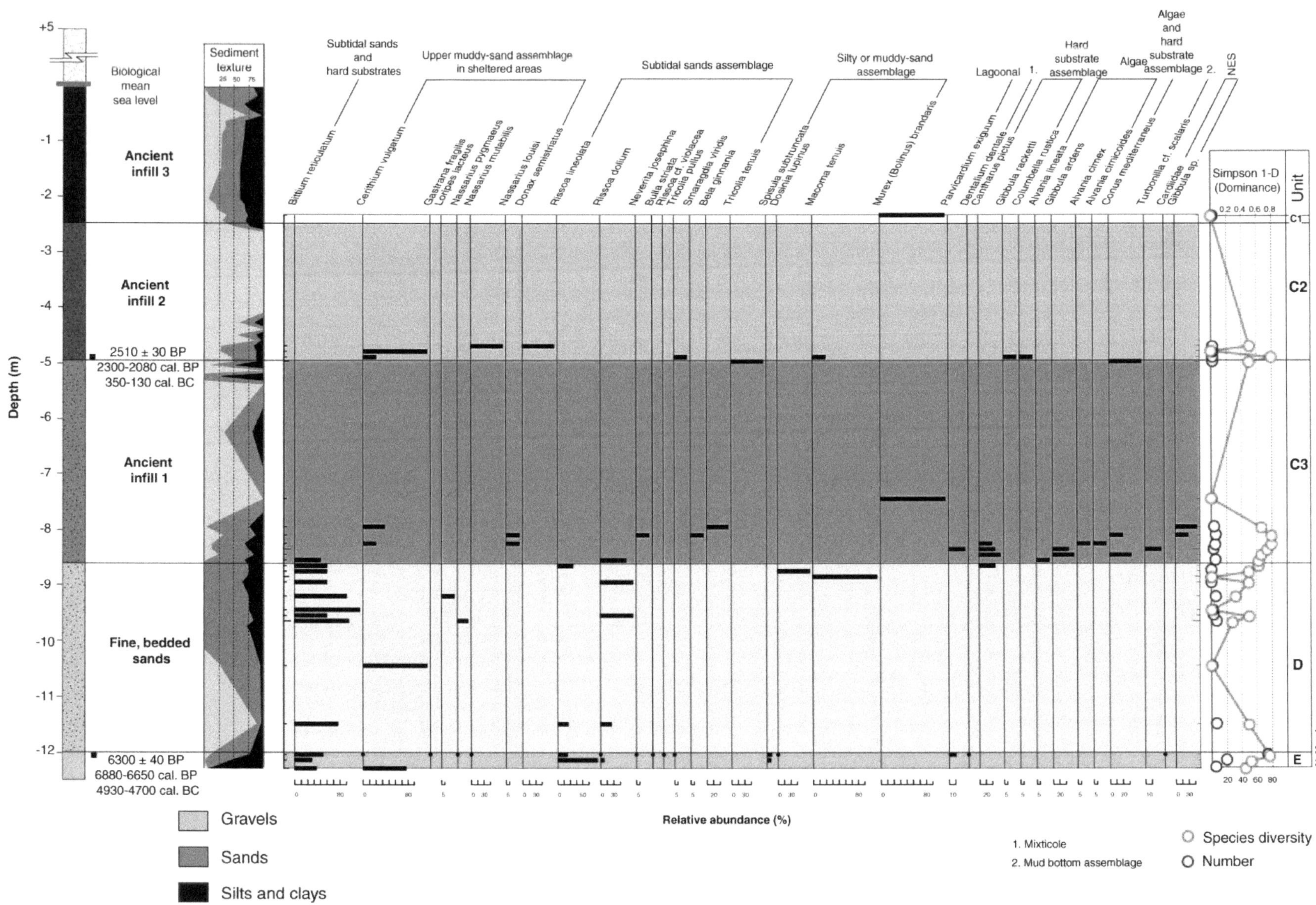

Figure 3.46: Macrofauna core TVIII. (1) Unit E: is dominated by three taxa, *Bittium reticulatum* (subtidal sands and hard substrates), *Cerithium vulgatum* (upper muddy-sand assemblage in sheltered areas) and *Rissoa lineolata* (subtidal sands assemblage). A protected environment is inferred from these data. (2) Unit D: comprises taxa from four main groups. These include *Bittium reticulatum* (subtidal sands and hard substrates), *Cerithium vulgatum*, *Loripes lacteus*, *Nassarius pygmaeus* (upper muddy-sand assemblage in sheltered areas), *Rissoa dolium* (subtidal sands assemblage) and *Macoma tenius* (silty or muddy-sand assemblage). These data evoke a low energy beach environment. (3) Units C: Low faunal densities and the juxtaposition of species from diverse ecological environments are consistent with artificially deposited polder sediments.

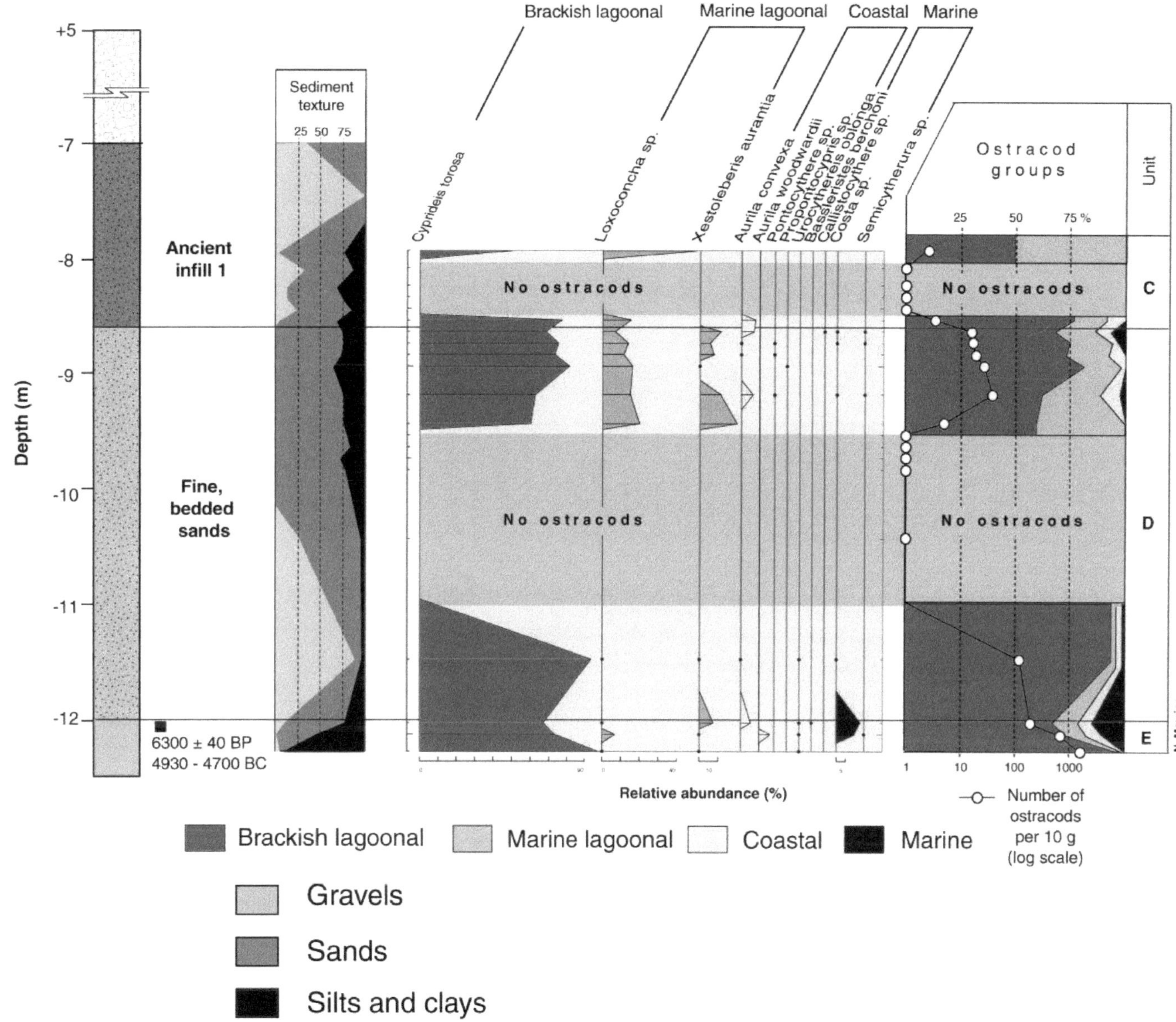

Figure 3.47: Ostracods core TVIII. There is very little biostratigraphic variation in units E and D, dominated by the brackish lagoonal species *Cyprideis torosa*. Secondary species include *Loxoconcha* sp. and *Xestoleberis aurantia*. The overlying polder deposits are void of ostracod fauna. These data corroborate a low energy environment during antiquity, in the wave shadow of Tyre island.

3.5.2.1.2 Low energy marine environment

In unit D, transition to a fine-bedded sands unit corroborates a sheltered marine environment. Fine-grained sands and sorting indices of ~0.4 are consistent with a well-sorted subtidal beach environment. Four assemblages dominate the molluscan faunas: subtidal sands and hard substrates (*Bittium reticulatum*), upper muddy-sand assemblage in sheltered areas (*Loripes lacteus*, *Gastrana fragilis*, *Cerithium vulgatum*), silty or muddy-sand assemblage (*Rissoa dolium*, *Rissoa lineolata*) and the subtidal sands assemblage (*Macoma tenuis*, *Dosinia lupinus*). The bottom part of unit C contained no ostracod faunas. The brackish water species *Cyprideis torosa* dominates at the top of the unit, with secondary peaks of *Loxoconcha* spp., *Xestoleberis aurantia* (marine lagoonal) and *Aurila convexa* (coastal).

The litho- and biostratigraphical proxies point to a low energy environment protected by the offshore island. Such a depositional context would have been ideal for the establishment of an anchorage haven from the Bronze Age onwards.

3.5.2.1.3 Infill phases

Units C1-3 comprise coarse grained gravels and sands. Poorly developed histograms and poor sorting indices of ~1.1 to 1.3 are consistent with polder deposits. The sedimentological analyses evidence three different phases and/or sources of infill deposits, beginning during the Hellenistic period (2510 ± 30, 2300-2080 cal. BP). These are clearly differentiated by the modal values of the three units 160 μm (unit A), 63 μm (unit B) and 250-63 μm (unit C).

The relative absence of molluscan and ostracod faunas further corroborates these infill phases. The molluscan fauna juxtapose taxa from diverse ecological environments, spanning the muddy-sand assemblage to the hard substrate assemblage. The low number of tests, their poor taphonomic condition and the absence of juvenile individuals are not consistent with the development of an *in situ* biocenosis. Similar facies have been elucidated in core TXVI and indicate that part of the ancient Hellenistic and Roman cities were founded upon artificial polder deposits.

3.5.2.2 What can we say about the location of a southern harbour at Tyre?

New archaeological data and the two short cores drilled in Poidebard's supposed southern basin are not consistent with a seaport as long hypothesised by many scholars. The modelling data suggest that this southern area was partially protected by a series of small sandstone islets, but not sufficiently enough to be favourable to the establishment of a second anchorage on the island. We evoke three reasons to explain Poidebard's misinterpretations: (1) The majority of his photographs were taken from high-altitude, at around 1000 m. Although these were good at yielding a very general rendering of the near-surface archaeological remains, optical distortions meant that the images failed to penetrate the totality of the water column. (2) Secondly, many of the archaeological remains have been significantly eroded by coastal processes since their submergence during the late Roman period. Great areas of the vestiges are today partially covered in tracts of sand, which makes their interpretation problematic. (3) Finally, and perhaps most importantly, Poidebard did not dive himself, but relied on reports from hard-hat workers. His personal *a prioris* regarding the remains, allied with an absence of first hand observations, appear to have significantly biased his interpretations. In spite of these flaws, the quality of Poidebard's research must not be undermined. His photographic archives and cartography of the Tyrian coastal zone are of an exceptional quality (Denise and Nordiguian, 2005), and he remains one of the pioneers of underwater archaeology.

In the light of our new research, Renan's hypothesis of a southeastern harbour appears more plausible. The modelling and coastal stratigraphy datasets clearly demonstrate that this leeward coastal fringe was a well-protected façade from the Bronze Age onwards. Although no diagnostic harbour facies have been found, we hypothesise that this area was the most conducive environment for the establishment of a second anchorage haven at Tyre. However, in the absence of archaeological data it is impossible to precisely reconstruct where, when and how this second basin evolved. It is possible that an artificial basin *sensu stricto* did not exist, but that rather the whole of this sheltered south-eastern façade served as a semi-natural anchorage. Archaeological data from this now landlocked façade are clearly lacking, rendering a robust geoarchaeological reconstruction for this portion of the island problematic. Manifestly more work is needed on this important question. Large-scale construction works are evidenced during the Graeco-Roman period, characterised by infill deposits in this southern area (cores TVIII and TXVI), which must be put into relation with the drowned polder wall identified by El Amouri *et al.* (2005).

3.6 Satellite anchorages

3.6.1 Tyre's outer harbours

The existence of outer harbours at Tyre is corroborated by (1) our palaeogeographical reconstructions of the Bronze Age and Iron Age islands (**Figure 3.48**), and (2) maritime archaeological finds (stone anchors, pottery) on the leeward side of proximal reefs (Poidebard, 1939; Frost, 1971; El Amouri *et al.*, 2005 [**Figures 3.49-3.51**]). Archaeological prospections have failed to unearth any artificial harbourworks in these areas, however, used in tandem with lighter vessels, these outer anchorages greatly enhanced the capacity of Tyre to accommodate and transit goods from the hinterland and Mediterranean trade routes (**Figure 3.52**). To the north of the city ~1300 m of drowned reef have been evidenced, while to the south ~750 m would have been exposed during the second and first millennia BC. The elucidated Bronze

Age and Iron Age topographies suggest that future archaeological work looking to understand the city's maritime façade should concentrate upon these areas.

3.6.2 Continental harbours

The absence of fertile agricultural land on Tyre implies that the city was dependent upon its hinterland and neighbouring continental settlements for food and freshwater supplies. The continent was also used to bury the dead as have shown necropolis finds from the El Bass area (Aubet, 2004). Indeed a number of passages of the Bible (i.e. Ezek 26:8) evoke the 'daughters of Tyre' when referring to the Tyrian coast, a plurality that is consistent with three tell sites on the coastal seaboard parallel to the city: Tell Mashuk, Tell Chawakir and Tell Rachidiye. The existence of these three settlements indirectly implies a series of coastal hubs along the continental façade that assured the transportation of goods and services during the Bronze and Iron Ages. Although the spatial resolution of our data in these areas is not as good as for Tyre island, we nonetheless propose a number of plausible working hypotheses based on the chronostratigraphic data and geomorphology of these areas.

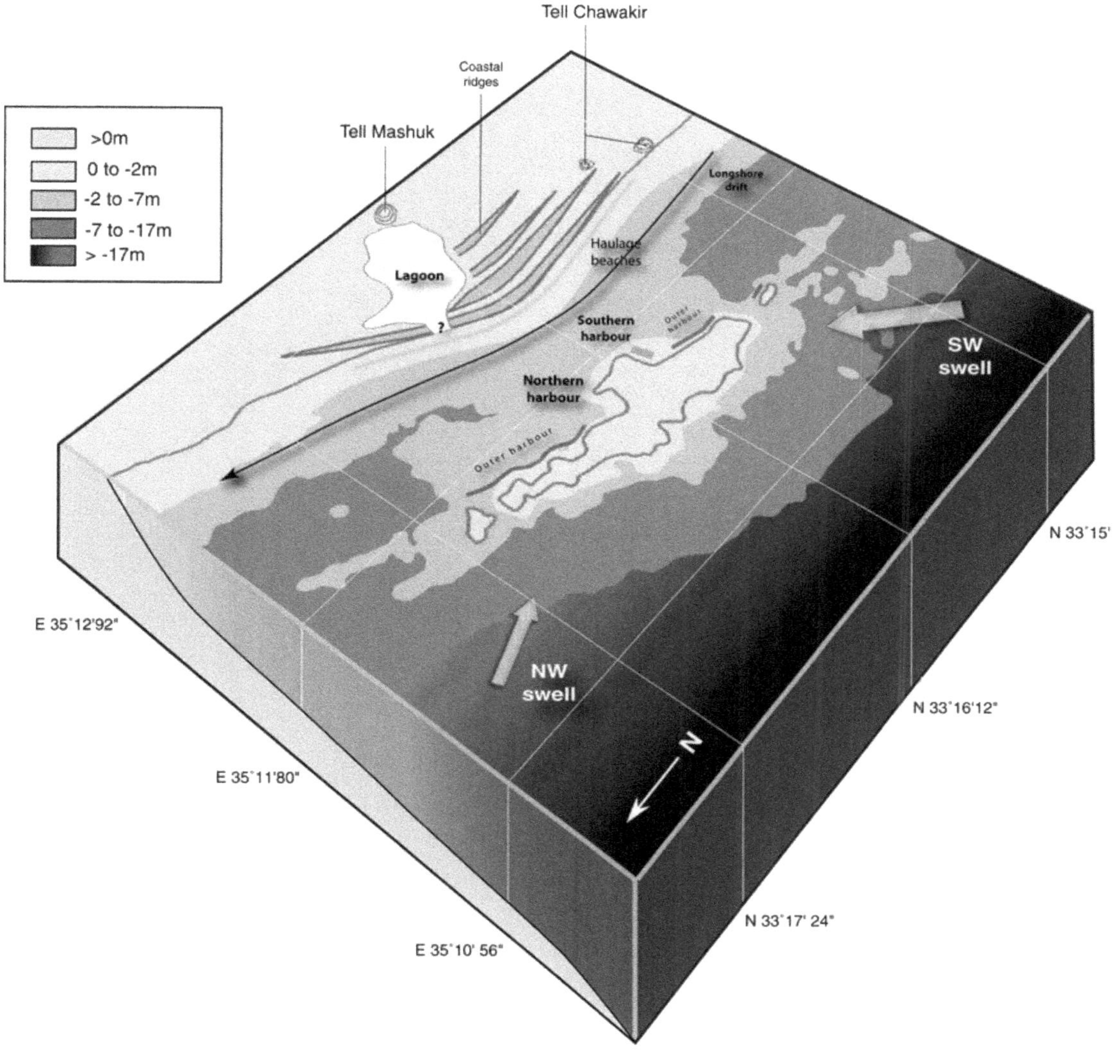

Figure 3.48: Proposed reconstruction of Tyre's ancient anchorage complexes during the first millennium BC. We identify four main harbour areas on this maritime façade: (1) the artificially protected northern harbour, presently buried beneath the city centre; (2) a second harbour complex on the south-eastern fringe of the island; (3) a series of outer harbours that exploited the extensive subaerial ridges and reefs to the north and south of the city; and (4) the continental harbour complexes that operated in tandem with Tell Mashuk, Tell Chawakir and Tell Rachidiye.

3.6.2.1 Tell Mashuk and a possible lagoonal harbour

Although it is equivocal which of the three tell sites was Palaeotyrus or Old Tyre *sensu stricto*, for many scholars the most eligible candidate is Tell Mashuk (Kenrick, 1855; Katzenstein, 1997). The site lies in the wave shadow of Tyre on the eastern axis of the city and has been occupied since the Bronze Age. It is likely to have served as a proto-settlement before the later colonisation of Tyre itself. At present, the tell remains are landlocked 1600 to 1800 m from the sea due to the significant progradation of the tombolo during the mid- to late Holocene (see chapter 2). Our stratigraphic data demonstrate, however, that the coastal area to the west of the tell was flooded around 6000 BP and gradually prograded seawards. Although we have not cored immediately at the base of Tell Mashuk, the topography and our geomorphological prospections in this vicinity suggest that the maximum marine ingression lay in proximity to the tell. Low accommodation space coupled with high sediment supply and the pronounced wave shadow generated by Tyre island led to the rapid accretion of a salient strandplain. Diagnositic beach sedimentology at the base of the stratigraphy shallows up to a fine-grained plastic clays facies dominated by *Cerastoderma glaucum*, a lagoon tolerant taxa (>80 % silts and clays). This stratigraphy suggests that by 5500 BP the area to the west of the tell formed a lagoon in communication with the open sea (**Figure 3.53**). The mid-zone of the lagoon facies is dated 4180 ± 30 BP, or 2430-2200 cal. BC.

Although the precise dimensions and phases of infilling of this lagoon are not clearly known, we hypothesise that it was used as an anchorage haven during the Bronze Age (draught depth <1 m). Use of the lower reaches of wadis and lagoons as natural harbours was widespread in the Levant during the third and second millennia BC (Raban, 1987a, 1990). The natural low energy conditions meant that there was no need for artificial harbourworks. Loading and unloading of cargo could take place on shallow draught boats hauled onto the lagoon shoreline. This lagoonal harbour complex would also have been used in tandem with the sandy beach ridges that enclosed the lagoon to the west.

At what stage the lagoon became isolated from the sea is unclear. The existence of a lagoon and a rapidly prograding salient strandplain raises the question of the inlet. In the absence of a flushing fluvial system, any inlet would gradually have been blocked by a beach ridge accumulation. It is possible that human societies maintained an artificial link with the open sea until the Iron Age, although the wide haulage beaches meant that the viability of the settlement was not totally dependent on this shallow anchorage. By Roman times, the lagoon had significantly silted up and we observe a transition to marshland deposits characterised by plant macrorestes and fossil snail tests. These marshlands persisted in the El Bass area until the mid-nineteenth century (Carmona and Ruiz, 2004).

3.6.2.2 Tell Chawakir

Tell Chawakir lies at the southern limit of the tombolo salient. The area was transgressed around 6000 BP and prograded rapidly to form a beach ridge strandplain (**Figure 3.54**). None of our cores in proximity to the tell have yielded lagoonal lithofacies analogous to those at Tell Mashuk. Conversely, the stratigraphy evokes sandy beaches at the base of the tells that could have been used as natural haulage ramps. It seems plausible that some of the cargo from Tyre's southern outer harbours was ferried directly to and from the shoreline by lighter vessels based at Tell Chawakir, without necessarily transiting via Tyre island *sensu stricto*.

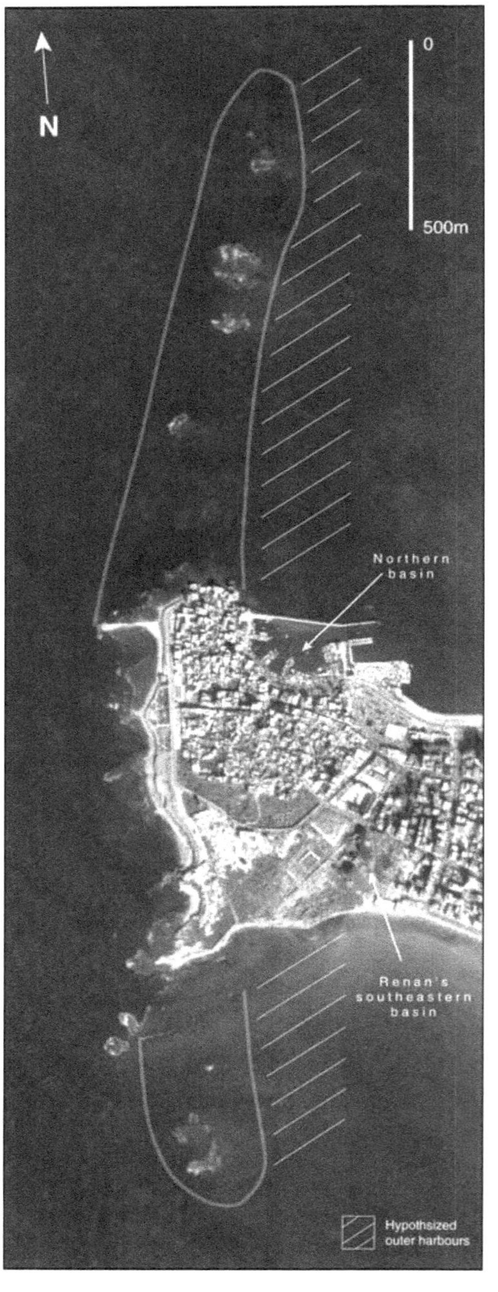

Figure 3.49: Drowned areas of the Tyrian coastal ridge visible from the air. Archaeological finds from around these areas suggests that they served as outer harbours during the Bronze and Iron Ages. Base image: DigitalGlobe, 2006.

Figure 3.50: Drowned coastal reef to the south of Tyre. We hypothesise that these drowned reefs were used as outer harbours during the Bronze and Iron Ages. Photograph: M. El Amouri, 2002.

Figure 3.51: Stone anchor found in proximity to submerged coastal reefs to the south of Tyre. Photograph: M. El Amouri, 2002.

Figure 3.52: Nineteenth century lithograph engraving of lighter vessels being used to unload a sailing craft offshore at Acre in Palestine. The technique has probably evolved very little since the Bronze Age. The use of outer harbours in tandem with lighters greatly increased Tyre's capacity to receive and transit trade goods (full title of lithograph: Saint Jean d'Acre, from the sea, in David Roberts The Holy Land, 1842).

3.6.2.3 Tell Rachidiye

Not all of Tyre's commerce was conducted by the sea. Terrestrial trade routes also existed with northern Syria, where the Tyrians traded with Uratu and Mesopotamia. Evidence of these terrestrial trade routes can be found at Rachidiye (Doumet-Serhal, 2004b). Rachidiye lies in proximity to outcropping springs at Ras El-Ain and was therefore important in supplying Tyre with its provision in freshwater. The Romans, exploiting the now subaerial tombolo, later built an aqueduct linking Tyre directly with this freshwater spring. Material from its necropolis has been shown to attest to close trade relations between Cyprus and Phoenicia during the Iron Age. The coastal area to the south of Tell Chawakir and north of Tell Rachidiye has been relatively stable during the Holocene. Our core stratigraphies from these areas (see chapter 2) reveal late Pleistocene palaeosols over a sandstone bedrock. No Holocene marine deposits have been found in this area. A series of drowned sandstone reefs are also manifest in aerial photographs to the west of the settlement. We hypothesise that these were used as outer anchorages for the larger vessels, although clearly greater work is required (**Figure 3.55**).

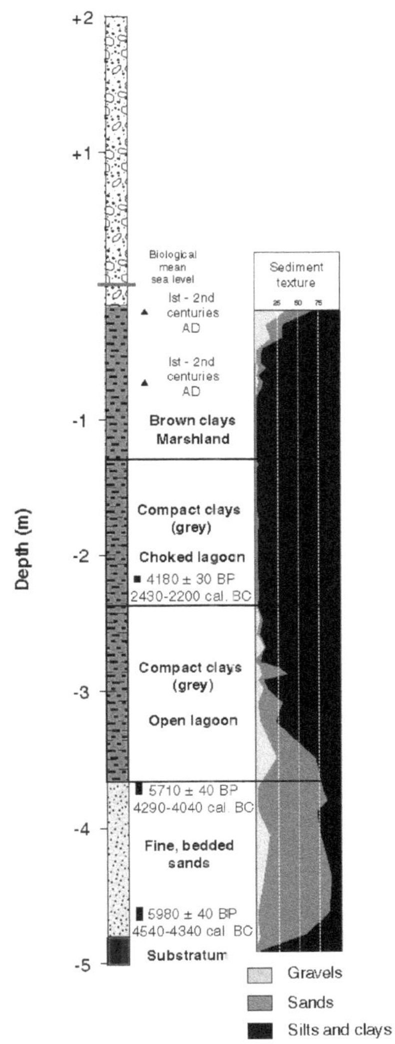

Figure 3.53: Stratigraphy of core TXVIII.

Figure 3.54: Photographs of the strandplain in proximity to Tell Chawakir. Above: looking south towards Tell Rachidiye. Below: looking west towards the sea. Photographs: N. Marriner.

3.7 Concluding remarks

While our work reveals the importance of Tyre's northern harbour as the city's main transport hub from the Bronze Age onwards, the role of its satellite infrastructures must not be underestimated. Traditionally, research has focused on Tyre's northern and southern harbours which catered for merchants on the trade routes during the Bronze and Iron Ages. Our new work shows that Tyre also possessed a series of secondary seaport complexes that were integral parts of this network, notably assuring the day-to-day running of Tyre (provision in water, food, building material etc.). In total we identify four harbour complexes on the Tyrian coastline (1) the artificially protected northern harbour, presently buried beneath the city centre; (2) a second harbour complex on the south-eastern fringe of the island; (3) a series of outer harbours that exploited the extensive subaerial ridges and reefs to the north and south of the city; and (4) the continental harbour complexes that operated in tandem with Tell Mashuk, Tell Chawakir and Tell Rachidiye. Lighter vessels would have interlinked these harbour complexes. The partially drowned coastal ridge upon which the ancient city lay served as the protective umbrella structure for these harbour complexes. The evolution of the site since 6000 BP means that the main seaports are today buried beneath tracts of coastal sediments, while the outer harbours have been drowned by subsidence of the Tyrian horst during the late Roman period (**see chapter 6**).

Future archaeological work at Tyre needs to focus on better understanding the evolution of maritime infrastructure since the Bronze Age. Although modern urbanisation renders this task difficult around the fringes of the northern harbour, El Amouri *et al.* (2005) and Descamps *et al.* (personal communication) have shown the rich archaeological potential of ancient land surfaces presently drowned to the north and south of Tyre. In geoarchaeological terms, our research has not only reconstructed the palaeogeographical evolution of the city's ancient harbours since 6000 BP, but has also highlighted the chronostratigraphic impacts of seaport dredging, largely neglected in previous studies of this nature. In chapters 4 and 5, we critically compare and contrast findings from Tyre with Sidon and Beirut.

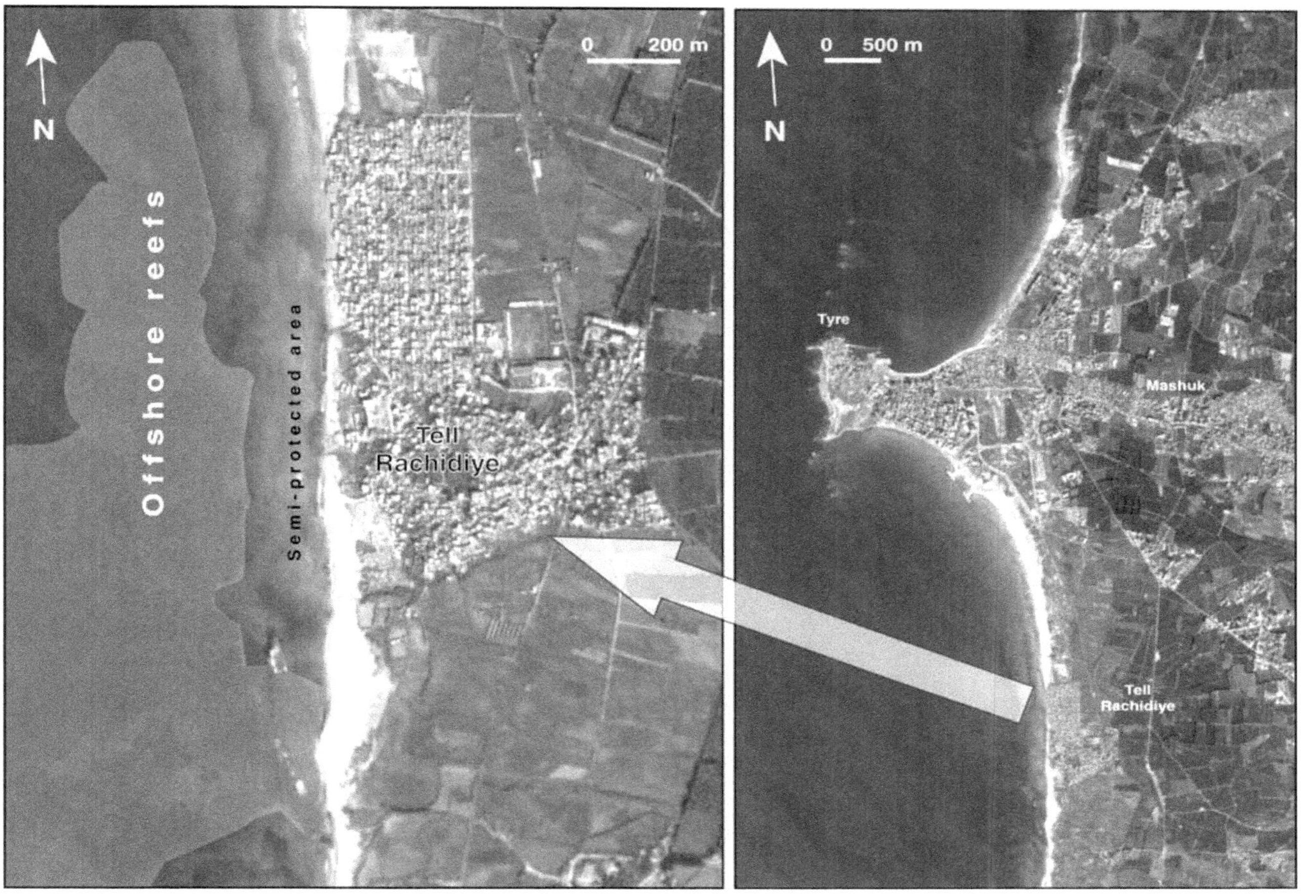

Figure 3.55: Aerial photograph of Tell Rachidiye. Base images: DigitalGlobe, 2006.

Chapter 4

Geoarchaeology of Sidon's ancient harbours

Under the auspices of the British Museum, terrestrial excavations have been underway at Sidon since 1998 (Doumet-Serhal, 2003, 2004c). In tandem with these archaeological surveys, 15 cores were drilled in and around Sidon's ancient port areas, with three main objectives: (1) to elucidate the evolution of the city's maritime façade and investigate its coastal palaeogeography; (2) to compare and contrast these data with Sidon's sister harbour, Tyre (chapter 3; Marriner et al., 2005; Marriner et al., 2006a-b); and (3) to investigate human coastal impacts, and more specifically the problem of accelerated coastal sedimentation. As we have demonstrated in chapter 1, silting of the Mediterranean's ancient ports is a recurrent theme in coastal geoarchaeology, playing a significant role in littoral progradation and human exploitation of the anchorages (Blackman, 1982a-b; Morhange et al., 2003b; Raban, 1985a-c; Raban, 1987a). Ancient societies strived permanently with the silting problem, and indeed in areas of high sediment supply it was a constant endeavour to maintain a viable draught depth (Marriner and Morhange, 2006a). From a geoarchaeological standpoint, the silting provides a multiplicity of research possibilities, not least because the fine-grained sediments and high water table anoxically preserve otherwise perishable artefacts, but also the port sediments are a high-resolution sedimentary archive, recording much of the site's maritime and occupation histories. In this chapter, we look to precisely relocate Sidon's ancient anchorages and reconstruct the human-environment interactions along the city's maritime façade.

4.1 Introduction

The great antiquity of the Sidon-Dakerman area has long attracted the interest of scholars (Renan, 1864; Contenau, 1920; Contenau, 1924a-b; Dunand, 1939, 1940, 1941, 1942-43, 1967; Doumet-Serhal, 2003, 2004c, 2006a). During the nineteenth century spectacular discoveries at Sidon, Murex Tell, Tell Dakerman and the surrounding lower foothills of the Lebanon range yielded a great multiplicity of archaeological material chronicling a long history of human occupation stretching back to the Neolithic (Saidah, 1979; see **Figure 4.1**). Canaan's oldest city according to Genesis (Genesis 10:15), Sidon tell occupies a modest rocky promontory that overlooks a partially drowned sandstone ridge and two marine embayments (Doumet-Serhal, 2003; Marriner *et al.*, 2006b; **Figures 4.2 and 4.3**). The southern basin, or *Crique Ronde*, is closed by the Dakerman promontory. This geomorphological duality segmented the territory of Sidon into two well-defined spatial entities, the tell or 'Greater Sidon', and the port city or 'Little Sidon', centred on the ancient northern harbour (Doumet-Serhal, 2004c). The outlying Land and Sea Castles mark the northern and southern limits of the ancient city, which covers some 16 hectares (Doumet-Serhal, 2003). During the Iron Age, this geomorphological endowment allowed Sidon to evolve into one of Phoenicia's key city-states, producing and transiting wealthy commodities to trading partners in Assyria, Egypt, Cyprus and the Aegean (Doumet-Serhal, 2006b; Forstner-Müller *et al.*, 2006; Griffiths and Ownby, 2006). This trading ascendancy is corroborated by the Old Testament's use of the term Sidonian to encapsulate all Phoenicians. Sidon enjoyed its apogee during the sixth to fifth centuries BC, at which time it superseded Tyre as Phoenicia's principal naval base.

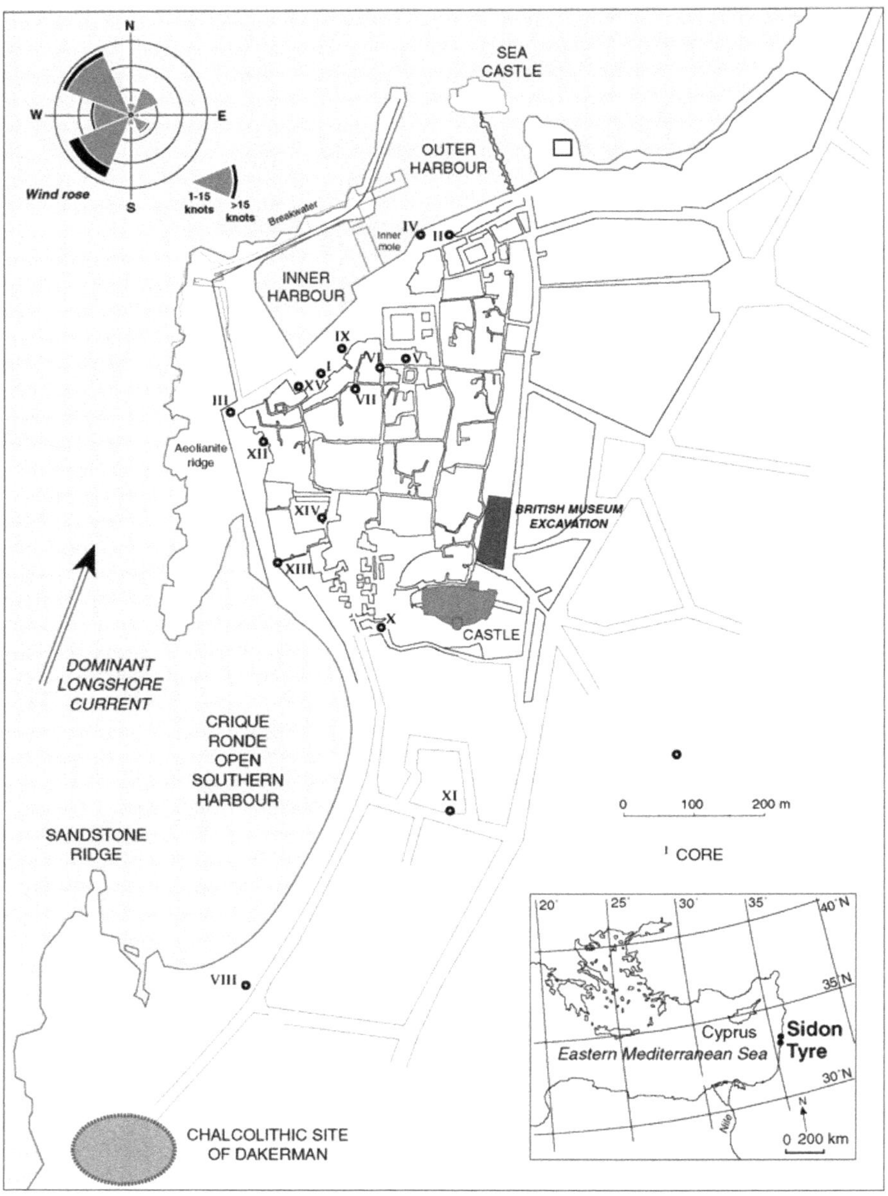

Figure 4.1: Sidon's ancient harbour areas and location of cores.

Figure 4.2: British Museum excavations have been underway on Sidon's tell since 1998. The work has revealed a long and complex history of occupation stretching back to the Early Bronze Age. The land and sea castles mark the northern and southern limits of the ancient settlement. Base image: DigitalGlobe, 2006.

Figure 4.3: 1830s lithograph of Sidon by David Roberts (in Roberts, 2000). The two ancient tells are clearly discernible in the landscape depiction.

Although Sidon has a long history of archaeological research (Renan, 1864; Contenau, 1920; Contenau, 1924a-b; Dunand, 1939, 1940, 1941, 1942-43, 1967) the ancient city had, until very recently, never been systematically explored. In light of the difficult geopolitical context, it was only in 1998 that the Lebanese Department of Antiquities authorised the British Museum to begin systematic excavations of the ancient tell (Curtis, 2000). The current excavations are led by Doumet-Serhal at the College site, one of three parcels of land acquired by the Department of Antiquities during the 1960s (**Figure 4.2**). Eight years on, a continuous stratigraphy spanning the third millennium BC through to the Iron Age has been established for the city (Doumet-Serhal, 2003, 2004c; **Figure 4.4**). The excavations since 1998 show the archaeological complexity and richness of each stage of Sidon's development, from the Early Bronze Age onwards. Indeed Sidon is the only ancient city in Lebanon where third millennium BC layers are easily accessible to excavation (Doumet-Serhal, 2003, 2004c).

In tandem with the terrestrial excavations, a series of 15 cores was drilled around Sidon's two marine embayments (**Figure 4.5**) looking to accurately reconstruct the evolution of the city's maritime façade and attempt to correlate geoarchaeological evidence from the coastal area with findings from the tell (Espic *et al.*, 2002; Morhange *et al.*, 2003b).

Sidon's coastal plain runs from the Litani river in the south, northwards towards the Awali (**Figure 4.5**). This low-lying topography, up to 2 km wide in places, comprises a rectilinear coastline. In the Sidon area, a series of faults has oriented the valleys and talwegs NW-SE (Dubertret, 1955, 1975; Sanlaville, 1977). The most important regional watercourses include the Litani, with headwaters in the Beqaa valley, and the Awali, which flows from the Jurassic anticlinal of Barouk-Niha. These watercourses alone transit ~280 x 10^6 m^3 and 130 x 10^6 m^3 of sediment per year (Abd-el-Al, 1948; Tayara, 1991; Soffer, 1994). Two modest fluvial systems, the el-Kamleh and the el-Barghout, flow north and south of the ancient tell respectively.

Figure 4.4: A-B: The six levels of occupation at Sidon during the third millennium BC. The Bronze Age stratigraphy of Sidon comprises six occupation layers from the end of Early Bronze I to the end of Early Bronze III (around 3000 to 2300 BC). A short hiatus is observed between the end of Early Bronze I and the beginning of Early Bronze II. The overall character that emerged from the third millennium installations was of a long-lived settlement with domestic structures such as ovens, querns and basins. C-D: Bronze Age and Iron Age stratigraphy at the College site. Photographs from Doumet-Serhal (2004).

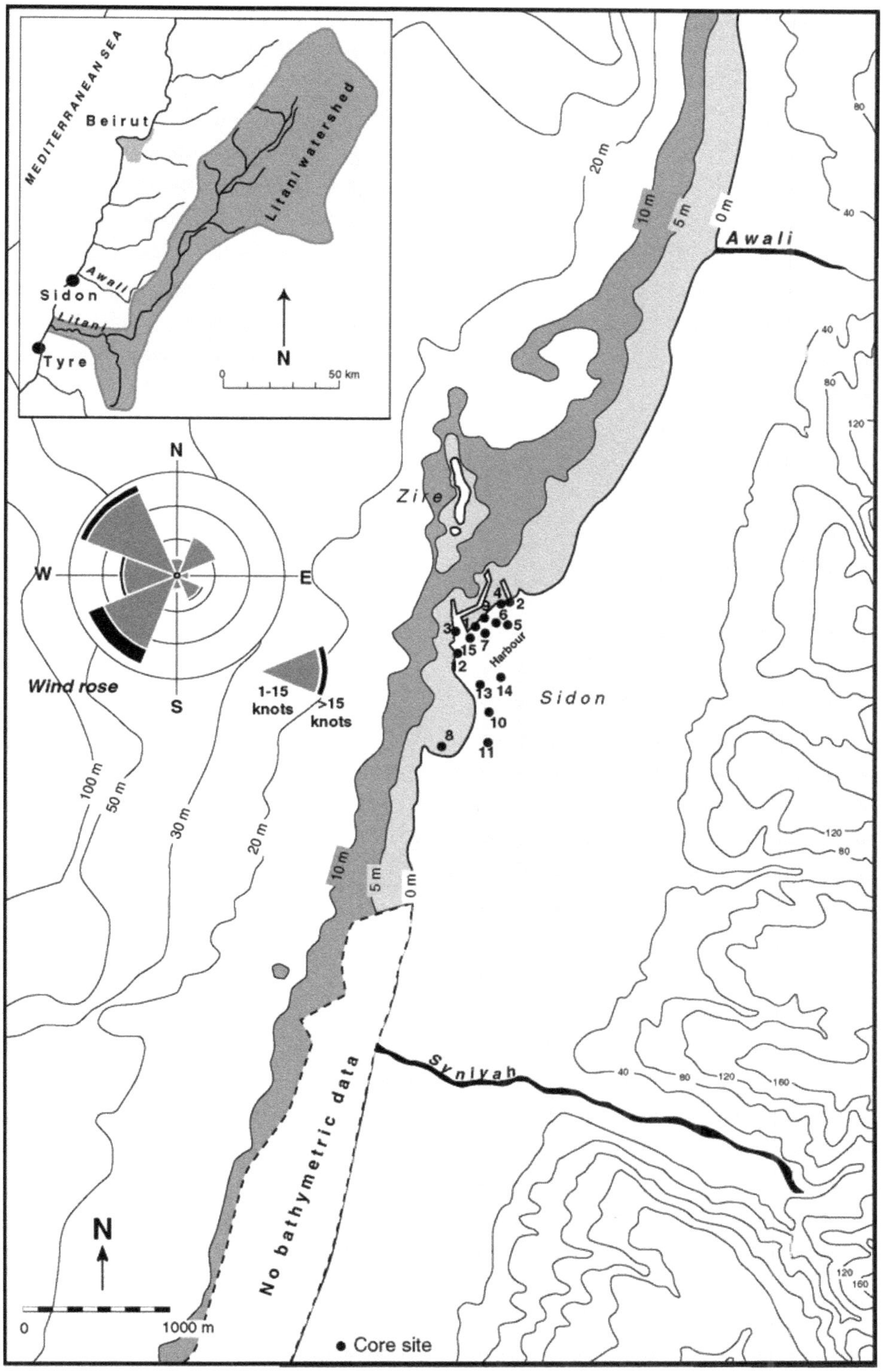

Figure 4.5: Sidon's coastal bathymetry.

4.2 Sidon's maritime façade: geoarchaeological context

Sidon's coastal physiography makes it an ideal location for the establishment of three natural anchorage havens. Two pocket beaches lie leeward of a Quaternary sandstone ridge, partially drowned by the Holocene marine transgression (**Figure 4.6**). To the south of the ancient city this ridge has been breached by the sea to form a large semi-circular embayment. Named the Egyptian harbour by Renan (1864) and later the *Crique Ronde* by Poidebard and Lauffray (Poidebard and Lauffray, 1951), this coastal zone presently comprises a sandy beach. It is overlooked to the south by the Dakerman promontory. This semi-protected bay is inferred to have been used as a proto-harbour by the latter tell's Chalcolithic inhabitants (Saidah, 1979).

Whether or not it was ever artificially protected by harbourworks has never been unequivocally demonstrated (Poidebard and Lauffray, 1951), a question we elucidate later in this chapter.

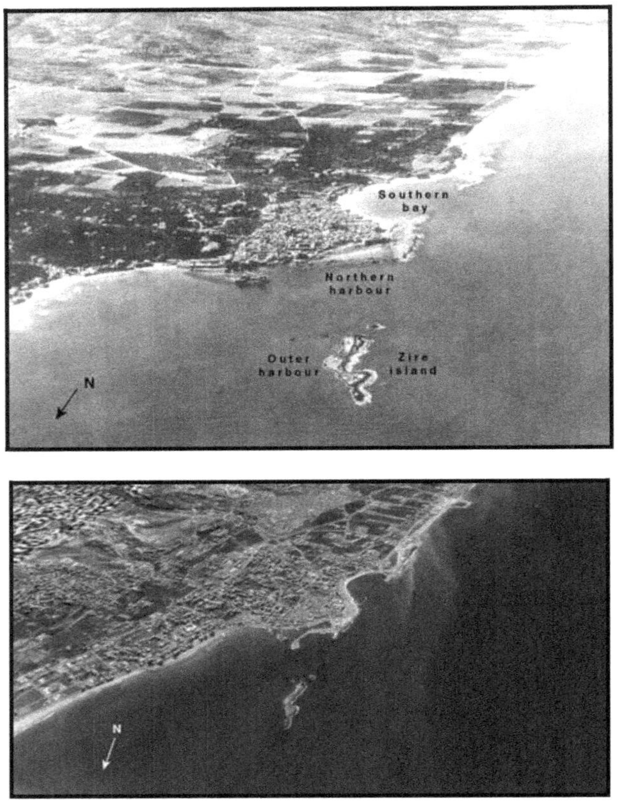

Figure 4.6: Above: Sidon and Zire during the 1940s (from Poidebard and Lauffray, 1951). Below: Sidon and Zire in 2005 (DigitalGlobe, 2006). Note the extensive redevelopment of the coastal front in both the northern harbour and southern bay. In the foreground, Sidon's outer harbour lies in the shadow zone of Zire island. The promontory of Sidon separates two coves, the northern harbour and Poidebard's *Crique Ronde*.

Northwest of the promontory lies a second bay, protected from the open sea by a prominent sandstone ridge. 580 m in length, this coastal edifice shields a shallow basin still used to this day; a medieval sea castle, built upon a small islet, closes the northern portion of the basin. This northern harbour, the centre of Sidon's economic and military activity in antiquity, is mentioned for the first time by Pseudo-Scylax who describes it as a closed harbour. Much of Poidebard and Lauffray's work was centred on this area where they identified a series of juxtaposed harbourworks (**Figures 4.7 and 4.8**). From their research emerged the vestiges of a closed ancient port comprising: (1) a reinforced sandstone ridge; and (2) an artificial inner harbour mole, perpendicular to the ridge, and separating two basins.

A third harbour area, the offshore island of Zire, is a unique feature of the Sidonian coastline (**Figure 4.6**). First described by Renan (1864), it was not until the work of Poidebard and Lauffray (1951) that a preliminary plan of the island, with its quarries and harbourworks, was drawn up. They identified a double seawall sheltering a series of quarries and harbour quays on its leeward side. In 1973, underwater surveys by Frost complemented her predecessors' work, uncovering a collapsed jetty and numerous scattered masonary blocks on the sea bottom in proximity to the island (Frost, 1973). She concluded that the island had not only served as a quarry and harbour but also supported a number of constructions. Carayon (2003) undertook the most recent archaeological work of note, in which he describes six quarry zones, detailing the cartography of Poidebard and Lauffray (1951). During our field investigations we surveyed and dated an uplifted marine notch (+50 cm) on these quarry faces pertaining to a short-lived sea-level oscillation around 2210 $\pm$ 50 BP (**Figure 4.9**; Marriner and Morhange, 2005a; Morhange *et al.*, 2006b). These data are in contrast with Tyre, where submergence of ~3 m is recorded since late antiquity by coastal stratigraphy, submerged urban quarters and harbourworks (chapters 3 and 6; El Amouri *et al.*, 2005; Marriner *et al.*, 2005; Marriner *et al.*, 2006a).

4.3 Methods and data acquisition

A series of 15 cores was drilled around the two marine embayments (**Figures 4.1 and 4.5**). In depth discussion of the techniques employed can be reviewed in chapters 1 and 2. These include multi-proxy litho- (sedimentology and grain size analyses) and biostratigraphical (marine molluscan faunas and ostracods) lines of investigation. Radiocarbon datings provide a working chronostratigraphic framework. Material included marine shells, wood fragments, charcoal and seeds (**see Table 4.1**).

4.4 Results and discussion

Detailed descriptions of the litho- and biostratigraphical data from Sidon's closed northern harbour and the *Crique Ronde* follows.

4.4.1 Where was Sidon's ancient northern harbour?

Although Sidon had three port complexes during antiquity, the northern basin, sheltered by an extensive aeolianite ridge, was naturally predisposed to become the main seaport from the Bronze Age onwards. As with many of the ancient harbours on the Levantine façade, Sidon's northern harbour is an example of a buried urban harbour *par excellence* (Marriner and Morhange, 2007). Diagnostic harbour facies - including fine-grained silts and plastic clays, and lagoonal and marine lagoonal fauna - have been elucidated around the fringes of the present fishing harbour for a distance of 100 m inland (**Figure 4.10**). Coastal progradation of the port coastline after the Byzantine period diminished the size of the harbour to its present dimensions. As at Tyre, this land encroachment accommodated urban growth during the medieval and modern periods. The reconstructed basin was approximately 50 % larger than present (**Figure 4.10**).

Figure 4.7: Aerial photographs of Sidon's maritime façade from Poidebard and Lauffray (1951). After the Second World War, Poidebard was called upon at Sidon to better comprehend the silting up of the northern seaport, which had not harboured deep draught boats since the seventeenth century. The work at Sidon was undertaken for the most part between 1946 and 1949, before being published in 1951. The research benefited from Poidebard's now rich *savoir faire* of harbour archaeology. Aerial photography, underwater prospections and dredging of the northern harbour brought to light a suite of advanced harbourworks. Unfortunately many of the structures visible on the shielding aeolianite ridge were destroyed during recent modernisation of the harbour area.

Figure 4.8: Example of artificial modification of the aeolianite ridge photographed by Poidebard in the 1940s. There are no chronological constraints upon this construction due to the inherent difficulties in dating rock-cut structures.

Figure 4.9: Uplifted notch on Zire island at +50 cm pertaining to a short sea-level oscillation around 2210 ± 50 BP (photograph: C. Morhange).

Sample	Depth below MSL	Lab code	Material dated	$^{13}C/^{12}C$ (‰)	^{14}C BP	±	Cal. BP	Cal. BC/AD
BH I 6	437.5	Ly-9470	Marine shells	1.9	4931	62	5440-5040	3490 - 3090 BC
BHIX 8	167.5	Lyon-1798 (GrA 20857)	Marine shells	est 0	2370	50	2130-1860	180 BC - 90 AD
BHIX 10	157.5	Lyon-1879 (Poz 998)	Marine shells	-2.26	2285	35	2000-1800	50 BC - 150 AD
BHIX 24	86.5	Lyon-1878 (Poz 1016)	Marine shells	-0.35	2350	35	2090-1870	140 BC - 80 AD
BHIX 36	236.5	Lyon-1796 (GrA 20859)	Marine shells	-1.61	2340	80	2180-1750	230 BC - 200 AD
BHIX 35	273	Lyon-1797 (GrA 20858)	Marine shells	est 0	2240	60	1990-1690	40 BC - 260 AD
BHIX 44	402.5	Lyon-1876 (Poz 1004)	Marine shells	1.37	3640	50	3680-3400	1730 - 1450 BC
BHIX 47	477.5	Lyon-1877 (Poz 1002)	Marine shells	1.58	4410	40	4720-4420	2770 - 2470 BC
BH VIII 6	350	Lyon-1799 (GrA 20809)	Marine shells	1.88	4220	50	4450-4140	2500 - 2190 BC
BH VIII 10	472.5	Lyon-1728 (OxA)	Marine shells	1.16	4060	40	4230-3960	2280 - 2010 BC
BH VIII 14	737.5	Lyon-1729 (OxA)	Marine shells	3.57	5955	45	6470-6270	4520 - 4320 BC
BH VIII 16	780	Lyon-1730 (OxA)	Marine shells	1.11	6030	45	6580-6310	4630 - 4360 BC
BH XV 3	161	Poz-13012	1 *Venerupis rhomboides*	-1.9	2255	35	1970-1760	20 BC - 190 AD
BH XV 12	247.5	Poz-13374	2 grape seeds	-27.3	2515	30	2740-2480	790 - 530 BC
BH XV 17	315	Poz-13375	Charcoal	-29.5	3385	35	3720-3550	1770 - 1600 BC
BH XV 24	367.5	Poz-13013	1 *Nassarius reticulatus*	4.4	3670	40	3690-3450	1740 - 1500 BC
BH XV 28	425	Poz-13006	2 *Glycymeris glycymeris* (juvs.) 1 *Lucinella divaricata*	8	4840	40	5280-5000	3330 - 3050 BC
BH XV 31	555	Poz-13007	5 *Bittium reticulatum*	-3.6	5180	50	5650-5430	3700 - 3480 BC

Table 4.1: Radiocarbon determinations and calibration.

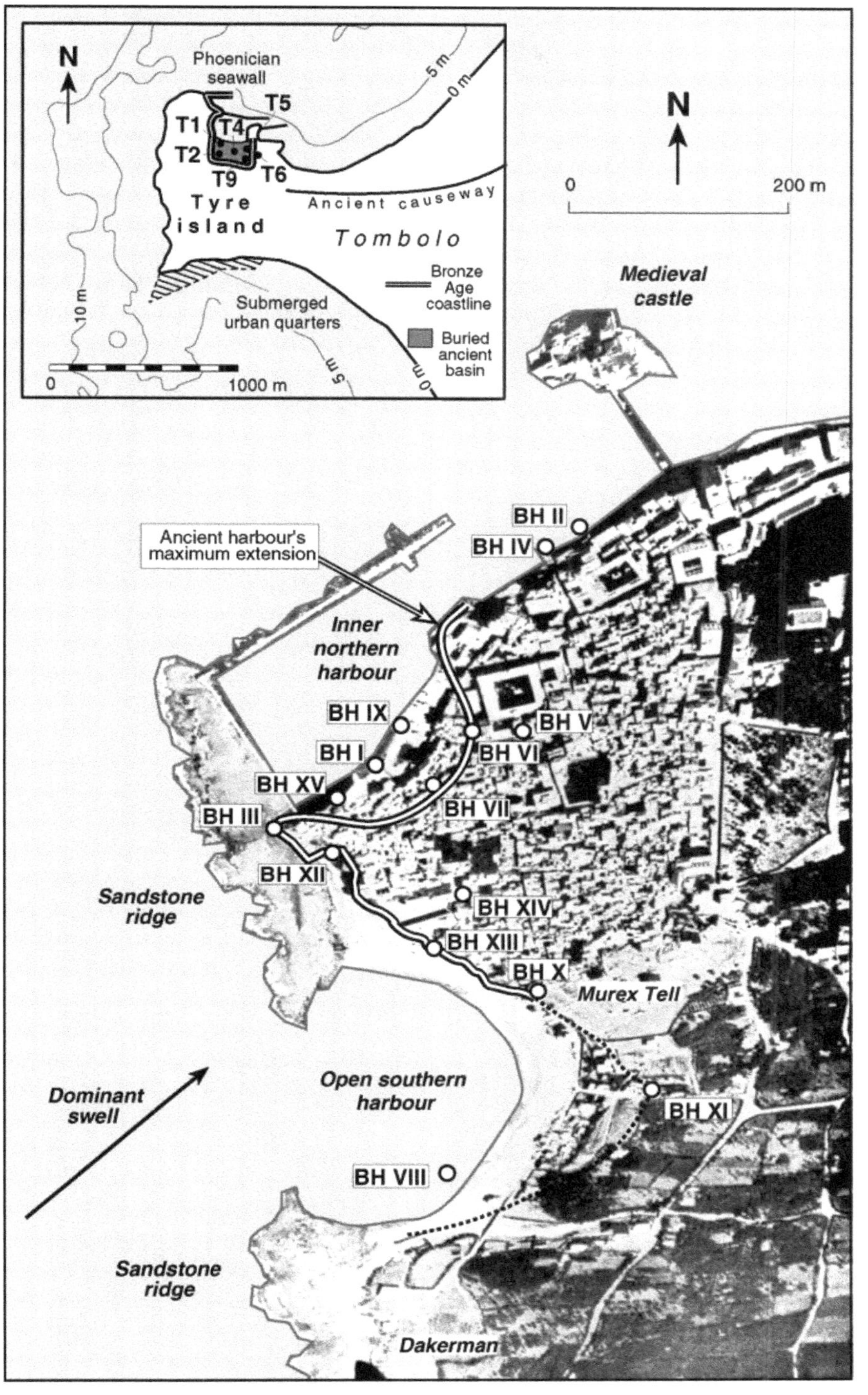

Figure 4.10: Sidon's reconstructed harbour limits in antiquity. Inset: Tyre's northern harbour limits.

Our geoarchaeological datasets elucidate a complex history of coastal change and human occupation. Investigation of the coastal archives has expounded understanding of Sidon's maritime history between the Bronze Age and medieval periods. Multidisciplinary investigations demonstrate that harbour management advances are clearly translated by distinctive sedimentary facies and faunal suites. The sedimentary history of the harbour details six periods.

4.4.2 When and how did Sidon's northern seaport evolve?

4.4.2.1 Transgressive unit

Description and interpretation: Unit D comprises a marine pebble unit which overlies the sandstone substratum and marks the Holocene marine transgression of the harbour area (**Figures 4.11, 4.14 and 4.17**). Many of the pebbles are encrusted with marine fauna such as

Serpulae. In core BH I, this grades into a silty sand with a medium sorting index of 0.88.

4.4.2.2 Pocket beach/Bronze Age proto-harbour

Description: Unit C2 is characterised by a shelly sand unit with poor sorting indices (1.17-1.27) and sand modal values of 250-200 µm. The sedimentological data concur a middle energy beach environment. In core BH IX, the bottom of the unit is constrained to 4410 ± 40 BP (2750-2480 cal. BC). Two *Loripes lacteus* shells from unit C of BH I yielded a radiocarbon age of 4931 ± 62 BP (3475-3070 cal. BC). Molluscan taxa from the following assemblages are attributed to this unit (**Figures 4.12, 4.15 and 4.18**): subtidal sands assemblage (*Bulla striata* and *Mysia undata*), upper clean-sand assemblage (*Smaragdia viridis*, *Nassarius pygmaeus*, *Neverita josephinia* and *Chamela gallina*) and the upper muddy-sand assemblage in sheltered areas (*Loripes lacteus* and *Cerithium vulgatum*). The ostracod fauna comprises taxa from the marine lagoonal (*Loxoconcha* sp. and *Xestoleberis* spp.) and coastal (*Aurila* spp., *Urocythereis* sp. and *Heterocythereis albomaculata*) domains, with some marine taxa being drifted in (**Figures 4.13, 4.16 and 4.19**). These biostratigraphic data are analogous to a pocket beach sheltered by the sandstone ridge.

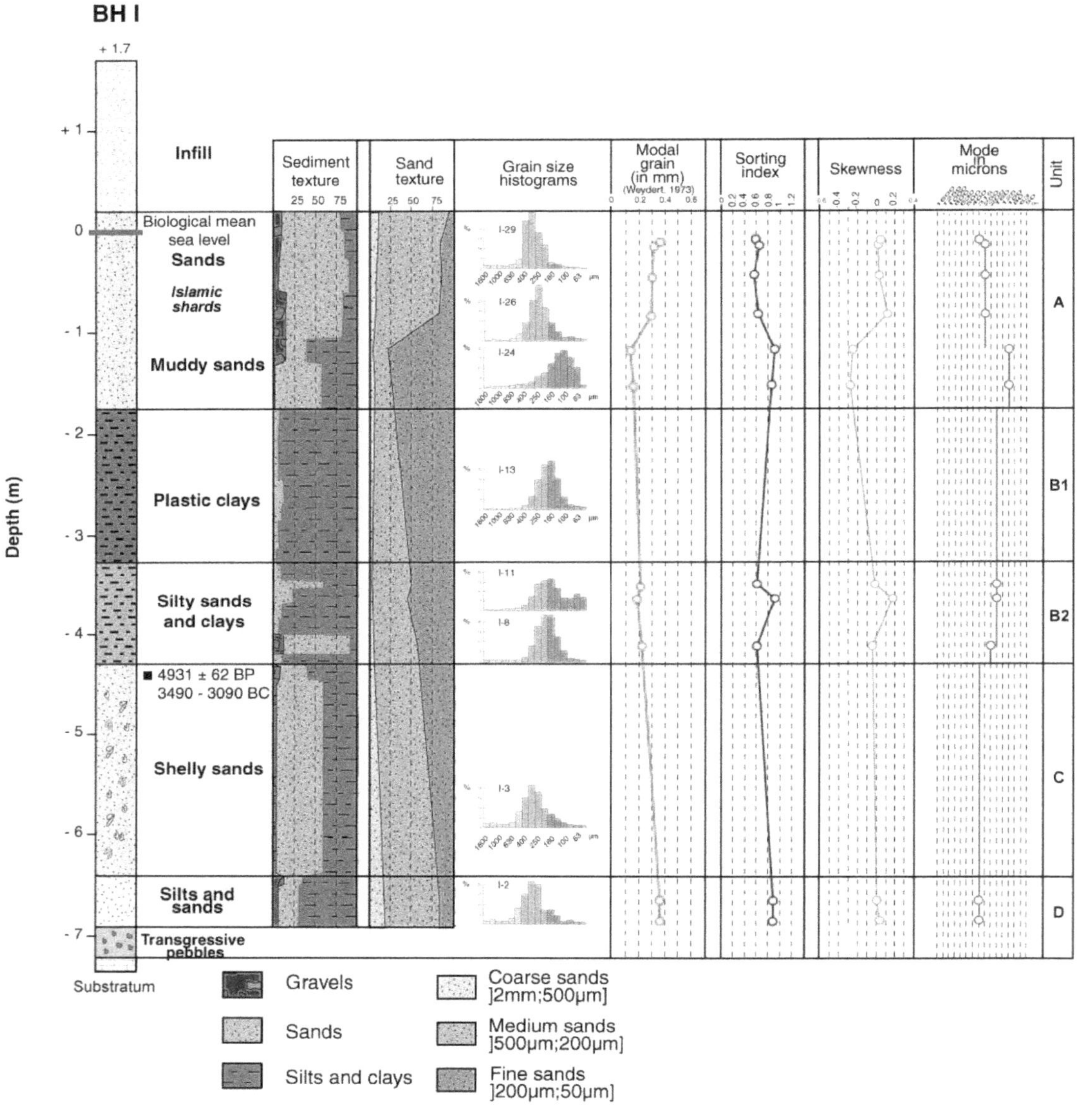

Figure 4.11: Sedimentology of core BH I (northern harbour). (1) Unit D marks the Holocene marine transgression, with a marine pebble unit which overlies the sandstone substratum. This grades into a silty sand (comprising 7 % gravels, 23 % sands and 69 % silts and clays). A sorting index of 0.88 is consistent with a mediocre sorting of the sediment. (2) A rise in energy dynamics in unit C is marked by transition to a shelly sand unit. (3) In unit B2, there is a net change in sedimentary conditions with a shift to fine grained sands and silts. Modal grain size values of 160 µm for the sand fraction and mediocre sorting indices (0.6-0.91) are concomitant with a low energy environment. (4) Accentuation of these low energy conditions is manifested in unit B1 by a plastic clays unit. The silts and clays fraction dominate at between 90-98 %. (3) The base of unit A is marked by a sudden rise in the sands fraction to the detriment of the silts and clays. The influence of the gravels and sands fractions gradually increases up the unit from 2 % for the gravels and 46 % for the sands at the base, to 11 % and 67 % respectively at the top. A shift from lower energy modal sand values at the base (100 µm) to middle energy values (315 µm) at the top concurs a gradual demise in harbour maintenance.

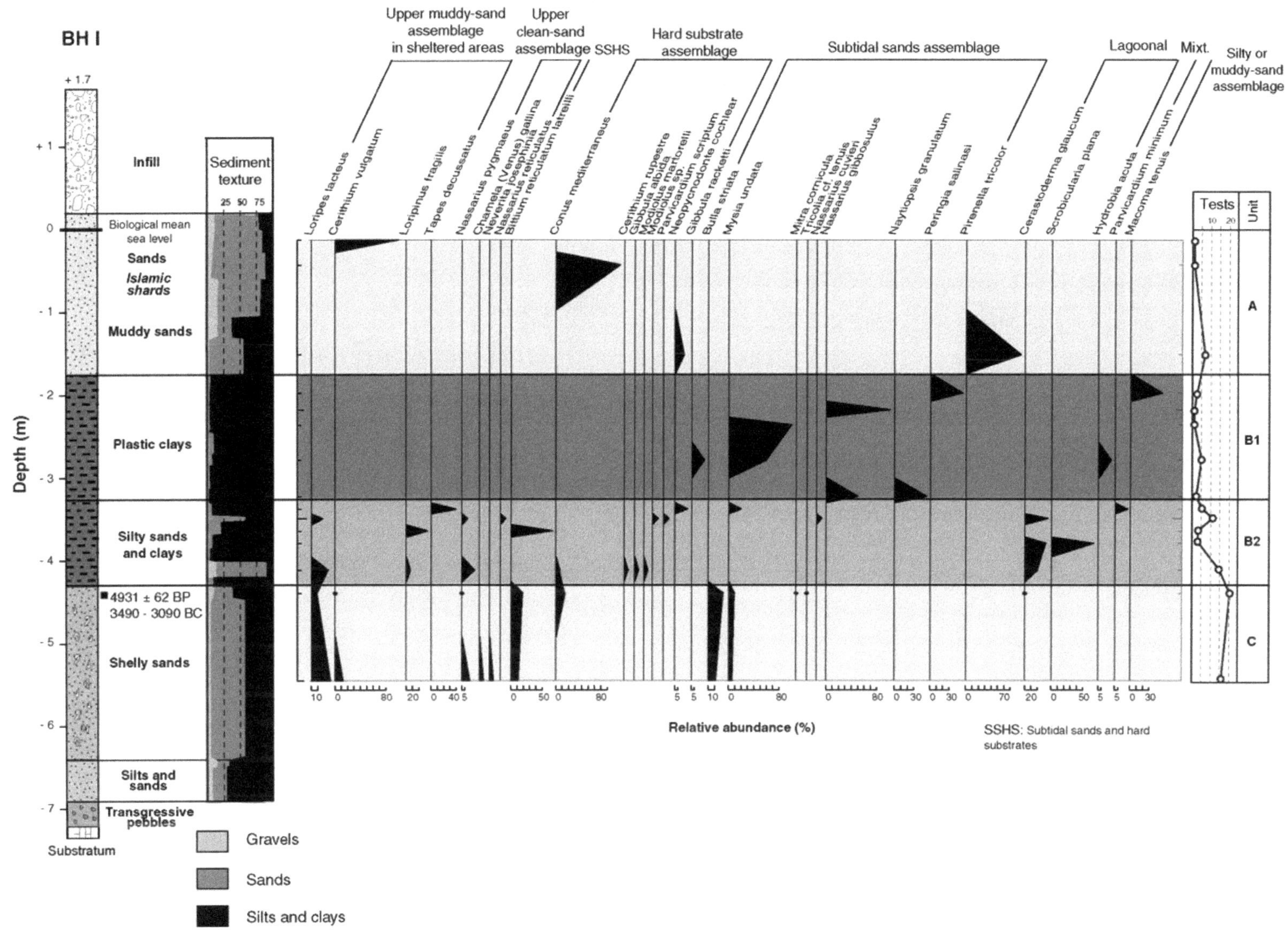

Figure 4.12: Molluscan macrofauna from core BH I (northern harbour). (1) Unit C is dominated by taxa from the upper muddy-sand assemblage in sheltered areas (*Loripes lacteus* and *Cerithium vulgatum*), upper clean-sand assemblage (*Neverita josephinia*, *Nassarius pygmaeus* and *Chamela gallina*) and the infralittoral sands assemblage (*Bulla striata* and *Mysia undata*). These groups are consistent with a natural semi-sheltered beach environment, which existed during the Bronze Age. (2) In unit B2 the dominant groups include the lagoonal assemblage, the upper clean-sand assemblage and the upper muddy-sand assemblage in sheltered areas. Rise in lagoonal taxa is consistent with artificial sheltering of the environment (i.e. man-made harbourworks). (3) Taxa from the subtidal sands assemblage (*Peringia salinasi*, *Nassarius gibbosulus* and *Mysia undata*) dominate unit B1. Secondary groups include lagoonal and the silty or muddy-sand assemblage, with a low species number indicating little interaction with the open sea. (4) Unit A is characterised by species from the subtidal sands assemblage, the hard substrate assemblage and the upper muddy-sand assemblage in sheltered areas. It translates a reopening of the environment to the influence of offshore marine dynamics, during the Middle Age.

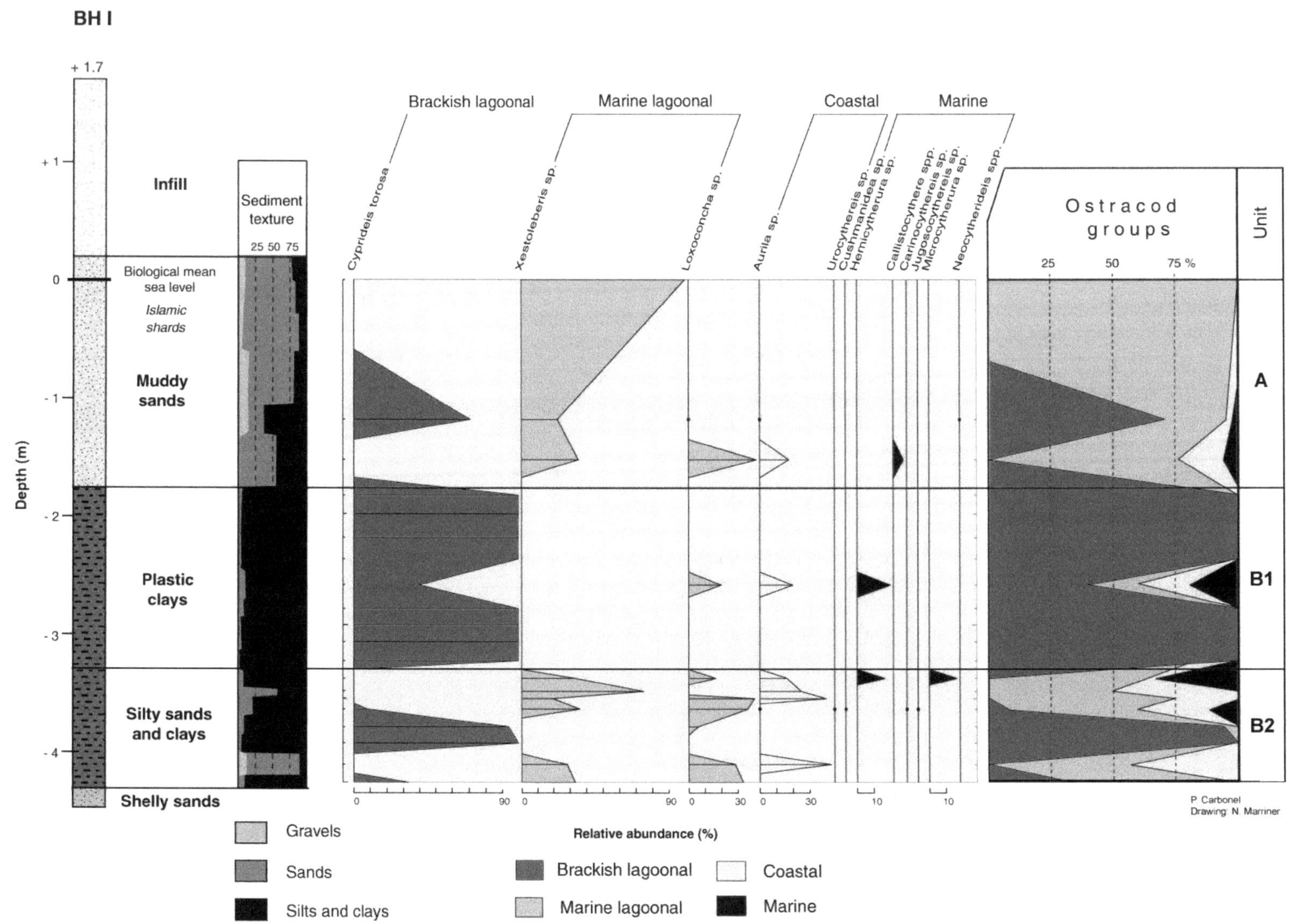

Figure 4.13: Ostracod microfauna from core BH I (northern harbour). Ostracods were only extracted from units B2, B1 and A. (1) Unit B2 is dominated by taxa from the brackish lagoonal (*Cyprideis torosa*) and marine lagoonal (*Xestoloberis* sp. and *Loxoconcha* sp.) ecological groups, indicating a sheltered harbour environment. Drifted in valves of coastal (*Aurila* sp.) and marine (*Microcytherura* sp. and *Hemicytherura* sp.) taxa attest to continued exposure to offshore marine dynamics and swell. (2) With the exception of a probable storm event in level 258 cm, importing coastal and marine taxa, unit B1 is characterised by the monospecific dominance of *Cyprideis torosa*. This taxon bears witness to a very well-protected harbour throughout the Byzantine period. As in Tyre, we interpret this as the harbour apogee. (3) Gradual reopening of the environment is indicated in unit A, by an increase in marine lagoonal (*Loxoconcha* sp. and *Xestoleberis* sp.) and coastal taxa (*Aurila* sp.), to the detriment of *Cyprideis torosa*. We infer a gradual demise of Sidon as a commercial centre during the Middle Age.

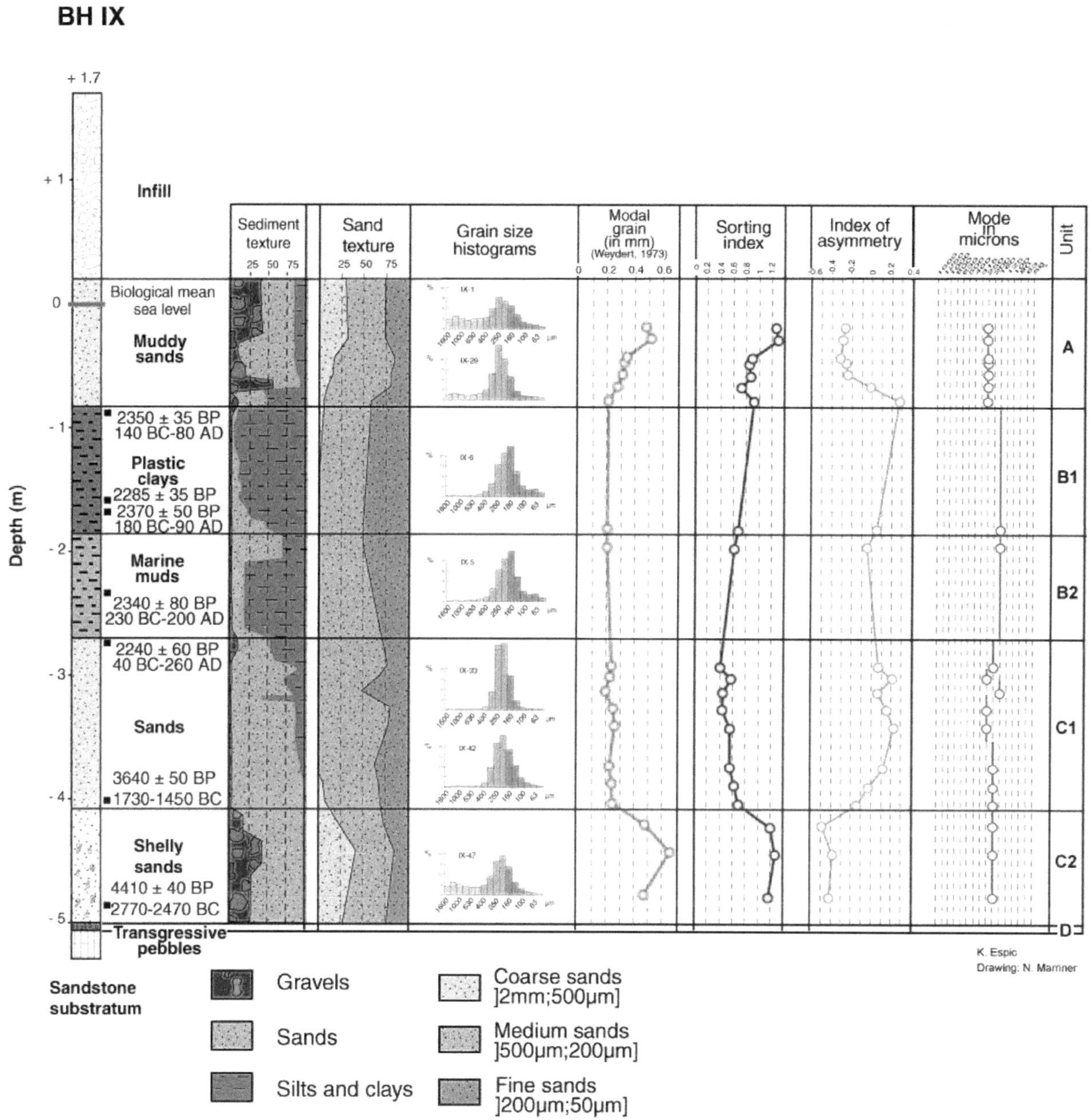

Figure 4.14: Sedimentology of core BH IX (northern harbour). (1) Unit D comprises a layer of sandstone pebbles encrusted with marine fauna (for example *Serpulae*), and is consistent with the marine transgression of the environment. (2) Unit C2 is characterised by a shelly sand unit, with poor sorting indices (1.17-1.27) and sand modal values of 200 µm. The sedimentological data concur a middle energy beach environment. The bottom of the unit is constrained to 4410 ± 40 BP. (3) A fall in energy dynamics in unit C1 is marked by a rise in the silts fraction (up to 59 %). The medium sands fraction of the histograms are well-developed, and sorting indices of 0.4-0.68 indicate a well to medium sorted sediment. The base of the unit is constrained to 3640 ± 50 BP. We interpret this unit as corresponding to the Middle Bronze Age proto-harbour, with the possible reinforcement of the sandstone ridge to improve shelter in the anchorage haven (i.e. cothon style harbour infrastructure). (4) Increased shelter in unit B2 is indicated by a sharp rise in the silts and clays fractions to between 31 % and 82 %. Grain size histograms are well developed in the fine sands fraction. The unit is dated ~2340 ± 80 BP, corresponding to the Roman period. As at Tyre, we interpret unit B1 and B2 as having been dredged, removing Phoenician and Hellenistic harbour layers. (5) Accentuated marine shelter is characterised in unit B1 by a plastic clay unit exceeding 75 % of the total dry sediment weight. This unit corresponds to the harbour apogee, with extensive harbour infrastructure. (6) The gradual demise of Sidon is represented in unit A by a rise in the gravels and sands fractions to the detriment of the silts and clays. A shift to higher energy modal values (250 µm) marks increased exposure of the environment to outer marine dynamics.

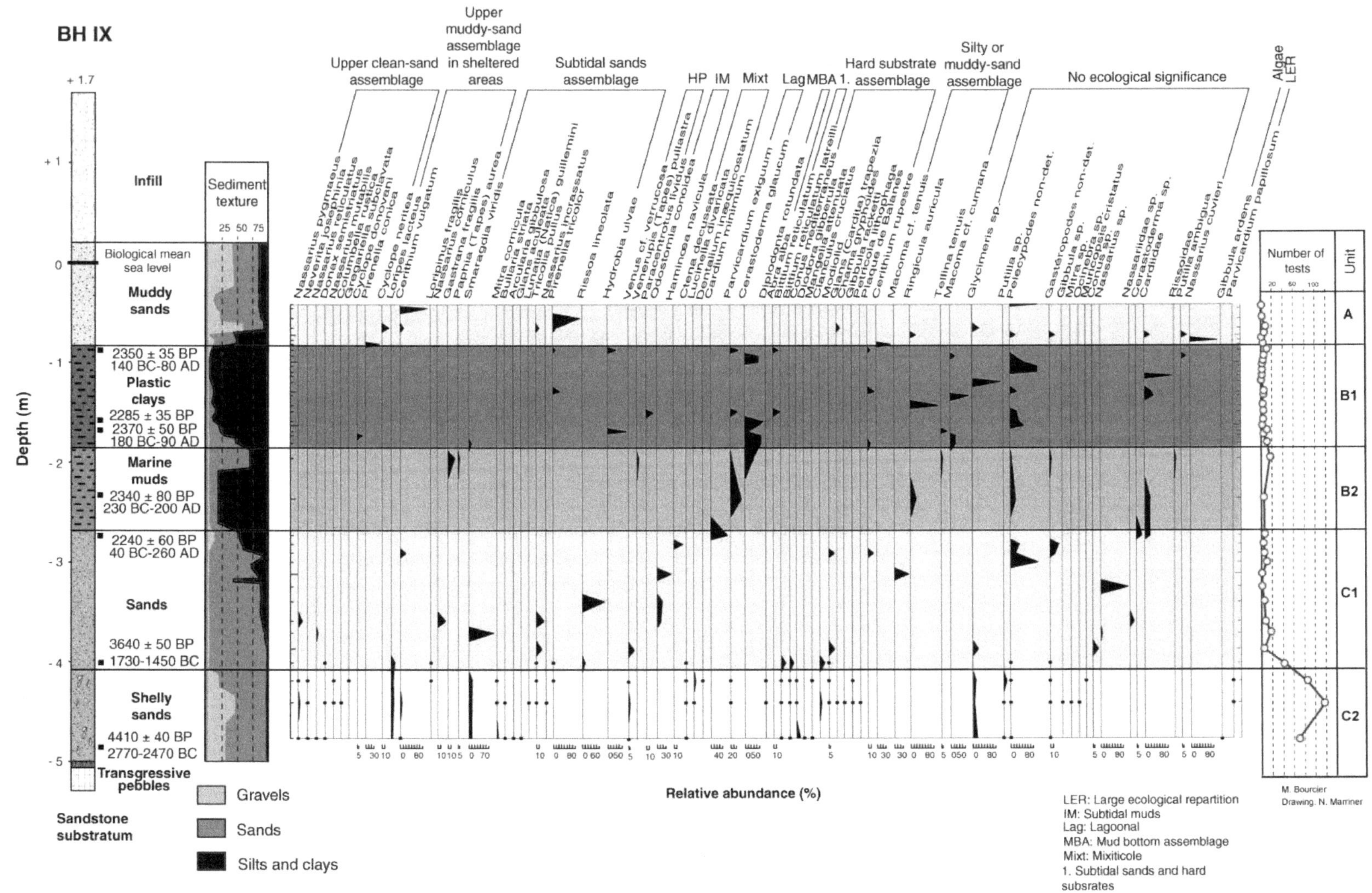

Figure 4.15: Molluscan macrofauna from core BH IX (northern harbour). (1) Unit C2 is characterised by taxa from the following assemblages: subtidal sands assemblage, upper clean-sand assemblage (*Smaragdia viridis* and *Nassarius pygmaeus*) and the upper muddy-sand assemblage in sheltered areas (*Loripes lacteus* and *Cerithium vulgatum*). These suggest a semi-open beach environment. Interaction with the sea is corroborated by low levels of tests from diverse *extra situ* ecological environments. (2) Unit C1 is characterised by taxa from the subtidal muds assemblage (*Odostomia conoidea* and *Haminœa navicula*), subtidal sands assemblage, upper muddy-sand assemblage in sheltered areas, upper clean-sand assemblage and silty or muddy-sand assemblage. These all corroborate a natural semi-sheltered environment, which served as a proto-harbour during the Middle Bronze Age. (3) Unit B2 is dominated by lagoonal taxa (*Parvicardium exiguum* and *Cerastoderma glaucum*) and silty or muddy-sand assemblage taxa. These groups attest to an artificial sheltering of the environment, linked to harbourworks. (4) The lagoonal and silty or muddy-sand assemblages continue to characterise the *in situ* taxa of unit B1, concurrent with confined harbour conditions during the Byzantine period. (5) Unit A is characterised by diverse taxa from the following groups: upper muddy-sand assemblage in sheltered areas, upper clean-sand assemblage, silty or muddy-sand assemblage and infralittoral sands assemblage.

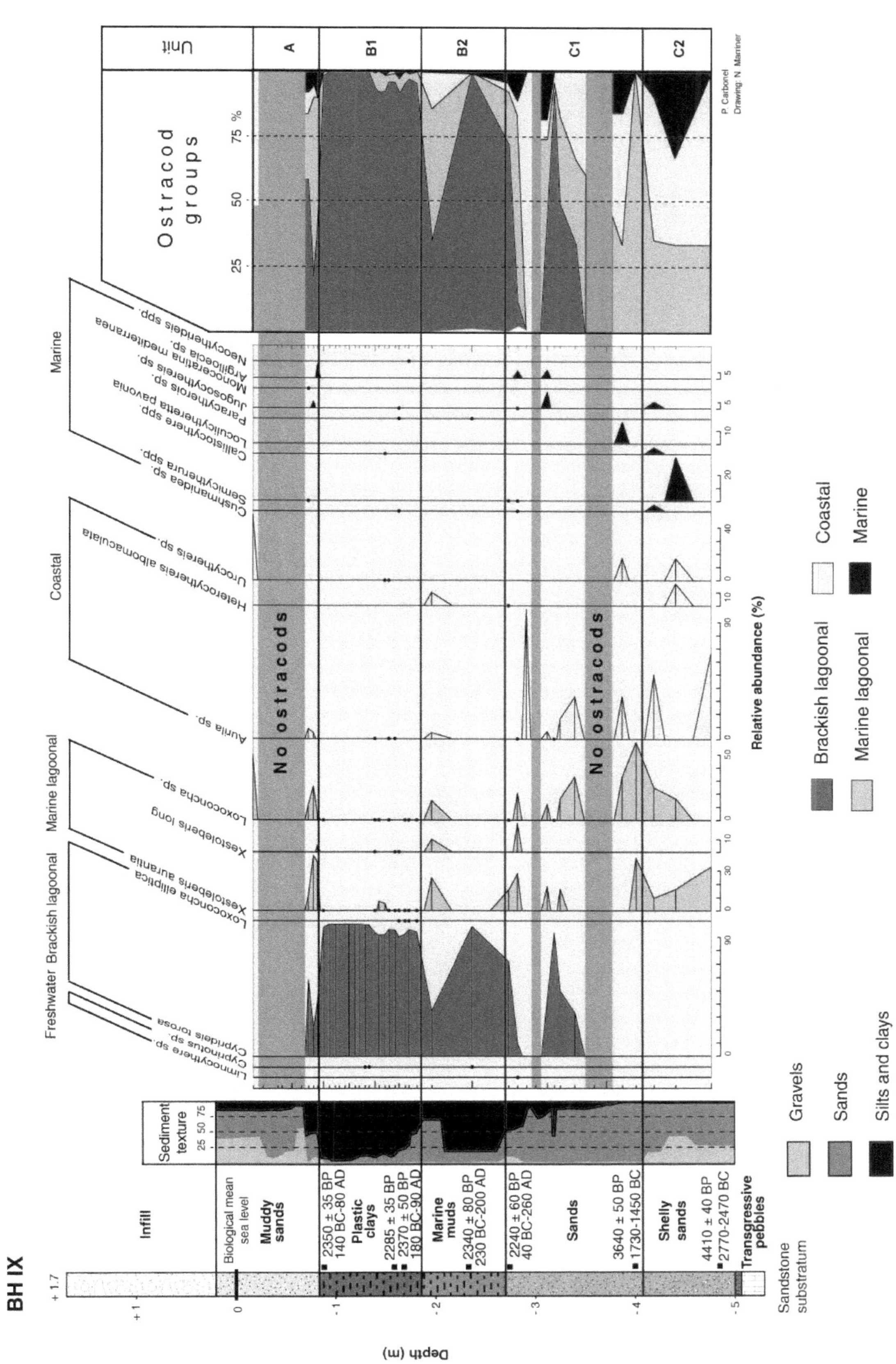

Figure 4.16: Ostracod microfauna from core BH IX (northern harbour). (1) Unit C2 is dominated by taxa from the marine lagoonal (*Loxoconcha* sp. and *Xestoleberis aurantia*) and coastal (*Urocythereis* sp., *Heterocythereis albomaculata*, *Aurila* sp.) domains with some marine taxa being drifted in. These attest to a semi-protected natural environment, sheltered by the sandstone ridge. (2) The bottom of unit C1 continues to be dominated by marine lagoonal and coastal taxa, with a rise in the brackish lagoonal taxon *Cyprideis torosa* towards the top of the unit. We infer a Middle Bronze Age harbour basin (the base of the unit is dated 3640 ± 50 BP), with the possible beginnings of a cothon style reinforcement of the sandstone ridge. (3) Continued dominance of *Cyprideis torosa* in unit B2 indicates an artificially sheltered harbour environment during the Roman period. Interaction with the open sea is indicated by secondary taxa from coastal and marine environments. (4) Quasi-dominance of *Cyprideis torosa* in unit B1 (>95 % relative abundance) indicates an extremely sheltered harbour, acting like a leaky lagoon with poor interaction with the open sea during the Byzantine period. (5) Exposure of the environment in unit A is indicated by a rise in marine lagoonal and coastal taxa, to the detriment of *Cyprideis torosa*.

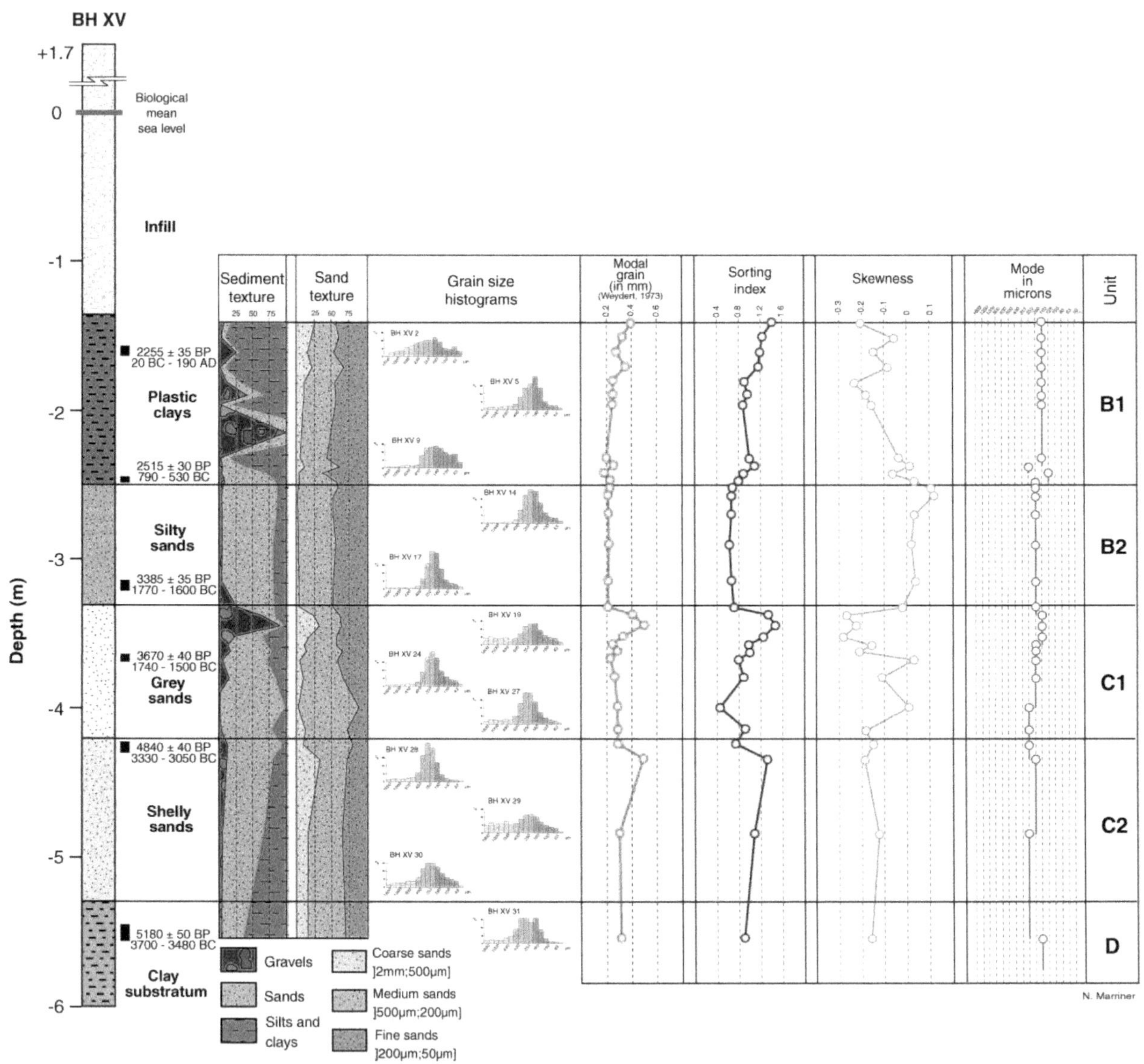

Figure 4.17: Sedimentology of core BH XV (northern harbour). (1) Unit D: comprises the transgressive contact, characterised by a reworking of the underlying Pleistocene clay substratum. (2) Unit C2: is characterised by medium-grained shelly sands. (3) Unit C1: A fall in modal values from 250 µm to 160 µm is concurrent with a drop in energy dynamics. We interpret this as early evidence for Bronze Age port infrastructure. (4) Unit B2: Transition to a silty sand unit, dominated by medium to fine sands, is consistent with a further fall in energy dynamics. (5) Unit B1: Poorly sorted, low energy plastic clays corroborate a low energy artificial harbour, protected by extensive port infrastructure. Pronounced peaks in the gravels fraction are associated with large wood fragments, seeds and ceramics, waste trapped in the harbour by the prevailing low energy conditions.

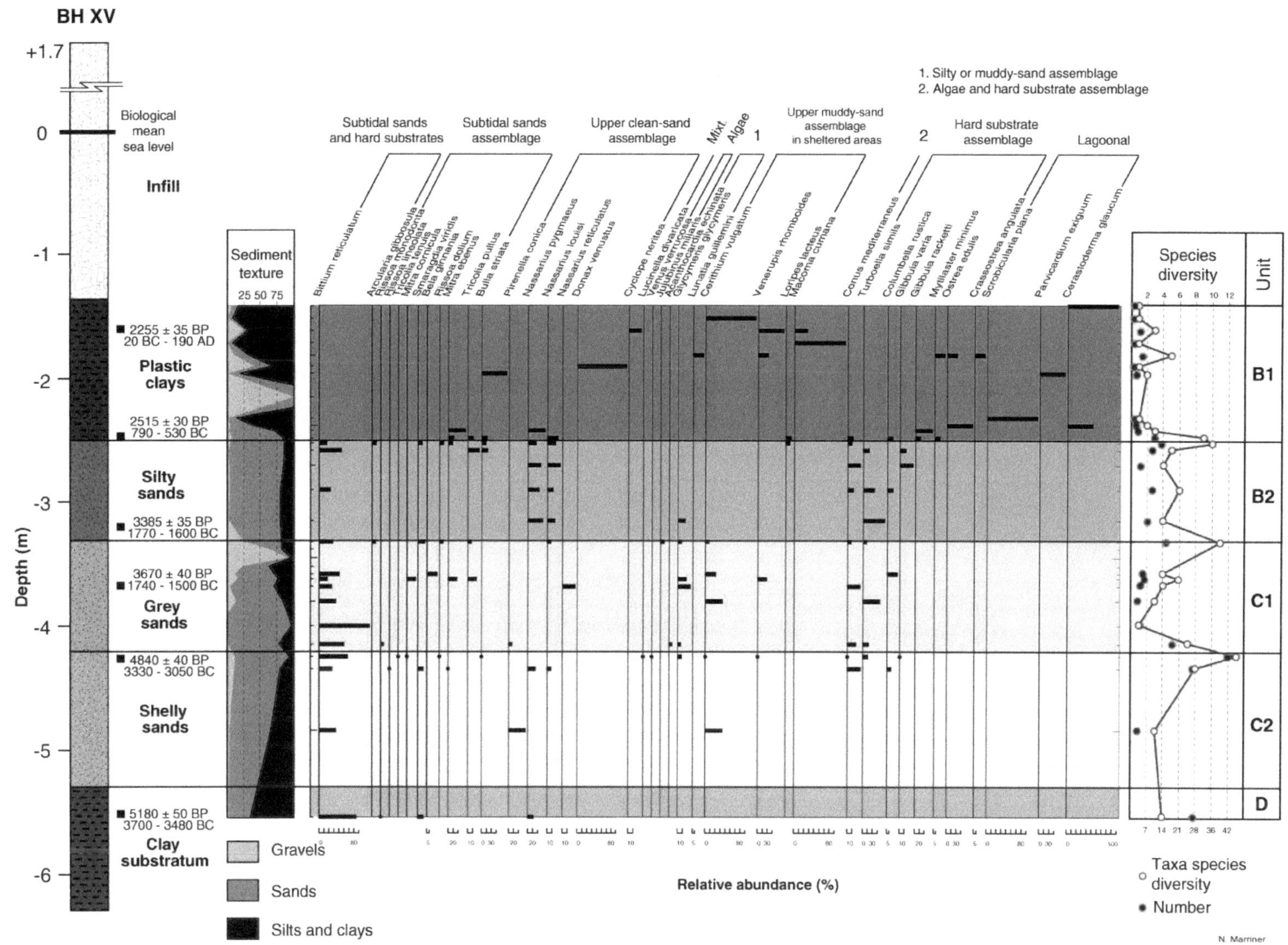

Figure 4.18: Molluscan macrofauna from core BH XV (northern harbour). (1) Unit D: The macrofauna suite is dominated by *Bittium reticulatum*, characteristic of subtidal sands and hard substrates. Secondary taxa include *Smaragdia viridis* (subtidal sands assemblage) and *Nassarius pygmaeus* (upper clean-sand assemblage). (2) Unit C2: A shift to medium-grained shelly sands in unit C2 is concurrent with subtidal sands and hard substrates, subtidal sands assemblage and upper clean-sand assemblage molluscan taxa. (3) Unit C1: Dominant molluscan groups include subtidal sands and hard substrates assembage, subtidal sands assemblage, upper clean-sand assemblage, silty or muddy-sand assemblage (*Glycymeris glycymeris*) and upper muddy-sand assemblage in sheltered areas (*Venerupis rhomboides* and *Cerithium vulgatum*). (4) Unit B2: Molluscan groups analogous of unit C1 persist into this facies and attest to a proximal shoreface. (5) Unit B1: A sharp rise in lagoonal molluscs (*Scrobicularia plana*, *Parvicardium exiguum* and *Cerastoderma glaucum*) concurs a well-protected port basin.

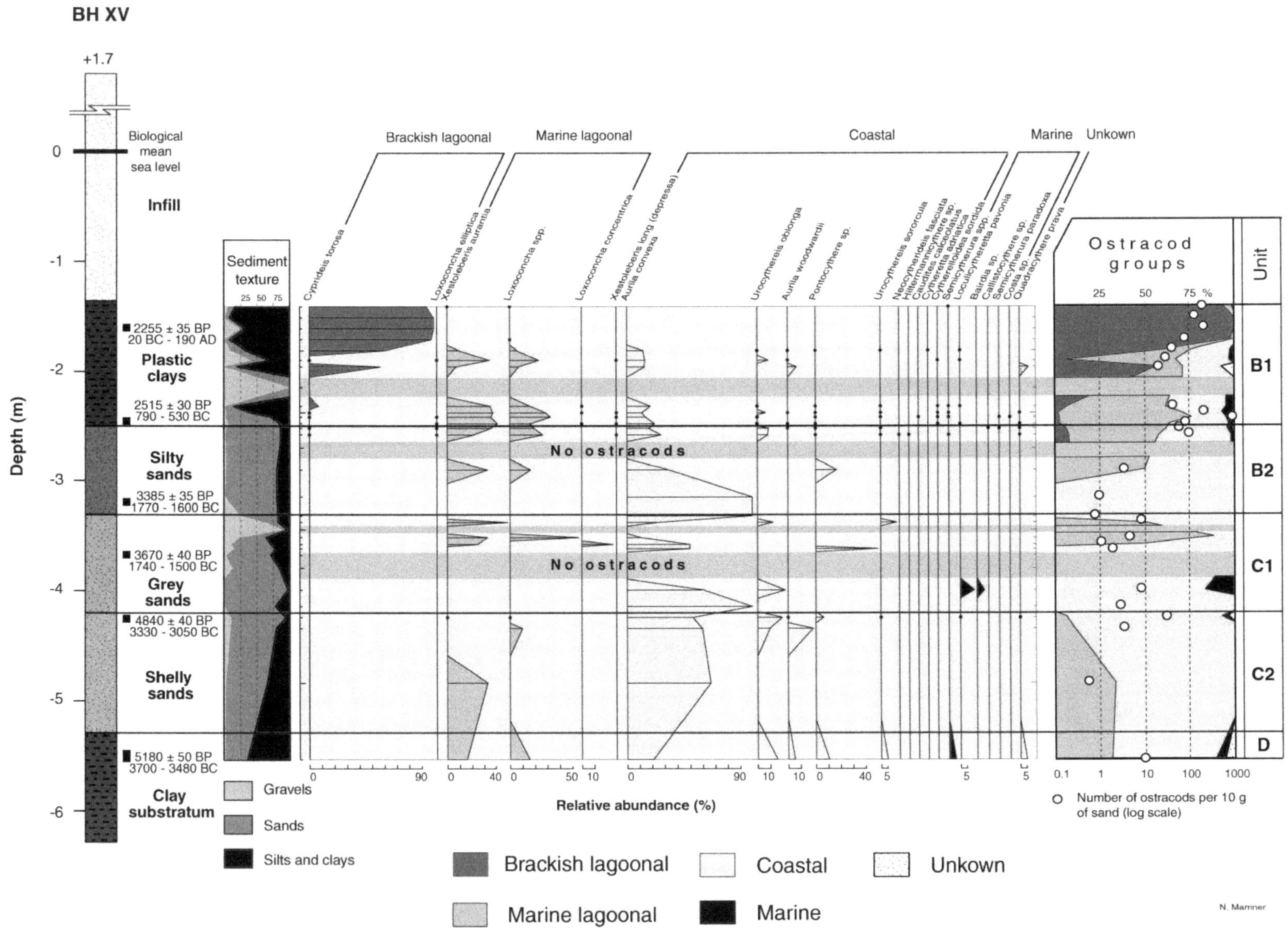

Figure 4.19: Ostracod microfauna from core BH XV (northern harbour). (1) Unit D: Marine lagoonal (*Xestoleberis aurantia* and *Loxoconcha* spp.) and coastal species (*Aurila convexa*, *Urocythereis oblonga*, *Aurila woodwardii*, *Pontocythere* sp. and *Urocythereis sororcula*) constitute the ostracod fauna. (2) Unit C2: Coastal ostracods dominate the unit. (3) Unit C1: *Aurila convexa* continues to dominate the ostracod fauna with other secondary coastal taxa including *Urocythereis oblonga*, *Aurila woodwardii* and *Pontocythere* sp. (4) Unit B2: Coastal ostracods persist into this facies and attest to a proximal shoreface. (5) Unit B1: The monospecific domination of the brackish water species *Cyprideis torosa* concurs a well-protected harbour basin.

Interpretation: At the time of Sidon's foundation, during the third millennium BC, maritime harbour technology was still very primitive (Marcus, 2002a; Fabre, 2004/2005). Existing Bronze Age evidence from the Levant shows a clear pattern of environmental determinism, where coastal populations founded settlements in proximity to naturally occurring anchorages such as leaky lagoons, estuaries and pocket beaches (Raban, 1987a; Raban, 1990). At Bronze Age Sidon, the northern pocket beach was ideally predisposed to serve as a proto-harbour. This northern cove, with its shelly medium sands facies, afforded the best natural shelter for larger merchant boats during storms.

4.4.2.3 Semi-artificial Bronze Age cove

Description: A fall in energy dynamics is translated by a rise in the silts fraction (up to 59 %). Medium sands dominate, with a well to medium sorted sediment. In BH IX, the base of the unit is constrained to 3640 ± 50 BP (1730-1450 cal. BC), a date confirmed by data from BH XV (3670 ± 40 BP or 1740-1500 cal. BC). We interpret this unit as corresponding to the Middle Bronze Age to Late Bronze Age proto-harbour, with possible reinforcement of the sandstone ridge improving the anchorage quality at this time. Small boats would have been hauled onto the beach face, with larger vessels being anchored in the embayment. Molluscan taxa include tests from the subtidal muds assemblage (*Odostomia conoidea* and *Haminœa navicula*), subtidal sands assemblage (*Tricolia pullus*, *Mitra ebenus*, *Rissoa dolium*, *Bela ginnania* and *Mitra cornicula*), upper muddy-sand assemblage in sheltered areas (*Loripinus fragilis*, *Nassarius corniculus* and *Cerithium vulgatum*), upper clean-sand assemblage (*Nassarius reticulatus*) and the silty or muddy-sand assemblage (*Glycymeris glycymeris*). The ostracod data evince continued domination of marine lagoonal and coastal taxa, with a rise in the brackish lagoonal taxon, *Cyprideis torosa*, towards the top of the unit in BH IX. These all corroborate a semi-sheltered environment that served as a proto-harbour during the Bronze Age (Frost, 1995; Raban, 1995).

Interpretation: Standardisation in ceramic vessel production and an increase in the number of sherds unearthed by the British Museum excavations are significant indicators of developments in urbanization during the late Early Bronze and early Middle Bronze (Doumet-Serhal, 2004c). Cretan and Egyptian archaeological material attests to trade relations between the Levantine coast and the Aegean and Egypt during the Middle Bronze Age (**Figure 4.20**; MacGillivray, 2003; Doumet-Serhal, 2006b; Forstner-Müller *et al.*, 2006; Griffiths and Ownby, 2006). Aegean imports became more numerous during the Late Bronze, indicative of expanding Mediterranean trade.

Concurrent with these terrestrial data, the coastal stratigraphy suggests that towards the end of the Middle Bronze Age and the Early Iron Age (~1200-1000 BC), growing Mediterranean trade prompted coastal populations into modifying these natural anchorages (Marcus, 2002a-b). For a detailed discussion on early artificial harbourworks the reader is referred to chapter 3. Le Roux *et al.* (2003a) have also evidenced Bronze Age lead contamination in Sidon's ancient harbour sediments, concomitant with the growth of the settlement at this time (**Figure 4.21**). Following a Middle Bronze Age pollution signal, lead pollution attains peak levels of ~60-80 mg/kg during the Roman period. We discuss these data in more detail in chapter 6.

Figure 4.20: A Minoan cup from a Middle Bronze Age context found at the College site in 2002 (in Doumet-Serhal, 2004a). Minoan, Mycenean and Euboean imports unearthed in Bronze Age and Iron Age habitation layers underline the importance of the sea and inter-state trade contacts at Sidon. Clear patterns are observed between the archaeological finds on the tell and the harbour stratigraphy, which translates a concomitant shift to more complex seaport installations to accommodate this expanding trade.

In Sidon's northern harbour, transition from shelly to fine-grained sands appears to be the earliest granulometric manifestation of human coastal modification. A single radiocarbon date from core BH XV constrains this alteration to the Middle Bronze Age (~1700 cal. BC) and must be confirmed by further data. The Phoenicians cleverly quarried sandstone ridges to form artificial quays and reinforce breakwaters (Frost, 1995). Surplus blocks were frequently reemployed to either construct (i.e. Arwad) or reinforce (i.e. Sidon and Tripoli) a seawall (Carayon, 2003; Carayon and Viret, 2004; Viret, 1999-2000, 2004, 2005) (**Figure 4.22**). In Sidon's northern harbour, urban redevelopment means that these vestiges are no longer visible (**Figure 4.23**), but they have been described by Arvieux (1735), Renan (1864), Lortet (1884) and surveyed by Poidebard and Lauffray (1951).

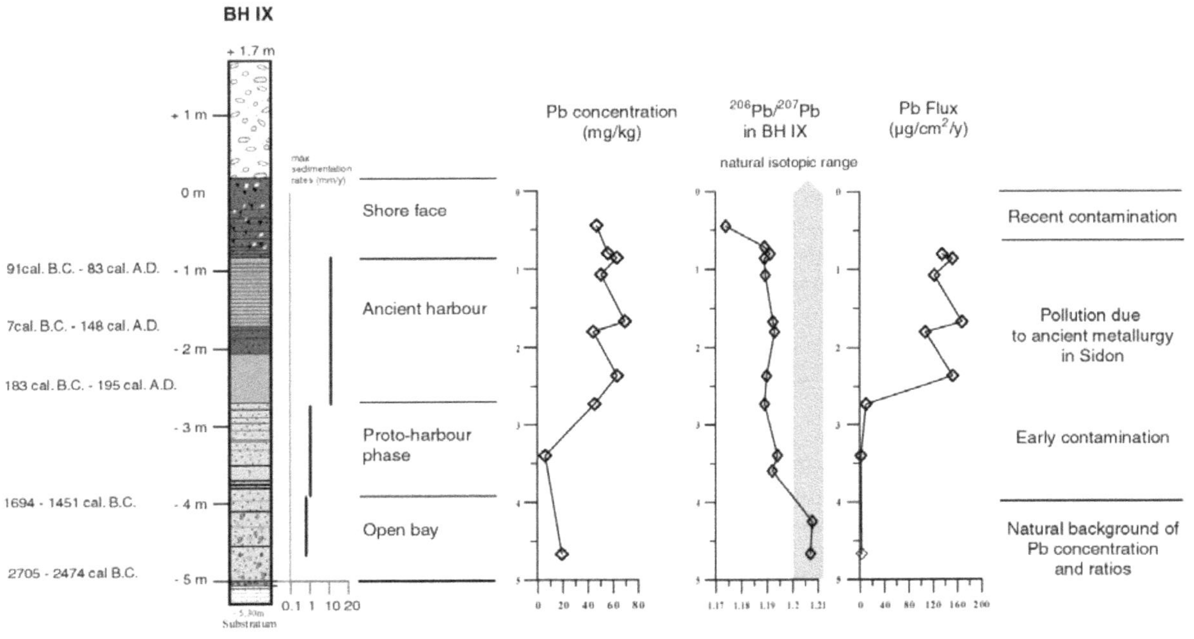

Figure 4.21: Pollutant lead contamination in Sidon's ancient harbour (adapted from Le Roux *et al.*, 2003). Although there is not a pronounced change in lead concentrations during the Bronze Age, the lead isotope signature attests to a clear shift in $^{206}Pb/^{207}Pb$ ratios. Isotope signatures are a more reliable means of reconstructing anthropogenic activities due to the dilution processes involved in measuring lead concentrations. The early shift in $^{206}Pb/^{207}Pb$ indicates the initial growth and expansion of the settlement before the well-defined foundation of the harbour. The Roman-Byzantine apogee is characterised by lead pollution levels equivalent to those observed in modern day harbours and estuaries.

Figure 4.22: Sidon's northern harbour and its ancient seawall. An uplifted marine notch is observed at ~50 cm above MSL. These are the last archaeological photographs available of the aeolianite ridge, concreted over during development of the northern fishing harbour (photographs: C. Morhange, 1998).

The lee of Zire was also exploited as a deep-water anchorage, or outer harbour, at this time although the island's two jetties date from the Persian period. In fair to medium weather large merchant vessels would have loaded and unloaded goods at this geomorphologically predisposed hub, their cargos ferried to and from the shoreline by lighters (Marcus, 2002a-b; Wachsmann, 1998). A whole suite of harbourworks, including seawalls, quays and mooring bits have been dug into the Quaternary sandstone, rendering Zire an integral component of Sidon's port system (Carayon, 2003; **Figure 4.23**).

Figure 4.23: Since 1998, Sidon's northern harbour has been extensively modernised, destroying great tracts of the ancient harbourworks (Marriner and Morhange, 2005). Better urban planning would have saved this unique cultural resource, which could potentially have been converted into an open air museum with obvious benefits to the tourist industry (image: DigitalGlobe, 2006).

The biosedimentological datasets show, however, that it was the northern harbour, naturally protected from the open sea by a sandstone ridge, which became the city's primary port basin. Difficulty in dating the first phase of artificial confinement, at both Sidon and Tyre, appears concurrent with two complimentary dynamics: (1) modest artificial harbourworks during the late Middle Bronze Age and Late Bronze Age; and/or (2) intense dredging during the Roman and Byzantine occupations.

4.4.2.4 Closed Iron Age to Roman harbours

Description: In unit B2 there is a net change in sedimentary conditions with a shift to silts and fine grained sands. Modal grain size values of 200-160 µm and medium sorting indices are concomitant with a low energy environment (**Figures 4.11, 4.14 and 4.17**). Dominant molluscan groups include the lagoonal assemblage (*Cerastoderma glaucum*, *Parvicardium exiguum* and *Scrobicularia plana*), the upper clean-sand assemblage (*Nassarius louisi* and *Nassarius pygmaeus*) and the upper muddy-sand assemblage in sheltered areas (*Gastrana fragilis*). A rise in lagoonal taxa is in compliance with anthropogenic sheltering of the environment by harbourworks (**Figures 4.13, 4.16 and 4.19**). Ostracod fauna is poor, with less than 50 tests per 10 g of sand, and dominated by taxa from the brackish lagoonal (*Cyprideis torosa*) and marine lagoonal (*Xestoloberis* sp. and *Loxoconcha* sp.) ecological groups, all indicative of a protected environment. In BH XV, peaks of *Aurila convexa* are concomitant with a proximal shoreface. Drifted-in valves of marine taxa (*Semicytherura* sp., *Microcytherura* sp. and *Hemicytherura* sp.) attest to continued connection with the open sea and offshore marine dynamics. In cores BH IX and BH XV, radiocarbon dates constrain the chronology of the unit to 2515 ± 30 (790 - 530 cal. BC) and 2340 ± 80 BP (230 cal. BC - 200 cal. AD), or the Phoenician/Persian to Roman periods. The lower foundation courses of Zire island's jetties have also been dated to the Persian period (Frost, 2000b; Carayon, 2003; **Figure 4.24**). Persistent age-depth anomalies concur analogous data in Tyre's ancient harbour, where strong chronostratigraphic evidence for dredging has been detailed from the Roman period onwards.

Interpretation: The British Museum excavations have shown that there was regular trade between Phoenician merchants and an Aegean clientele during the first millennium BC (Doumet-Serhal, 2003). Greek merchants at Sidon are also attested to by numerous finds of Euboean pottery (Doumet-Serhal, 2004a). Such important trade clearly required advanced seaport infrastructure and docking facilities. Unfortunately, given the relative absence of pre-Hellenistic deposits, high-resolution reconstruction of the northern harbour's Phoenician history is problematic. Advanced harbour management techniques during the Roman and Byzantine periods culminated in the repeated dredging of Sidon's northern harbour, removing this stratum from the geological record (**Figure 4.25**). In cores BH I and BH IX, there is an unambiguous sediment hiatus between ~1700-1500 cal. BC and the Roman period. As at Tyre, reworking of the marine bottom is delineated by the chronostratigraphic dataset and supports unequivocal evidence from Marseilles (Hesnard, 2004a-b) and Naples (Giampaola *et al.*, 2004; Giampaola and Carsana, 2005). In BH XV, a more coherent and continuous chronology suggests that this area was less affected by dredging activity. The data evoke advanced harbourworks during the Phoenician and Persian periods.

Siltation, notably under deltaic and urban contexts, was a well-recognised problem in antiquity with four sedimentary sources of note: (1) local watercourses (Ribes *et al.*, 2003); (2) biogenic sands reworked by longdrift currents; (3) erosion of adobe constructions and urban runoff (Rosen, 1986; Doumet-Serhal, 2006c); and (4) use of the basin as a base-level waste dump (Morhange, 1994). Sidon's gravels fraction from the Roman period comprises a suite of discarded objects trapped at the bottom of the basin, including ceramics, wood, seeds, leather artefacts etc. Indeed, an inscription from Roman Ephesus, demanding citizens not to throw waste into the port, underlines that ancient societies must have been acutely aware of this problem (Die Inschriften von Ephesos, Bonn, I, 1979, no. 23).

It is postulated that extensive dredging during the Roman and Byzantine period explains (1) the observed stratigraphic hiatus; and (2) dating inversions. Previously, these problems, in the absence of robust chronological frameworks, were most often ignored or left unexplained.

4.4.2.5 Closed late Roman and early Byzantine harbours

Description: Accentuation of these low energy conditions is manifested in unit B1 by a plastic clays unit. Throughout much of this facies, the silts and clays fraction comprises >90 % of the total sediment. Radiocarbon ages are often stratigraphically incoherent for this unit. Lagoonal (*Cerastoderma glaucum* and *Parvicardium exiguum*) and upper muddy-sand assemblage in sheltered areas taxa (*Cerithium vulgatum*, *Venerupis rhomboides*, *Loripes lacteus* and *Macoma cumana*) continue to characterise the *in situ* species, attesting to confined harbour conditions. Quasi-dominance of *Cyprideis torosa*, with a rich faunal density, indicates an extremely sheltered, lagoon-like harbour during the late Roman and early Byzantine periods.

Interpretation: By Roman times, the discovery and use of hydraulic concrete greatly enhanced engineering possibilities (cf. Vitruve, V, 12), locally deforming coastal landscapes (Brandon, 1996; Oleson, *et al.*, 2004a). At this time we observe pronounced transition from environmental to anthropogenic determinism. The Romans were able to conceive long breakwaters or offshore harbour basins, Caesarea Maritima and Anzio being examples *par excellence* (Felici, 1993; Blackman, 1996). All-weather basins could be constructed at locations where no natural roadstead existed. During this technological revolution, Sidon's northern harbour underwent significant changes with the formation of an inner artificial mole perpendicular to the sandstone ridge (Poidebard and Lauffray, 1951). This yielded an extremely well-protected basin, translated in the geological archive by a silt facies containing lagoonal molluscs and microfossils.

Under these closed conditions sedimentation rates rose significantly - ~1 mm/yr during the mid-Holocene compared to ~10 mm/yr for the Roman period - not least because of the overriding confinement, but also linked to increasing human use and abuse of the surrounding watershed that flushed sediment into coastal depocentres (Devillers, 2005). This dual phenomenon, increased confinement coupled with a rise in sediment yields, is consistently observed in base-level harbour basins throughout the circum Mediterranean. For example sedimentation rates of 20 mm/yr are recorded at Archaic Marseilles (Morhange, 1994), 15 mm/yr at Roman Alexandria (Goiran, 2001) and 10 mm/yr at Roman and Byzantine Tyre (Marriner and Morhange, 2006a). In its most acute form the ensuing coastal progradation led to harbour landlocking, isolating basins many kilometres from the sea. Celebrated examples are known from the Anatolian deltas of the Aegean and include the sites of Troy, Ephesus, Priene and Miletos (Brückner *et al.*, 2002; Kraft *et al.*, 2003; Kraft *et al.*, 2006).

In addition to artificial dredging, engineering solutions to the siltation problem have been asserted, although many of these remain speculative (Blackman, 1982b). At Sidon, Poidebard and Lauffray (1951) identified a flushing channel, carved into the sandstone to link the northern basin with the open sea. Undated, the two scholars hypothesised this to be an *ad hoc* desilting channel, dug to generate current through the inner harbour and alleviate the effects of sediment deposition. The stratigraphy shows dredging and coeval desilting infrastructure to have been insufficient in completely eradicating the problem; two thousand years later, rapid silting up means that the majority of the ancient basin is now buried beneath the modern city centre (Marriner *et al.*, 2006b).

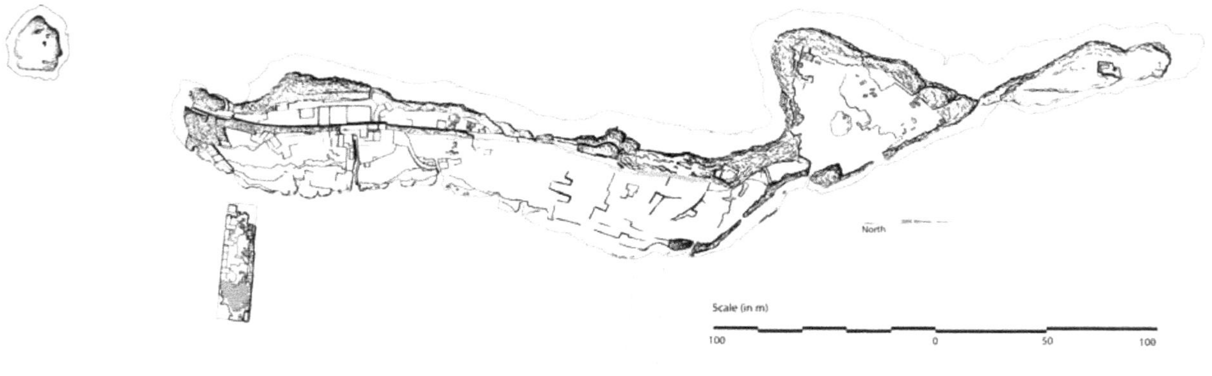

Figure 4.24: Zire island from Carayon (2003).

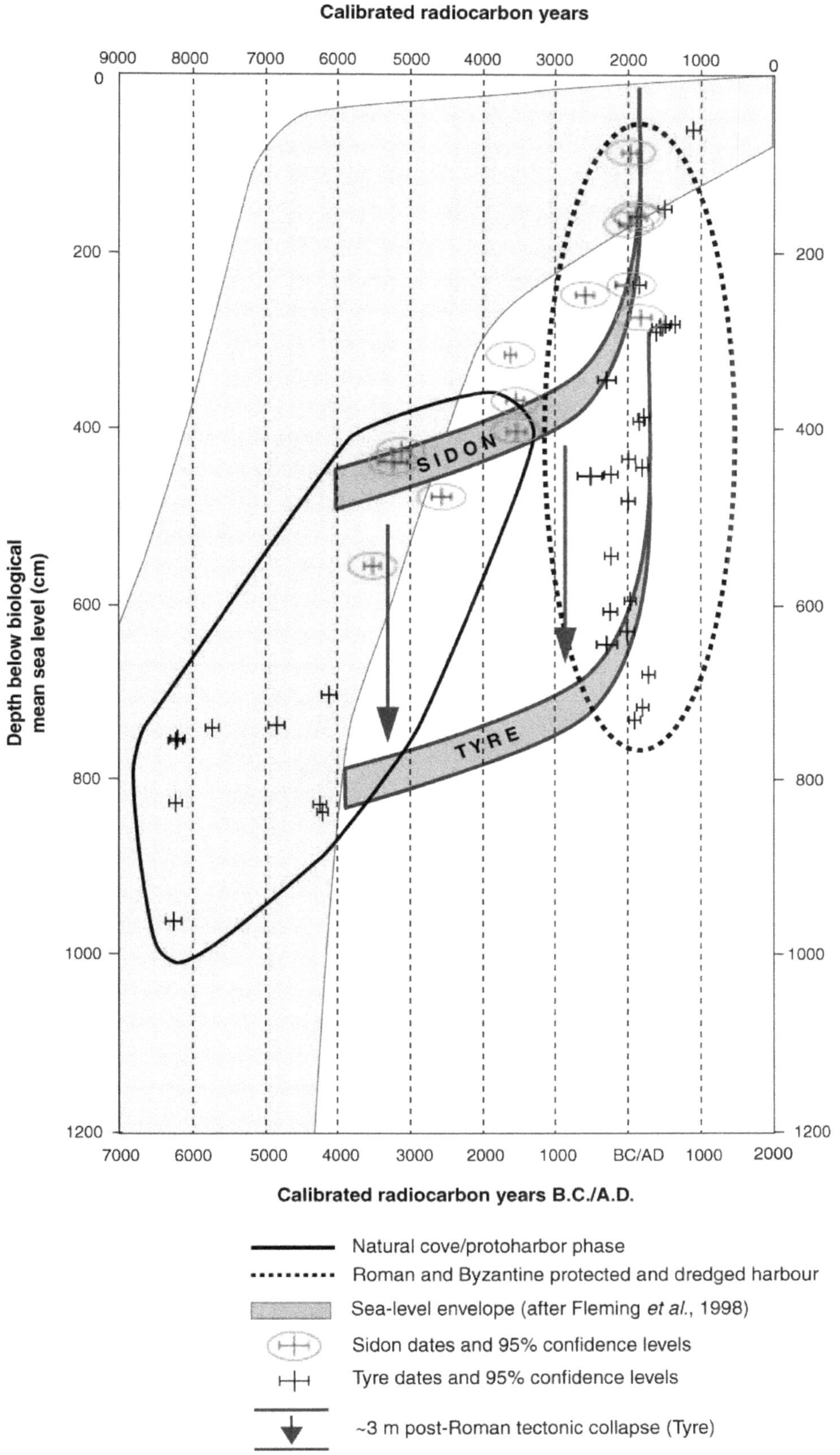

Figure 4.25: Chronostratigraphic evidence for Roman and Byzantine dredging of Sidon and Tyre's ancient harbours. The older radiocarbon group corresponds to naturally aggrading marine bottoms. Quasi-absence of chronostratigraphic record between BC 4000 to 500, coupled with persistent age depth inversions, are interpreted as evidence of harbour dredging.

While Bronze Age populations benefited from Sidon's geological endowment, Byzantine societies inherited the Romans' rich maritime *savoir faire*. The Byzantine period in Sidon is marked by advanced reinforcement of the antecedent port infrastructure, with a notable persistence of Roman technology and its opulent legacy of engineering achievements (Hohlfelder, 1997). This is corroborated by a plastic clay unit with diagnostic lagoonal macro- and microfossils, typical of pronounced confinement. These geological data support archaeological evidence from Beirut's Byzantine harbour suggesting that the Levantine coast was still an important trade zone at this time (Saghieh-Beydoun, 2005). Such a trade apex, coeval with advanced port infrastructure and management techniques, leads us to propose a harbour apogee for Sidon during the Byzantine period. This advanced port infrastructure is consistent with a pronounced lead pollution peak linked to extensive ancient metallurgy at Sidon during the Roman and Byzantine periods (Le Roux *et al.*, 2003a). Such high concentrations of lead (80 ppm and >100μg/cm^2/yr) are similar to levels (50 to 200 ppm) observed in modern contaminated harbours (Buckley *et al.*, 1995; Croudace and Cundy, 1995).

4.4.2.6 Exposed Islamic harbour

Description: The base of unit A is marked by a sudden rise in the sands fraction to the detriment of the silts and clays. The relative abundance of the gravels and sands fractions gradually increases up the unit (**Figures 4.11, 4.14 and 4.17**). Ceramics constrain this facies to the Islamic period. A shift from lower energy modal sand values at the base (100 μm) to middle energy values (up to 315 μm) at the top of the facies, could substantiate a gradual demise in harbour maintenance. Biostratigraphic data affirm a reopening of the environment to the influence of offshore marine dynamics, with taxa from the subtidal sands assemblage, the hard substrate assemblage and the upper muddy-sand assemblage (**Figures 4.12, 4.15 and 4.18**). The ostracod suites manifest an increase in marine lagoonal (*Loxoconcha* sp. and *Xestoleberis* sp.) and coastal taxa (*Aurila* spp.) to the detriment of *Cyprideis torosa* (**Figures 4.13, 4.16 and 4.19**). We posit that this corresponds to a fall in maintenance levels during the Islamic period.

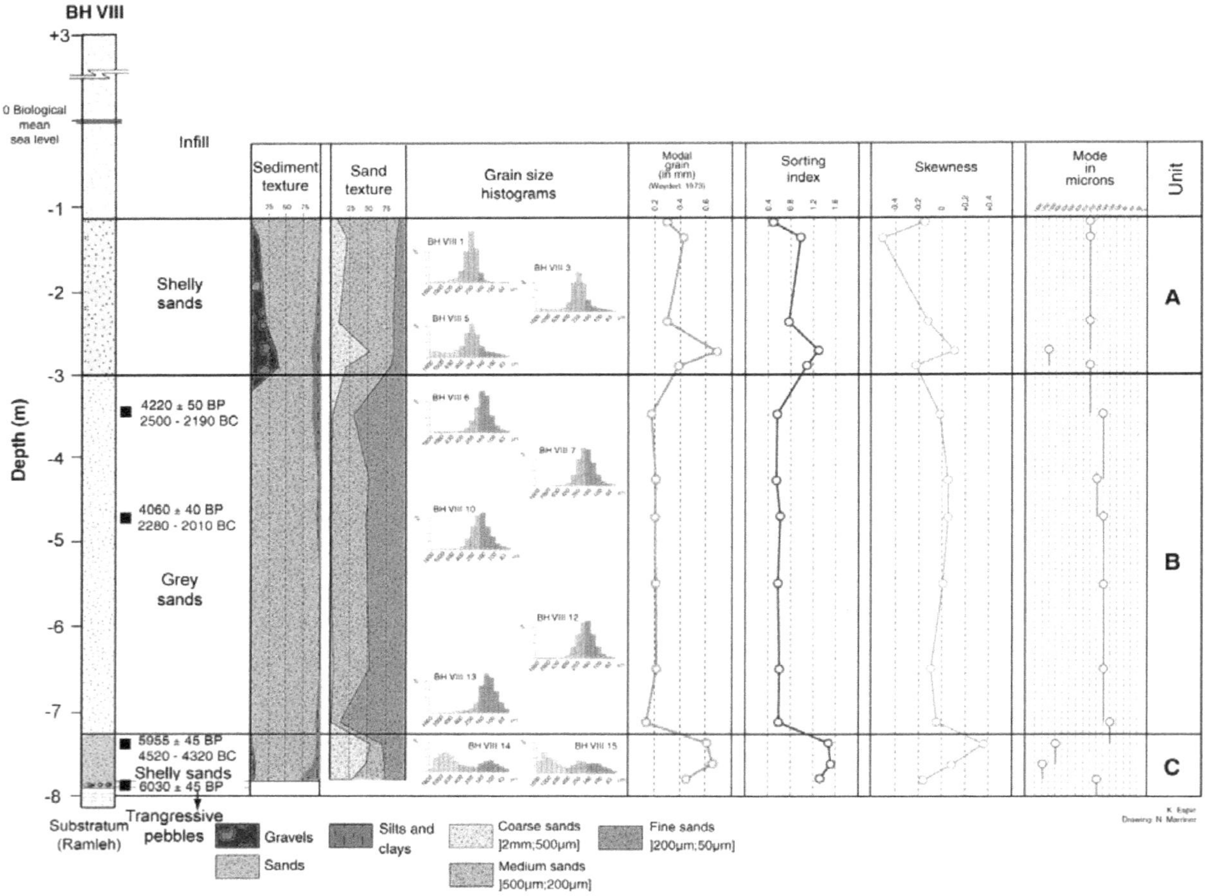

Figure 4.26: Sedimentology of core BH VIII (southern cove). (1) Unit C: is the Holocene flooding surface and is dated 6000 BP. The unit comprises poorly sorted coarse-grained shelly sands. (2) Unit B: is characterised by medium to well sorted sands. These are dominated by medium and fine sands consistent with a well-sheltered medium to low energy beach cove. We infer this cove to have been used as a proto-harbour during the Bronze Age. (3) Unit A: comprises coarser grained sands, consistent with a prograding shoreline. This progradation is linked to a loss of accommodation space under a context of stable RSL and high sediment supply.

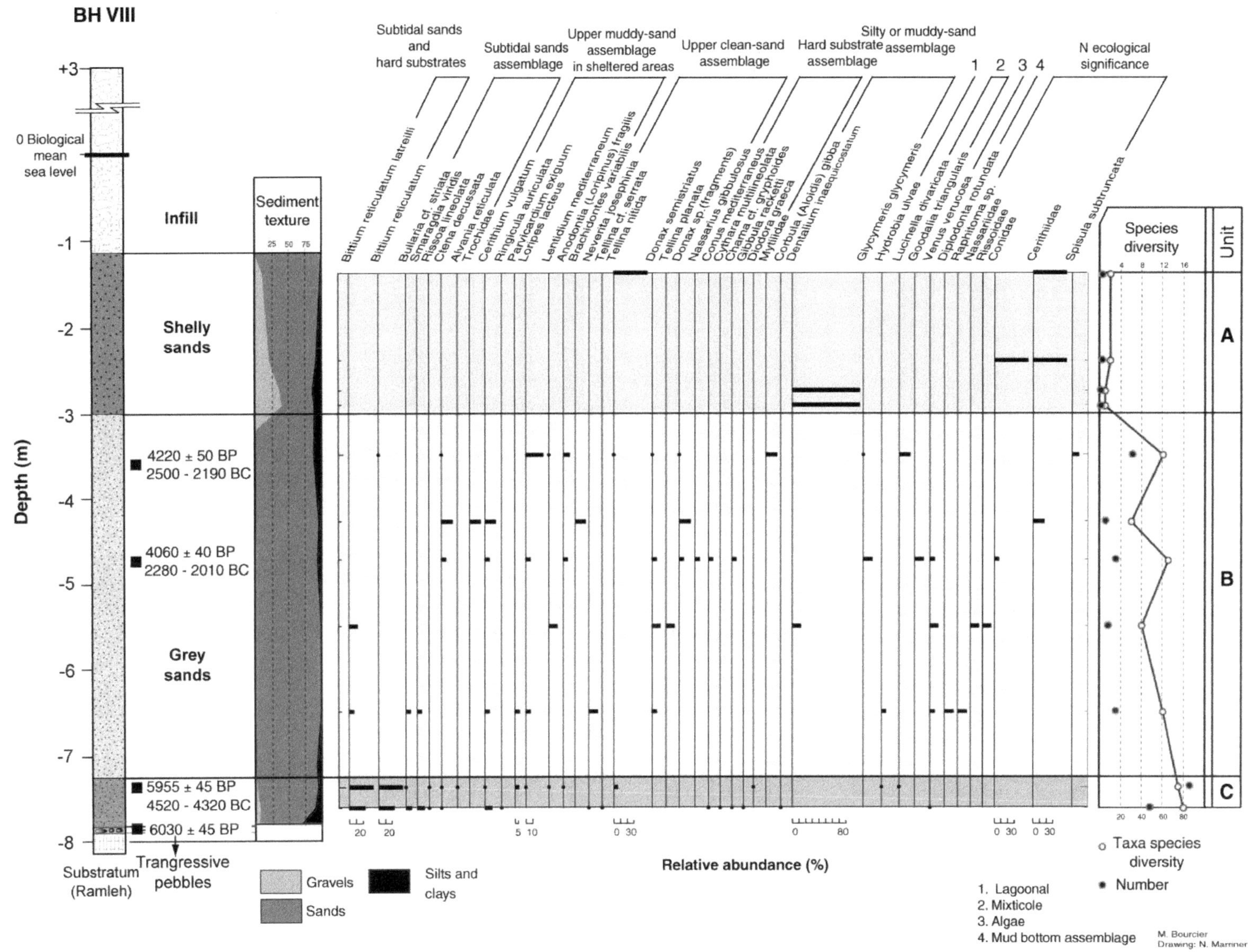

Figure 4.27: Molluscan macrofauna from core BH VIII (southern cove). (1) Unit C: This unit's molluscan suite is concomitant with the sedimentological data. The facies is dominated by taxa from the subtidal sands and hard substrates assemblage (*Bittium* spp.), the subtidal sands assemblage (*Bulla striata*, *Smaragdia viridis*) and the upper muddy-sand assemblage in sheltered areas (*Parvicardium exiguum*, *Loripes lacteus*). (2) Unit B: comprises taxa from diverse ecological assemblages including the subtidal sands and hard substrates assemblage, the subtidal sands assemblage and the silty or muddy-sand assemblage. These palaeoecological data support a medium to low energy marine cove. (3) Unit A: comprises a poor faunal suite, concurrent with relatively high energy beach dynamics on the beach face. This hydrodynamism was not conducive the fossilisation of tests.

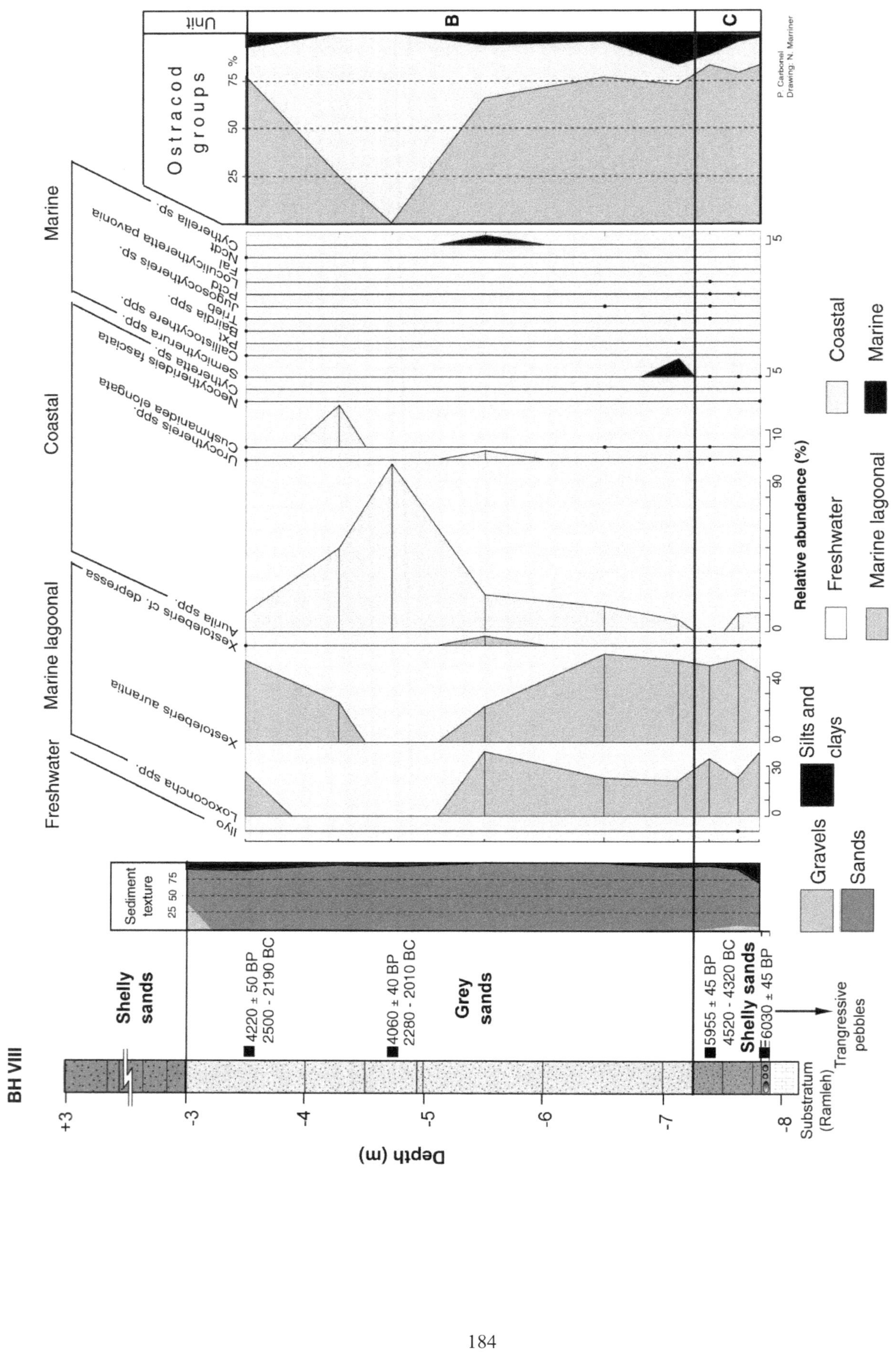

Figure 4.28: Ostracod microfauna from core BH VIII (southern cove). BH VIII's ostracod suite is dominated by three taxa, *Loxoconcha* spp. (marine lagoonal), *Xestoleberis aurantia* (marine lagoonal) and *Aurila* spp. (coastal). These species support a medium to low energy coastal environment since the Holocene marine flooding of the cove. Influence of sporadic higher energy dynamics is attested to by drifted in marine taxa (i.e. *Semicytherura* sp.).

Interpretation: Three hypotheses are advanced to explain the demise of Sidon's port areas during the Islamic period, namely: (1) cultural; (2) tectonic; and/or (3) tsunamogenic. These are explored in chapter 6. In part 4.4.3, we discuss the geoarchaeological evidence from the *Crique Ronde*, around Tell Dakerman. Human exploitation of this southern cove as a natural anchorage predates the northern harbour, attested to by rich Neolithic and Chalcolithic finds at Dakerman (Saidah, 1979).

4.4.3 When and how did Sidon's southern cove evolve?

In keeping with the classic Iron Age model of twin harbours (Lehmann-Hartleben, 1923), the existence of an artificial southern harbour has attracted the attentions of archaeologists since the 1940s and 1950s (Poidebard and Lauffray, 1951). Poidebard and Lauffray's surveys in this southern bay did not, however, yield any structures pertaining to the existence of an artificial harbour basin. Three periods are distinguished in the stratigraphic record.

4.4.3.1 Holocene transgression

Description and interpretation: The sandstone substratum is overlain by a thin pebbly sand unit, dated 6030 ± 45 BP (4630-4360 cal. BC) and marking the Holocene marine transgression of the cove (**Figure 4.26**). This subsequently grades into a coarse shelly sand fraction, constrained to 5955 ± 45 BP (4520-4320 cal. BC) and typical of the subtidal zone. *Bittium* spp., analogous of subtidal sands, characterises the molluscan fauna (**Figure 4.27**). *Loxoconcha* spp. and *Xestoleberis aurantia* dominate the ostracoda and concur a shallow marine embayment (**Figure 4.28**).

4.4.3.2 Bronze Age pocket beach

Description: The onset of unit B is dated to after 5955 ± 45 BP, and comprises a medium to fine-grained sand unit. Sample BH VIII 10 yielded a radiocarbon age of 4060 ± 40 BP (2280-2010 cal. BC). The molluscan fauna is diverse with tests from a range of ecological contexts including subtidal sands, the upper muddy-sand assemblage in sheltered areas, the upper clean-sand assemblage, the silty or muddy-sand assemblage and the lagoonal assemblage. Marine lagoonal ostracod taxa persist into the unit, and are accompanied by a gradual rise in coastal taxa (*Aurila* spp., *Urocythereis* spp. and *Cushmanidea elongata*). Sporadic tests of marine species (*Semicytherura* spp. and *Loculicytheretta pavonia*) indicate continued communication with the open sea.

Interpretation: At no point during antiquity do our sedimentological and palaeoecological data show evidence for artificial harbourworks in the southern cove. During the Bronze Age this embayment would have served as a fair weather harbour for the inhabitants of Sidon and Dakerman. The sandy beaches prevalent in the southern cove would have accommodated smaller vessels, drawn from the water onto the beachface. This phenomenon is still practiced today throughout the Mediterranean by fishermen with light, shallow draught vessels (**Figure 4.29**). Larger vessels would have been sheltered in the better protected northern harbour.

Figure 4.29: Beaching of small, shallow draught fishing vessels on the coast of Nerja in Spain, 2006 (photographs: W. Iredale). This practice is still commonplace throughout the Mediterranean.

4.4.3.3 Prograding shoreline

Description: Unit A comprises a shelly sands unit. An important rise in the gravels fraction (21-40 %) and the much coarser nature of the sands concurs a prograding shoreline in proximity to the mid-littoral zone. The molluscan fauna is poor and essentially comprises badly preserved tests and shell debris, reworked by the action of the swash. The higher energy dynamics of the swash zone were not conducive to ostracod test preservation.

Interpretation: Under a context of stable relative sea level and moderate to high sediment supply Sidon's southern cove gradually prograded seawards, diminishing the cove to its present disposition. The absence of artificial harbourworks and relative persistence of middle energy coastal dynamics means that this littoral deformation is less pronounced than the northern harbour. At the end of the 1990s, the bay's seaboard fringe was heavily concreted, burying the vast majority of the Holocene deposits (**Figure 4.30**).

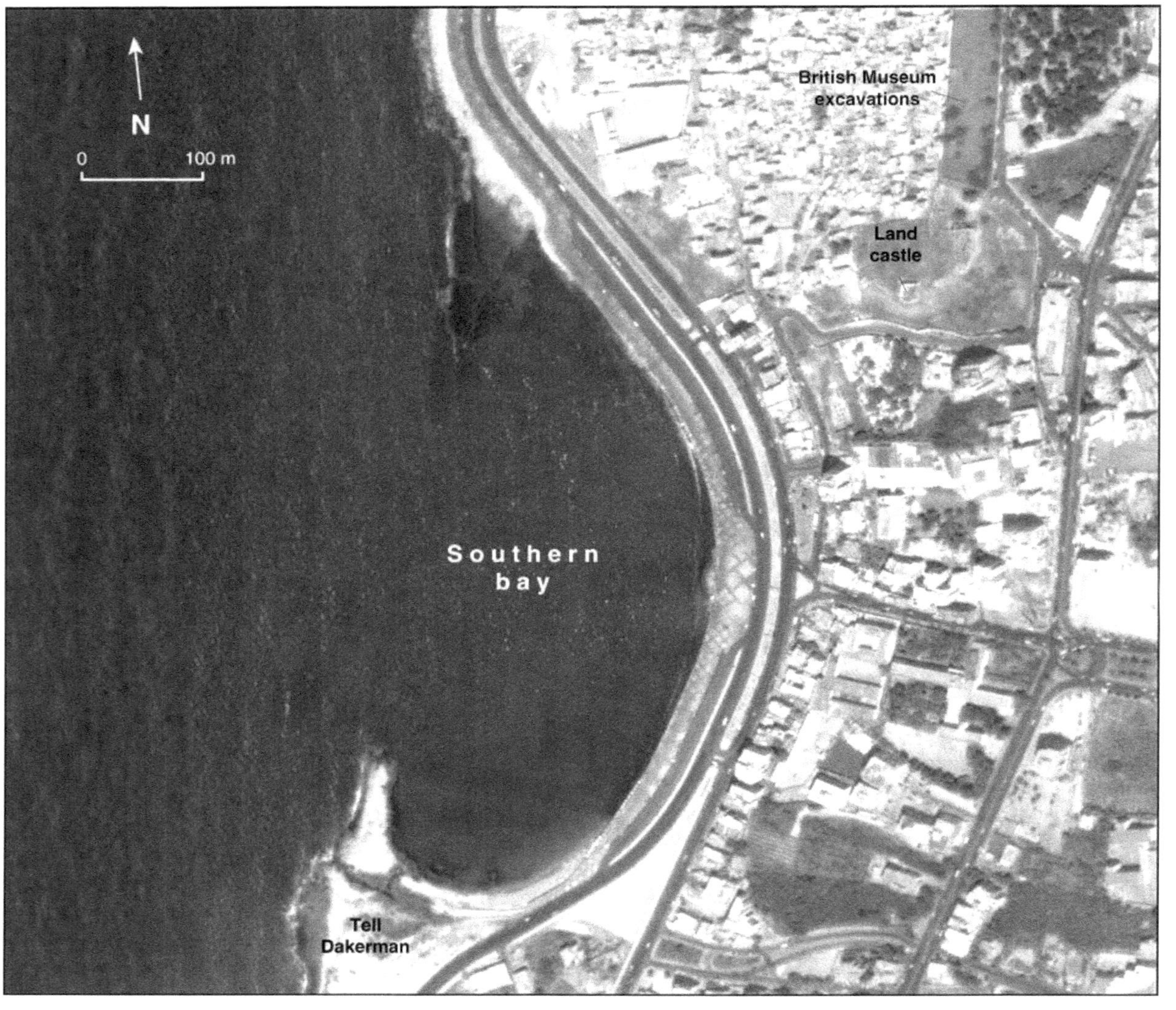

Figure 4.30: Aerial photograph of Sidon's southern bay in 2006 (image: DigitalGlobe). Redevelopment since the 1990s means that much of the Holocene coastal deposits have been concreted.

4.5 Concluding remarks

Our geoarchaeological data at Sidon demonstrate that harbour history can be clearly chronicled by diagnostic litho- and biostratigraphies. The difficulty in dating the first phase of artificial harbour confinement translates modest harbourworks during the Middle and Late Bronze Age. Early seafarers used natural semi-protected pocket beaches with little need to modify the natural coastline. The weight of natural factors in influencing human exploitation of the coastal environment prevailed until the Phoenician period (Marriner *et al.*, 2006a-b). During the Iron Age human societies at Sidon adapted the natural environment using semi-artificial infrastructure, including reinforcement and adaptation of the shielding sandstone ridge. The Roman and Byzantine periods mark a watershed in seaport technology, notably with the discovery of hydraulic concrete, translated stratigraphically at Sidon by a plastic clay unit.

Historical coastal progradation has led to the partial isolation of the ancient harbours beneath the modern city centres. Around half of the northern basin is now buried beneath thick tracts of marine sediments, and the historical coastline presently located 100 to 150 m inland. Such progradation has fossilised Sidon's archaeological heritage, offering significant scope for future research and our understanding of Phoenicia's maritime history (Marriner and Morhange, 2005a). In chapter 6, data from Sidon will be compared and contrasted with the findings from Beirut and Tyre.

Chapter 5

Geoarchaeology of Beirut's ancient harbour

Beirut was the third and final seaport complex to be investigated in this project. As at Sidon, our knowledge of Beirut's ancient tell has advanced significantly over the past decade, thanks namely to redevelopment of the city centre and excavations centred around the modern port. In spite of this research, understanding of the city's coastal palaeoenvironments during antiquity is poor. Buried Iron Age harbourworks presently 300 m from the sea attest to pronounced coastal deformation during the past 3000 years. This process has been significantly accentuated during the last two centuries by redevelopment of the port, which remains in use some 5000 years after its foundation. In this chapter, we elucidate the coastal stratigraphy east and west of the Bronze Age tell to yield new insights into the evolution of the Beirut seaboard, in addition to the complex history of human-environment interactions. We use these chronostratigraphic data to (1) precisely locate the main anchorage haven during antiquity; and (2) propose a chronology for its evolution.

5.1 Introduction

While archaeological discovery in Lebanon has a long and productive history, research in the nation's capital has been hampered by enduring demographic, geographic and geopolitical factors (Renan, 1864; Chéhab, 1939; Mouterde, 1942-43; Lauffray 1944-45, 1946-48). Although we know from written accounts that Beirut played an important role in world affairs, especially during the Roman and Byzantine periods (Mouterde and Lauffray, 1952; Hall, 2004), precise archaeological and topographical data regarding the city has long been missing. Despite the fact that the identification of ancient Bêruta/Berytos/Berytus with modern Beirut had never been called into question, the exact location of the ancient city and tell within the modern agglomeration was open to fervent debate (Vaumas, 1946; Davie, 1987). Sporadic surveys by twentieth century scholars tended to suggest the ancient city lay in an area between the present port seaboard, delimited by rue Foch to the east, rue Allenby to the west and the place de l'Etoile to the south (**Figure 5.1**). Beirut has, however, been built up progressively on ancient habitation layers rendering extensive archaeological excavations in such a dense urban fabric politically and logistically difficult (Forest and Forest, 1977).

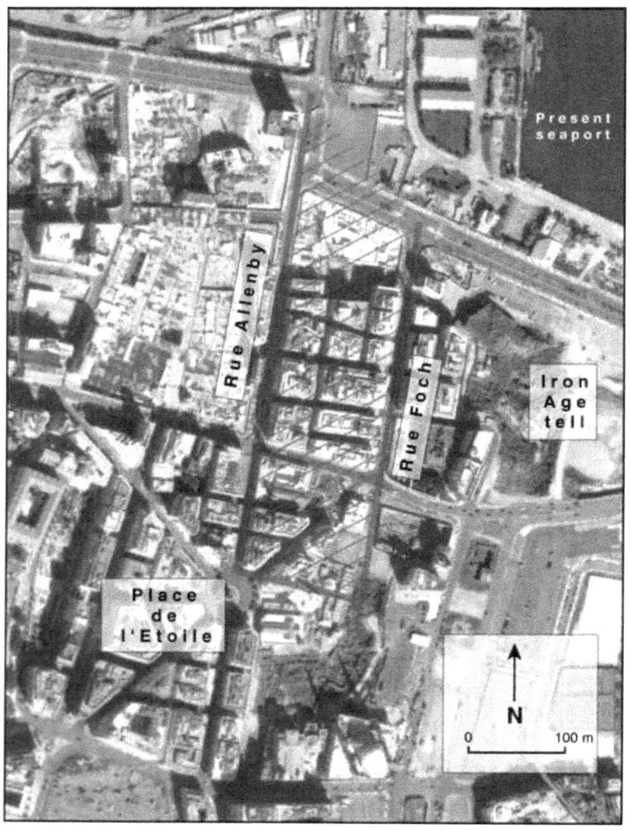

Figure 5.1: Early archaeological surveys tended to suggest that the ancient city lay in an area between the present port seaboard, delimited by rue Foch to the east, rue Allenby to the west and the place de l'Etoile to the south (Renan, 1864; Chéhab, 1939; Mouterde, 1942-43; Lauffray 1944-45, 1946-48; Vaumas, 1946; Mouterde and Lauffray, 1952; Davie, 1987). Base image: DigitalGlobe, 2006.

Against this backcloth of data paucity, plans to reconstruct and modernise Beirut's city centre during the early 1990s offered exciting opportunities to explore the evolution of this important site on an unprecedented scale (Lauffray, 1995; Lefèvre, 1995a-b). The area surrounding the present seaport is rich in buried archaeological monuments and relics, bearing witness to a complex history of human occupation spanning some 5000 years (Gavin and Maluf, 1996; Elayi and Sayegh, 2000; Curvers and Stuart, 2004; Doumet-Serhal, 2004a). Since 1993, national and international institutions have supported a project involving hundreds of archaeologists covering the quasi-totality of ancient Beirut, and notably the ancient centre delimited by the medieval walls.

For *Solidère*, the reconstruction agency, one of the early concerns was to marry urban development with archaeology so that the rich historical heritage of the city centre could be integrated into the rebuilding process. Unfortunately, as wealthy developers and politicians vied with archaeologists, the project received criticism from some quarters for falling short of many of these initial goals, this in spite of rigid government legislation designed to avoid bygone errors (e.g. the Lebanese Antiquity Law; Lauffray, 1995; Karam, 1996; Naccache, 1996, 1998; Seeden, 1999; Raschka, 2006). Nonetheless, the unique urban excavation has produced a great mass of data since its inception in 1993 (**Figure 5.2**; Perring *et al.*, 1996; Saghieh, 1996; Butcher and Thorpe, 1997; Cumberpatch, 1997; Curvers and Stuart, 1997; Finkbeiner and Sader, 1997; Heinze and Bartl, 1997; Thorpe, 1998-1999; Thorpe *et al.*, 1998-1999; Elayi and Sayegh, 2000; Faraldo Victorica and Curvers, 2002; Doumet-Serhal, 2004a).

Without doubt, the main positive outcome of the project has been the spatial coverage of archaeological and topographical data obtained for both the ancient and historical periods (Elayi and Sayegh, 2000; Curvers and Stuart, 2004). From a geoarchaeologist's standpoint, therefore, coupling this data with a study of the coastal stratigraphy offered unprecedented opportunities to: (1) accurately relocate the city's ancient harbour(s); (2) precisely reconstruct 5000 years of coastal deformation; and (3) better comprehend human-environment interactions at both the local and regional scales, through a comparison with Sidon and Tyre (Marriner *et al.*, 2005; Marriner *et al.*, 2006a-b).

5.2 Archaeological and geomorphological contexts

Beirut's rocky promontory, 6 km long by 2 km wide, is one of the most defining geomorphological traits of the otherwise rectilinear Lebanese coast (Dubertret, 1940). Wedged between the Mount Lebanon range to the east and the Mediterranean to the west, the promontory is surrounded by the sea on two of its three sides; it is intersected by a network of transverse faults which cut across the Lebanon chain (**Figures 5.3 and 5.4**; Dubertret, 1955).

The peninsula attests to a long history of human occupation beginning in the middle Palaeolithic (Fleisch, 1946; Copeland and Wescombe, 1965, 1966). Twenty sites are known from the Neolithic, the oldest of which was discovered in 1930 at Tell Arslan, 8.5 km south of Beirut, underneath a Roman habitation layer (Bergy, 1932). For later periods, Beirut's tell and city centre bear out 5000 years of continuous human occupation spanning the Bronze Age and Iron Age, in addition to the traces left by the Persian, Seleucid, Roman and Byzantine empires (Mouterde, 1966; Hall, 2004). At its greatest extent, archaeological evidence suggests that metropolitan Beirut covered an area of at least 1.2 km by 0.8 km (Mikati and Perring, 2006).

Figure 5.2: Excavations in Beirut city centre (photograph: Richard Barnes). Archaeological work has been underway since 1993, yielding insights into great tracts of the ancient city from its Bronze Age beginnings through to the medieval period. Under the auspices of both national and international institutions, the project has involved hundreds of archaeologists from around the world.

The ancient city was founded on the northern part of the peninsula, in a depression between two hills, Ashrafieh to the east (102 m above MSL) and Ras Beirut (95 m) to the west (Vaumas, 1946; **Figures 5.3 and 5.4**). Ashrafieh and the eastern portion of the Beirut promontory comprise Neogene marls, while Ras Beirut has been fashioned in Cenomanian limestones (Dubertret, 1955, 1975). Three periods of Quaternary sea-level still-stand have cut abrasion platforms into the northern façades of Ashrafieh and Ras Beirut, at 85-100 m, 40-65 m and 6-20 m (Vaumas, 1944; Sanlaville, 1977). The vast majority of the archaeology is concentrated upon the lower, northward facing terrace which reaches 800 to 1000 m width in the Nahr Beirut area (Sanlaville, 1977). The Ashrafieh-Ras Beirut depression included a shallow bay open to the north, formed by the drowning of a talweg system after 6000 BP. It is this natural cove that later became the ancient seaport. In the centre was a small island, Burg al-Mina, while to the east and west two elevated land spurs closed the basin (**Figure 5.4,** see **Figure 5.10** for the coastal reconstruction of Beirut). A SSW-NNE trending fault cuts from the port to Basta et-Tahta (**Figure 5.4**), and puts into contact the Neogene and Cenomanian formations (Dubertret, 1955, 1975).

This ancient city was very easy to defend: to the west access was blocked by the abrupt cliffs of Raouchey whilst the river Nahr Beirut blocked access from the east (**Figure 5.4**). The Nahr Beirut, fed by its two tributaries Wadi Hammana and Wadi Salima, is 23 km long and flows in a narrow, deeply incised valley which diminishes as it approaches the sea (**Figure 5.3**). To the south, defence was afforded by the hills of Ashrafieh. In ancient times, the only means to enter the ancient city was either by sea, via the port, or through a narrow talweg corridor between the hills of Ashrafieh and Ras Beirut (which presently runs between the roads of Damas and Basta; **Figure 5.4**).

During the Bronze and Iron Ages it appears probable that, as at Sidon, two urban nuclei existed, an "upper city" centred on the tell and a "lower city" on the banks of the ancient harbour (Elayi and Sayegh, 2000; Curvers, 2002). The Hellenistic and Roman-Byzantine cities were founded south, southwest and west of the ancient tell. Much of the city centre has preserved this rigid grid-like morphology, and features such as the Roman hippodrome and forum are still clearly discernible in the urban fabric (**Figure 5.5**).

Despite its former glory, Beirut was little more than a small fishing harbour by the seventeenth century. During the nineteenth century, Laorty-Hadji (1855) laments the demise of the harbour "Çà et là des débris et des fûts de colonnes, misérables restes de l'antique Béryte, servent dans le port à amarrer les bateaux. Formé par une jetée, ce port autrefois profond et commode, les habitants l'avaient laissé encombrer de ruines et de sables." Jessup (1910) compounds his predecessor's words "[...] there was no harbour, only an open roadstead, and boats landing from ships anchored outside would strike bottom before reaching the beach [...]".

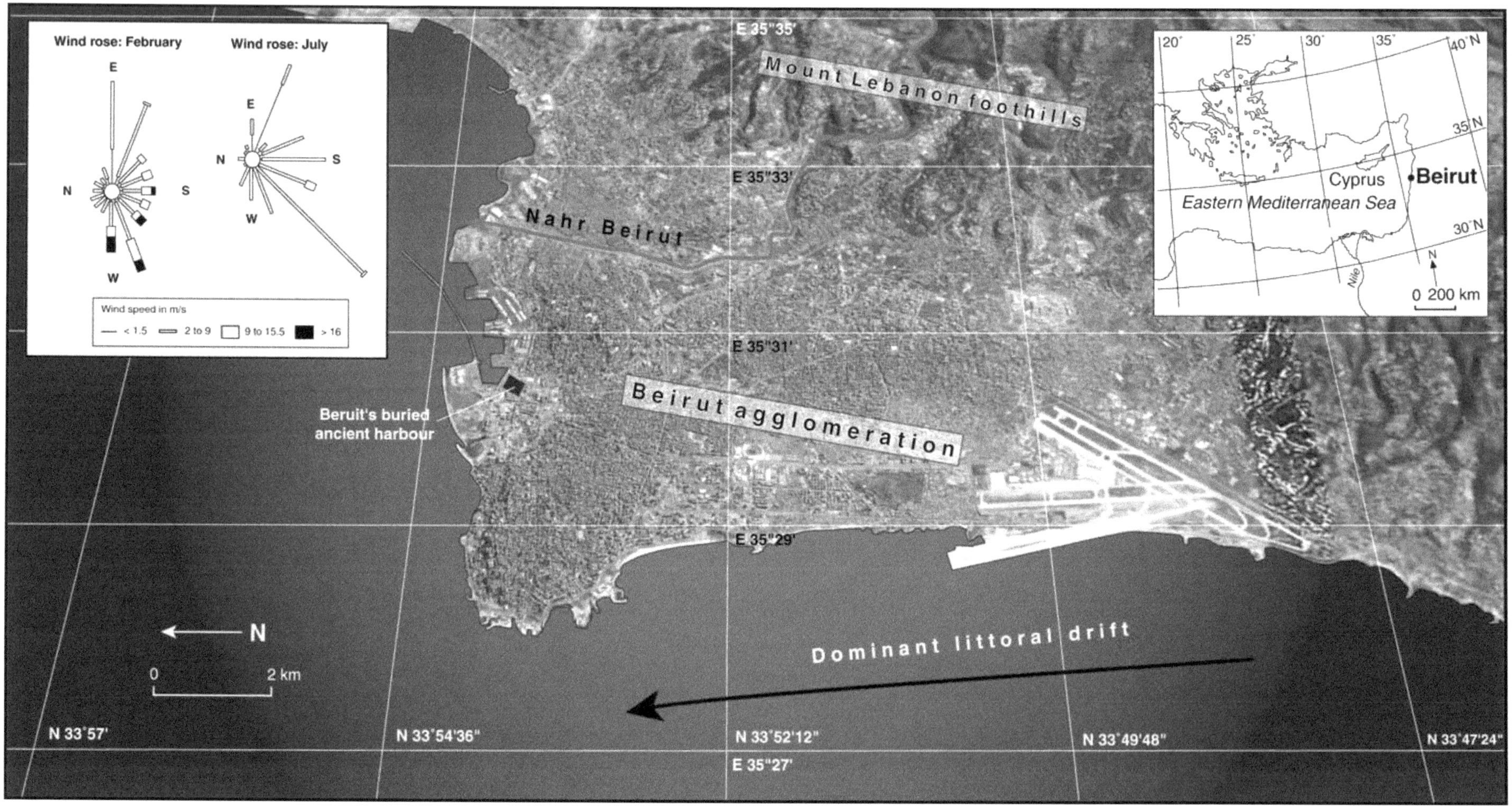

Figure 5.3: Oblique aerial photograph of the Beirut agglomeration in 2006. The city occupies a peninsula sandwiched between the Mount Lebanon range to the east and the Mediterranean sea to the west. The ancient city lies on the northern flank of the peninsula, where its ancient anchorage was protected from the dominant south-westerly wind and swell (base image: DigitalGlobe, 2006).

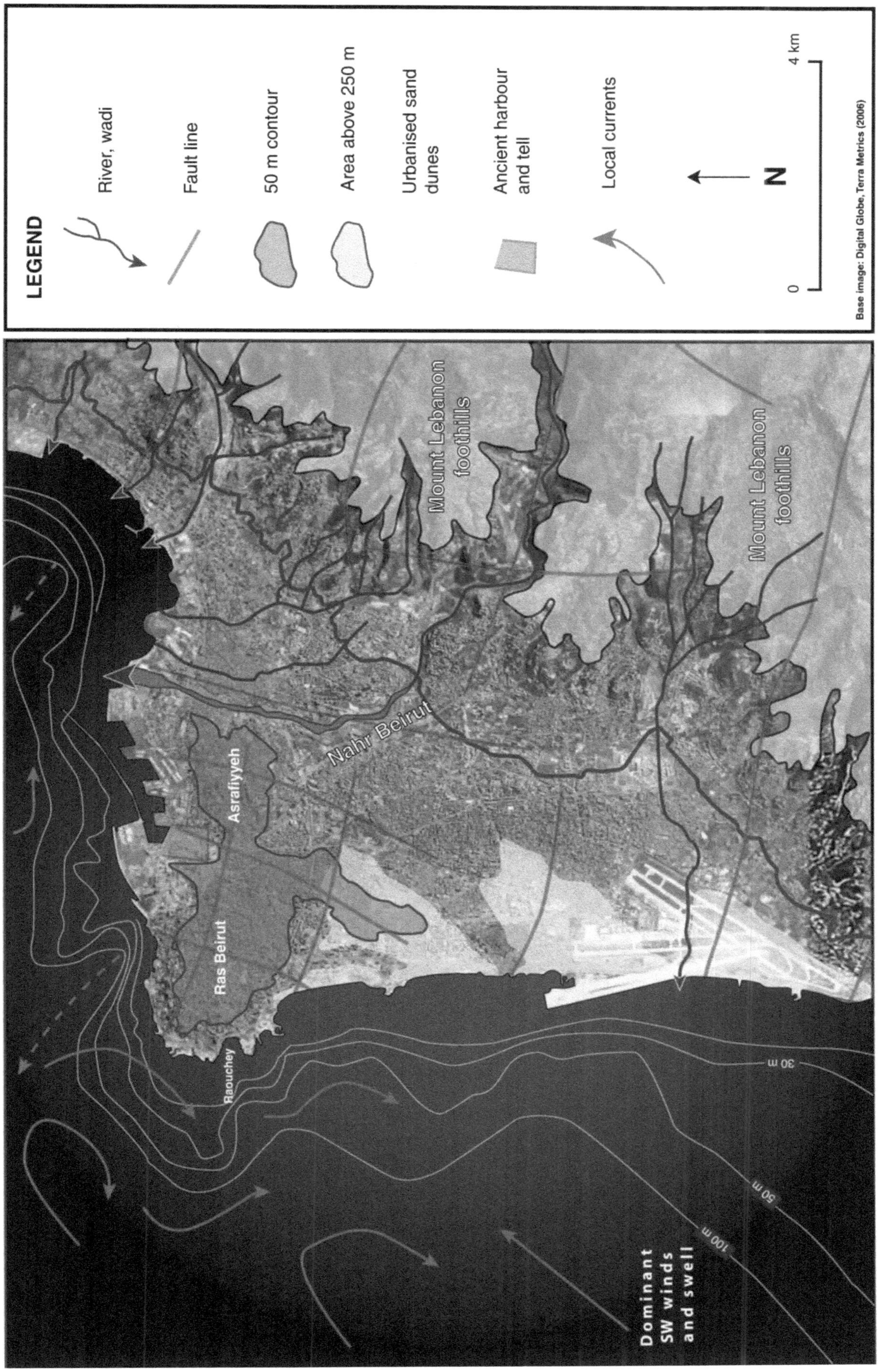

Figure 5.4: Geography of the Beirut peninsula (base image: DigitalGlobe, 2006).

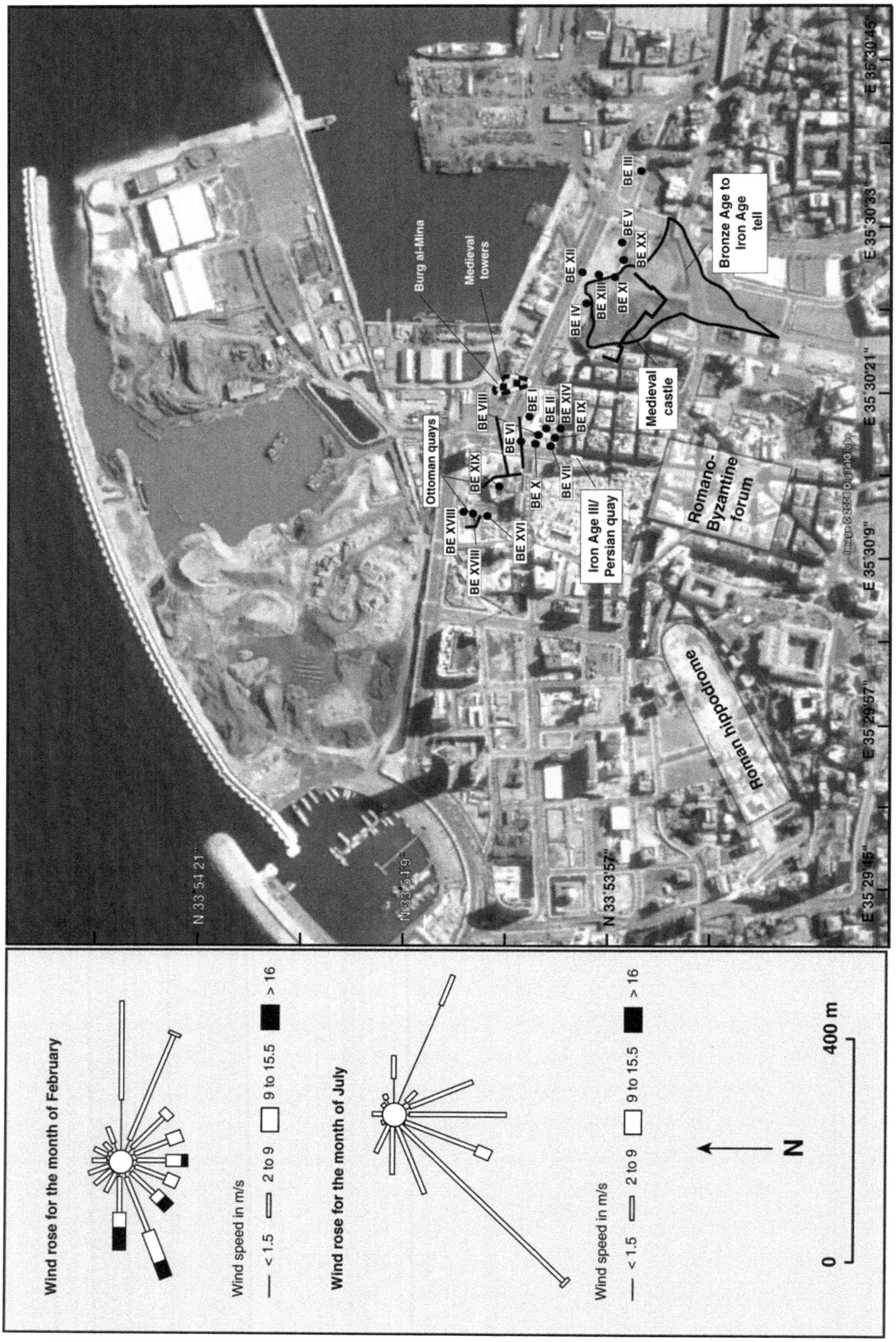

Figure 5.5: Location of core sites at Beirut. Archaeological data from Elayi and Sayegh (2000) and Marquis (2004).

5.3 Methods and data acquisition

A total of twenty cores was drilled at sites around Beirut's ancient tell (**Figure 5.5**). We refer the reader to earlier chapters on Sidon and Tyre for a detailed review of the equipment and methods used. These include numerous sedimentological, biostratigraphical and chronostratigraphical proxies. Old city maps and engravings of Beirut have been studied to precisely reconstruct changes in the port coastline since the nineteenth century (Ormsby, 1839; Wyld, 1840; Löytved, 1876; Baedeker, 1912; Royal Engineers, 1841; Scott, 1841; Skyring, 1841). All maps were georeferenced before being superimposed to form a single chart. This composite map was subsequently married with the archaeological and stratigraphic data to yield insights into coastal deformations during the ancient and historic periods (Francou, 2002).

5.4 Where was Beirut's ancient harbour?

Before the recent excavations of the city centre, the exact location of Beirut's ancient harbour had been open to ardent speculation. In the absence of solid archaeological evidence, four main hypotheses dominated the literature. (1) For Mesnil du Buisson (1921), the harbour lay between Bab es-Santiye, Burg al-Mina island and the rue de la Marseillaise, just north of the present silted up basin (**Figure 5.6**). (2) Like his homologue, Dussaud (1927) constrained the ancient port to the northern façade of the Beirut peninsula but is very vague in attributing a precise geographical location: "[...] la population de marins dût demeurer le long des anses, aujourd'hui en partie comblées, qui ouvrent vers le nord, disposition qui signale les meilleurs abris de la côte syrienne". (3) For Vaumas (1946) the ancient harbour lay in the bay of Saint-Andre, the cove east of the Iron Age tell (**Figure 5.6**). (4) Finally, it was not until the 1980s that Davie (1987) proposed a plausible and well-argued hypothesis for an ancient seaport in the area between Zaytuneh and the Rivoli cinema, at the foot of the tell's south-western flank (**Figure 5.7**). In order to test these early hypotheses, we drilled 20 cores to the east and west of the ancient tell (**Figure 5.5**).

Ancient Beirut lies in an area that was fertile, well-drained, flat and wide, and where the sea could be exploited for food and transportation. Whilst the internal structure of the city has remained relatively stable since antiquity, the biggest topographical changes have occurred along the southern coastal flank of the settlement. Iron Age harbourworks at excavation plot BEY 039 (rue Allenby; **Figure 5.8**), more than 300 m from the present artificialised coastline, attest to significant coastal deformations during the past 3000 years. A study of nineteenth and twentieth century maps, coupled with our elucidated coastal stratigraphy and harbour infrastructure, has yielded a precise reconstruction of port coastal changes since the Bronze Age (**Figures 5.9 to 5.12**).

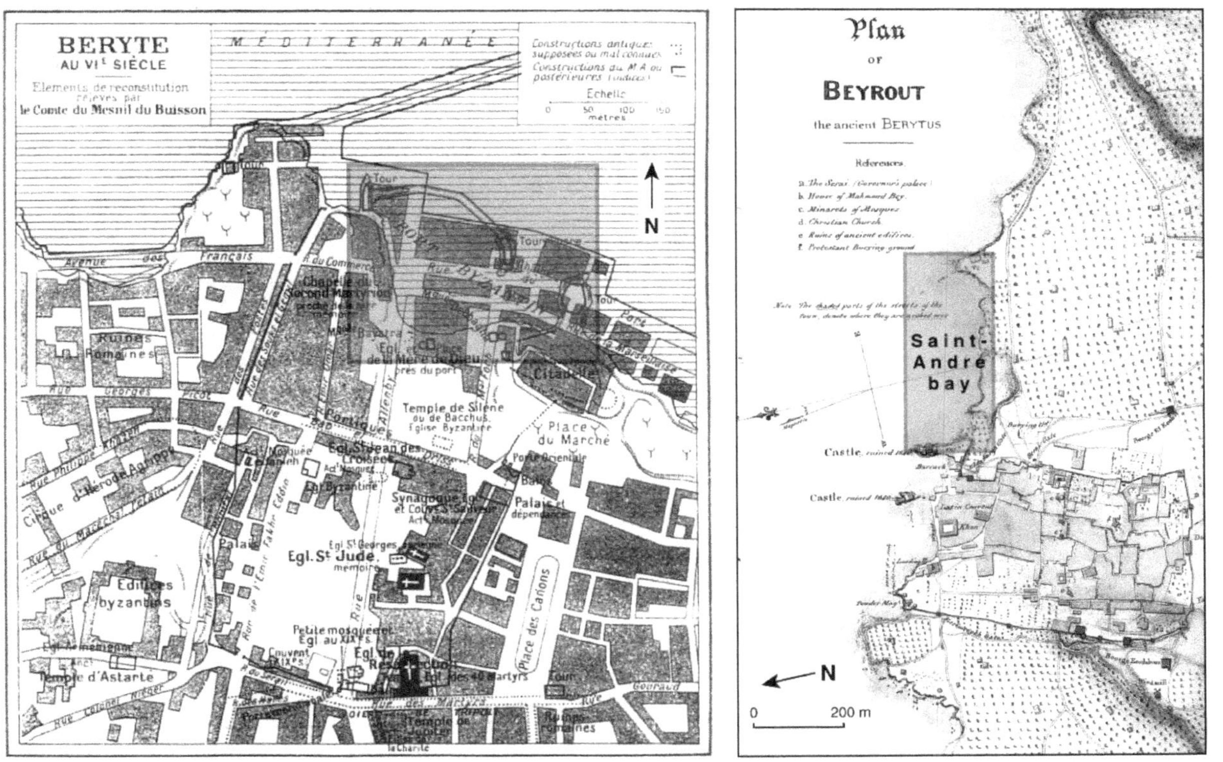

Mesnil du Buisson (1921)

Vaumas (1946) on Royal Engineers map of Beirut (1841)

Figure 5.6: Mesnil de Buisson (1921) and Vaumus' (1946) hypotheses for the location of Beirut's ancient harbour (blue shading).

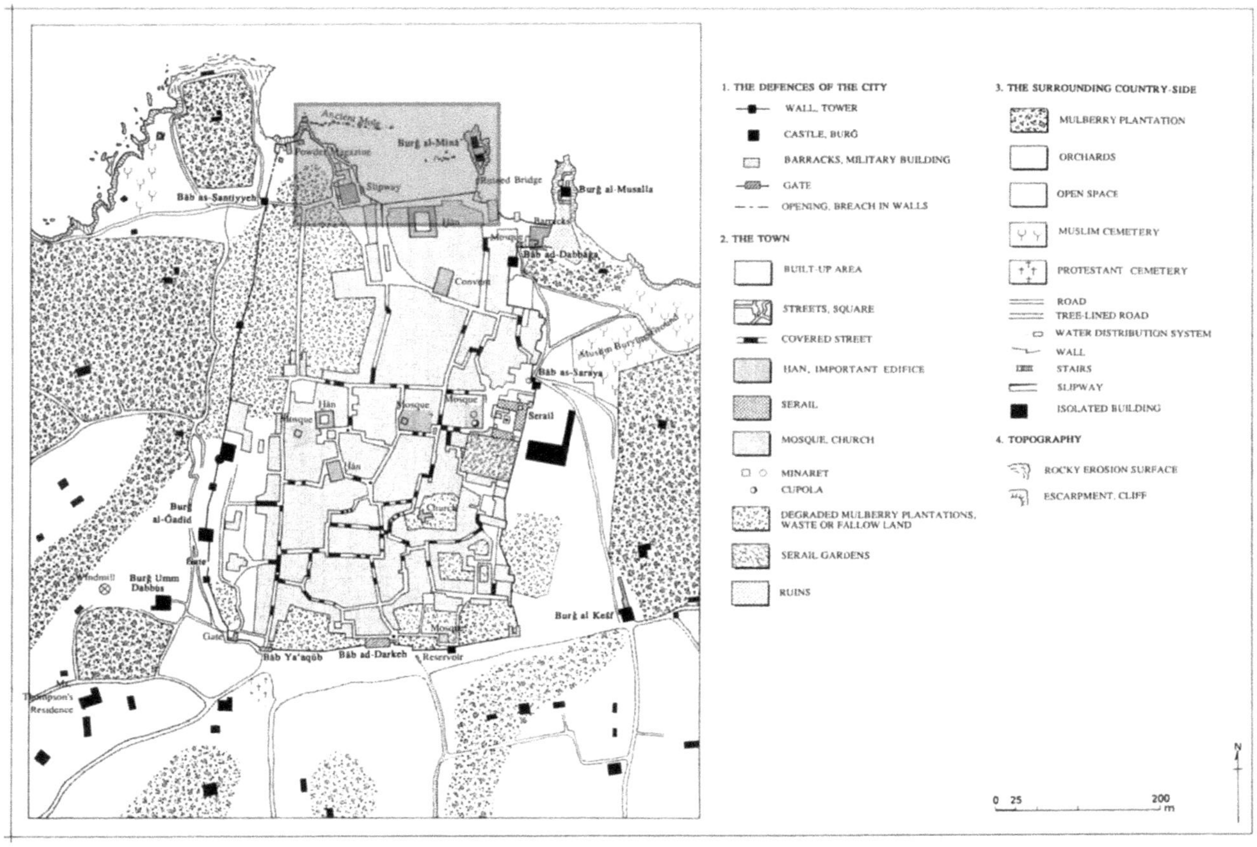

Figure 5.7: Davie's (1987) proposed location for Beirut's ancient harbour.

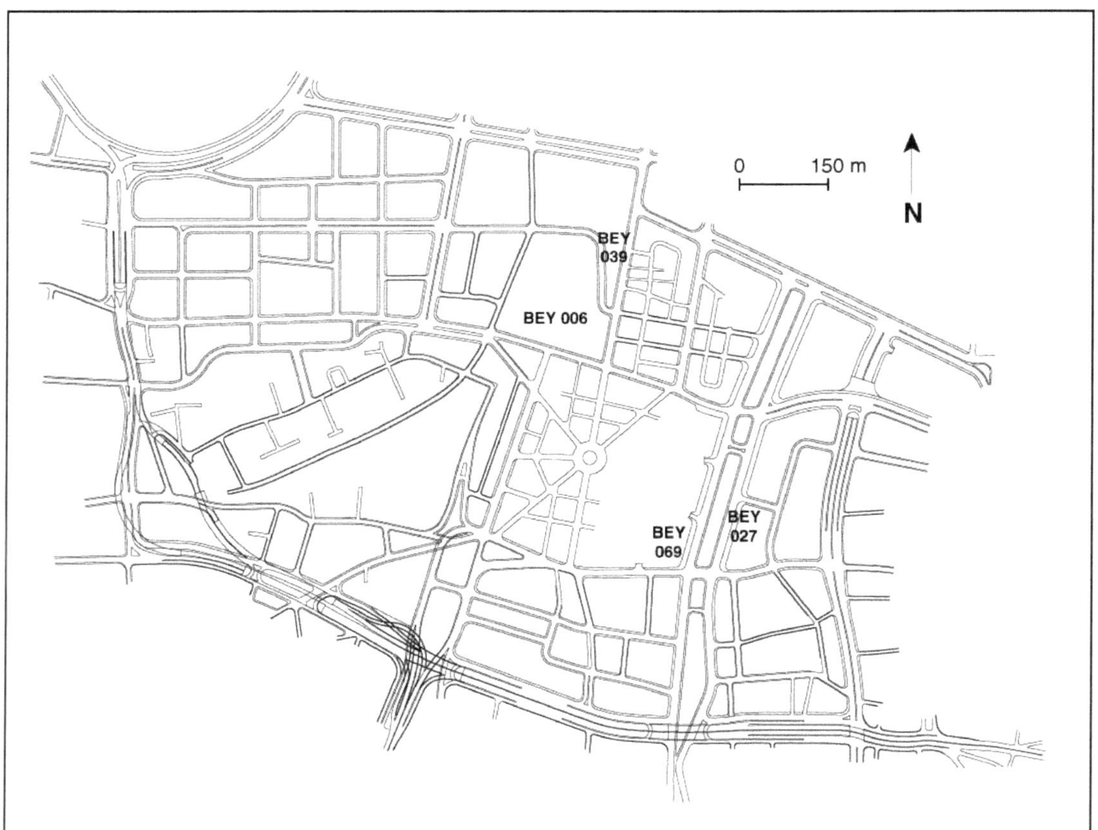

Figure 5.8: Location of BEY archaeological sites discussed in the text (base map from Curvers and Stuart, 2004).

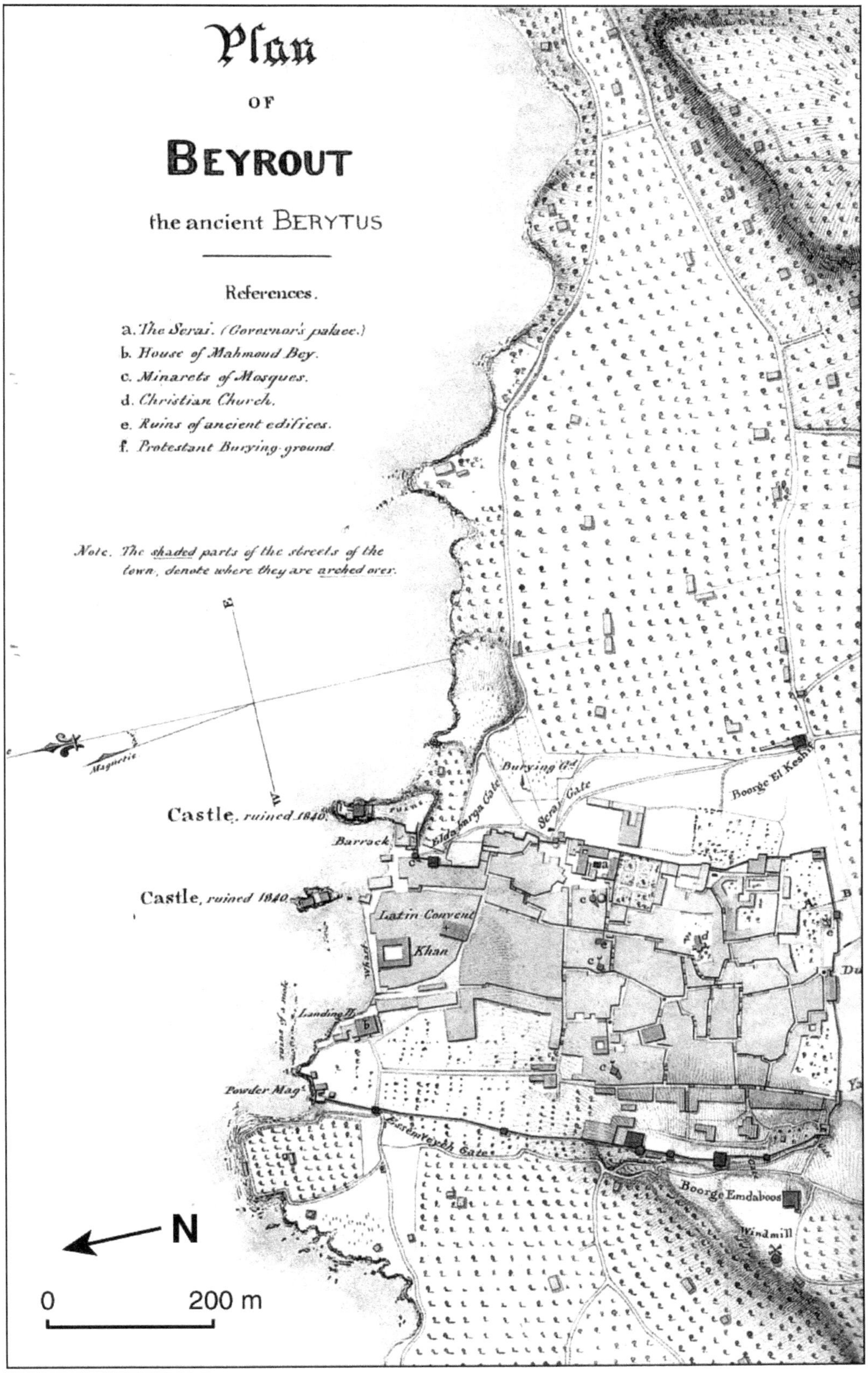

Figure 5.9: Map of Beirut as it appeared in 1841 (Royal Engineers, 1841) before the significant construction works undertaken to modernise the ancient harbour. The port coastline has been significantly artificialised since this time (see figures overleaf).

Figure 5.10: Löytved's 1876 map of Beirut in which the first phase of coastal redevelopment is clearly manifest.

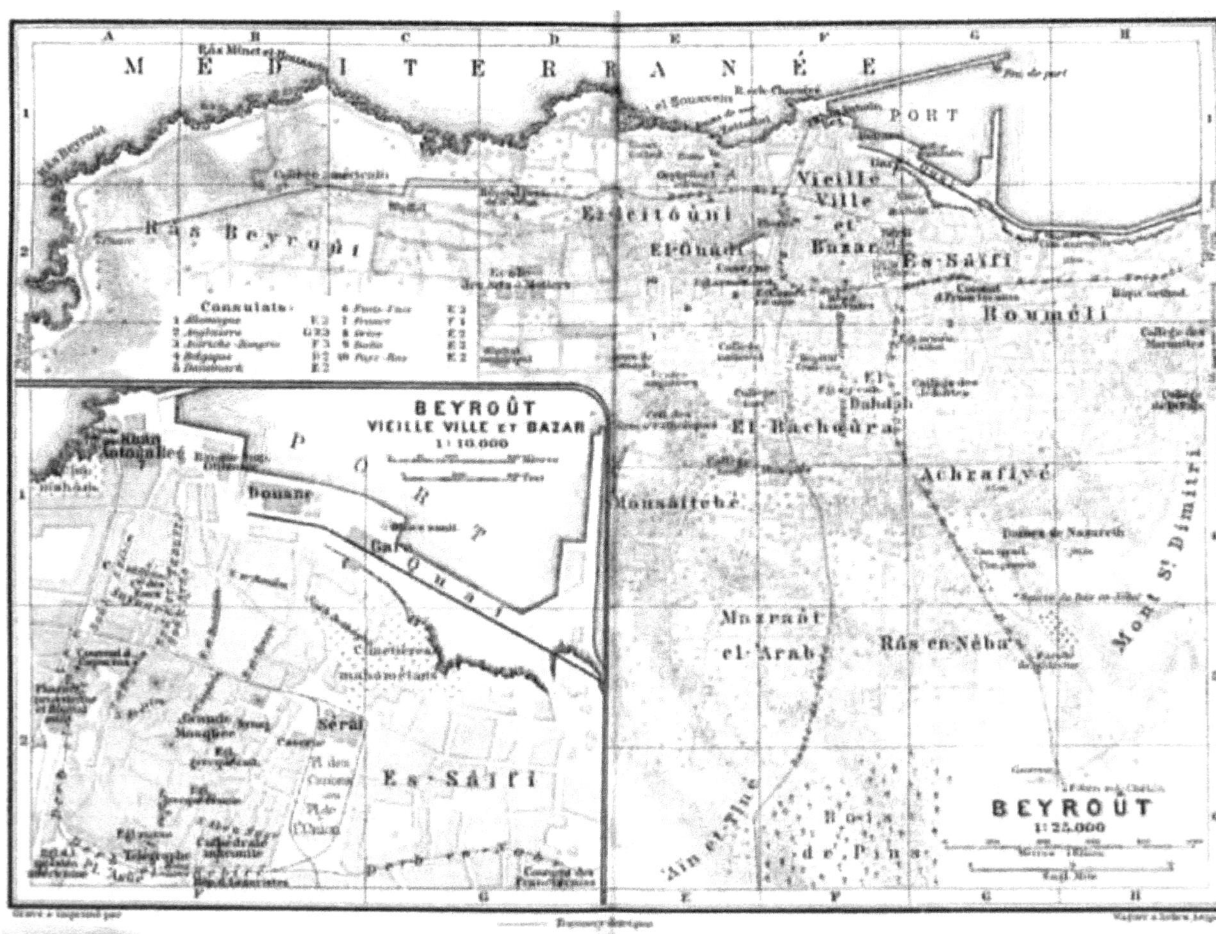

Figure 5.11: Beirut [Beyrout] city centre and coastline in 1912 (from Baedeker, 1912).

Figure 5.12: 5000 years of coastal deformation in Beirut's ancient harbour. Note the progressive regularisation of the coastline from a natural indented morphology to an increasingly rectilinear disposition during later periods (base image: DigitalGlobe, 2006). Archaeological data from Elayi and Sayegh (2000) and Marquis (2004).

Unlike Sidon and Tyre, the entirety of the former basin at Beirut today lies landlocked beneath the city centre. This is due not only to silting up since antiquity, but also the considerable construction works undertaken since the nineteenth century. Only the western cove has yielded fine-grained sediments diagnostic of a sheltered environment, although we suggest that sandy beaches to the east of the tell could have been used as a fair-weather anchorage for shallow draught vessels from the Bronze Age onwards. Like the southern bay at Sidon, cores Be III, Be V, Be XII and Be XX only manifest medium grain marine sands. Due to the dense urban fabric, the majority of our cores in the western harbour are concentrated in the northwest portion of the basin between Burg al-Mina and the Ottoman quays (**Figure 5.12**). Archaeological surveys south of this core network indicate that the maximum extension of the ancient harbour during the Bronze Age lay some 300 m from the present coastline (Elayi and Sayegh, 2000).

We present and discuss the sedimentological datasets in part 5.5. The coastal deformations are characterised by two periods:

(1) **Bronze Age to medieval period coastal changes.** The basin was formed by the drowning of an active talweg ~6000 BP; our calculations suggest that the ancient cove had an approximate surface area of ~50,000 m² at this time (**Figure 5.12**). The archaeological data support a ~70 m progradation of the coastline between the Early Bronze and the Roman periods (Elayi and Sayegh, 2000; Marquis, 2004). A Middle to Late Bronze Age shoreline has been unearthed at BEY 069 (Bouzek, 1996). This find suggests that the lower reaches of the drowned wadi formed a creek at the base of the tell, which was exploited as an anchorage haven at this time. Such a scenario is consistent with data from southern Levant, where wadis were widely exploited as natural anchorages during the Bronze Age (Raban, 1987a; Morhange *et al.*, 2005b). Data from BEY 027 infers that by the Late Bronze Age, the creek had silted up to become inactive, before complete urbanisation during the Hellenistic period (Arnaud *et al.*, 1996; Mendleson, 1996).

For later periods, an Iron Age III/Persian quay was unearthed at excavation site BEY 039 (**Figure 5.13**). Located in rue Allenby, the quays indicate the contemporary extension and north-south orientation of the port. Although the exact location of the southern Iron Age coastline is not known exactly, the pattern of Iron Age archaeological finds shows that it lies just north of rue Weygand (Elayi and Sayegh, 2000).

As at Sidon and Tyre, we hypothesise that repeated dredging during the Roman and Byzantine periods reduced the natural progradation tendency by removing sediment from silted up portions of the basin. This is partially confirmed by the ceramics data, and we are presently undertaking a series of radiocarbon datings to establish a more complete chronology. From the onset of the Byzantine crisis and the beginning of the nineteenth century, settlement demise accounts for rapid silting up; the medieval archaeology attests to a 30 m coastal progradation between 660 AD and 1600 AD (Marquis, 2004).

Figure 5.13: A-C: Iron Age III/Persian harbour quay at excavation plot 039. The excavations at this plot covered ~3000 m². This quay presently lies ~300 m from the coastline. It shows the western bank of the harbour at this time, and also the south-north trending façade of the seaport in this area. The quay comprises ramleh sandstone blocks, measuring ~60 cm by 30 cm, and fixed together by a grey mortar. The two underlying strata constitute larger blocks (60 by 100 cm) stacked without the use of a mortar. This type of construction is typical of the Phoenician period. D. Quay mooring bit 140 cm from the quayside, 45 cm in diameter and fashioned in a ramleh sandstone block (all photographs from Elayi and Sayegh, 2000).

(2) **Renaissance to modern period coastal changes.** Between the Renaissance period and today, five sets of maps have been used to document the evolution of the coastline. This work builds upon earlier research by Davie (1987). The maps exploited include those of Ormsby (drawn in 1831 and published in 1839), Wyld (1840), Royal Engineers (1841), Scott (1841), Skyring (1841), Löytved (1876) and Baedeker (1912).

At the end of the eighteenth century, Beirut was a small fishing settlement of 4000 inhabitants, playing a very minor role in the transport of goods to Syria (Monicault, 1936). The rise of the city as a major commercial port goes back to the coastal revival initiated by steamship navigation, which triggered a shift in the economic activity from inland caravan cities like Damascus (Hastaoglou-Martinidis, 1998; Davie, 2000). In 1832, Beirut was established as the capital of Vilayet Sidon, a measure that attracted consular representation and foreign traders. After this time, the seaport expanded in four major phases, clearly documented by the coastal artificialisation. (1) **Between 1867 and 1876**, Beirut underwent a significant phase of seaport redevelopment. Low import duties, the building of the wharf, and the construction of the Beirut-Damascus cross mountain road opened Beirut to the Syrian/Arabian interior and made it the region's principal entrepot, supplanting Acre and Sidon as the Levant's primary maritime hub. (2) **Between 1920 and 1943**, Beirut fell under French colonial rule and underwent a renewed phase of seaport redevelopment (Monicault, 1936). (3) **Post-World War II**, Beirut also saw a series of new port constructions, notably with an expansion of the docking capacities. (4) Finally, **since the early 1990s**, following a number of decades of geopolitical turmoil, the Lebanese government has set about rebuilding and modernising central Beirut and its port infrastructure. These ongoing works notably include a large artificial port to the northwest of the present basin (**Figure 5.12**).

In totality, Beirut's port area bears witness to over 5000 years of human modification and artificialisation, characterised by a ~1 km dislocation of the coastline. Research shows increasing levels of coastal regularisation since the Neolithic, with the most profound changes taking place after the eighteenth century.

5.5 When and how did Beirut's ancient harbour evolve?

Only the cores between Burg al-Mina and the western land spur yielded facies analogous with a well-protected ancient harbour (**Figure 5.12**). Litho- and biostratigraphical studies from three of these cores, Be VIII, Be IX and Be X have established five phases in the evolution of Beirut's seaport. It is important to stress that the chronological interpretations are based upon preliminary ceramic and radiocarbon chronologies and cross-correlation with the well-dated sequences at Sidon and Tyre. Further radiocarbon dates are presently underway.

5.5.1 Early to mid-Holocene low energy lagoon

Description: The limestone substratum in core Be VIII is overlain by a dark plastic clays unit (3/2 5Y olive black) comprising over 75 % silts and clays. Medium sorting indices of ~0.8 and skewness values of -0.11 to -0.17 are consistent with the predominance of medium to fine sands (**Figures 5.14, 5.17 and 5.20**). The molluscan suite is relatively poor, with very few tests being represented. Those present comprise species from the upper muddy-sand assemblage in sheltered areas (*Loripes lacteus*, *Cerithium vulgatum*), lagoonal (*Cerastoderma glaucum*) and the upper clean-sand assemblage (*Nassarius mutabilis*; **Figures 5.15, 5.18 and 5.21**). The ostracod fauna has a relatively low faunal density of between 20 and 40 tests per 10 g of sand. The suite is dominated by the brackish lagoonal species *Cyprideis torosa* (**Figures 5.16, 5.19 and 5.22**). Tests of marine lagoonal and coastal taxa are concomitant with proximity to the coastline.

Interpretation: We interpret this plastic clays unit as a naturally sheltered coastal environment which began accreting as transgressing Holocene sea levels reached the basin vicinity. Analogous facies have been observed at both Sidon and Tyre and support data from the northern coast of Israel (Galili and Weinstein-Evron, 1985; Cohen-Seffer *et al.*, 2005). This lagoon environment would have been protected by a series of proximal beach ridges, breached during periods of storm and high swell. Full marine conditions are only recorded in the cove by a sublittoral sands unit at ~6000 BP (see 5.5.2). Further radiocarbon datings are presently underway to establish more robust chronologies.

5.5.2 Pocket beach/Bronze Age proto-harbour

Description: Unit C comprises a shelly sands unit with medium sorting indices of between 0.5 and 1. Medium sands dominate the unit and constitute 50 to 80 % of the total sands fraction. The sedimentological proxies all point to a medium to low energy marine environment. The molluscan faunas are dominated by taxa from four assemblages, the upper clean-sand assemblage (*Cerithium vulgatum*), the subtidal sands assemblage (*Rissoa* spp.), the upper muddy-sand assemblage in sheltered areas and the subtidal sands and hard substrates assemblage (*Bittium reticulatum*). Medium to low energy marine dynamics are further corroborated by the marine lagoonal (*Loxoconcha* spp., *Xestoleberis aurantia*) and coastal (*Pontocythere elongata*, *Hiltermannicythere* sp., *Aurila convexa*) ostracod species.

Interpretation: The proxy data evoke the accretion of a sublittoral sand facies within a naturally protected marine cove, which began accreting after ~6000 BP. This stratigraphy is reminiscent of the pocket beach units observed at both Sidon and Tyre (chapters 3 and 4; Marriner *et al.*, 2005, 2006a-b). Such a sheltered environment, looking north and protected from the dominant winds and swell by the southwestern flank of the peninsula, explains the decision by human societies to settle on the tell overlooking this marine embayment. The settlement's topography would have been known to sea-faring crews who hopped between natural anchorages during the early days of maritime trade (Marcus, 2002a-b). Bronze Age settlers used the fertile land on this wide terrace for food. Interaction with the sea, through this natural cove, permitted trade and the exploitation of marine food resources.

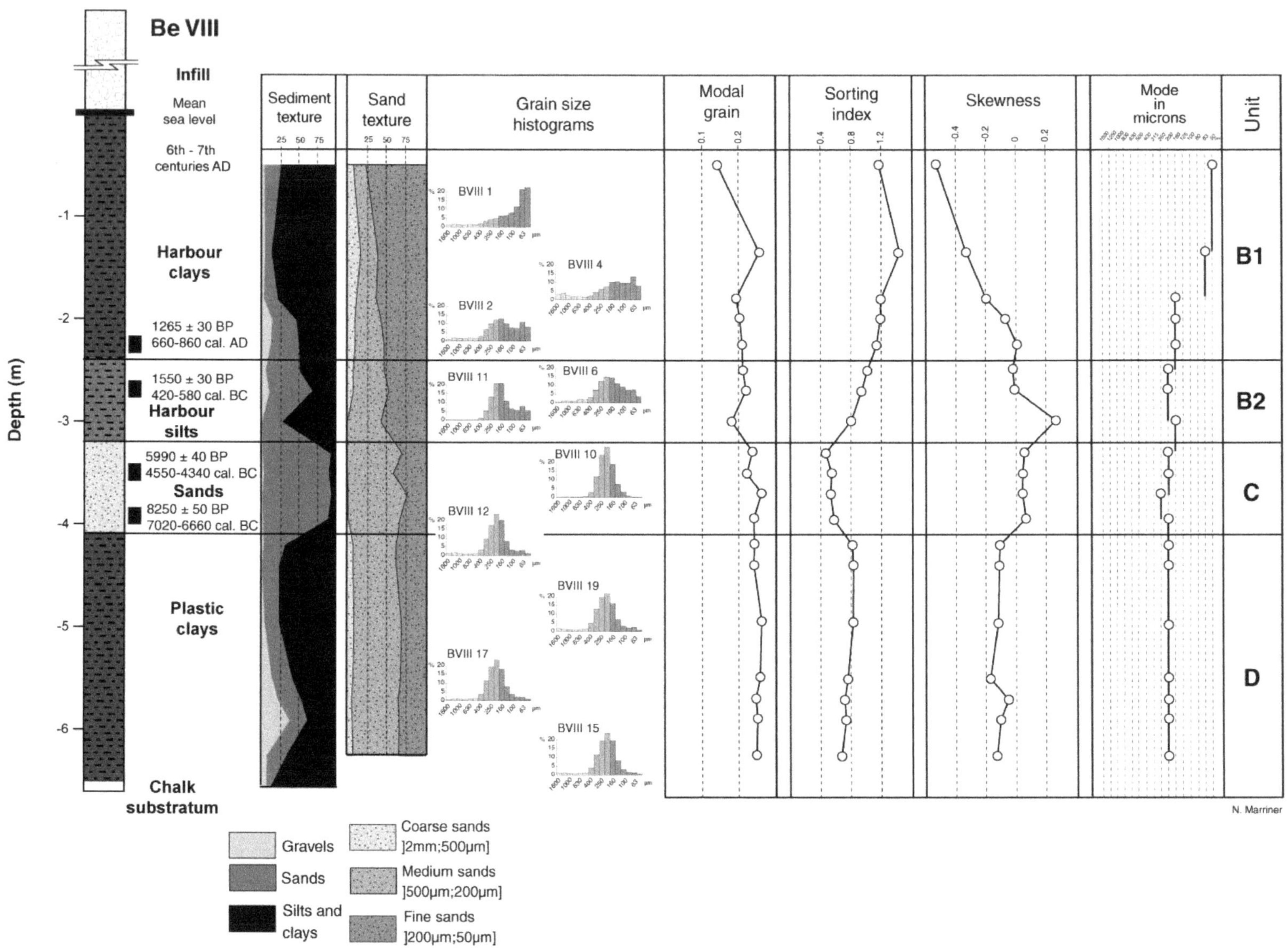

Figure 5.14: Sedimentology of core Be VIII. (1) Unit D: is a plastic olive black (3/2 5Y) clay lithofacies comprising 46-86 % silts and clays with medium sorting indices of 0.68 to 0.82. We infer a low energy lagoon environment. (2) Unit C: is a medium to fine-grained sand unit. Slightly positive (0-0.1) skewness values attest to a subtidal beach. The base of the unit has been dated 8250 ± 50 BP. This early Holocene date appears consistent with the reworking of underlying lagoon deposits during the Holocene transgression ~6000 BP, as corroborated by a radiocarbon date in core Be X. We are presently undertaking further radiocarbon dates on this unit. (3) Unit B2: comprises a fine grained silty sand unit. The gravels fraction comprises ceramics, seeds and wood fragments typical of a harbour unit. The top of the unit has been dated to the late Roman/Byzantine period. This facies is consistent with the Iron Age/Roman harbour sedimentology elucidated at both Sidon and Tyre. (4) Unit B1: constitutes a plastic clays units (49-86 %) characterised by poor sorting indices (1.14-1.43). This sedimentology is typical of a well-protected harbour basin during the Byzantine period, existing at Beirut until the 6th-7th centuries AD.

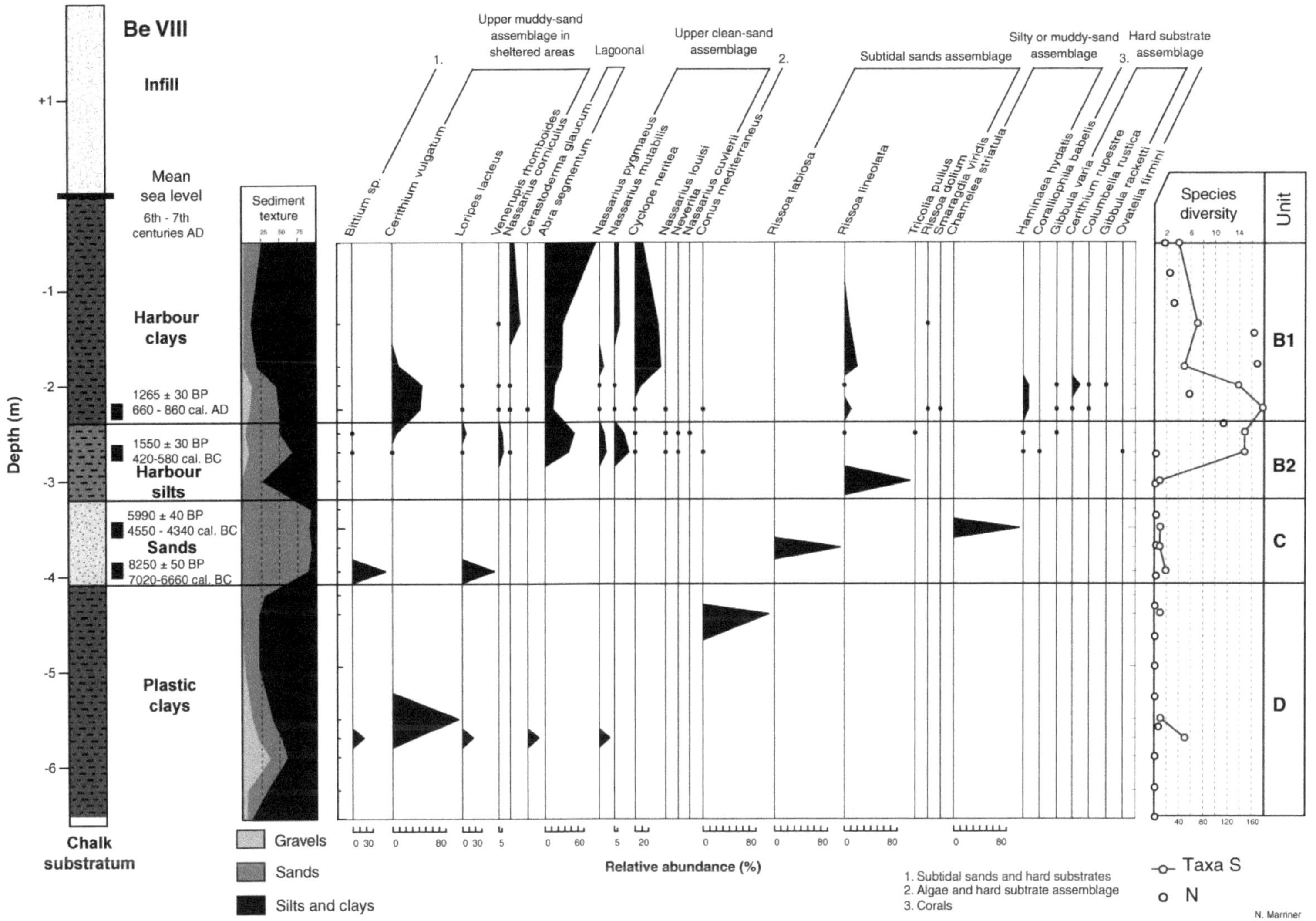

Figure 5.15: Molluscan macrofauna of core Be VIII. (1) Unit D: The unit's poor molluscan fauna comprises individuals typical of low energy coastal environments including *Cerithiumm vulgatum*, *Loripes lacteus* (upper muddy sand assemblage in sheltered areas) and *Cerastoderma glaucum* (lagoonal). We infer a lagoon environment from these data. (2) Unit C: has a very poor molluscan suite. It includes the following taxa, *Bittium reticulatum* (subtidal sands and hard substrates), *Loripes lacteus* (upper muddy-sand assemblage in sheltered areas) and *Rissoa labiosa* (subtidal sands assemblage). (3) Unit B2: has a rich molluscan fauna with taxa typical of a protected harbour environment, including *Cerithium vulgatum*, *Loripes lacteus* (upper muddy-sand assemblage in sheltered areas), *Abra segmentum* (lagoonal), *Nassarius pygmaeus*, *Nassarius mutabilis* (upper clean-sand assemblage) and *Rissoa lineolata* (subtidal sands assemblage). (4) Unit B1: The diagonistic harbour species persist into this lithofacies and attest to a well-protected lagoon environment. The unit is dominated by the lagoonal species *Abra segmentum*, *Cyclope neritea* (upper clean-sand assemblage) and *Cerithium vulgatum* (upper muddy-sand assemblage in sheltered areas). Secondary taxa include *Nassarius corniculus* (upper muddy-sand assemblage in sheltered areas), *Nassarius mutabilis* (upper clean-sand assemblage) and *Rissoa lineolata* (subtidal sands assemblage).

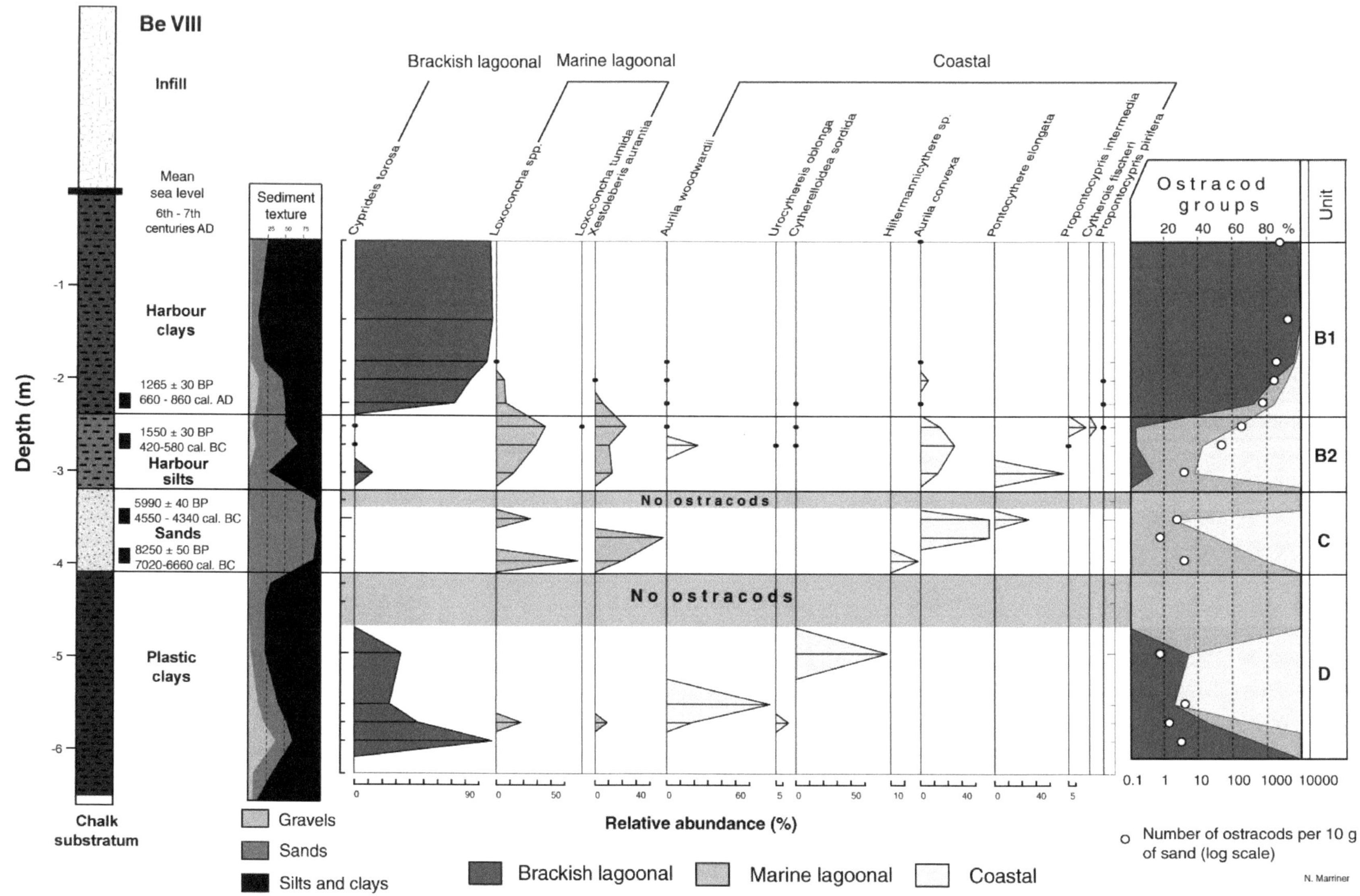

Figure 5.16: Ostracod microfauna of core Be VIII. (1) Unit D: comprises a poor ostracod fauna with fewer than 10 tests per 10 g of sand. The unit is dominated by the brackish lagoonal species *Cyprideis torosa* and *Aurila woodwardii* (coastal). Secondary species include *Loxoconcha* spp. and *Xestoleberis aurantia* (marine lagoonal). (2) Unit C: is dominated by marine lagoonal (*Loxoconcha* spp., *Xestoleberis aurantia*) and coastal (*Aurila convexa*, *Pontocythere elongata*) taxa. We infer a medium to low energy beach environment. (3) Unit B2: The ostracod fauna translate a well-protected environment. The dominant taxa include *Loxoconcha* spp., *Xestoleberis aurantia* (marine lagoonal), *Aurila convexa* and *Pontocythere elongata* (coastal). Secondary species include *Cyprideis torosa* (brackish lagoonal), *Aurila woodwardii*, *Cytherois fischeri* and *Propontocypris pirifera* (coastal). (4) Unit B1: constitutes a high faunal density (>1000 tests per 10 g sediment) dominated almost exclusively by *Cyprideis torosa*. We infer a well-protected lagoon environment consistent with the technological apogee of the harbour during the Byzantine period.

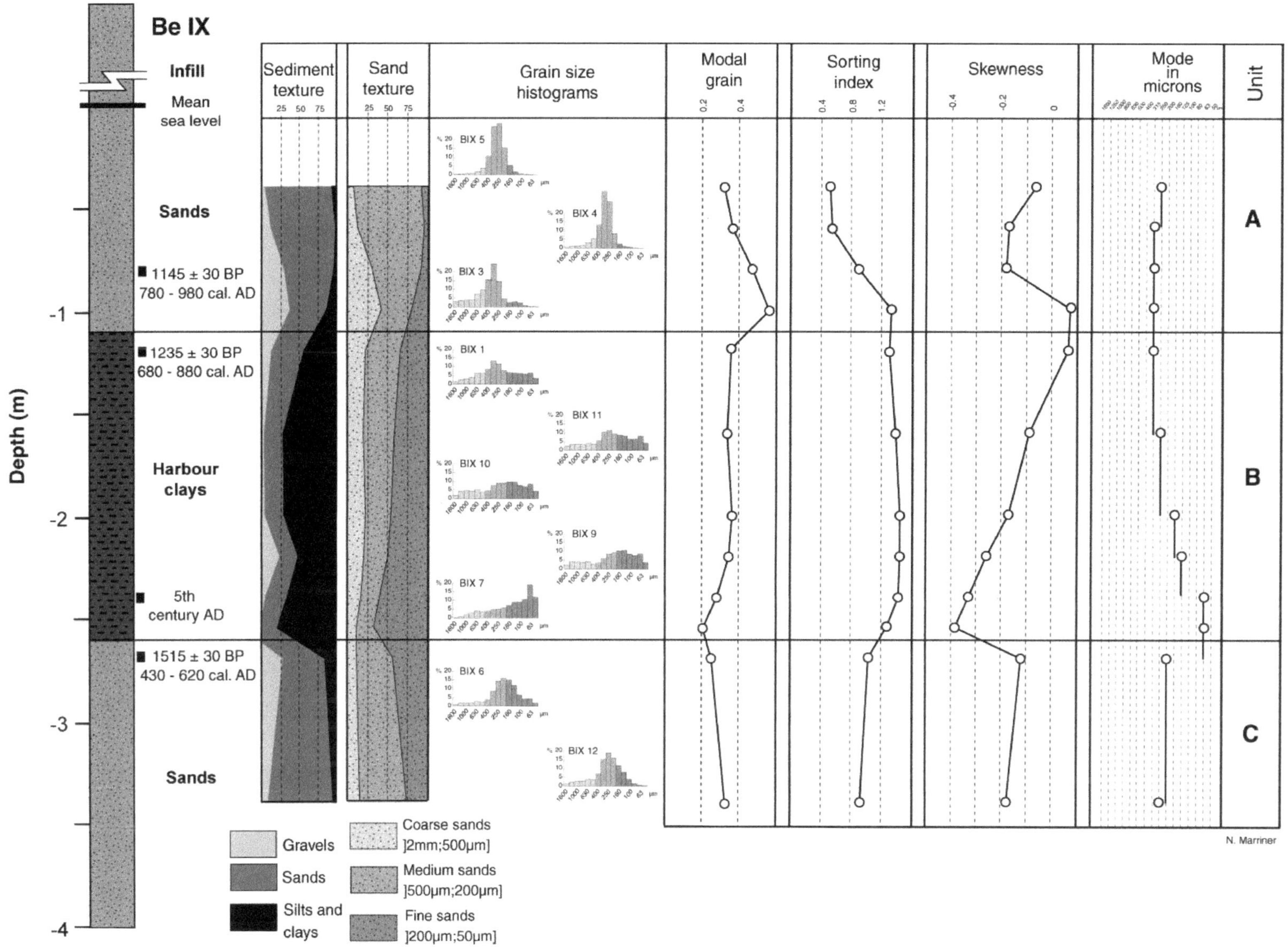

Figure 5.17: Sedimentology of core Be IX. (1) Unit C: is a medium to fine-grained sand facies with sorting indices between 0.9 and 1. We infer a middle to low energy beach environment. (2) Unit B: is a plastic clays lithofacies. Poorly developed grain size histograms are typical of an artificially protected environment. The chronology of the unit is consistent with the Byzantine period. A chronological hiatus of Iron Age and Roman deposits suggests that these have been removed through dredging. (3) Unit C: Transition to a coarse to medium grain sand unit corroborates a relative demise in harbour infrastructure. Cross-correlation with a similar stratigraphic unit in core Be X indicates that this shift in sedimentary environments occurred during the 6th-7th centuries AD.

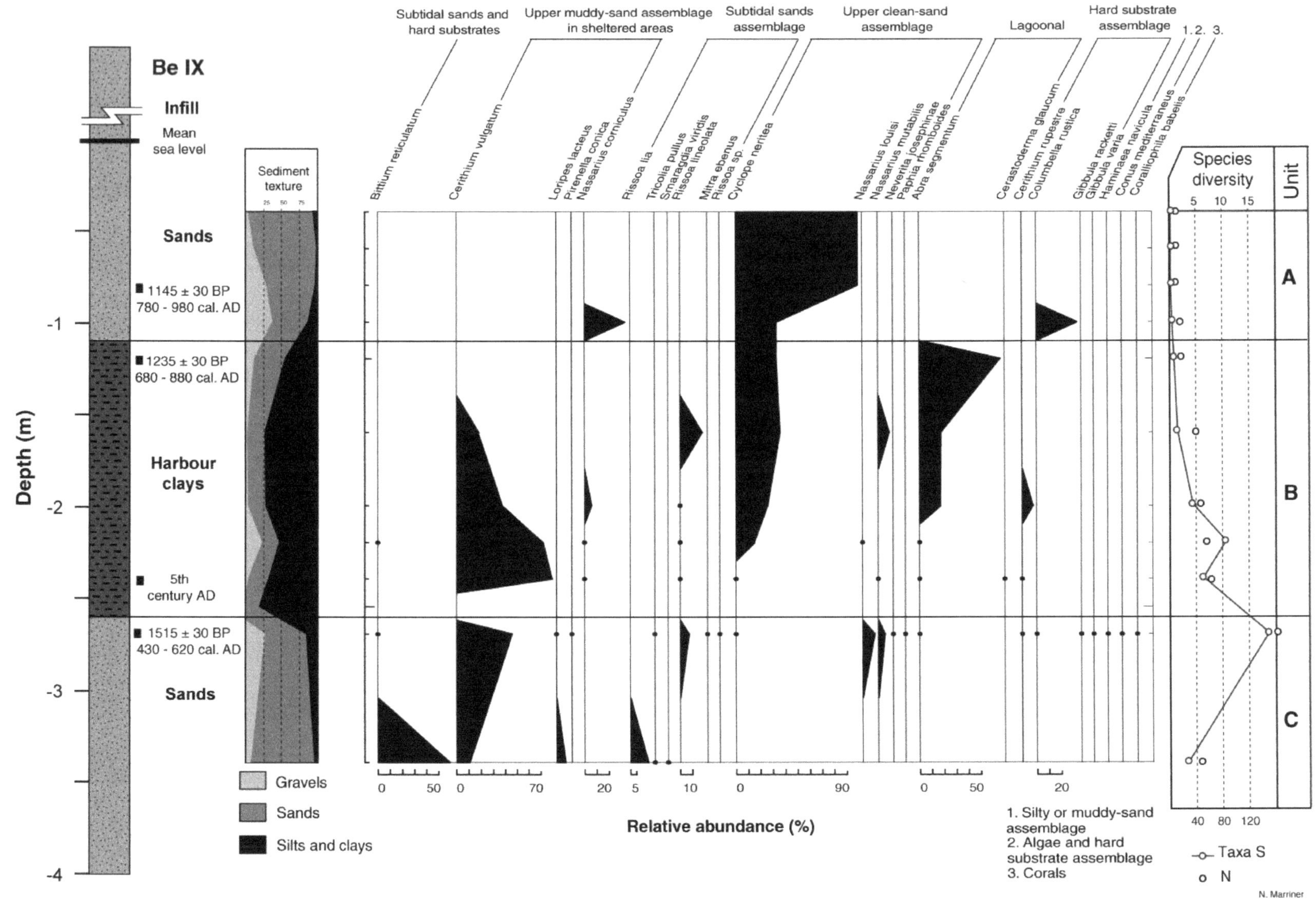

Figure 5.18: Molluscan macrofauna of core Be IX. (1) Unit C: constitutes taxa typical of a medium to low energy beach environment including *Bittium reticulatum* (subtidal sands and hard substrates), *Cerithium vulgatum*, *Loripes lacteus* (upper muddy-sand assemblage in sheltered areas), *Rissoa lia*, *Rissoa lineolata* (subtidal sands assemblage), *Nassarius louisi* and *Nassarius mutabilis* (upper clean-sand assemblage). (2) Unit B: Three taxa diagnostic of a well-protected harbour environment dominate this unit. They are *Cerithium vulgatum* (upper muddy-sand assemblage in sheltered areas), *Cyclope neritea* (upper clean-sand assemblage) and *Abra segmentum* (lagoonal). (3) Unit C: comprises a very poor molluscan suite. The unit is dominated by the upper clean-sand assemblage species *Cyclope neritea* indicative of a prograding beach unit.

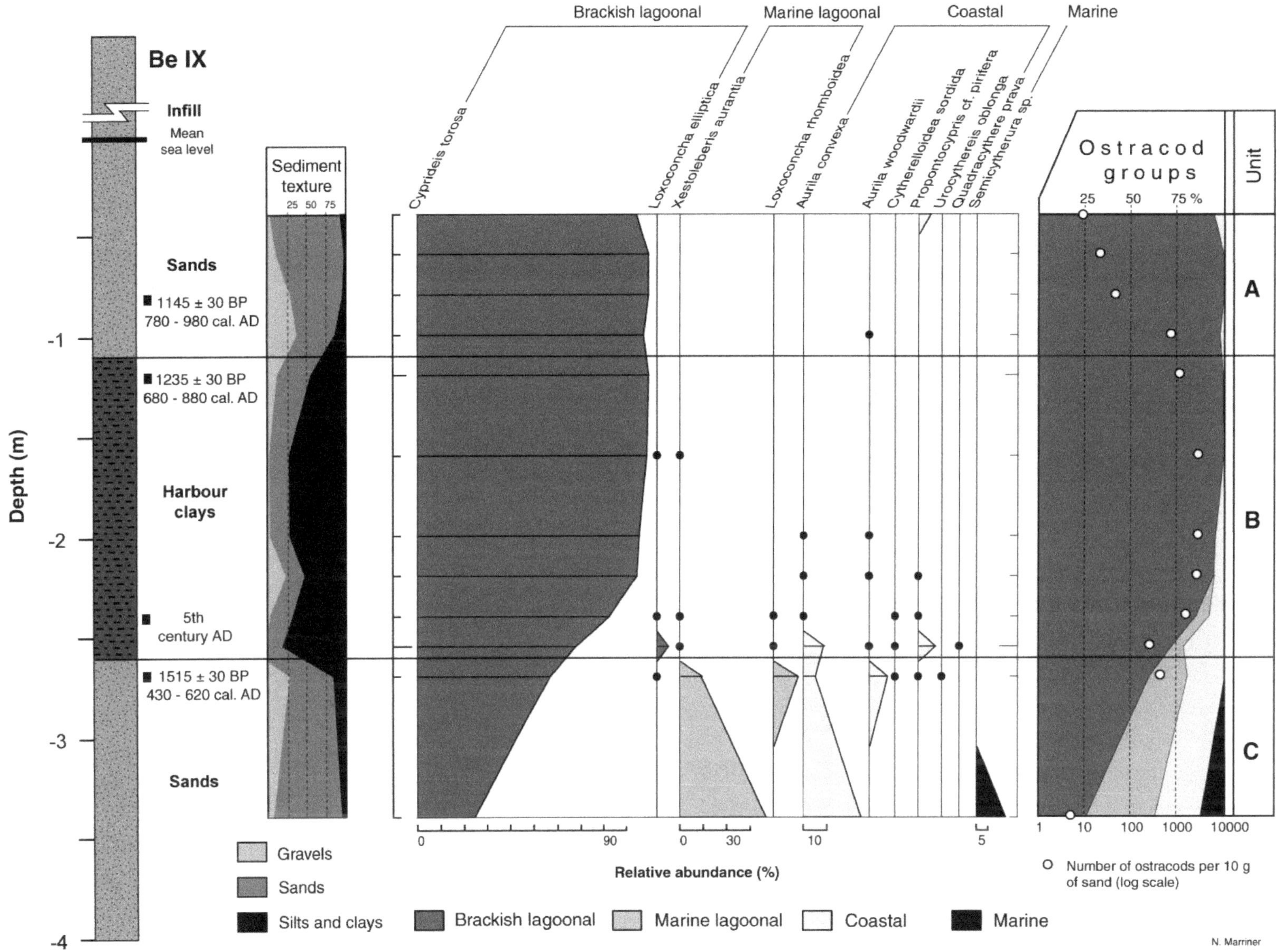

Figure 5.19: Ostracod microfauna of core Be IX. (1) Unit C: constitutes taxa from all four ecological groups including, *Cyprideis torosa* (brackish lagoonal), *Xestoleberis aurantia*, *Loxoconcha rhomboidea* (marine lagoonal), *Aurila* spp. (coastal) and *Semicytherura* sp. (marine). (2) Unit B: has a rich faunal density of over 1000 tests per 10 g of sand. The facies is dominated by *Cyprideis torosa* which attains relative abundance levels >90 %. The mono-specific predominance of this species indicates a hyposaline environment typical of a well-protected harbour. (3) Unit A: despite a shift in sedimentary conditions, *Cyprideis torosa* persists into this unit. We note a fall in faunal densities to >100 tests per 10 g of sand.

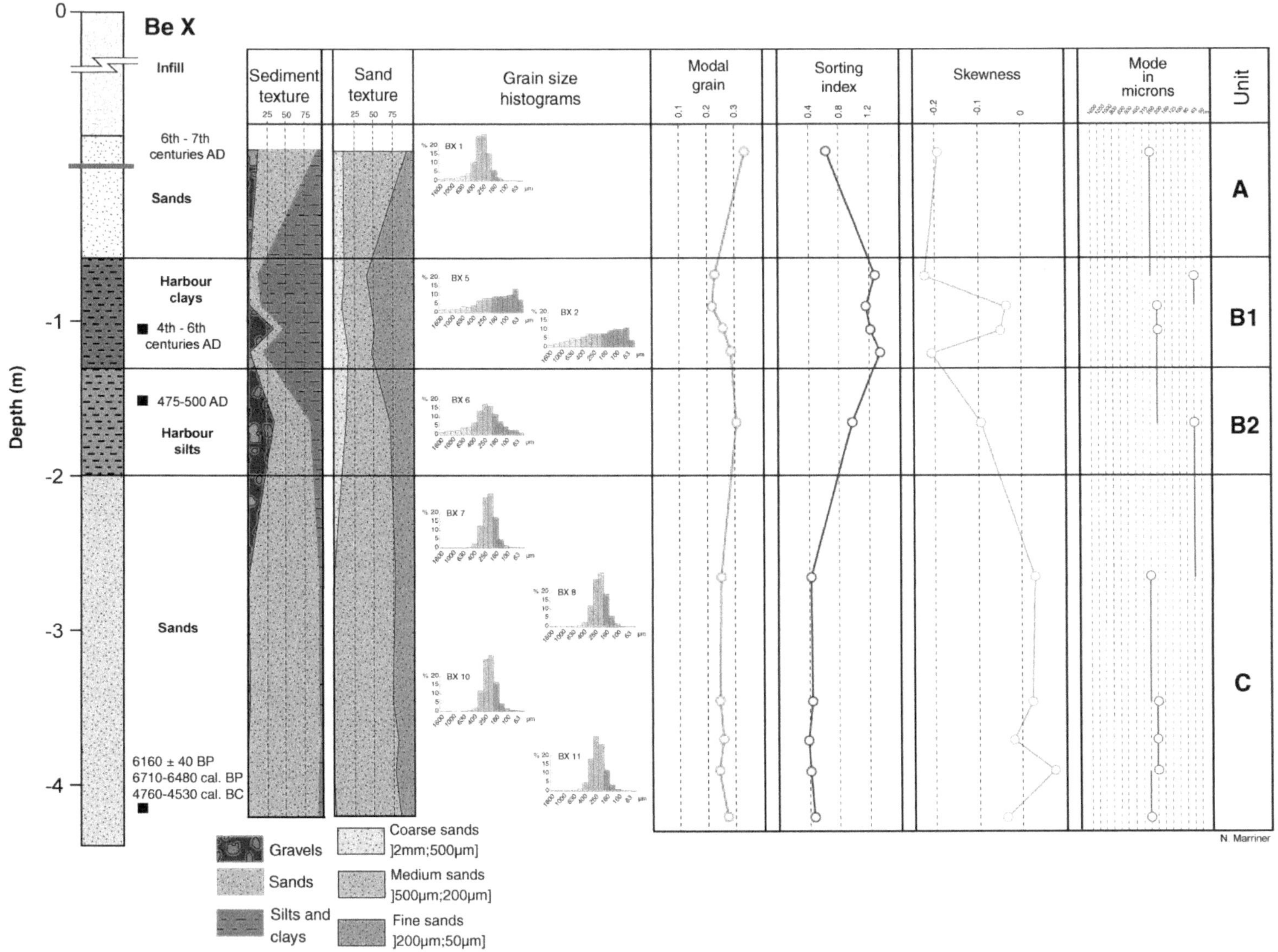

Figure 5.20: Sedimentology of core Be X. (1) Unit C: comprises a medium grain, well-sorted (~0.4) sand unit. The base of the unit is dated 6160 ± 40 BP (6710-6480 cal. BP), and is consistent with the accretion of a medium to low energy beach unit after the Holocene transgression. We suggest that this naturally protected cove served as a proto-harbour during the Bronze Age. (2) Unit B2: constitutes a medium sorted silty sand unit, rich in cultural debris including ceramic sherds, seeds and wood fragments. The ceramics have been dated to 475-500 AD corroborating a sheltered harbour during the late Roman/Byzantine period. The absence of earlier Iron Age and Roman deposits is consistent with late Roman/Byzantine dredging. (3) Unit B1: comprises a poorly sorted (~1.2) plastic clays unit. The poorly developed grain size histograms are diagnostic of an ancient harbour depocentre. The unit is dated by ceramics to the Byzantine period. (4) Unit C: is a medium grain sand unit. It attests to the relative demise of the Byzantine harbour during the 6th-7th centuries AD.

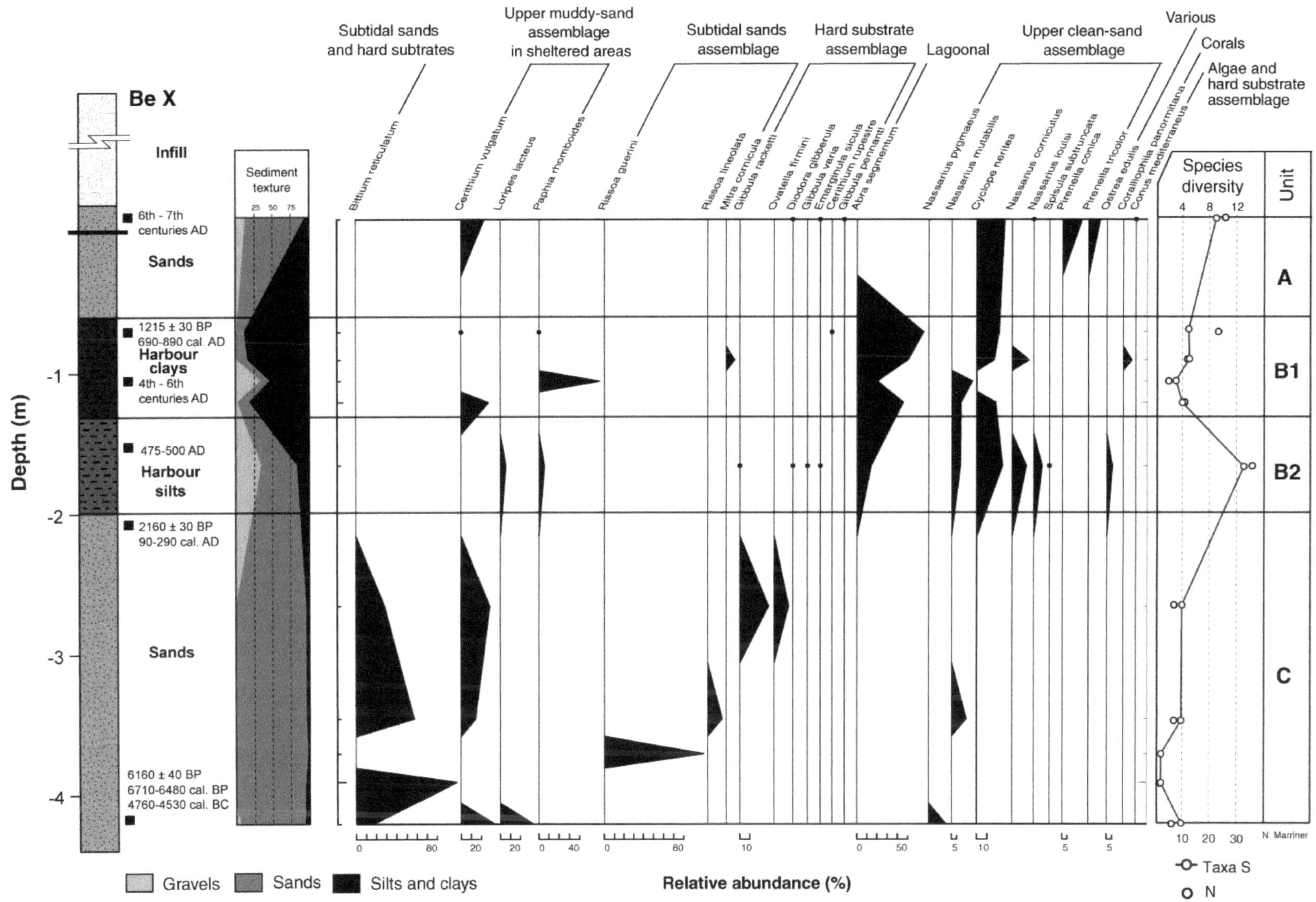

Figure 5.21: Molluscan macrofauna of core Be X. (1) Unit C: juxtaposes taxa from a number of different ecological groups, including *Bittium reticulatum* (subtidal sands and hard substrates), *Cerithium vulgatum*, *Loripes lacteus* (upper muddy-sand assemblage in sheltered areas), *Rissoa guerini* (subtidal sands assemblage), *Rissoa lineolata* (subtidal sands assemblage), *Gibbula racketti* (hard substrate assemblage), *Nassarius pygmaeus* and *Nassarius mutabilis* (upper clean-sand assemblage). We infer a medium energy beach environment. (2) Unit B2: comprises molluscs indicative of a sheltered low energy environment. These include *Loripes lacteus* (upper muddy-sand assemblage in sheltered areas), *Abra segmentum* (lagoonal), *Nassarius mutabilis*, *Cyclope neritea* and *Nassarius corniculus* (upper clean-sand assemblage). (3) Unit B1: Hypersaline lagoon conditions are attested to by the dominance of *Abra segmentum*. Secondary taxa include *Cerithium vulgatum*, *Paphia rhomboides* (upper muddy-sand assemblage in sheltered areas), *Nassarius mutabilis*, *Cyclope neritea* and *Nassarius corniculus* (upper clean-sand assemblage). (4) Unit C: Taxa typical of a prograding beach characterise this unit. These notably include *Cyclope neritea* and *Pirenella* spp.

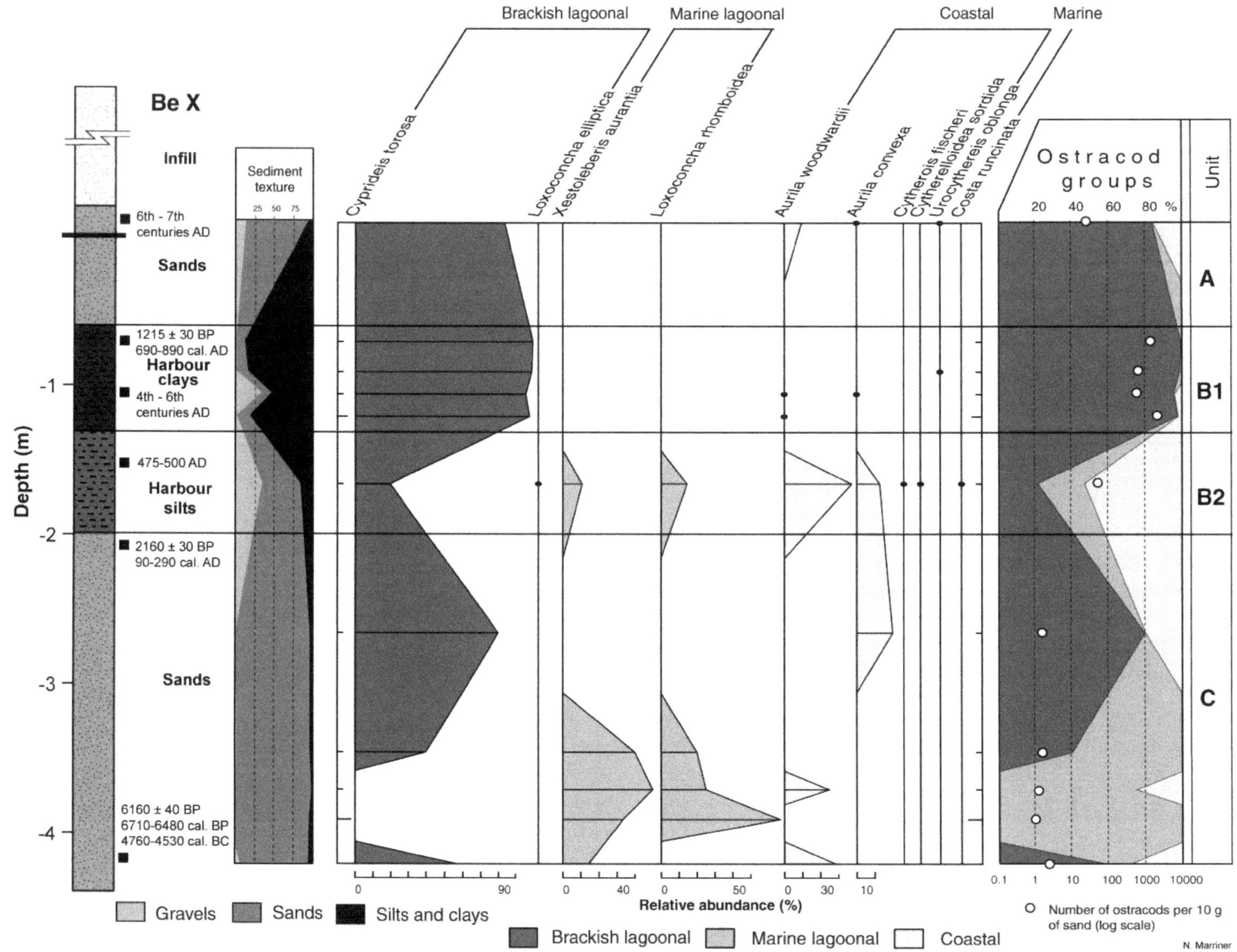

Figure 5.22: Ostracod microfauna of core Be X. (1) Unit C: A semi-protected beach environment is attested to by brackish lagoonal (*Cyprideis torosa*) and marine lagoonal (*Xestoleberis aurantia* and *Loxoconcha rhomboidea*) taxa. (2) Unit B2: juxtaposes taxa from all four ecological groups, including *Cyprideis torosa* (brackish lagoonal), *Xestoleberis aurantia*, *Loxoconcha rhomboidea* (marine lagoonal), *Aurila* spp. (coastal) and *Costa runcinata* (marine). (3) Unit B1: is almost exclusively dominated by *Cyprideis torosa*. We infer a protected hypersaline environment, typical of a well-protected harbour. (4) Unit C: Despite a shift in the sedimentary environment, *Cyprideis torosa* persists into this unit, concurrent with a gradual demise of the harbour.

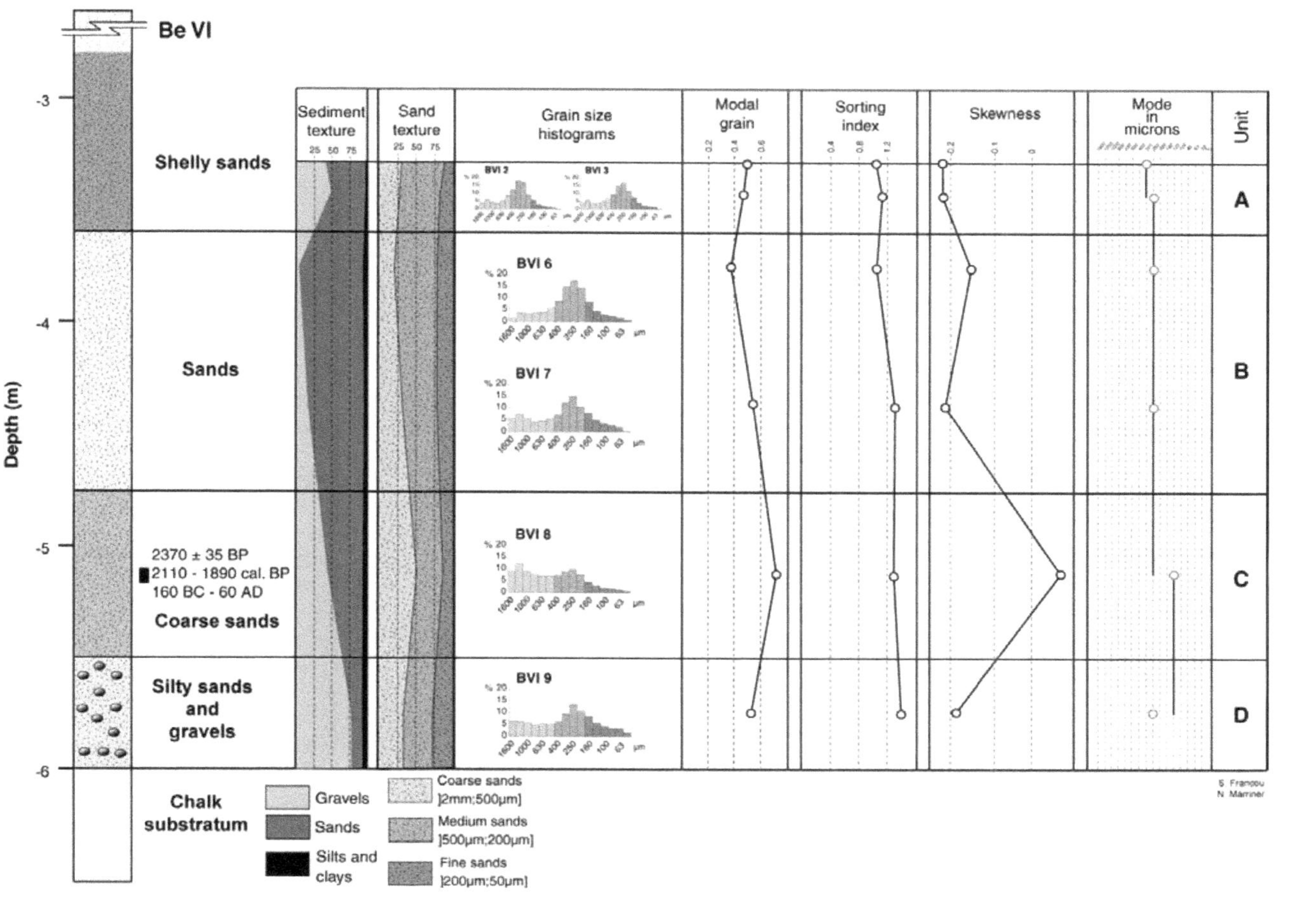

Figure 5.23: Sedimentology of core Be VI. Poorly sorted sands and the dominance of coarse and medium sands confer an exposed marine environment in this area since 6000 BP. There are no stratigraphic units diagnostic of a harbour environment.

Beirut, although not a prominent Bronze Age town, benefited from its geographical position halfway between Sidon and Byblos to serve as a stop-over for traders *en route* to and from Egypt, southern Levant, Syria and Anatolia. It followed the general model of urban development, namely the construction of palaces, temples and city walls designed to protect and consolidate this maritime-derived wealth (Curvers and Stuart, 2004). Although overshadowed by its sister city-states Byblos, Sidon and Tyre, Beirut is mentioned in numerous Egyptian cuneiforms from the second millennium BC as a thriving Bronze Age settlement (Arnaud and Salvini, 2000).

Unlike Sidon, there is no sedimentological evidence in support of primitive Bronze Age harbourworks. This could be for two reasons: (1) the absence of such infrastructure during the Bronze Age at the site; and/or (2) the failure to pick up a diagnostic signature due to the proximal position of the cores to the cove entrance. The heart of the Bronze Age basin appears to have been located ~120-150 m to the south (**Figure 5.12**), although archaeological investigations have thus far not unearthed any artificial infrastructures dating to this period.

5.5.3 Closed Iron Age (?) to Roman/Byzantine harbours

Description: In unit B1, the onset of artificial harbourworks is characterised by a sharp rise in the silts and clays unit (25 to 75 % of the total sediment). Sorting indices of 0.8 to 1 attest to a medium sorted sand fraction. The sediment juxtaposes fine-grained sands and silts with a rich coarse-grained debris constituting wood fragments, glass, seeds and ceramic sherds. Sample BVIII 5 has been dated 1550 ± 30 BP (420 to 580 cal. AD), consistent with the Roman period. Such a juxtaposition of coarse-grained versus fine-grained material is a classic feature of artificial harbours in proximity to an urban settlement. The macrofauna is dominated by five assemblages, the upper muddy-sand assemblage in sheltered areas (*Cerithium vulgatum*, *Loripes lacteus*, *Venerupis rhomboides*), the upper clean-sand assemblage (*Cyclope neritea*, *Nassarius louisi*, *Nassarius mutabilis*, *Nassarius pygmaeus*), the lagoonal assemblage (*Abra segmentum*) and the subtidal sands assemblage (*Rissoa lineolata*). Species from the lagoonal (*Cyprideis torosa*), marine lagoonal (*Loxoconcha* spp., *Xestoleberis aurantia*) and coastal (*Aurila convexa*, *Pontocythere elongata*) domains constitute the ostracod suite. This is accompanied by a rise in faunal densities to between 100-300 tests per 10 g of sand.

Interpretation: The presence of artificial harbourworks is clearly manifest in the basin stratigraphy. Unfortunately, unequivocally attributing these changes to the Iron Age is problematic given repeated dredging phases during the Roman and Byzantine periods. Paradoxically, Roman and Byzantine dredging has culminated in an archiveless harbour for the Iron Age period. Only at Sidon have we been able to localise and date pockets of fine-grained Iron Age sediments (chapter 4; Marriner *et al.*, 2006b). Research has shown that the Roman period marks a technological watershed in many ancient harbours of the Mediterranean (Oleson, 1988; Rickman, 1988). At Marseilles and Naples, for example, wide-reaching infrastructure enterprises were undertaken not only to overdeepen the rapidly silting up port, but also to completely refashion harbour layouts (Hesnard, 2004a-b).

Although we record a gradual progradation of the harbour coastline between the Bronze Age and Roman periods, we have very little information on the nature of harbourworks employed (**Figure 5.12**). Thus far, only an Iron Age III/Persian quay has been unearthed (Elayi and Sayegh, 2000; Sayegh and Elayi, 2000; **Figure 5.13**). This is buried ~60-70 m south of the core network. Just north of two nineteenth century Ottoman quays, no fine-grained harbour sediments have been unearthed indicating the harbour was not more extensive in this northerly direction (**Figures 5.23 and 5.24**). It seems likely that the nineteenth century quay reemployed foundation courses from a more ancient harbour structure. However, in the absence of strong archaeological data, we can only tentatively hypothesise that this edifice was adapted and recycled from the Iron Age onwards. Unlike the southern prograding harbour coastline, the northern limits of the port basin have probably varied very little over the past 6000 years.

During the Phoenician period archaeological finds in and around the basin suggest that the harbour was used as a commercial and fishing hub (Curvers and Stuart, 1996). By the second half of the eighth century BC, the Phoenician city-states were providing Assyria with much of its maritime resources (Elayi, 1984, 1990). Abundant imported ceramics and transport containers are consistent with a rich Mediterranean trade (Curvers and Stuart, 2004). Archaeological research on Phoenician and Persian urban quarters bordering the harbour shows the importance of murex (*Bolinus brandaris* and *Hexaplex trunculus*) finds, which can be put into relation with purple-dye production (Doumet, 1980; 2004; Marquis, 2004). Finds of nets and fish bones on a large scale also indicate the importance of the fishing industry (Elayi and Sayegh, 2000). Fish would have been dried and exported to local and regional markets in the hinterland, and possibly further afield. A harbour of this size was also capable of assuring the transportation of cedar wood, prized in construction throughout the Levantine basin (Elayi, 1984, 1990). The relative absence of military finds appears to suggest that the city did not support a large military float as was the case at Sidon and Tyre during the same period (Elayi and Sayegh, 2000).

Figure 5.24: Ottoman quays unearthed during the EDRAFOR construction works (photographs courtesy of EDRAFOR). These quays lie just outside the ancient basin and attest to a gradual dislocation of the harbour coastline since the Bronze Age.

The existence of smaller secondary and/or outer harbours must not be excluded. Two coves appear conducive to the accommodation of shallow draught boats. (1) The ~400 m by ~150 m palaeo-embayment immediately east of the tell (**Figure 5.12**) has yielded medium to fine-grained sand facies at the base. Although it is difficult to make precise interpretations on the geomorphology of this cove because the upper parts of the coastal stratigraphy have been reworked by the nineteenth century redevelopment, the archaeological remains and topographic data in this area do allow the disposition of the cove to be reconstructed (Ormsby, 1839; Wyld, 1840; Royal Engineers, 1841; Scott, 1841; Skyring, 1841; Löytved, 1876; Baedeker, 1912; Davie, 1987). (2) ~350 m west of Beirut's ancient harbour a second marine cove, still manifest in nineteenth century maps, might have served as a fair-weather harbour (**Figure 5.12**). Clearly further work is required to corroborate these working hypotheses.

The archaeological data demonstrate that the sixth to fifth centuries BC were a period of urban expansion of the lower city, which spread out and beyond the two overlooking promontories (Sader, 1999; Curvers, 2002;

Curvers and Stuart, 1998-1999). The pottery found in Beirut from the earliest Hellenistic contexts includes amphorae originating from Sidon, Tyre, Cyprus and the Aegean. The presence of these imported wares indicates the importance of Beirut and its harbour in Mediterranean trade at this time. On the basis of trends in pottery finds, it has been hypothesised that Beirut was a dependency of Sidon during the Hellenistic period (Reynolds, 1999), becoming independent after the second century BC. At this time, there was a shift from Ptolemaic dependence to the new imperial policy of the Seleucid rulers. During the Roman-Byzantine period, the monumental centre of the city shifted from the Iron Age tell and harbour area, to a zone outside the borders of the Iron Age city (Lauffray, 1944-1945; Davie, 1987; Faraldo Victorica and Curvers, 2002).

5.5.4 Closed late Roman and early Byzantine harbours

Description: Transition to plastic harbour clays concurs the technological apogee of Beirut's ancient harbour during the late Roman and Byzantine periods. The unit is constrained by numerous ceramic finds dating to the fifth to seventh centuries AD. Sorting indices of between 1.2 and 1.5 are concomitant with a poorly sorted sediment (**Figures 5.14, 5.17 and 5.20**). Fine sands comprise over 50 % of the total sand fraction. Poorly developed histograms are consistent with the juxtaposition of fine-grained deposits trapped in the basin by a fall in water competence and coarse-grained material deriving from slope wash and urban waste. Five groups dominate the molluscan fauna, namely the upper muddy-sand assemblage in sheltered areas, the upper clean-sand assemblage, the silty or muddy-sand assemblage, the lagoonal assemblage and the subtidal sands assemblage (**Figures 5.15, 5.18 and 5.21**). A sharp rise in the lagoonal species *Abra segmentum* indicates a heavily artificialised basin. This tendency is corroborated by the ostracod data, characterised by the monospecific domination of *Cyprideis torosa* and accompanied by a sharp rise in faunal densities (>1000 tests per 10 g of sand; **Figures 5.16, 5.19 and 5.22**).

Interpretation: Urban development which started in the Roman period continued under Byzantine rule and the hinterland too witnessed an era of prosperity with the development of agriculture, mainly oil and wine production, and of the silk, glass and purple industries (Hall, 2004). Archaeological data suggest that the early Roman city grew up on the tracks of the previous Hellenistic city wall (Saghieh-Beydoun *et al.*, 1998-1999). Between the second and fourth centuries AD, major changes in the urban fabric and layout are recorded, consistent with the foundation of the new imperial capital at Constantinople in 330 AD (Sader, 1999). Against this backcloth of economic and cultural opulence Beirut's harbour was developed to accommodate increased mercantile traffic. Archaeological data from the port area record a significant overhaul in the basin topography (Curvers, personal communication). Production of Beirut amphorae was expanded to collect and export goods from the hinterland, in response to the city's economic growth (Reynolds, 1999). This concentration of wealth was associated with a number of new public and monumental constructions, including a hippodrome, two large bath complexes and new forums (Curvers and Stuart, 2004; Doumet-Serhal, 2004a).

Consolidation of Roman engineering techniques and infrastructure is translated in the ancient harbour by the B1 plastic clays unit. Although we have very little archaeological data pertaining to the nature of harbourworks during the late Roman and Byzantine periods, research at sites such as Caesarea, Ostia and Cosa shed light on the technology and techniques employed at Beirut (Oleson, 1988; Oleson and Branton, 1992; Brandon, 1996, 1997; Hohlfelder, 1997; Humphrey *et al.*, 1998; Oleson *et al.*, 2004a-b).

We attribute the relative absence of seaport infrastructure unearthed from this and earlier periods to three factors: (1) repeated destruction of port edifices during catastrophic earthquakes (Guidoboni *et al.*, 1994). Beirut's peninsula is located in a highly seismic area (Dubertret, 1955, 1975); the ancient city lies upon three fault lines, one of which intersects the harbour basin (Darawcheh *et al.*, 2000; **Figure 5.4**). Calculations indicate that +7 earthquakes (Mercalli scale) have an average return period of 50 years in this area (Plassard and Kogol, 1981). Iron Age, Roman and Byzantine habitation layers show evidence of anomalies and structural fracture lines, degradation that has been linked to seismic activity (Saghieh Beidoun, 1997). The most devastating historical earthquake to have struck Berytus occurred in 551 AD, and was associated with massive urban destruction (see 5.5.5); (2) the recycling of port infrastructures throughout history; and (3) finally, the absence of archaeological excavations in these areas. Until the harbour area has been satisfactorily surveyed, it is impossible to assign more weight to one factor over another.

5.5.5 Islamic and Medieval harbours

Description: Transition to a well-sorted medium-grained sand after the sixth to seventh centuries AD is consistent with an opening up of the harbour basin to coastal dynamics (**Figures 5.14, 5.17 and 5.20**). A rapid fall in the silts and clays fraction from +75 % in unit B1 to <20 % in unit A suggests that this change was relatively rapid. The transformation is accompanied by an increase in molluscan species from the upper clean-sand assemblage (*Cyclope neritea*) and the hard substrate assemblage (*Columbella rustica*; **Figures 5.15, 5.18 and 5.21**). *Cyprideis torosa* continues into this unit, although a pronounced decline in faunal densities is observed from 1000/10g sand to <100/10g sand (**Figures 5.16, 5.19 and 5.22**).

Interpretation: The transition from fine-grained to coarser-grained sedimentation is a classic stratigraphic feature of ancient harbours. It is usually linked to a partial/total abandonment of the basin and/or the economic and political decline of a site (Goiran and Morhange, 2003; Marriner and Morhange, 2007). Unlike Sidon and Tyre, the widespread archaeological excavations undertaken since 1993 mean that there is no shortage of evidence to explain the sudden demise of Beirut's seaport, centred on the sixth to seventh centuries AD (Curvers and Stuart, 2004). As we have outlined in earlier chapters, there appear to be four complementary dynamics at work: (1) historical; (2) tectonic; (3) tsunamogenic; and (4) climatic. Although these points will be addressed in more detail in chapter 6, it is nonetheless important to insist upon two points.

(1) **Archaeological:** The seventh century AD marks a decline in Byzantine hegemony on the circum Levant and an expansion of Islamic influence (Bonner, 2005). At Beirut, two lines of archaeological evidence document economic decline after the sixth to seventh centuries AD. (i) **Urban**: excavations suggest that great tracts of the city were left in ruin after the 551 AD earthquake, with patchy evidence for urban rebuilding (Seeden and Thorpe, 1997-1998; Curvers and Stuart, 2004; Mikati and Perring, 2006). Many parts of the city show a mixed pattern of abandonment and continued occupation. (ii) **Ceramic ware**: the pottery recovered from the forum (BEY 006) of this period manifests a significant change in the range and character of goods being imported into the city, consistent with a decline in Mediterranean traffic (Perring *et al.*, 1996; Perring, 1999).

(2) **Lithostratigraphical:** Excavations undertaken in Beirut's harbour by Curvers have revealed the presence of tree branches and considerable amounts of unabraded Roman pottery and rubble in sixth to seventh century AD layers. Surveys in the Ottoman harbour have unearthed harbours muds and silts which lie unconformably above sea-scoured bedrock (Curvers and Stuart, 2004). These have been attributed to tsunami action and indirectly infer considerable damage to the city's seaport infrastructure. This archaeological evidence, coupled with our stratigraphic data, support major changes in the port's configuration at this time. At no point during the Islamic and medieval periods do we record such a well-protected harbour. In light of this, there appears to be a clear link between the retraction of the Byzantine Empire to its Anatolian core and the catastrophic destruction of many parts of Beirut, including its harbour area, during the 551 AD earthquake and tsunami.

The 551 AD earthquake is the best documented earthquake to have struck ancient Beirut in antiquity (Guidoboni *et al.*, 1994; Soloviev *et al.*, 2000). Described as the highest magnitude event to have affected the eastern Mediterranean (Darawcheh *et al.*, 2000), its seismic repercussions were felt as far afield as Mesopotamia. At a more general level, the period is characterised by significant and catastrophic coastal deformations throughout the region (Goiran, 2001; Pirazzoli, 2005; Morhange *et al.*, 2006b; see chapter six for more detailed discussion).

The 551 AD earthquake is mentioned by several pre-1900 AD catalogues including those of Bonito (1691) and Perrey (1850). Recent research has postulated two hypotheses for the event's epicentre: (1) Guidoboni *et al.* (1994) have presented three ancient texts which they believe attest to an epicentre off the coast of Beirut; (2) Ambraseys *et al.* (1994), on the other hand, have suggested that the earthquake's epicentre lay in the Jordan rift valley. Tsunami destruction along the Lebanese coast from Tripoli to Tyre would tend to favour the former.

The accounts of a number of chroniclers who lived at the time of the event have survived to present, including Malalas (491-578 AD), Agathias (532-580 AD) and John of Ephesus (507-586 AD), and an itinerary dated 560-570 AD written by a traveller named Antoninus Placentinus. According to Antoninus Placentinus, the Bishop of Beirut – an eyewitness of the earthquake – estimates that there were 30,000 deaths due to the earthquake (Darawcheh *et al.*, 2000). Although this figure is difficult to corroborate, it seems plausible as Beirut was considered to be the pearl of the Phoenician coast around this time. Fires and disease probably also contributed to this number in the aftermath of the event.

The accompanying tsunami is documented by John of Ephesus' account: "before the earthquake happened, the sea retired roughly two miles, then the people were rushed in the seabed to find wealth at the sunken ships, then an immense wave returned, flooding the shore and drowning ships as well as the people who were in the seabed and along the coast" (in Guidoboni *et al.*, 1994).

5.6 Concluding remarks

Coastal artificialisation during the past 200 years means that Beirut's ancient harbour is completely landlocked beneath the city centre. The ancient harbour lies in a talweg depression between the Iron Age tell to the east and a land promontory to the west (**Figure 5.12**). The ancient anchorage took full advantage of these natural endowments, which were gradually reinforced from the Iron Age onwards. The harbour comprised a deep bay looking north and sheltered from the dominant south-westerly winds. It appears that the basin was closed by a belt of partially drowned reefs, gradually reinforced to form an artificial mole (although it should be noted that no traces of this mole have been uncovered during the recent excavations). It is hypothesised that Borj al-Mina served as an outer harbour during the Bronze and Iron Ages (see **Figure 5.12**), in the same way as Zire island at Sidon.

According to Raban (1995), Levantine harbours were preferentially located in three areas: (1) on coastal peninsulas (Akko, Athlit, Sidon); (2) on offshore islands in proximity to the coastline (Tyre, Arwad); and (3) in coastal lagoons and wadis (Dor). The Beirut model appears to be a unique composite of these, comprising a drowned wadi – whose upper creek was manifestly used as an anchorage during the Bronze Age – and a small drowned island sandwiched between two enveloping land promontories.

The coastal stratigraphy elucidates five phases in the evolution of the basin spanning the last 6000 years. The core network lies in the northern portion of the ancient basin; clearly complementary stratigraphic data are needed further south, where the heart of the Bronze Age and Iron Age anchorages is to be found. Within this context, an interesting area for future research is the rue Allenby, where Iron Age III/Persian quays have been uncovered (Elayi and Sayegh, 2000). Sediments in the immediate vicinity of this cove have the potential to reveal more precise data on the Bronze Age and Iron Age histories. Despite the significant areas excavated, the opportunity to survey the ancient harbour using classical dig techniques, *sensu* Marseilles and Naples, could be rued in years to come (Lauffray, 1995; Karam, 1996; Naccache, 1996, 1998; Seeden, 1999; Raschka, 2006).

Chapter 6

Beirut, Sidon and Tyre: 5000 years of human-environment interactions

In this chapter we compare and contrast the stratigraphic data from all three ports to establish the relative role of natural and anthropogenic factors in forcing ancient harbour coastal stratigraphy and progradation since the mid-Holocene. The geomorphology of Phoenicia's buried harbours has been significantly affected by the flooded Pleistocene landscape. First, the shape and size of the flooded basins has influenced the available accommodation space. Second, the pattern of infilling has been heavily controlled by the interaction of environmental (river and coastal processes) and human factors. Here, we attempt to (1) link diagnostic coastal stratigraphies with harbour technology; and (2) better understand the origin of sediments accumulating in the basins. The chapter also endeavours to correlate the Beirut, Sidon and Tyre data with the wider Levantine context.

6.1 What are the similarities between Phoenicia's harbours?

The type stratigraphy manifest at Beirut, Sidon and Tyre shows considerable chronostratigraphic similarities (**Figure 6.1**). Although differences are observed in the granularity of sediments (see chapter 6.2), there are clear patterns in both the vertical (stratigraphic) and lateral evolutions of the three basins, changing from natural coves to increasingly artificialised seaports. Four main periods have been identified during the past 5000 years. In the absence of conspicuous archaeological data, it is argued that such a chronostratigraphic approach can be used to make informed inferences about the nature of harbour technology used (Marriner and Morhange, 2007).

6.1.1 Bronze Age proto-harbours

Bronze Age proto-harbours have been largely discussed in preceding chapters 3, 4 and 5. A number of research advances have nonetheless emerged during the course of our research and merit being recapitulated. One of the Levantine coast's key defining traits is the series of south-north trending Pleistocene shorelines that rim tracts of the region's seaboard (Durbertret, 1955; Sanlaville, 1977; Frechen *et al.*, 2001, 2002, 2004; Sivan and Porat, 2004). Referred to as 'kurkar' in Israel and 'ramleh' in the Lebanon, these sandstone outcrops have created a unique coastal geomorphology exploited by human societies for many millennia. At numerous localities on the Phoenician coastline, Holocene flooding of the shore-parallel ridges has given rise to partially drowned islands and reefs, many of which served as anchorage havens during the Bronze Age (Raban, 1985a-c, 1995; Marcus 2002a-b; Marriner *et al.*, 2006a; Haggai, 2006).

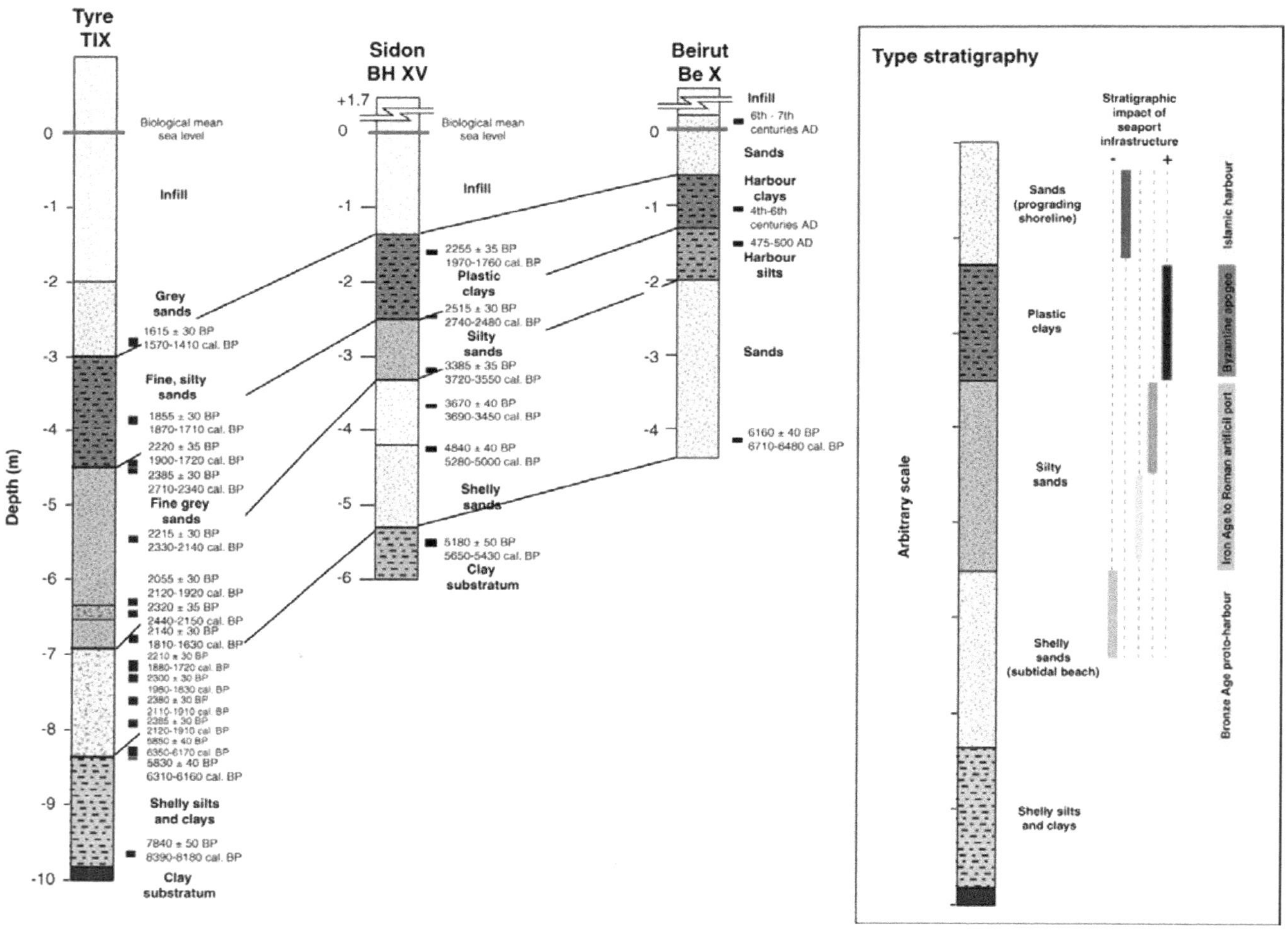

Figure 6.1: Comparison of the stratigraphy at Beirut, Sidon and Tyre. Inset: type stratigraphy from the three sites. The thicker sediment fill at Tyre is explained by the ~3 m tectonic collapse of the Tyrian horst during the late Roman period.

Stabilisation of Mediterranean sea levels around 6000 BP (Laborel *et al.*, 1994; Lambeck and Bard, 2000; Vella and Provansal, 2000; Morhange *et al.*, 2001; Sivan *et al.*, 2001) engendered the sedenterisation of human societies along present coastlines. In Syria, Lebanon and Israel, coastal settlements were preferentially founded around low energy coastal basins and the debouches of fluvial systems (Frost, 1995; Raban, 1995). Beirut, Sidon and Tyre, established during the third millennium BC, all integrate partially drowned islets into their harbour models. The primitive nature of seafaring at this time (Wachsmann, 1998) meant that the three basins required little or no human modification, and the natural defences are translated stratigraphically by medium to fine-grained sands that began accreting after 6000 BP. A lack of diagnostic harbour clays attests to absent or modest harbourworks. Given this, how did the three proto-harbours operate? Boats, or barquettes, from this period were generally small and hauled from the water onto the beachface (**Figures 6.2 and 6.3**). Larger trade vessels anchored in semi-protected bays and pocket coves, and goods were ferried to and from the shoreline by lighter craft (Frost, 2004). Offshore anchoring is corroborated by finds of Bronze Age stone anchors at Sidon and Tyre, concurrent with the extensive discoveries from Byblos (**Figure 6.4**, Dunand, 1936, 1954; Frost, 1969a, 1970).

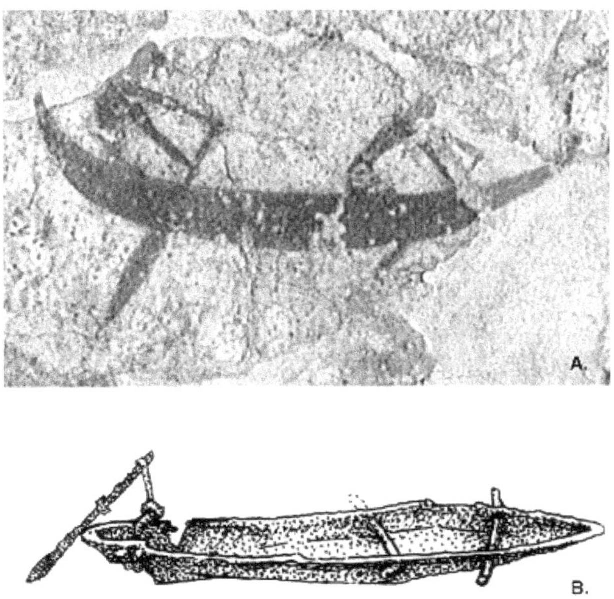

Figure 6.2: Artistic representations of Bronze Age barquettes or lighters. These small vessels were widely used during the Bronze and Iron Ages to load and unload large cargo-carriers anchored offshore (from Frost, 2004). A: Simple service craft depicted in the 'Admiral's house' at Thera (Santorini). B: Votive bronze boat found in the Temple of the Obelsiks at Byblos. The craft clearly has a removable steering oar (from Frost, 2004).

Towards the end of the Middle Bronze Age and early Late Bronze Age, archaeological evidence from the southern Levant attests to human modification of these natural anchorages (Raban, 1984, 1985a-c, 1987a-b, 1990, 1995; Marcus, 2002a-b). With the expansion of Bronze Age trade, Phoenicia's major ports of call had to be safe in all seasons, including facilities for docking, repair, maintenance and entrepot. A rich literature shows that early engineers carved installations out of the sheltered landward side of reefs (Frost, 1995; Raban, 1984, 1995; **Figure 6.5**). Although these infrastructures are notoriously difficult to date, a Bronze Age origin has long been attributed to them by many scholars (Poidebard, 1939; Frost, 1995; Viret, 2005). It should be noted, however, that these could equally be ascribed to the Iron Age (Haggai, 2006).

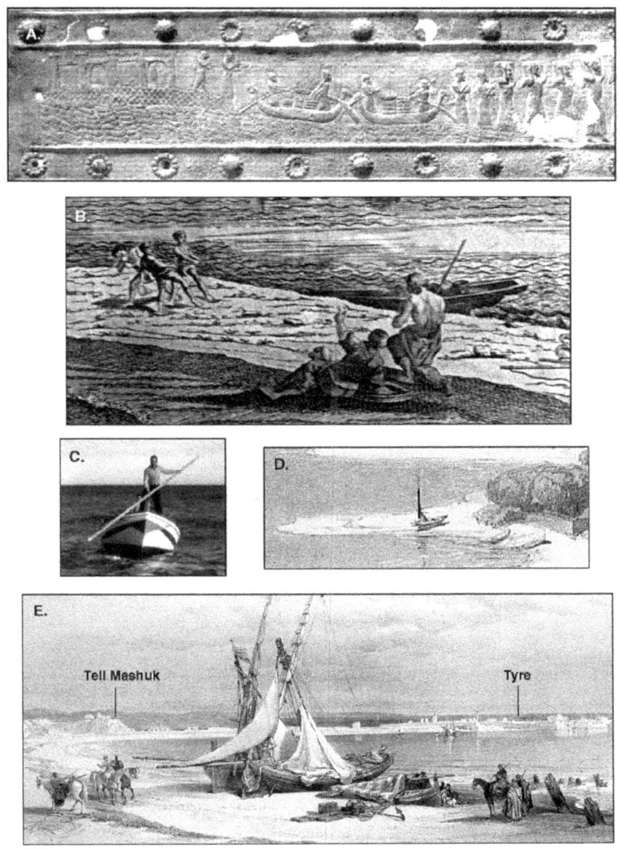

Figure 6.3: Evidence of beach boating during the past 3000 years. This technique was particularly commonplace during the Bronze and Iron Ages because no artificial modification of the coastline was required. A. Beach hauling at Palaeo-Tyre (from the Gates of Balawat). B. The basic rudiments of the technique changed very little during the following millennia. This example is taken from an eighteenth century engraving of Rhodes by Georg Proust (photograph, N. Marriner). C. Beaching of an 8 m long vessel at Arwad (in Frost, 2004). D-E. Evidence of beaching at Sidon and Tyre during the nineteenth century (lithographs by David Roberts, 1839, in Roberts, 2000).

On purely stratigraphic grounds, the existence of such artificial harbourworks at Beirut, Sidon and Tyre has been difficult to prove unequivocally. Only at Sidon have we been able to date a moderate shift from medium to finer-grained sedimentary conditions during the Middle Bronze Age (~1500 cal. BC). This is by no means conclusive and clearly requires much greater work. At Beirut and Tyre, there is a paucity of stratigraphic evidence for Bronze Age harbourworks. This would tend to suggest either: (1) the absence of such infrastructure at this time; and/or (2) the moderate stratigraphic impacts of these edifices. A dearth of diagnostic harbour units at Byblos, the pearl of the Phoenician coast during the Bronze Age, further corroborates the findings from Beirut, Sidon and Tyre (Dunand, 1936, 1954; Frost, 1969a, 2002a, 2004; Frost and Morhange, 2000; Stefaniuk *et al.*, 2005). The difficulties presented by maritime engineering at this time appear to have been overcome by the extensive use of lighter vessels to load and unload larger trade crafts anchored offshore (Wachsmann, 1998; Marcus, 2002a-b).

Figure 6.4: Bronze Age anchors. (A) A stone anchor of the quintessentially Byblian type. The shape is designed for use on rocky shores and the apical groove to prevent rope slippage during raising and lowering (photograph in Frost, 2004). B. Anchor finds from a Bronze Age ship at Newe Yam. (C-E) Frost's stone anchor typology: (C) the rocky coast anchor uses weight alone; (D) coral anchor commonly used among the brittle corals of the Red Sea; (E) sand anchor with wooden spikes used to fix the anchor to the clastic sediment bottom.

From this first harbour phase, therefore, it is noted that Beirut, Sidon and Tyre, in addition to the other major Bronze Age ports of the Phoenician coast, coincided with areas of partially drowned sandstone. Natural anchorages were the rule with very little need for human modification of the original geomorphological endowments (Frost, 1995; Marcus, 2002a-b).

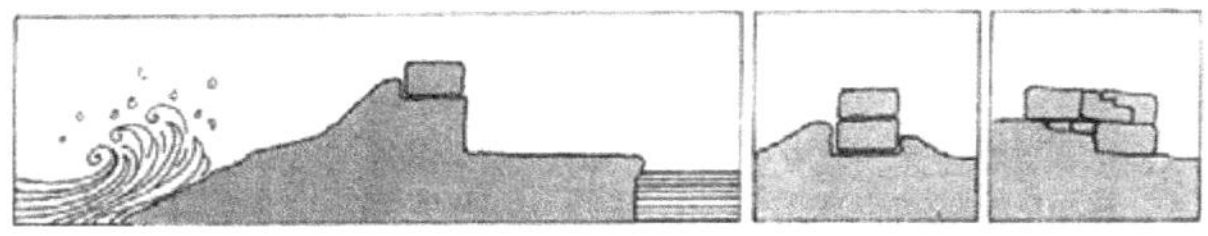

Figure 6.5: Schematic representation of Bronze Age/Iron Age rock cut harbour quays (in Frost, 1995). The blocks quarried on the leeward side are used to strengthen and heighten the natural reef.

6.1.2 Artificial Iron Age to Roman harbours

Archaeological evidence from the three sites attests to a pattern of expanding Mediterranean trade that prompted coastal populations into artificialising the basins during the Iron Age. Beirut, Sidon and Tyre were key players in a cabotage network running from Ashkelon in the south up to Arwad in the north (**Figure 6.6**). There is a clear spatial pattern in the distribution of the Levant's anchorages, with intercalating distances of ~20 to 40 km between each site (average: 31.25 km). Seafaring on the Phoenician coast during the Bronze and Iron Ages was manifestly dominated by short hopping-type navigation (Barnett, 1958; McCaslin, 1980; Artzy, 1985; Knapp, 1993; Lipinski, 2004).

The increased use of iron at the turn of the first millennium BC meant that much larger shipping vessels could be constructed (Casson, 1994; Pomey and Rieth, 2005). Such boats had far greater exigencies than their Bronze Age counterparts (Wachsmann, 1998), with needs for docking, repair and entrepot. Unfortunately, Roman and Byzantine dredging means that there is very little stratigraphic evidence at Beirut, Sidon and Tyre pertaining to this shift during the Iron Age. A single core from Sidon (BH XV) concurs advanced harbour infrastructure during the Iron Age, although the data is not widespread enough to derive any generalised conclusions. Archaeological evidence from Athlit, a small Phoenician trading outpost 55 km south of Tyre, attests to an artificial mole which has been dated to the ninth/eighth centuries BC (Haggai, 2006). Ongoing stratigraphic studies within the site's basin will help to more transparently elucidate the sedimentological impact of Iron Age coastal infrastructure. Iron Age III/Persian quays at Beirut are also consistent with this model of well-developed port infrastructure during the Iron Age (Elayi and Sayegh, 2000).

Clearly dated fine-grained silts and clays are manifest during the Hellenistic and Roman periods at all three sites. Although this is not the technological apogee of the seaport trio, the sedimentology concurs advanced harbour infrastructure. Romano-Byzantine artificial moles have been evidenced at both Sidon and Tyre, their construction greatly facilitated by the discovery of hydraulic concrete (Poidebard and Lauffray, 1951; Hohlfelder, 1997; Noureddine and Helou, 2005; Descamps, personal communication). Indeed, recent studies have demonstrated that the quality of Roman hydraulic concrete was not replicated until the nineteenth century (Rickman, 1988; Oleson *et al.*, 2004a-b).

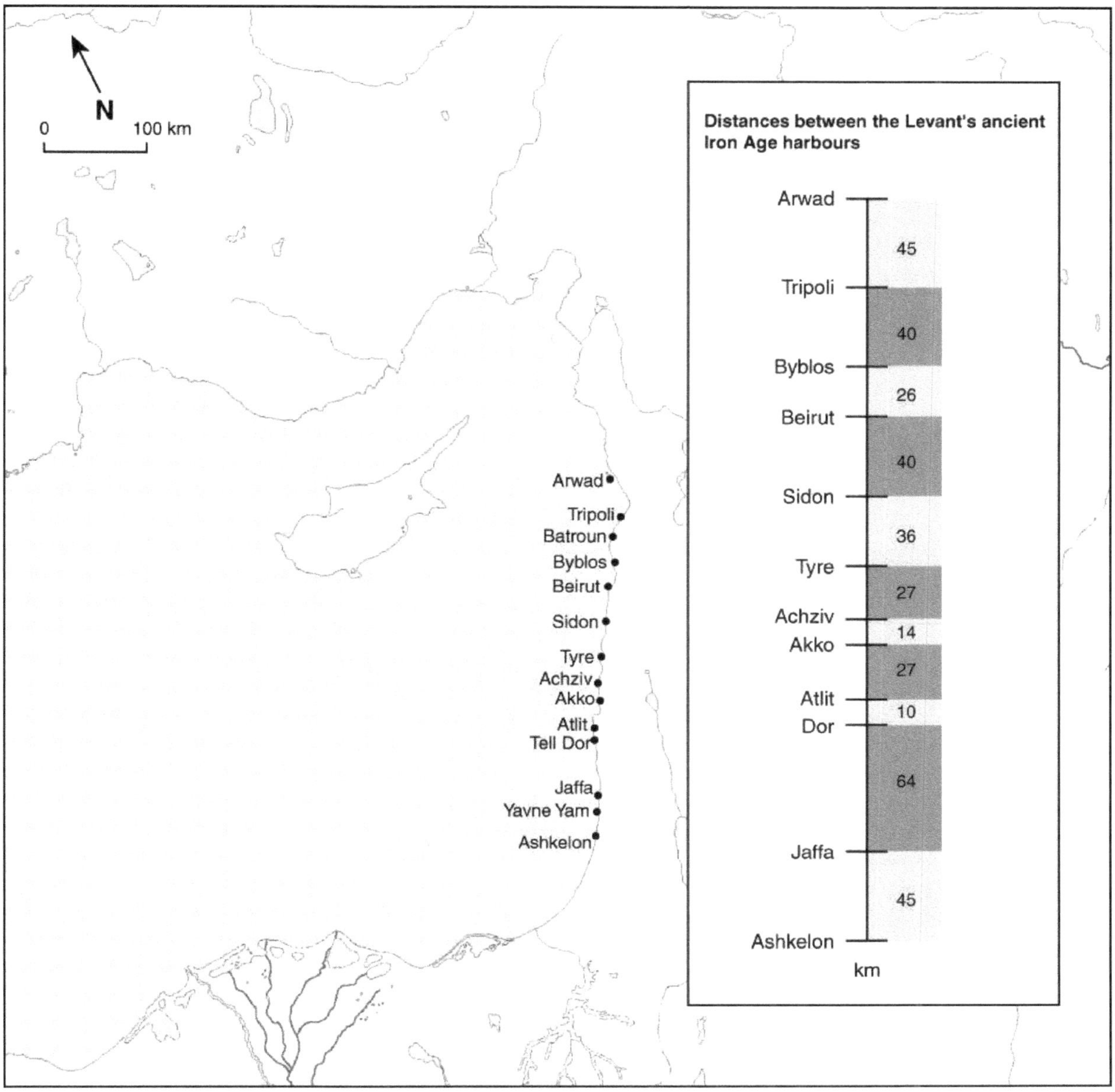

Figure 6.6: Iron Age harbours of the Levantine coast. Inset: distances between neighbouring ports of call. An average of 31.25 km was obtained, indicating short port-hopping navigation during this period.

Vitruvius elucidates three different types of construction: (1) Type 1 was the placement of concrete within a flooded containment system (**Figure 6.7**); (2) Type 2 was the casting of concrete blocks above water on the end of a pier or natural protruding feature (**Figure 6.8**); and (3) Type 3 was the placement of concrete within an evacuated watertight container or enclosure. For a detailed discussion of Roman and Byzantine harbour constructions we refer the reader to the extensive bibliography of Oleson (1988), Oleson and Branton (1992) and Oleson *et al.* (2004a-b).

Rapid rates of silting threatened the long-term viability of the harbours, culminating in repeated dredging of the basins (Marriner and Morhange, 2006a; Marriner *et al.* 2006b). Elsewhere in the Mediterranean, the transition to Roman rule is marked by the refashioning of many ancient harbours (Rickman, 1988). Although at some sites completely artificial roadsteads were carved or annexed onto the coastline, the enduring maritime trade routes meant that the major ports of call remained comparatively static (Arnaud, 2005). Instead, the Romans chose to reinforce the pre-existing ports and in many cases significantly remoulded harbour morphologies. Although we have very little architectural evidence pertaining to these port overhauls at Beirut, Sidon and Tyre, stratigraphic hiatuses and widespread dating inversions nonetheless point to considerable modification. The three settlements were all key cities of the Levantine seaboard at this time, requiring effective harbour management strategies for trade and transport purposes.

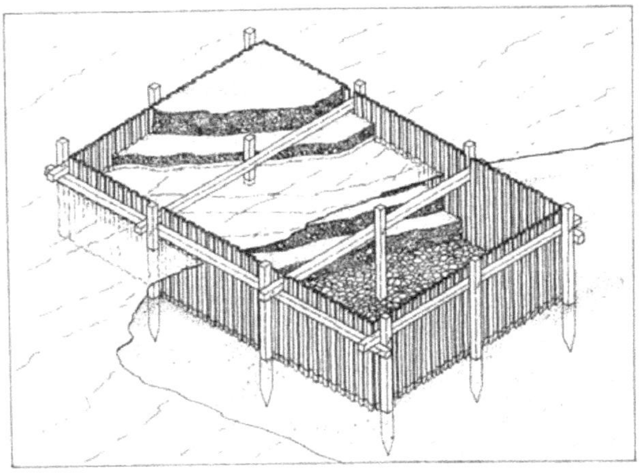

Figure 6.7: Roman mole Type 1. Hypothetical reconstruction after Vitruvius in Brandon (1996). This technique was widely used at Cosa and Caesarea Maritima. A slight variant is the caisson or settling barge, which was floated into position before being weighted down with ballast and hydraulic concrete.

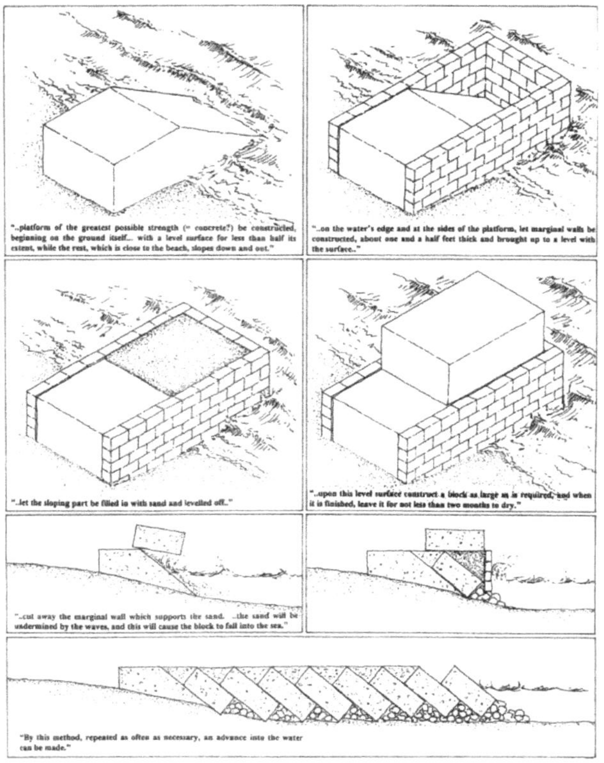

Figure 6.8: Roman mole Type 2. Hypothetical reconstruction from Brandon (1996).

6.1.3 Late Roman/Byzantine apogee

In 448-450 AD, Beirut acquired the title of metropolis, joining neighbouring Tyre as one of the Mediterranean's key cities (Mikati and Perring, 2006). At this point in time, the Byzantine Empire attained a zenith in prosperity (Hall, 2004) translated by the technological apogee of the three harbours. At Beirut and Sidon, the period is translated by a diagnostic plastic harbour clay. An analogous unit is observed at Tyre, although here the facies comprises fine-grained silts and sands. The consolidation of Roman construction techniques, coupled with the economic importance of the Levantine seaboard during the Byzantine period, culminated in lagoon-like harbours very well protected from the open sea.

Biostratigraphically, the late Roman-Byzantine apogee is marked by a sharp increase of Cyprideis torosa at all three sites, consistent with hyposaline basins. Lagoonal and fine-grained macrofaunal suites also characterise these facies. At no other point in the stratigraphic record do we observe such well-protected ports. For Beirut, the dramatic coastal artifialisation of the past 200 years means that there is a dearth of data pertaining to the nature of the enveloping infrastructure during antiquity.

6.1.4 Islamic and medieval harbours: transformation or decline?

The sixth to seventh centuries AD mark a turning point in the evolution of the three harbours. Coarse-grained sand units attest to a semi-abandonment of the harbours and silting up of the basins' landward fringe. Although we have briefly touched on this subject in chapters 3, 4 and 5, the asynchronous demise of all three seaports invokes a major regional crisis. Four hypotheses are advanced to explain the collapse of the seaports after the Byzantine apogee.

6.1.4.1 Historical

The sixth to eighth centuries AD were a period of significant geopolitical permutations, when the Byzantine, Persian and Arabic superpowers all vied for control of the Levant and its important maritime façade (Kaegi, 2005). During the sixth century AD, the Byzantine Empire was faced with a number of calamities including the Persian invasions, Justinian's over-expenditures on war and public building programs, and outbreak of the plague in 542 AD (Mitchell, 2006). Weakened by these internal malaises, the expansion of Arab forces during the seventh century AD presented a serious strain upon Byzantine control of the Levantine basin (Kaegi, 2005). The latter were eventually defeated in 636 AD at the Battle of Yarmuk. After this, Islamic forces sequestered both Damascus and Jerusalem, and gained control of the maritime ports on the Phoenician and Judean coasts (**Figure 6.9**). The archaeological evidence from Beirut and Sidon supports a contraction in inter-state trade during the sixth and seventh centuries AD, consistent with profound changes in the eastern Mediterranean's trade network (Pirenne, 1937; Elayi and Sayegh, 2000; Doumet-Serhal, 2003, 2004a).

During and after these permutations, it is speculated that harbour infrastructure fell into a state of disrepair (Marriner *et al.*, 2006a). A priori our sedimentological data are broadly consistent with these historical interpretations. Nevertheless, an alternative school of historical research has emerged tending to moderate the premise of a general decline of 'Syrian' harbours after the Byzantine period (Lewis, 1951; Fahmy, 1966; Borrut, 1999-2000; Sourdel and Sourdel, 1989). Some historians have gone so far as to say that the Levantine coast became the cradle of Islamic maritime development (Borrut, 2001). Historical sources evoke three pivotal Levantine harbours, Acre, Tyre and Tripoli, while the other ports seemingly disappear from the maritime map. Acre and Tyre were, for instance, important centres of naval construction during the ninth to tenth centuries AD. Beirut on the other hand, was arguably reduced to little more than a small agricultural town in a frontier zone between two warring empires (Mikati and Perring, 2006). Kennedy (1985a-b) offers an interesting alternative view, namely that the coastal cities of Phoenicia had suffered a period of decline prior to the Islamic conquest. He cites the abandonment of theatres, reduction in size of bath houses and changes in street layout as evidence of this transformation. For him, the post-Byzantine Levantine cities followed one of three trajectories: (1) some cities thrived such as Damascus and Aleppo; (2) others continued to decline (notably the coastal cities); and (3) a third group, comprising mainly inland settlements, continued unaffected. This evidence would tend to support a shift in the importance of inland caravan routes to the detriment of maritime cabotage, where just a few seaports (e.g. Acre, Tyre and Tripoli) continued to thrive.

Our stratigraphic data indicate rapid coastal progradation from the sixth century AD onwards, entraining the deformation and dislocation of Beirut, Tyre and Sidon's basins. Although manifestly smaller, it is difficult to accurately constrain the dimensions of these ports as the post-Byzantine sediment record lies beneath the present basin and is not readily accessible. In sum, while there is little dispute that change took place, there are divergent views on the chronology and scale of change. Our harbour chronostratigraphies do not allow us to precisely reconcile these chronological changes merely corroborate the general tendency of demise.

6.1.4.2 Tectonic

Beirut, Sidon and Tyre attest to significant seismic activity during the late Roman period (**Figures 6.10 and 6.11**). What is the relative sea-level evidence for tectonic movements at each site?

(1) At Tyre, new archaeological and stratigraphic data support tectonic collapse of ~3 m during a relatively narrow chronological window (**Figures 6.12-6.14**). Two types of evidence exist. (i) Archaeological: A priori, port opening at Tyre appears to have been amplified by the late Roman collapse of the island bastion. Its northern Roman mole is currently 2.5 m below present sea level (Descamps, personal communication), translating a subsidence of ~3-3.5 m. On the southern shore, walls and drowned quarries at -2.5 m below MSL have also been discovered (El Amouri *et al.*, 2005; **Figures 6.12 and 6.13**). (ii) Stratigraphic: Similar subsidence is translated in the city's coastal stratigraphy, notably a ~3 m offset in sediment accommodation space and radiocarbon dated strata between Tyre and Sidon (**Figure 6.14**).

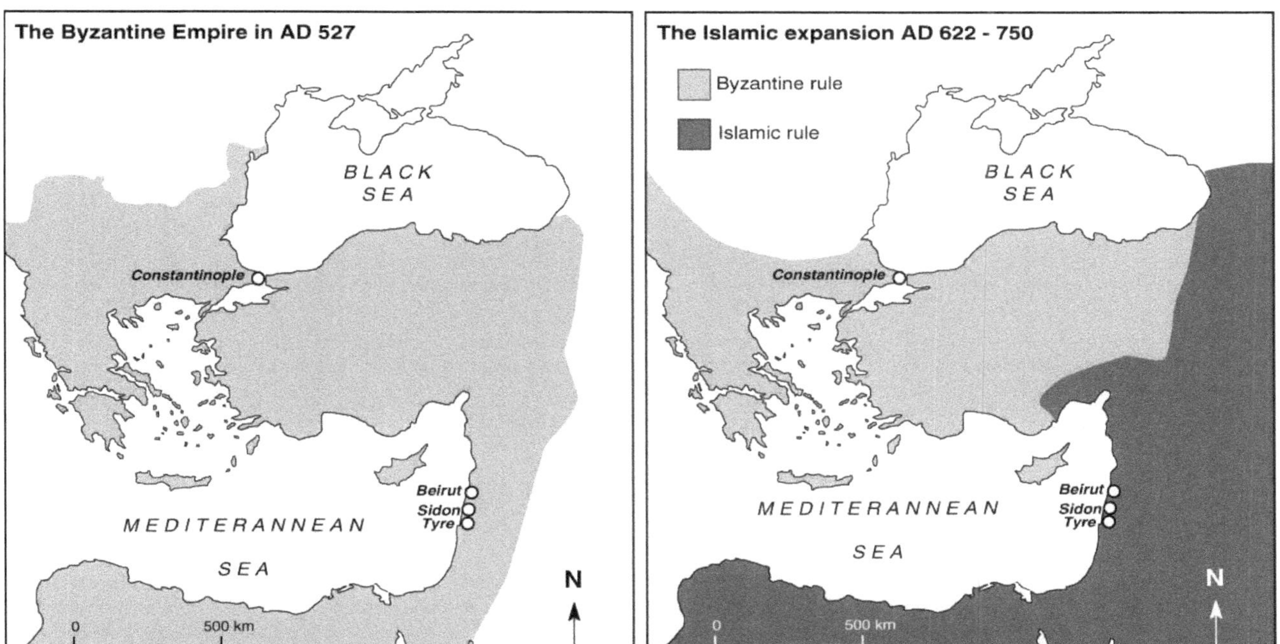

Figure 6.9: Retraction of the Byzantine Empire after the sixth century AD.

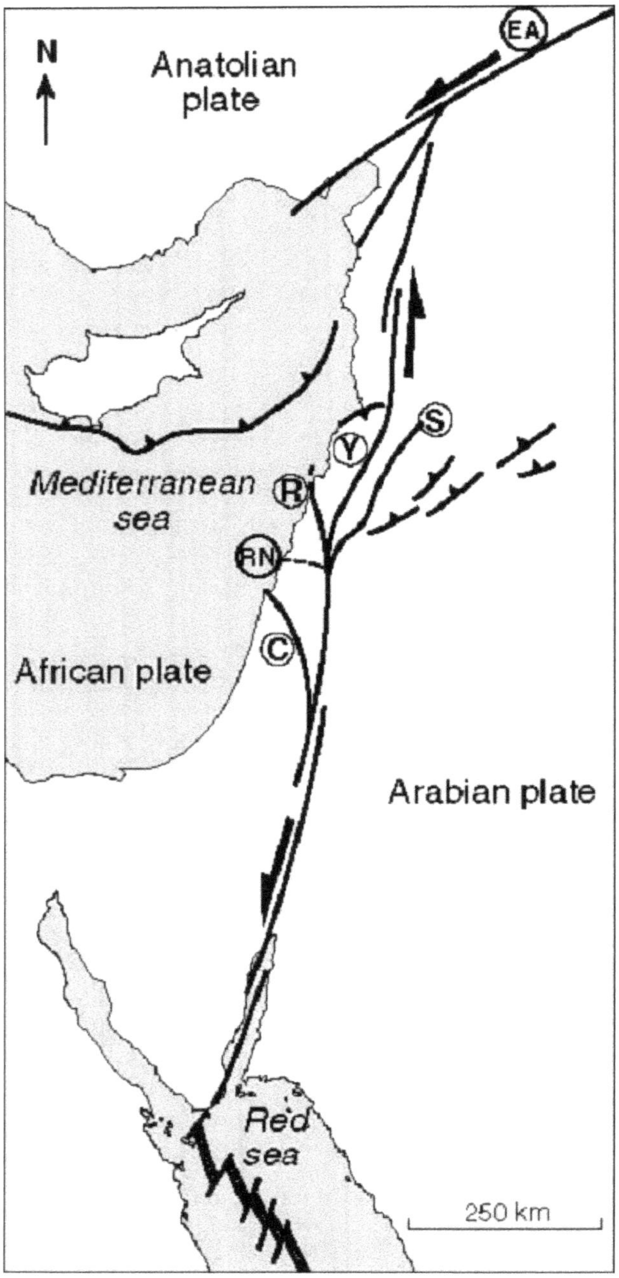

Figure 6.10: Tectonic setting of the Lebanese portion of the Dead Sea Transform fault system. C- Carmel fault; R - Roum fault; Y - Yammuneh fault; S - Serghaya fault; RN - Rosh Hanikra / Ras Nakoura fault, EA – East Anatolian fault.

(2) At Sidon, by contrast, sea-level data show that the magnitude of crustal mobility is inferior to Tyre, around +50 cm since antiquity (Marriner and Morhange, 2005), yet the same coarse sand facies is persistently observed. These contrasting data would tend to suggest that late Roman coastal tectonics cannot solely explain the demise of Phoenicia's ancient harbours.

(3) At Beirut, there is widespread textual and archaeological evidence for an earthquake event during the mid-sixth century AD (Russell, 1985; Guidoboni *et al.*, 1994; Saghieh Beidoun, 1997; Darawcheh *et al.*, 2000; Harajli *et al.*, 2002). This account by Agathias (in Mikati and Perring, 2006) provides an insight into the drama of the disaster. "[...] the lovely city of Berytus, the jewel of Phoenicia, was completely ruined and its world-famous architectural treasures were reduced to a heap of rubble, practically nothing but the bare pavements of the buildings being left. Many of the local inhabitants were crushed to death under the weight of the wreckage [...] The restored city was very different from what it had been in the past, though it was not changed beyond recognition, since it still preserved a few traces of its former self."

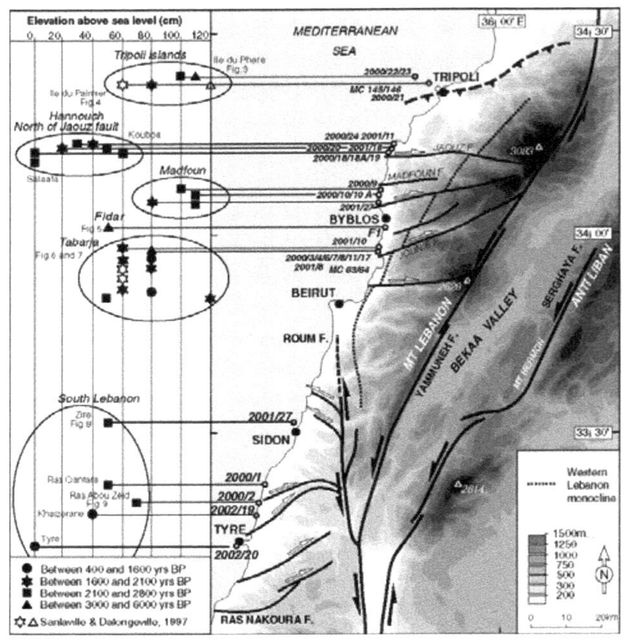

Figure 6.11: Structural map of Lebanon and adjacent areas. Sample locations and their corresponding elevation above sea level are shown on the left-hand side of the figure (from Morhange *et al.*, 2006).

Recent excavations intimate that the ancient sources did not exaggerate in their description of the destruction caused by the event (Curvers and Stuart, 2004). New research has yielded closely dated stratigraphic sequences at a number of dig sites that unequivocally corroborate the widespread earthquake damage (Elayi and Sayegh, 2000). In the aftermath, Beirut underwent altering patterns of trade, production and consumption. The archaeology also shows that many parts of the city were left in partial ruin or even abandoned, with limited evidence for reconstruction. Mikati and Perring (2006) present a model of 'continuity' but degradation of urban infrastructure at post-earthquake Beirut. Dating of raised shorelines north of Beirut affirms uplift of 50 to 80 cm (Morhange *et al.*, 2006b).

6.1.4.2.1 Regional shoreline deformations during the late Roman period

The presence of two raised Holocene shorelines along the Lebanese coast has attracted the interest of earth scientists for a number of decades (Sanlaville, 1977; Sanlaville *et al.*, 1997). Morhange *et al.* (2006b) have recently published new radiocarbon datings of these raised shorelines, elucidating two main Holocene sea levels: (1) an upper shoreline at ~+120 to +140 cm, which lasted from ~6000 to 3000 BP. On the nearby coasts of Syria and Turkey, this upper shoreline affirms the occurrence of seismotectonic displacement(s) around 3000 BP, decreasing in amplitude from Turkey to southern Lebanon (Pirazzoli, 2005); (2) a lower shoreline at ~+80 ± 40 cm, developed between 2700 BP and the sixth century AD. The main cluster of dates for this lower shoreline suggests that it was uplifted during the late Roman period (**Figure 6.15**). These uplifted shorelines have resulted from the activation of the Yammuneh and the Roum-Tripoli Thrust, as well as slip along transverse faults (**Figure 6.11**). Data from Tyre suggest that the Rosh Hanikra/Ras Nakoura fault marks the southern boundary of the Levantine vertical displacements, with no evidence for coastal uplift being reported from Israel during the Holocene (**Figure 6.14**; Sivan *et al.*, 2001). This is confirmed by the relative stability of its ancient harbours and coastal archaeology, including, for example, Achziv's Roman fish ponds at ~0 MSL (Raban and Galili, 1985; Galili *et al.*, 1988; Galili and Sharvit, 1998). Manifestly the Rosh Hanikra/Ras Nakoura tectonic hinge zone lies in the 27 km coastal area that separates Tyre and Achziv. What is the pattern of uplift in Lebanon during the late Roman period? From a high lower shoreline around the Tripoli area in north Lebanon (between 80 and 120 cm), uplift decreases in intensity in the Beirut area (Sanlaville, 1977)—although it is important to stress that the coast is completely urbanised — increasing again in the Sidon sector, where uplift remains nonetheless lower than in northern Lebanon. In contrast, the archaeological data from the Tyrian horst manifest ~3 m of tectonic collapse since Roman times (Marriner *et al.*, 2006a). This pattern of uplift and subsidence intimates tilting of fault bound panels and/or differential vertical movements across Jaouz, Madfoun, Jounieh and the ENE trending faults in the vicinity of Tyre. The local amount of uplift at each site appears to vary with (1) the spatial distribution of slip vector on the fault plain; (2) the distance from the site to the fault plain; and (3) local inhomogeneities in surface deformation. The precise role of the Lebanese tranverse faults therefore remains an open debate pending better spatial resolution of the uplift pattern.

It is difficult to ascertain whether uplift of the lower shoreline resulted from a single seismic event or from several large earthquakes between the fourth and the mid-seventh centuries AD (Stiros, 2001). The Levantine coast does seem to have been particularly affected by the 551 AD earthquake (Plassard, 1968; Russell, 1985; Saghieh Beidoun, 1997) however attributing the lower raised shoreline to a unique seismic event is contentious. Given the overall extent and diversity of the uplifts, more than 500 km coastwise, it is difficult to invoke a single-earthquake scenario. On the other hand, the lower shoreline episode manifestly corresponds to a narrow chronological window which evokes a relatively short period of intense seismic activity. According to data from various sources (Plassard and Kogoj, 1981; Russell, 1985) it is interesting to note that during Pirazzoli's (1986) Early Byzantine Tectonic Paroxysm a cluster of five earthquakes ≥8 are documented on the Levantine coast against a mere two during the period AD 600 to AD 1100 (scale sensu Plassard and Kogoj, 1981).

Figure 6.12: This ancient urban quarter on the southern fringe of Tyre was drowned by ~3 m during the late Roman period. Tyre is the only submerged ancient harbour on the Phoenician coast, a pattern evoking differential vertical movements along fault bound panels and local inhomogeneities in surface deformation. In spite of this pronounced seismic collapse, the type stratigraphy at Tyre, Sidon and Beirut remains the same. This leads us to conclude that relative sea-level variations are not important determinants in dictating harbour evolution (base image: DigitalGlobe, 2006). Archaeological data adapted from El Amouri *et al.* (2005).

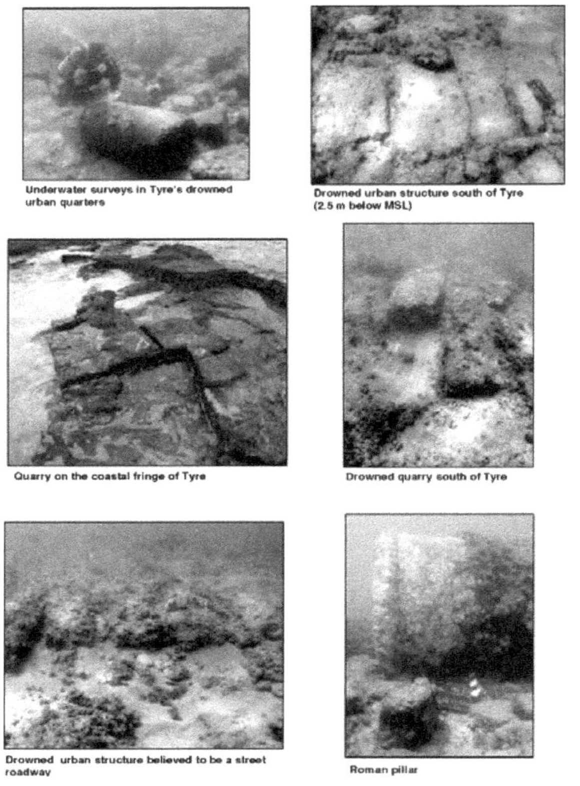

Figure 6.13: Drowned archaeological remains in Tyre's southern urban quarter (all photographs from the 2002 UNESCO diving project, headed by C. Morhange).

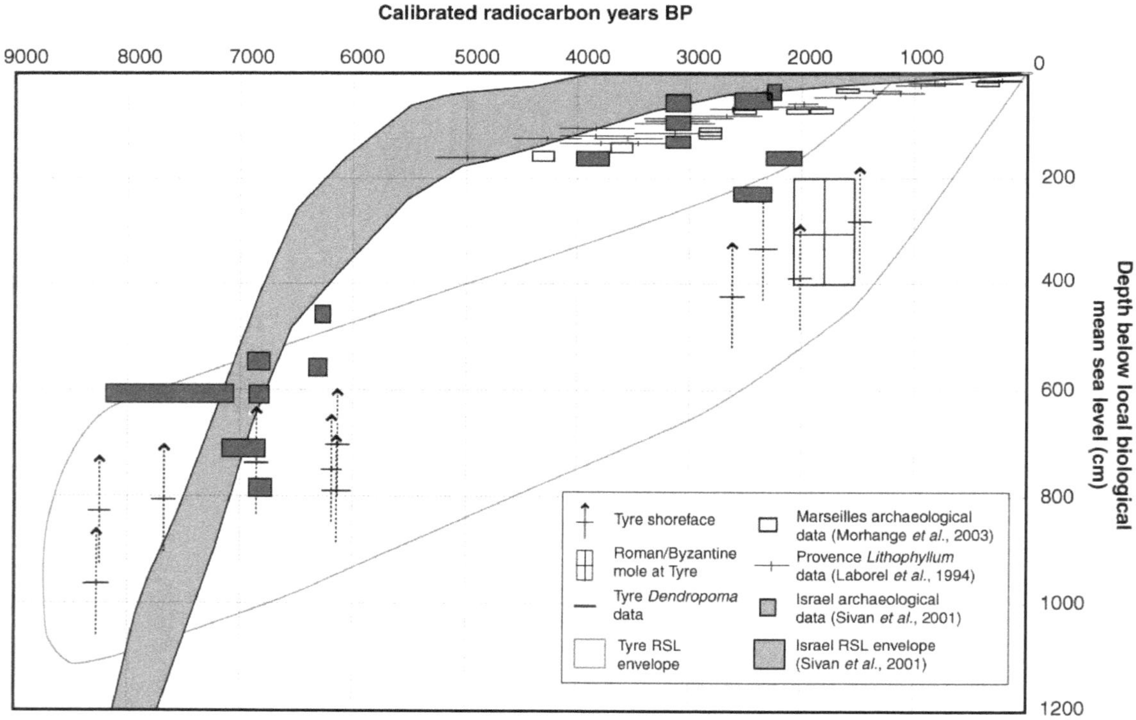

Figure 6.14: Sea-level tendencies at Tyre since 8000 BP. Although the results fit well with low error (± 5 cm) sea-level data from the stable coasts of Provence (Laborel *et al.*, 1994; Morhange *et al.*, 2001, 2003), the large error margins and absence of precise sea-level indicators found at the site means that the envelope cannot be used as a precise RSL curve. The 3 m collapse of the Tyrian horst during the late Roman period is clearly translated by offsets with the empirical data from Marseilles, and modelled scenarios for the Israeli coast (Sivan *et al.*, 2001). RSL data from the western margin of the Nile delta at Alexandria have also been plotted (Goiran, 2001). For Tyre, please note the paucity of data points between 6000 and 3000 BP.

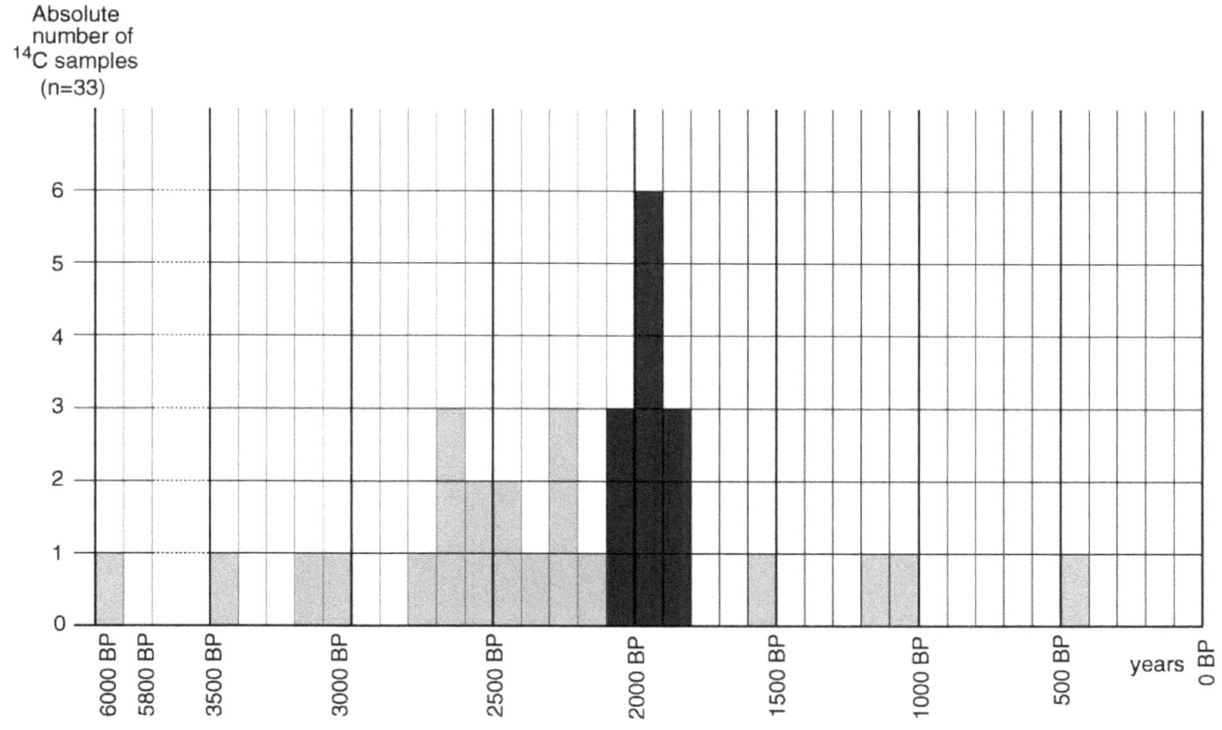

Figure 6.15: Histogram of the 33 radiocarbon dates from the Lebanese coast (after Morhange *et al.*, 2006). The late Roman (EBTP) seismic crisis is marked in dark grey.

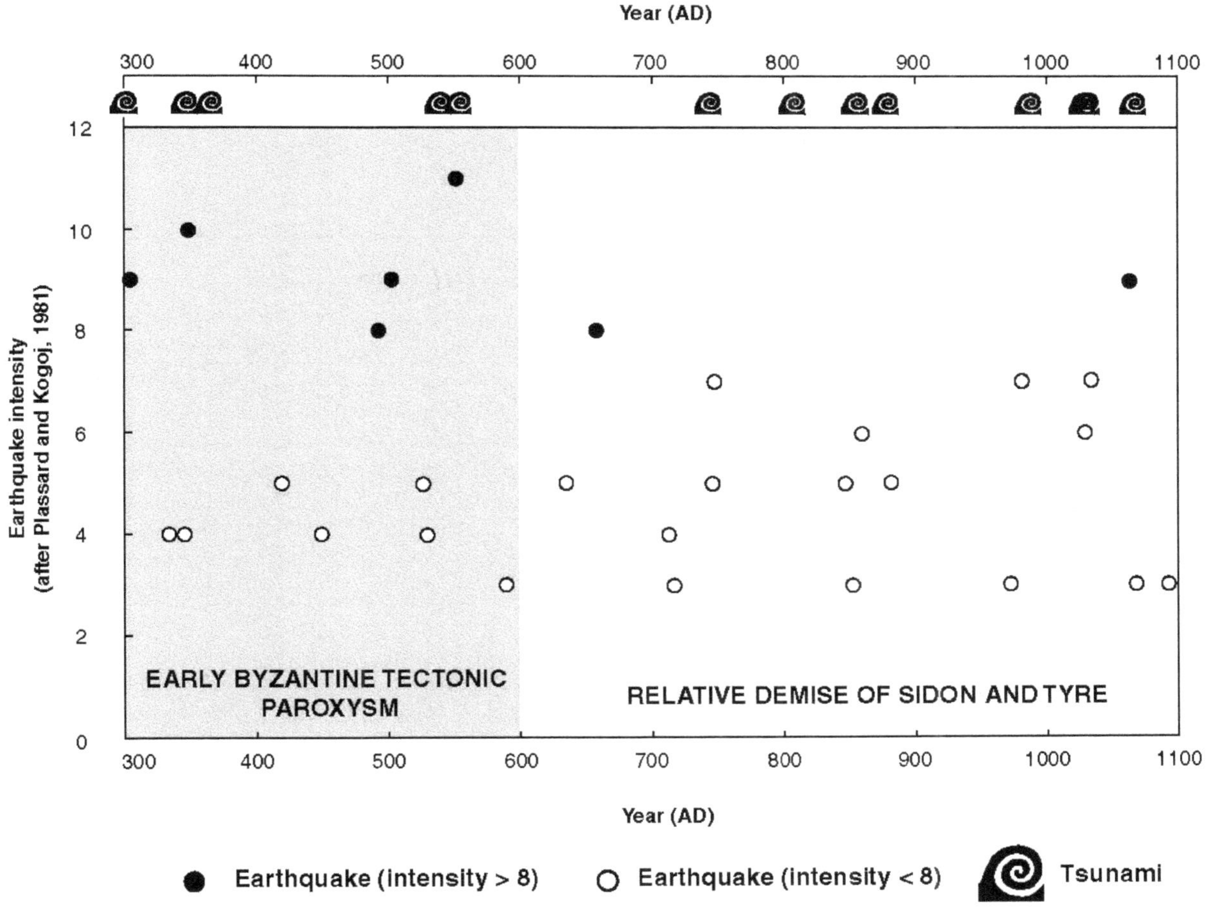

Figure 6.16: Chronology of earthquake and tsunami events affecting the Levantine coast between AD 300 and AD 1100. Data compiled from Ambraseys (1962), Plassard and Kogoj (1981), Russell (1985), Guidoboni *et al.*, (1994), Darawcheh *et al.* (2000) and Soloviev *et al.* (2000).

Morhange *et al.*'s (2006b) results constrain relative sea-level stability along the Lebanese coast to 1000 BP and later. This stability is documented by three types of indicators, archaeological (in situ fish tanks and sea walls at present MSL), biological (development of Dendropoma rims) and geomorphological (well-developed abrasion platforms at present MSL). Ongoing radiometric dating of the contact between the base of Dendropoma bioconstructions and the substratum consistently shows dates younger than 1000 BP along the Lebanese and Israeli coasts.

The stratigraphic evidence from the three harbours demonstrates that their demise was centred on the sixth to eighth centuries AD. These data indicate a slight chronological offset between the fourth to sixth century seismic crisis and the sixth to eighth century seaport opening (Pirazzoli, 1986, 2005; Pirazzoli *et al.*, 1996). **Figure 6.16** plots earthquake and tsunami events on the Levantine coast illustrating that the fourth to eleventh centuries were characterised by repeated seismic shocks, possibly provoking partial harbour damage.

6.1.4.3 Tsunamogenic

Given the quasi-identical stratigraphy at all three sites, we can conclude that local relative sea-level changes have played a relatively minor role in dictating harbour stratigraphy. This discrepancy leads us to moderate tectonic collapse in favour of documented tsunamogenic impacts. More than ten tsunamis struck the Levantine coast between the fourth to eleventh centuries AD, severely damaging harbour infrastructure (Ambraseys, 1962; Guidoboni *et al.*, 1994; Soloviev *et al.*, 2000). The best documented event is the 551 AD tsunami which struck the Phoenician coast at Beirut, Sidon and Tyre. Archaeological evidence from Beirut's ancient harbour indicates significant harbour damage and reworking of harbour silts and clays, concurrent with major changes in port configuration (Mikati and Perring, 2006).

6.1.4.4 Climatic

The Levant has yielded a number of high-resolution palaeoclimate records that provide an independent measure of the timing, amplitude and duration of past climate events (**Figure 6.17**; Bar-Matthews *et al.*, 1997, 1998; Schilman *et al.*, 2002; Enzel *et al.*, 2003; McGarry *et al.*, 2004; Devillers, 2005; Rosen, 2006). These time-series indicate that Levantine climate during the Holocene has been much more variable than previously believed (Schilman *et al.*, 2001). Significantly, palaeoclimate proxy data from the Byzantine-Islamic transition concur a significant shift from wet humid conditions to an arid climate. A sharp fall in precipitation is recorded in a number of local palaeohydrological data and oxygen isotope records (Stiller *et al.*, 1983; Frumkin *et al.*, 1991; Bar-Matthews *et al.*, 1997; Lemcke and Sturm, 1997; Schilman *et al.*, 2002; Enzel *et al.*, 2003; McGarry *et al.*, 2004). In the Dead Sea, for example, a pronounced drop in lake levels between the fifth and late eighth centuries AD corresponds to a period of prolonged regional drought (Enzel *et al.*, 2003). Sea and terrestrial isotope records between 1300 and 900 years BP show a similar increase in δ18O values (Δδ18O ~0.5‰), indicating the development of drier conditions (Bar-Matthews *et al.*, 1999; Schilman *et al.*, 2002). This increased aridity has been attributed to a lower number of cyclones affecting the southern Levant at this time.

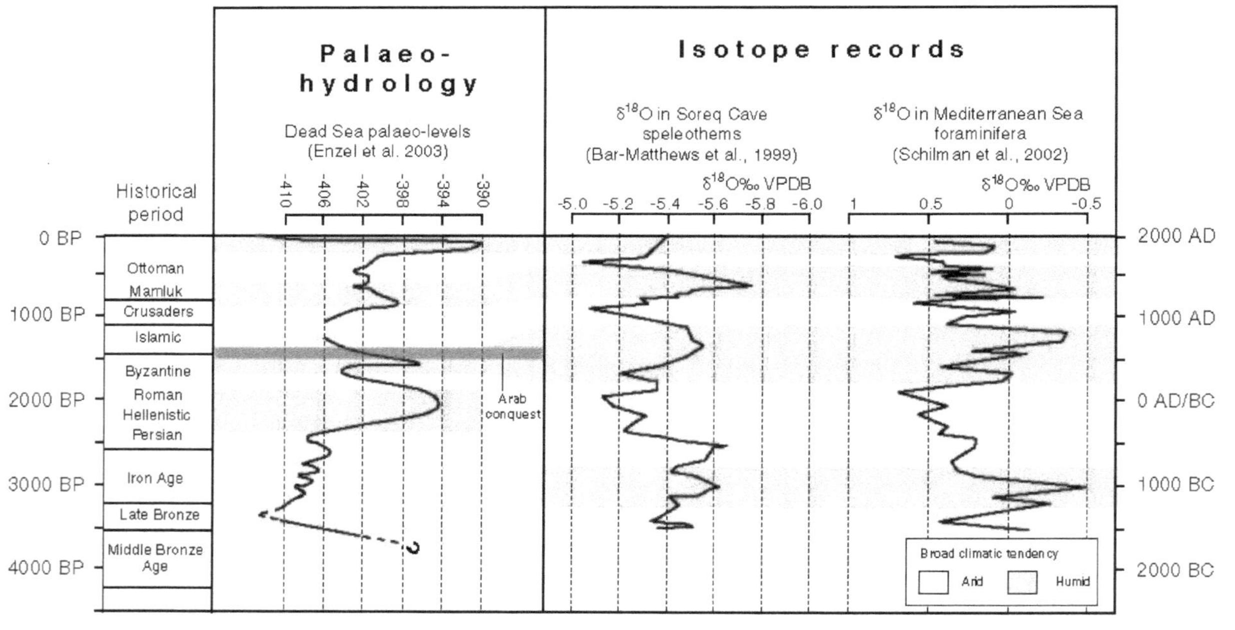

Figure 6.17: Palaeoclimate records from the Levantine basin (Bar-Matthews *et al.*, 1999; Schilman *et al.*, 2002; Enzel *et al.*, 2003). Increased aridity during the sixth century AD has been used as evidence to explain the Byzantine collapse. Offsets observed in the data are due to different sampling resolution and the slight dyssynchrony of climate changes between areas.

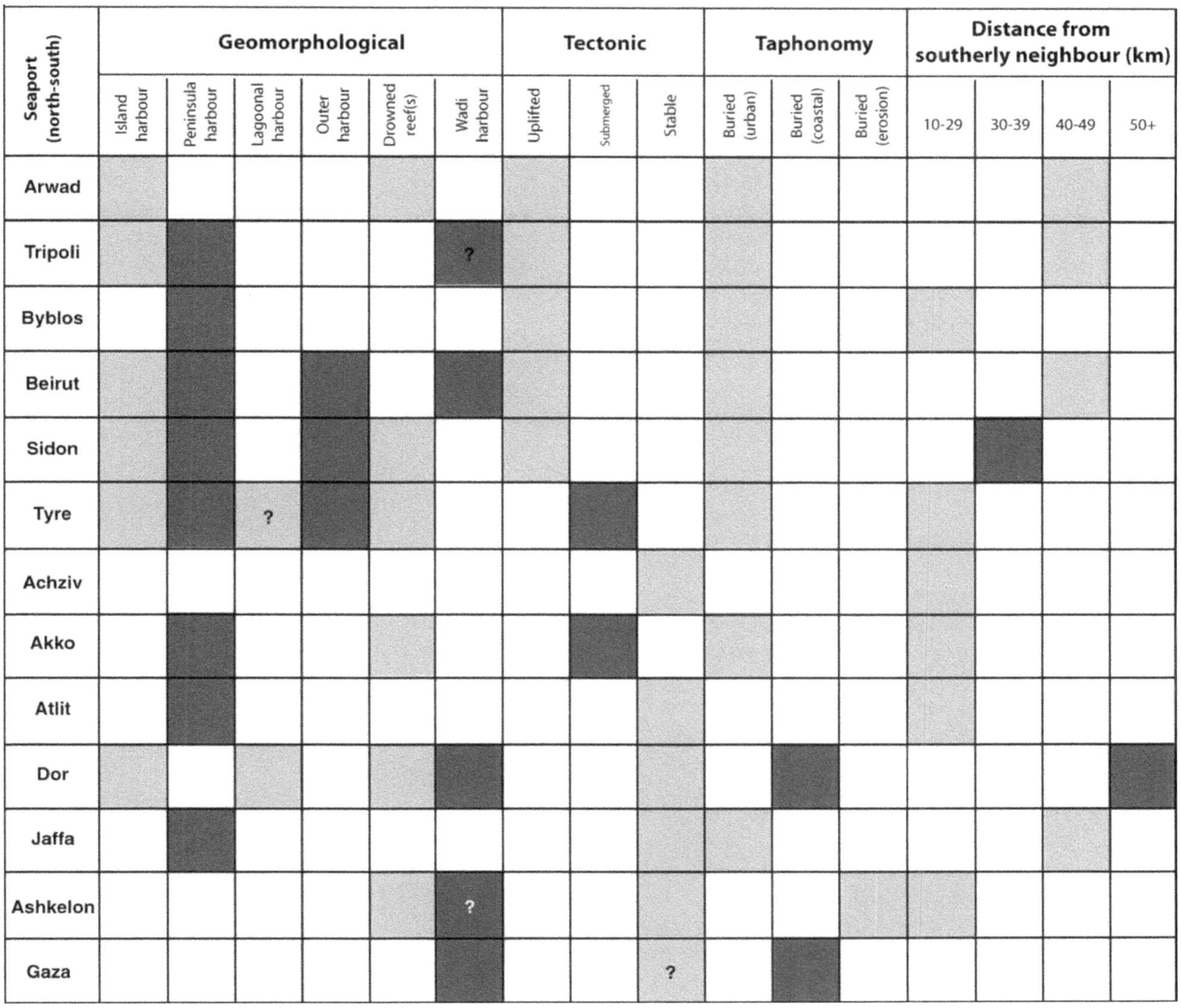

Figure 6.18: Comparison of the geomorphological, tectonic and taphonomic traits of the Levant's principal ancient anchorages.

Given that the stability of the agricultural based Byzantine society relied heavily on the constancy of climate, this deterioration has led some scholars to draw direct causal links between Byzantine cultural depression and increased aridity (Issar, 2003). Since agricultural production depended on factors such as fertile soils, precipitation and irrigation, prolonged climatic downturns have been evoked to explain the collapse of the agricultural, water-harvesting Byzantine society and Arabic expansion into the Levant from the arid area of current Saudi Arabia (deMenocal, 2001; Issar *et al.*, 1992; Nuzhet *et al.*, 1997; Weiss and Bradley, 2001). Although manifestly important in forcing cultural change (Yoffee and Cowgill, 1988; Weiss *et al.*, 1993; Butzer, 1996; Tainter, 1998; Peiser, 1998; Cullen *et al.*, 2000; Webster, 2002; Berglund, 2003; Haug *et al.*, 2003; Rosen and Rosen, 2001; Rosen, 2006), the present fashion arraigning climate change for societal downturn must be moderated. Episodes of climate deterioration are not necessarily detrimental to all, or negative in all of their aspects, and the temptation to attribute societal change solely to environmental determinism needs to be avoided.

6.1.5 Concluding remarks

The Levant was arguably one of the richest regions of the Byzantine Empire during the early centuries of its existence (Hall, 2004). It was at this time that the area witnessed an impressive demographic upswing, corroborated by settlement expansion, urban growth and agricultural development. From the material record, a broad picture of rural settlement and rich agricultural production is corroborated in the communities of Greater Syria until the mid-sixth century AD (Kennedy, 1985a). This apogee is clearly recorded in the harbour stratigraphy.

The demise of Phoenicia's ancient harbours, concurred by stratigraphic data from all three sites, needs to be nuanced on a number of grounds. We would like to insist upon two points. (1) The demise was relative. From a technological apogee during the late Roman and

Byzantine periods, we observe a medievalisation of the three cities along the lines of the Arabic medinas. The ascension of many inland sites during the sixth to eight centuries AD confers a rise in the importance of inland caravan routes. Archaeological evidence translates a decrease in Mediterranean imports at Beirut, Sidon and Tyre, intimating that the settlements looked inland towards the Arabian peninsula. (2) Attributing the demise to a precise historical or natural catastrophe is difficult. The decline of the Levantine maritime façade appears to have been engendered by decades of disease, natural disasters, poor climate and military conflict, though it is impossible to accord move weight to one factor over another.

On the basis of the evidence observed at the three harbours, can we advance a model of Phoenician harbours and their evolution? Raban (1995) has worked extensively on a number of Bronze and Iron Age harbour sites in Israel. Our data from the Lebanese coast complement this research and add an important vertical, or stratigraphic, dimension which was lacking in previous research. The data show a number of striking structural and stratigraphic trends (**Figure 6.18**).

(1) Structural. Four harbour types are manifest: (i) coastal peninsula harbours (Akko, Athlit, Sidon); (ii) offshore island harbours in proximity to the coastline (Tyre, Arwad, Zire); (iii) coastal lagoon and wadi harbours (Palaeo-Tyre, Dor); and (iv) composite harbours. The Beirut model appears to be a unique amalgam of these, comprising a drowned wadi – whose upper creek was manifestly used as an anchorage during the Bronze Age – and a small drowned island sandwiched between two enveloping land promontories.

(2) Stratigraphic. For coastal harbours, the evolution in technological techniques is well-documented by the stratigraphy. Three types of anchorages can be distinguished. (i) Proto-harbours. Bronze Age societies exploited coastal geomorphology to establish natural anchorage havens in low energy coves. The stratigraphic impacts of these harbours were limited. (ii) Iron Age semi-artificial harbours. This harbour type lies at the intersection between natural anchorages and completely artificial harbours. The type stratigraphy comprises a fine-grained sand/silt facies indicative of the artificial reinforcement of shielding reefs. (iii) Roman and Byzantine artificial harbours. At this time, natural roadsteads were no longer a prerequisite for port foundation. At Beirut, Sidon and Tyre this period is characterised by a significant phase of coastal artificialisation and port remodelling. The type stratigraphy is a plastic harbour clay.

6.2 Sediment sources

We ran a series of multivariate statistical analyses on the sediment texture, sand texture and ostracod groups to better distinguish the similarities and differences between the three sites. Despite the significant stratigraphic parallels elucidated at the three harbours, these statistical analyses indicate clear discrepancies in the granularity and biostratigraphy of sediments accumulating in the basins; this bears credence to the uniqueness of each site (**Figures 6.19 - 6.21**). The PCA and cluster analyses demonstrate that the most pronounced differences exist between Tyre and its two sister harbours, with a notable absence of plastic harbour clays during the Roman and Byzantine periods (**Figures 6.22 and 6.23**). There is very little overlap between the Tyrian data points and those from Sidon and Beirut. Conversely, the variance between Sidon and Beirut is more limited. These divergences are attributable to two factors: (1) the coastal processes operating within the harbour (degree of protection, proximity to the shoreline); and (2) the origin of infilling sediments.

For the latter, Morhange (2001) has demonstrated that four main sources play a role in harbour accretion: (1) clastic fluvial sediments from local and regional water courses; (2) marine inputs, e.g. biogenic sands; (3) use of the basin as an ad hoc waste dump by human societies; and (4) the erosion of adobe infrastructure (Rosen, 1986), subsequently sluiced into the harbour basin by runoff. While the cultural inputs are clearly attested to in the gravels fraction, including for example ceramic sherds, glass, seeds, and leather fragments, the origin of the clastic fraction is not unequivocal. In order to better understand the origin of sediments accumulating in Tyre's harbour we undertook a series of sedimentological and mineralogical analyses (**Figure 6.24**). To widen the scope of the study to Egypt and the southern Levant, these findings have been integrated into a much larger dataset comprising research by Stanley *et al.* (1998), Sandler and Herut (2000) and Ribes *et al.* (2003) (see Table 6.1).

Sediment supply is critical to understanding the coastal response of littoral zones during the Holocene (Anthony, 1995; Dubar and Anthony, 1995; Ercilla *et al.*, 1995; Stanley, and Galili, 1996). Against this backcloth, the origin of coastal sediments on the Levantine coast has attracted the attention of scholars for many decades (Venkatarathnam *et al.*, 1971; Maldonado and Stanley, 1981; Emelyanov, 1994; Stanley and Wingerath, 1996; Stanley *et al.*, 1997; Sandler and Herut, 2000). Broadly speaking, two schools of thought dominate the literature: (1) a first school asserts the dominant role of the eastern Mediterranean gyre as the important land-sea dispersal mechanism of Nilotic sediments. Increased erosion of Sinai and southern Levantine coasts since the construction of the Aswan High Dam has been used as evidence of the Nile's dominant role in sedimentary budgets to the east and northeast of the delta. While the majority of the sand and gravel fraction aliments the Nile pro-delta, the remaining suspended load is reworked and transported northeast to the Sinai and the southern Levant (Emery and Neev, 1960; Pomerancblum, 1966; Goldsmith and Golik, 1980; Golik, 1993). How influential this sediment source is beyond southern Israel is ambiguous; (2) a second school, which has come to the fore in recent years, asseverates that local Levantine water courses have had a more significant role to play in the region's coastal sedimentation (Stanley *et al.*, 1997, 1998; Sandler and Herut, 2000). Unfortunately, precise quantification is complicated by several sources of smectite-rich assemblages and similar mineralogical signatures for both the Nile and Levantine fluvial systems (IS>kaolinite>illite).

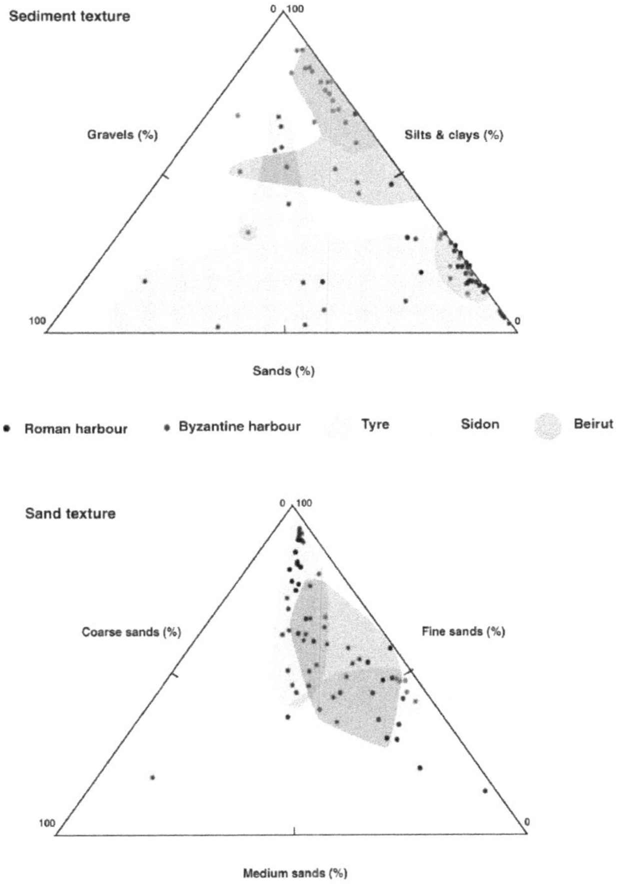

Figure 6.19: Ternary diagrams of Beirut, Sidon and Tyre's ancient Roman and Byzantine harbours. Beirut and Sidon's northern harbour facies consistently comprise fine-grained material. During the Roman period for example, Sidon's harbour constitutes >80 % silts and fine sands, whereas at Tyre the sand fraction still predominates. How should such a sharp disparity be construed? We advance two hypotheses, technological and geomorphological, both of which are not mutually exclusive: (1) Tyre's northern harbour is comparatively open, with a wide ~100-150 m channel entrance separating the two ancient moles. This created increased exposure to outer marine dynamics. In contrast, the harbours of Beirut and Sidon are quasi-landlocked by three main obstacles, scilicet the sandstone breakwater, the sea castle island and the inner mole. (2) Geographically, Tyre lies 9 km from the mouth of the Litani, Phoenicia's most important fluvial system. This river delivers coarse sediment inputs to the coastal zone, constituting mainly sands and gravels, trapped in base-level depocentres such as harbour basins. In contrast, Beirut and Sidon are situated near much smaller fluvial systems and watersheds, yielding mainly medium sands and silts.

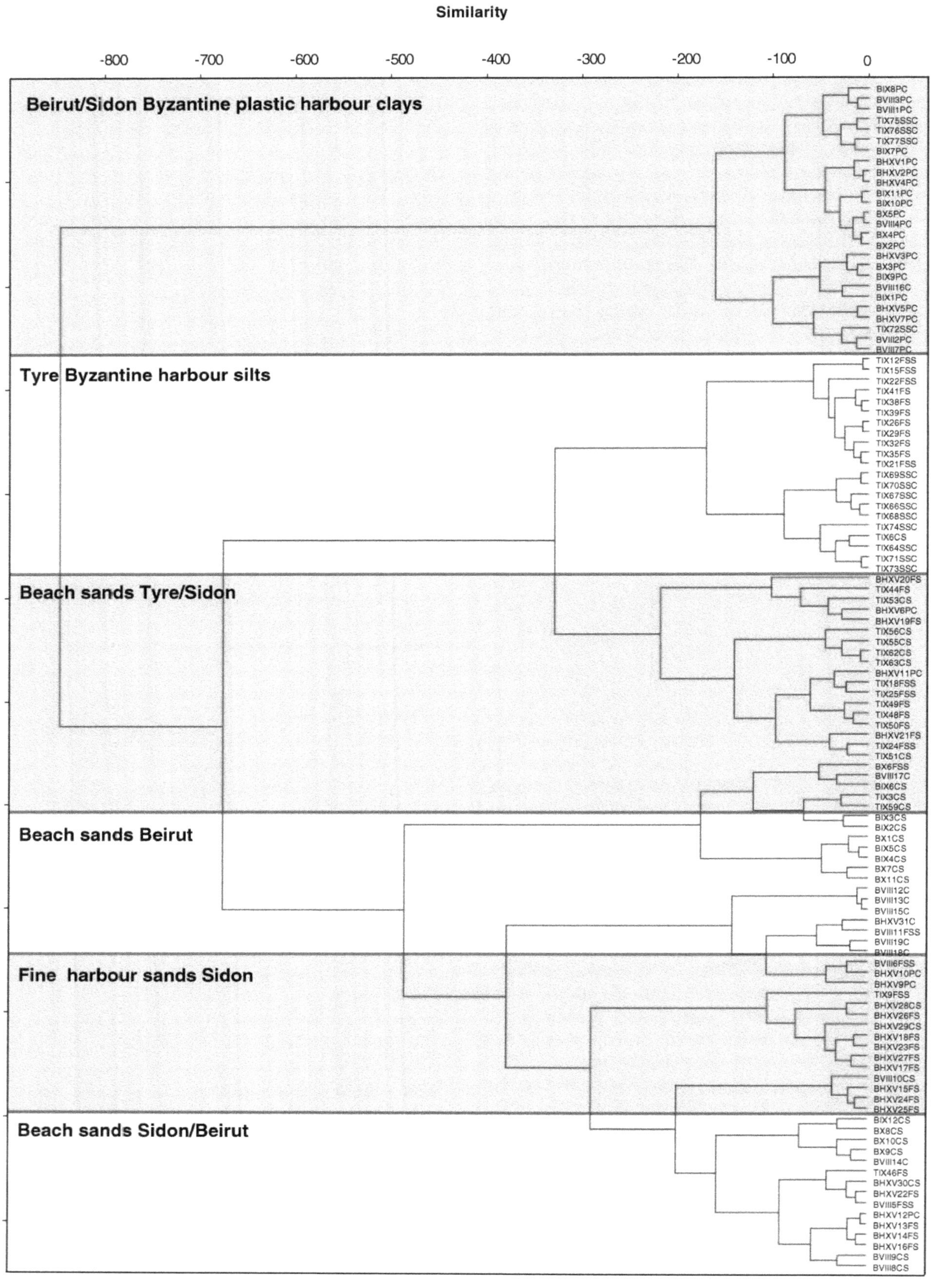

Figure 6.20: Cluster analysis of litho- and biostratigraphical datasets from Beirut, Sidon and Tyre (alogorithm: Ward's method, similarity measure: Euclidean). Despite the stratigraphic similarities elucidated at the three sites, the statistical analysis shows that the sites can be clearly differentiated.

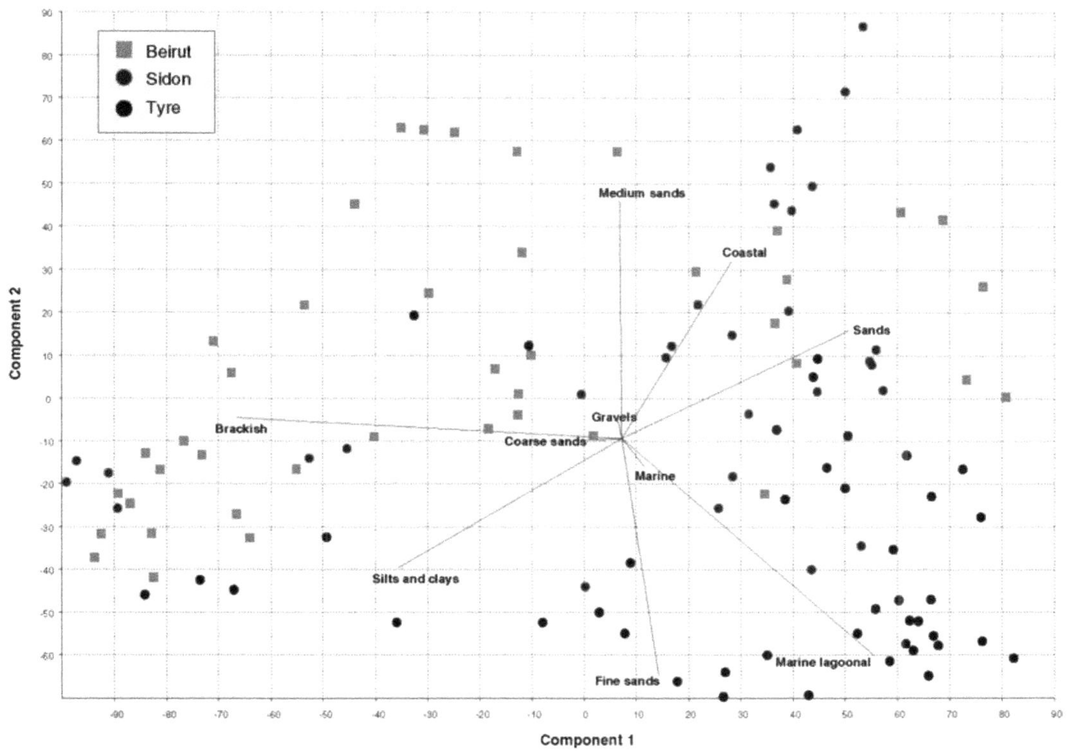

Figure 6.21: PCA scatter plot of axes 1+2 of a multivariate dataset comprising sediment texture, sand texture and the four ostracod ecological groups (Var-Covar matrix). 68 % of the total variation in the datasets is explained by components 1 and 2. The PCA allows us to correlate specific lithofacies with discrete ostracod groups. Broadly speaking, brackish lagoonal ostracods are associated with the silts and clays fraction (i.e. low energy hyposaline conditions), marine and marine lagoonal taxa with the fine sand fraction and coastal taxa with the sands and medium sands fraction. Internal data spread at sites are explained by variable lithostratigraphic and biostratigraphic factors (i.e. plastic harbour clays which contrast with prograding beach deposits). Please see below.

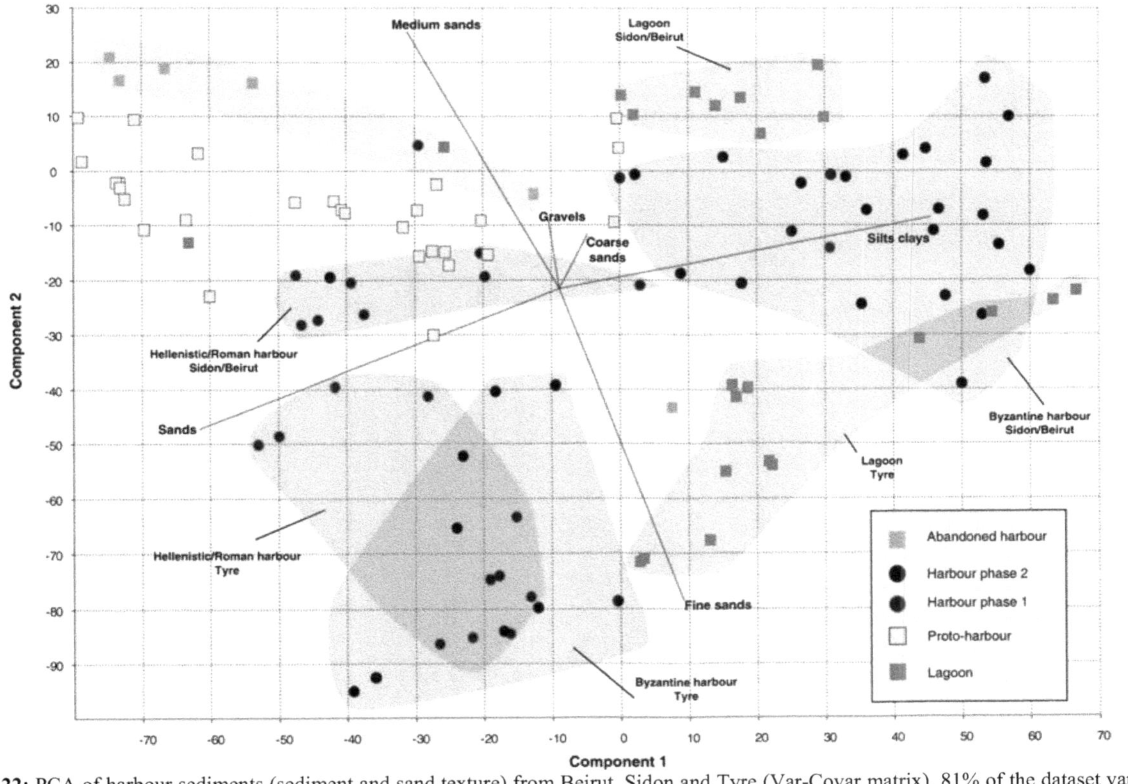

Figure 6.22: PCA of harbour sediments (sediment and sand texture) from Beirut, Sidon and Tyre (Var-Covar matrix). 81% of the dataset variation is explained by components 1 and 2. The different harbour sedimentary environments are clearly differentiated by the multivariate analyses. The most significant discrepancies are between Tyre and Sidon/Beirut. We attribute these contrasts to differences in sediment sources and the degree of harbour protection.

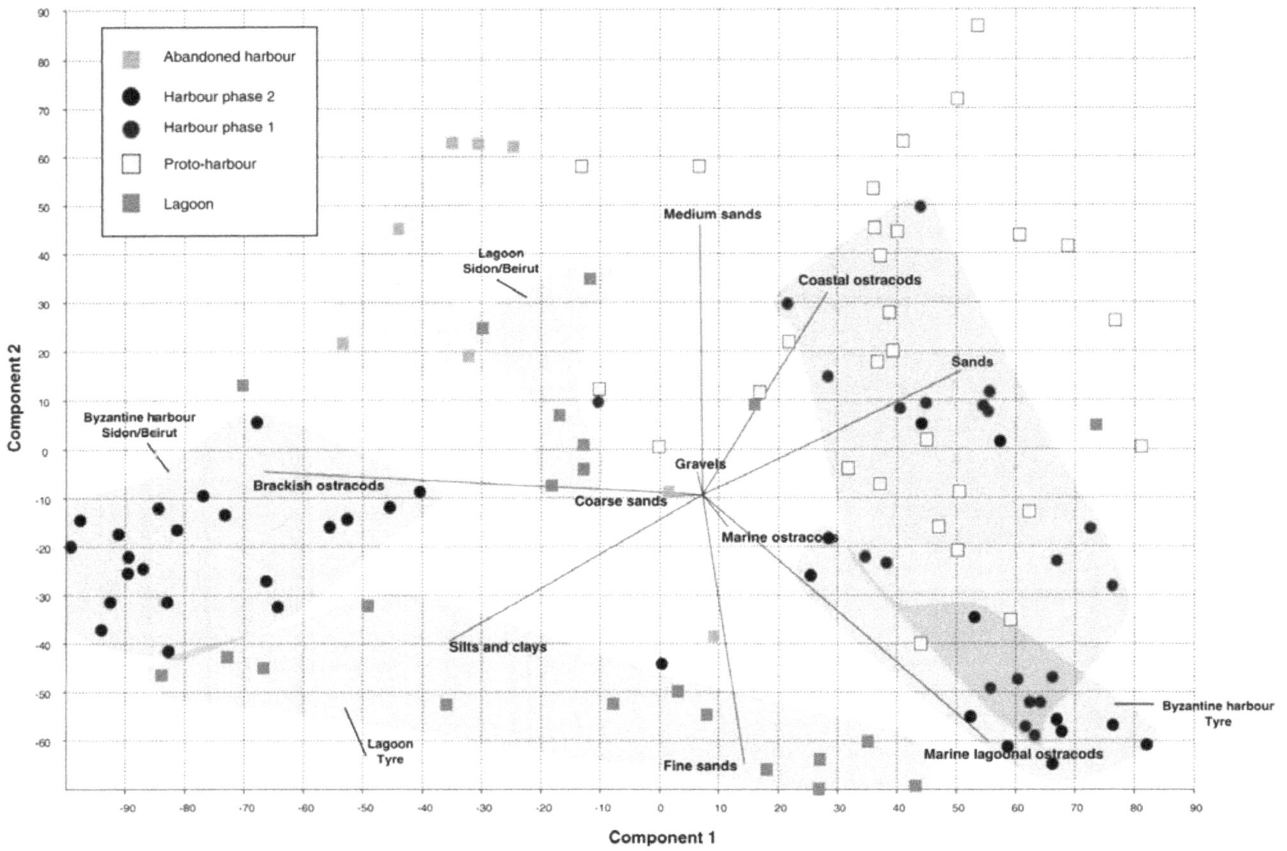

Figure 6.23: PCA of harbour sediments (sediment and sand texture) and ostracods from Beirut, Sidon and Tyre (Var-Covar matrix). 68% of the dataset variation is explained by components 1 and 2. On the basis of this multivariate analysis, we differentiate five discrete depositional environments at the three harbours. (1) An early Holocene lagoon environment, comprising fine-grained shelly sands and brackish ostracods. (2) A medium/low energy bay comprising medium sands and coastal ostracods. The data shows a high degree of variance consistent with the relatively open nature of the coastal environments at this time and the variability in coastal processes. (3) A fine-grained Iron Age/Roman artificial harbour unit, which is much better constrained than the preceding proto-harbour deposits. Less statistical variance translates an artificialisation of the coastal environments at all three sites. (4) A fine-grained Byzantine harbour unit. The increasingly confined conditions are clearly attested to by a sharp fall in the statistical data variance. Two clusters are apparent, the Beirut/Sidon cluster comprising plastic clays and brackish ostracods, and a Tyre cluster constituting fine sands and marine lagoonal ostracods. (5) A well-constrained semi-abandoned harbour.

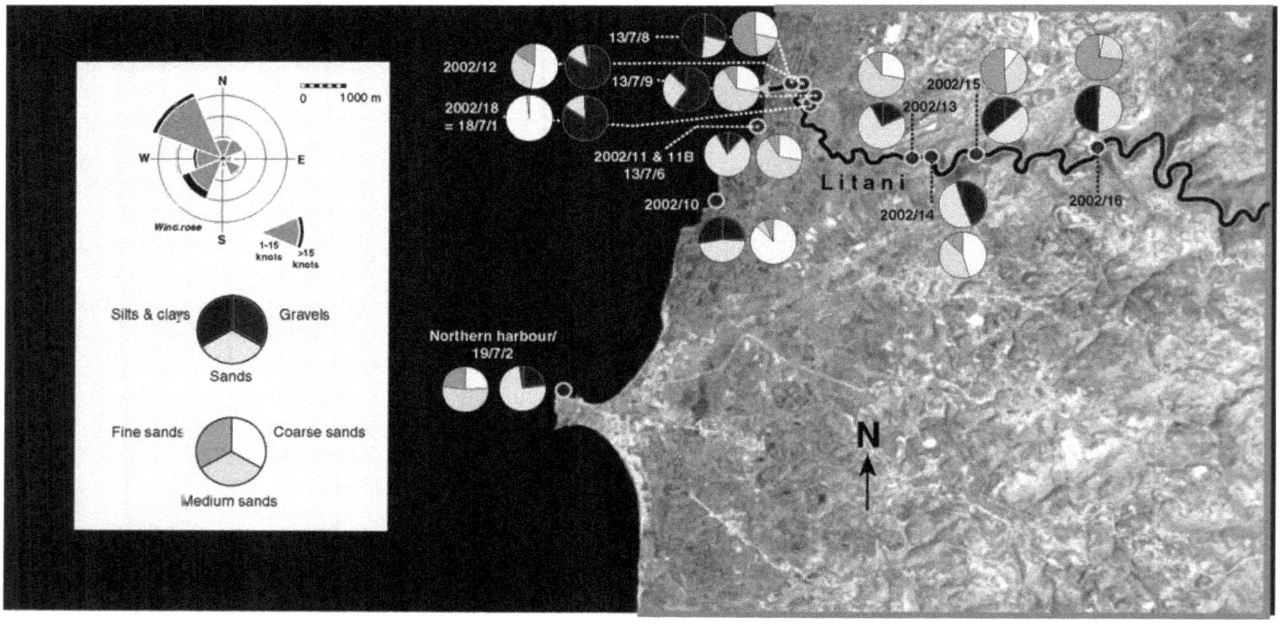

Figure 6.24: Sedimentology of Litani sediments. During the winter months, the system transports gravels and coarse sands to the coastal zone.

	Sample	IS	Kaolinite	Illite	Chlorite	Palygorskite	Clinophilite	Saddle value	Ka/IS
Tyre TIX	TIX 7g	75	15	10				0.69	0.2
	TIX 10g	60	15	15			10	0.74	0.25
	TIX 24g	60	15	20			5	0.77	0.25
	TIX 26g	75	15	10				0.65	0.2
	TIX 50g	70	10	20				0.65	0.14
	TIX 51g	50	15	35				0.77	0.3
	TIX 62g	80	15	5				0.63	0.19
	TIX 64g	80	15	5				0.68	0.19
	TIX 77g	65	25	10				0.82	0.390
	TIX 79g	80	15	5				0.52	0.19
Litani analogues	1378g	55	40	5				0.63	0.73
	1379g	55	45					0.6	0.82
	20028g	85	15					0.36	0.18
	200210g	90	10					0.43	0.11
	200211g	85	15					0.49	0.18
	200215g	50	50					0.58	1
	200216g	50	50					0.67	1
	swampg	95	5					0.33	0.05
Sidon BHI	BHI 24	75	25					-	0.33
	BHI 21	75	25					-	0.33
	BHI 20	75	20	5				-	0.27
	BHI 17	75	25					-	0.33
	BHI 16	80	20					0.75	0.25
	BHI 15	75	25					0.87	0.33
	BHI 14	75	20	5				0.75	0.27
	BHI 13	70	30					0.64	0.43
	BHI 11	75	20	5				-	0.27
	BHI 9	80	20					-	0.25
	BHI 3	60	30	10				-	0.5
	BHI 2C	95	5					-	0.05
Sidon archaeological site	Sc1	95	5					-	0.05
	Sc2	85	15					-	0.18
	Sc7	90	10					-	0.11
	Sc8	95	5					-	0.05
	Sc16	85	15					-	0.18
Awali	Awali 1	55	45					-	0.82
	Awali 2	50	50					-	1.00
	Awali 3	40	55	5				-	1.38
	Awali 4	40	60					-	1.50
	Bargouth	90	5	5				-	0.06
Litani	Litani 1	30	65	5				-	2.17
	Litani 2	45	55					-	1.22
	Litani 3	45	55					-	1.22
Sainiq	Sainiq 1	30	70					-	2.33
	Sainiq 2	40	60					-	1.50
	Sainiq 3	35	65					-	1.86

Table 6.1: Clay assemblages from the south Lebanon coast and intersecting waterways (grey: present study; white: Ribes *et al.*, 2003).

	Sample	IS	Kaolinite	Illite	Chlorite	Palygorskite	Clinophilite	Saddle value	Ka/IS
Nile	Nile 1	80	10	5		5		0.4	0.24
	Nile 2	85	15	0	0	0		0.44	0.16
Israel fluvial samples	El Arish	70	10	10	5	5		0.8	0.25
	Shiqma 92	75	15	10				0.79	0.37
	Evtah 92	75	10	10	5			0.84	0.17
	Lakhish 97	85	15					0.67	0.16
	Lakhish 92	85	10		5			0.73	0.16
	Soreq 97	80	10	10				0.8	0.35
	Soreq 92	85	10		5			0.77	0.16
	Yarqon 97	80	10	10				0.8	0.3
	Yarqon 92	85	5	10				0.92	0.18
	Poleg 97	70	20	10				0.75	0.35
	Poleg 92	80	10	10				0.65	0.19
	Alexander 97	80	10	10				0.53	0.26
	Alexander 92	80	5	10	5			0.73	0.17
	Taninim 97	85	5	10				0.72	0.14
	Oren 97	85	15					0.62	0.14
	Me'arot 97	85	15					0.68	0.23
	Qishon 97	80	10	10				0.85	0.21
	Qishon 98	85	10		5			0.76	0.13
	Na'aman 97	80	10	10				0.76	0.16
	Na'aman 92	85	15					0.73	0.24
	Bezet 97	85	15					0.67	0.21
	Bezet 92	85	15					0.83	0.2
	Glilot 1a	80	10	10				1.06	0.24
	Glilot 2	75	15	10				1.07	0.41
	Glilot 2a	80	10	10				1.04	0.25
	Ayalon 97	80	10	10				0.78	0.3
	Sh'alabim clay	90	10					0.55	0.19
	Shiqma reservoir	80	10	10				0.74	0.23
Israel offshore	C12	75	15	10				0.67	0.31
	C11	75	15	10				0.73	0.29
	C10	75	15	10				0.69	0.28
	C9	70	15	10	5			0.63	0.28
	C8	70	15	10	5			0.69	0.32
	C7	70	15	10	5			0.64	0.28
	S1	75	10	10	5			0.58	0.27
	C6	70	15	10	5			0.67	0.26
	C5	70	15	10	5			0.59	0.26
	C4	80	10	10				0.69	0.24
	H22	80	10	10				0.85	0.21
	H25	90	10					0.88	0.21
	B2	75	10	10	5			0.67	0.19
	B1A	75	10	10	5			0.63	0.21
	C3	70	15	10	5			0.95	0.47
	C13	70	20	10				0.62	0.33
	C14	75	15	10				0.58	0.34
	C15	75	15	10				0.61	0.4
	S2	75	5	10	10			0.59	0.26
	S3	80	10	10				0.6	0.29
	S6	70	15	10	5			0.63	0.33
	S7	70	15	10	5			0.63	0.3
	S8	15	10	5				0.6	0.33
	S9	70	15	10	5			0.53	0.29
	S10	75	15	10				0.57	0.2
	AT1-1	80	10	10				0.59	0.22

Table 6.1 continued: Clay assemblages from the Israeli coast and intersecting waterways (from Sandler and Herut, 2000).

6.2.1 Data acquisition

A series of twenty samples from core TIX (Tyre's northern harbour), the surrounding coastal zone and Litani watershed have been analysed for their clay mineralogical content. All samples were decarbonated in hydrochloric acid and rinsed; the organic content was removed using disodium peroxodisulphate. The remaining sediment was put into suspension using an ultra-sound bath and left to decant for two hours. The fraction inferior to 2 μm was subsequently pipetted off and concentrated into a clay paste using a centrifuge. This clay concentrate served to prepare three separate slides for each sample: an untreated control slide, a heated slide and a glycolated slide.

The slides were analysed using a Philips PW1050-81 diffractometre equipped with a cobalt anticathode. Identification of each sample's clay mineraology is based on changes in the spectral signature after physical or chemical treatment. Semi-quantification of the clay content was made using the glycolated signatures and peak areas. The intensities of different peaks were subsequently weighted and transformed into percentages (Moore and Reynolds, 1989). The saddle value of the Illite-Smectite (IS) phases was also calculated. Sandler and Herut (2000) have demonstrated that Nile IS phases are more expandable (>80 %) than Levantine derived IS, rendering this index a powerful tracer in differentiating between Nilotic and local sediment sources. These results were subsequently compared and contrasted with other regional datasets to better understand the role of the southeast Mediterranean conveyor in transporting and redistributing fluvial sediments to the coastal zone (Stanley *et al.*, 1998; Sandler and Herut, 2000; Ribes *et al.*, 2003).

6.2.2 Results and discussion

6.2.2.1 Litani and Tyre data

Description: The Litani yields a unique clay composition comprising IS phases (50-60 %) and kaolinite (40-50 %) with high saddle values (0.58-0.67, **Figure 6.25**). Very high kaolinite content is unique to south Lebanon, with all of the fluvial systems between the rivers Litani and Awali manifesting similar signatures (Ribes *et al.*, 2003). The assemblage is consistent with the erosion of kaolinite-rich red and brown hamra soils and recent colluvions in the Beqaa valley and surrounding watersheds. A second series of samples was analysed on the coastal zone south of the Litani delta. The assemblage is clearly differentiated and comprises high IS (85-90 %), low kaolinite content (10-15 %), negligible or nil illite and low saddle values (0.36-0.49). This signature is archetypal of the dark brown/black soils and sandy soils prevalent on the coastal seaboard up to the Awali river in the north (Lamouroux, 1967; Lamouroux *et al.*, 1967; Sanlaville, 1977, **Figure 6.26**). The deep brown/black soils are rich in expandable clays (low saddle values) which gives them a very compact and prismatic structure. An analogous clay assemblage is observed at the contact of core TIX's Maximum Flooding Surface.

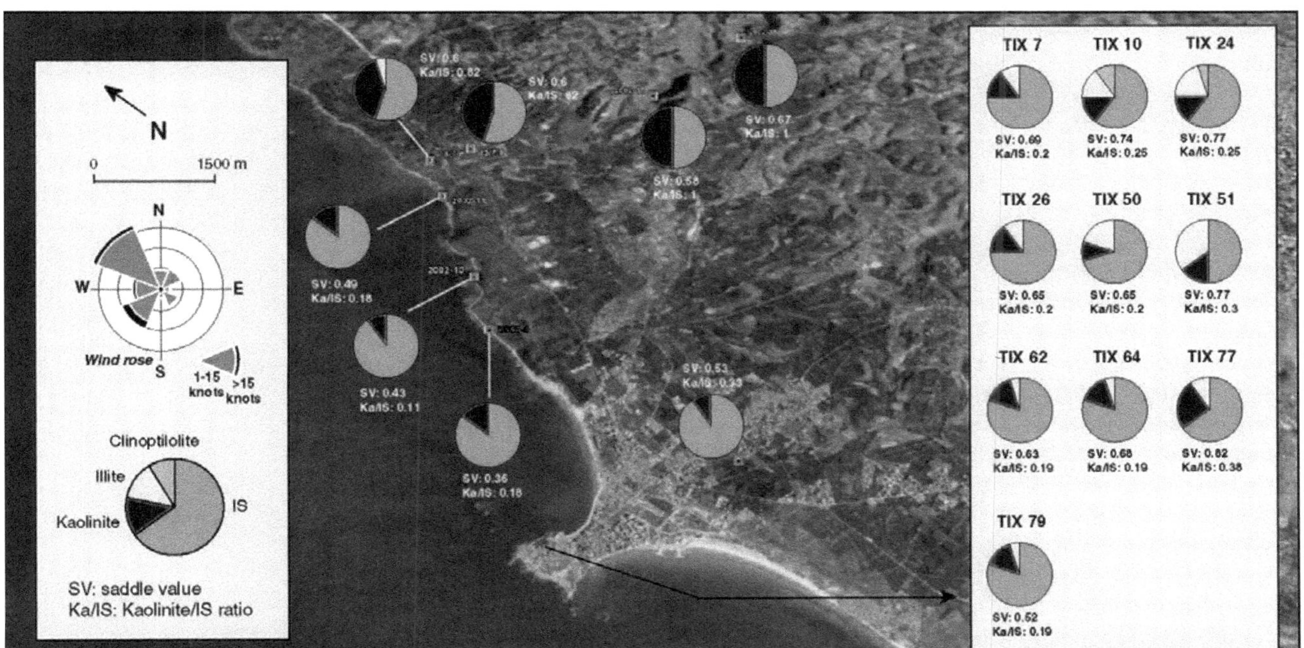

Figure 6.25: Clay mineralogy of fluvial and coastal deposits between Tyre and the Litani. Our data have evidenced three clay assemblages in the Tyre vicinity: (1) a Litani assemblage *sensu stricto* comprising IS phases (50-60 %), kaolinite (40-50 %) negligible or nil illite and high saddle values (0.58-0.67); (2) a coastal zone and littoral plain assemblage characterised by high IS (85-90 %), low kaolinite content (10-15 %), negligible or nil illite and low saddle values (0.36-0.49). This clay signature is typical of the tracts of sandy and brown/black soils prevalent on the coastal plain; and finally (3) Tyre's northern harbour assemblage manifests great variability in its clay assemblages consistent with local sediment sources and the influence of human societies on harbour sedimentation since 5000 BP.

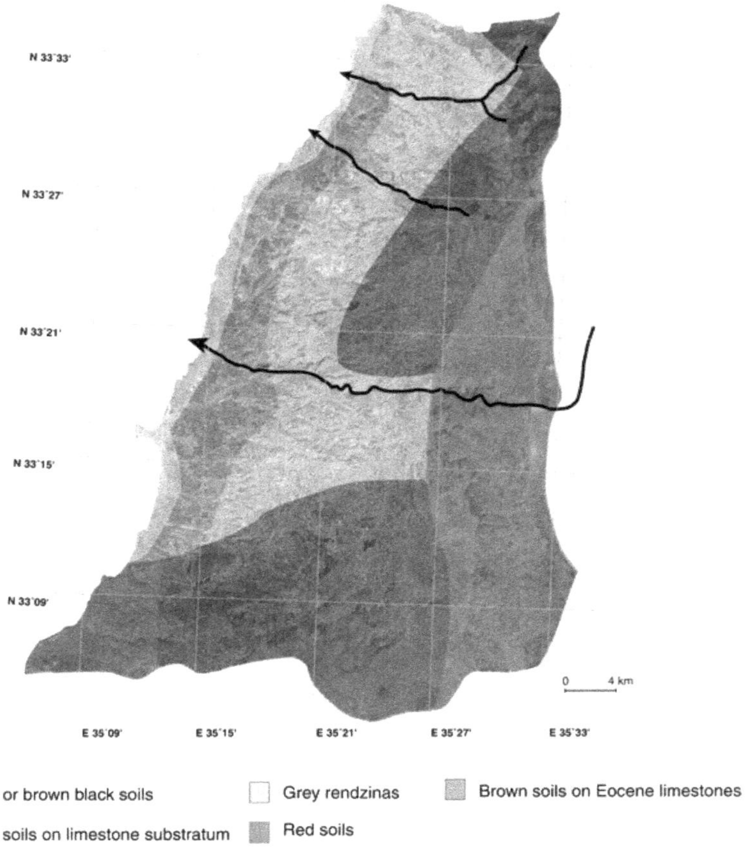

Figure 6.26: South Lebanese soils (after Sanlaville, 1977).

Surprisingly, neither of these two assemblages appears to contribute significantly to the Tyre harbour samples. The Tyre clay assemblages are variable and do not manifest any distinct source-sink pattern. IS values vary between 50 to 80 %, with 10 to 25 % kaolinite and 5 to 35 % illite. High saddle values of between 0.63 and 0.82 are recorded (**Figure 6.27**). This variability in the clay signatures evokes mixtures of local material and clays derived from the northern coast of Israel. The base of the core comprises an assemblage analogous to the brown/black soils observed on the coastal fringe and plain. During the early Holocene flooding of the cove, this signature evokes reworking of local sediment sources available in the basin. A rise in saddle values suggests an introduction of less expandable IS phases from the Litani and southern fluvial systems. The middle part of the core is dominated by illitic phases accompanying IS and low levels of kaolinite. These illitic phases are concurrent with the Cretaceous outcrops available in the Tyre vicinity (**Figure 6.28**). Samples TIX 10 and TIX 24 both contain clinoptilolite derived from Tyrian Eocene outcrops. Part of the source-sink variability might also be explained by post-sedimentary alterations in the clay composition, although at present this phenomenon is poorly understood (Chamley, 1989).

Interpretation: The infilling of Tyre's northern basin has been complicated by the influence of mankind, with marked variability in the up core clay signatures. Two significant anthropogenically forced changes are observed in the clay assemblages. (1) There is a notable rise in illite during the Roman and Byzantine periods. This is consistent with the increased weathering of Tyrian Eocene and Cretaceous outcrops. We evoke two hypotheses to explain this pattern: (i) deforestation of Eocene and Cretaceous outcrops during the Roman and Byzantine periods leading to accentuated erosion in the Tyre vicinity; and/or (ii) the use of this geology in Romano-Byzantine construction. The former hypothesis is corroborated by rapid rates of sedimentation (10 mm/yr) during the Roman and Byzantine periods, indicative not only of improved harbour protection but also an increase in watershed erosion. Compiled climate records from the Levant also indicate that this period was much cooler and wetter (**Figure 6.29**). Increased mountain flooding would have played a key role in contributing illite and kaolinite rich sediments to the coastal zone. Conversely, a fall in kaolinite content attests to more minor agricultural impacts during the Bronze Age and Islamic period. These periods were also drier and warmer than the intercalated wet phases. (2) The sudden rise in clinoptilolite during the Byzantine period is most logically explained by the quarrying of Eocene outcrops and their use in either harbour or urban construction. Although Eocene outcrops are extensive in the Tyre vicinity, we did not find a similar assemblage in any of our other samples.

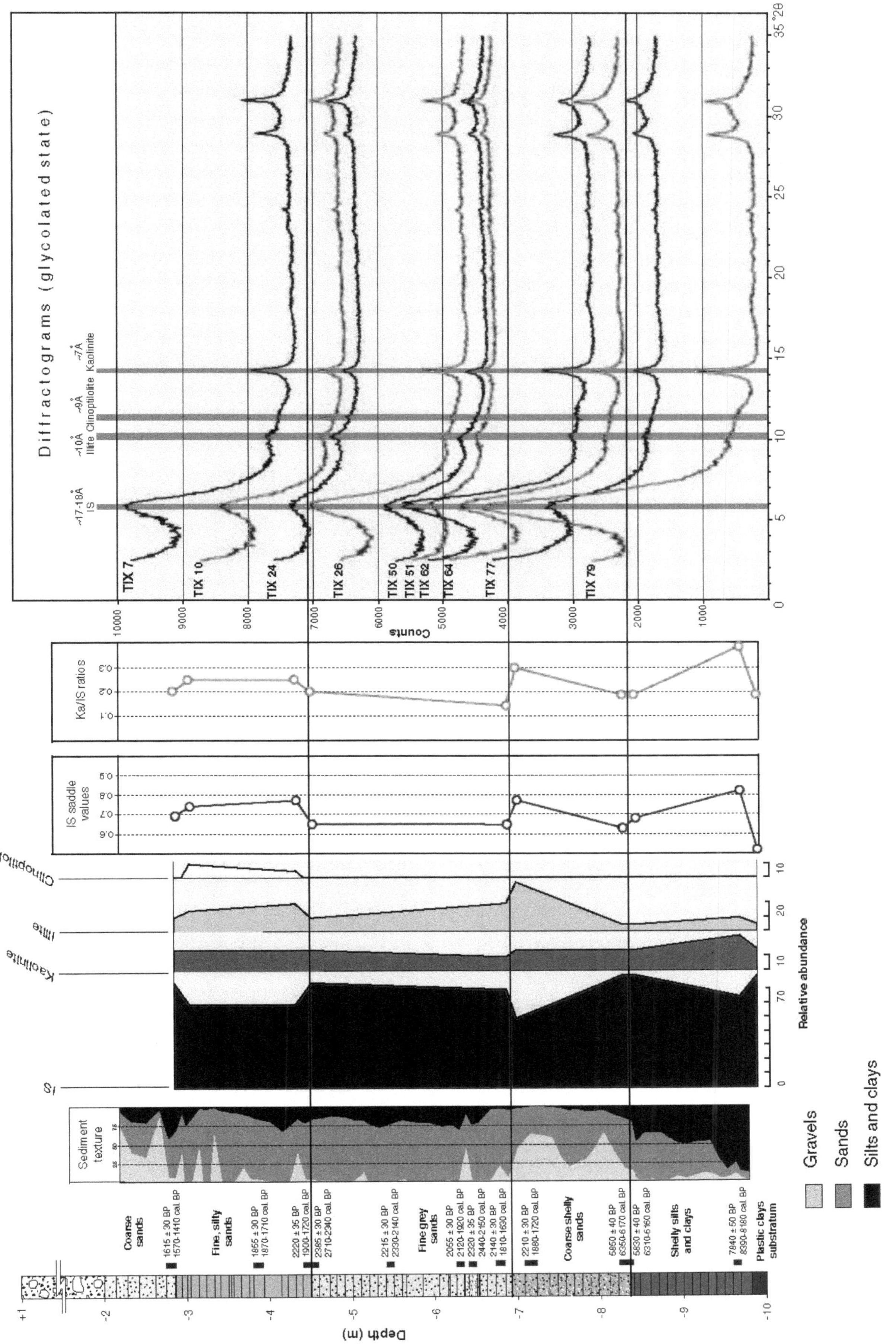

Figure 6.27: Clay assemblages from northern harbour core TIX and corresponding diffractograms (glycolated state).

The coastal area between Tyre and the Litani manifests a unique IS (80-90 %) and kaolinite assemblage (10-20 %) characterised by low saddle values and Ka/IS ratios. Surprisingly, there is no clear link between the Litani source area and coastal sedimentation to the south of the river. Geomorphological prospections in this area reveal that the coastline is presently undergoing significant erosion, with a reworking of sandy and brown/black soils yielding this unique clay signature (**Figure 6.30**). Although data is lacking, a fall in Litani sediment yields has manifestly entrained an erosion of coastal zones in the Tyre vicinity. Because this erosion of older sedimentary stocks is dated to the twentieth century, it would be ambiguous to directly extrapolate this pattern to the Holocene. For example, in Israel Nir (1984) has shown a ten to twenty fold increase in erosion rates along the Levantine seaboard during the past 100 years. In light of this, we hypothesise that the modern coastal samples are not good analogues for the Holocene record.

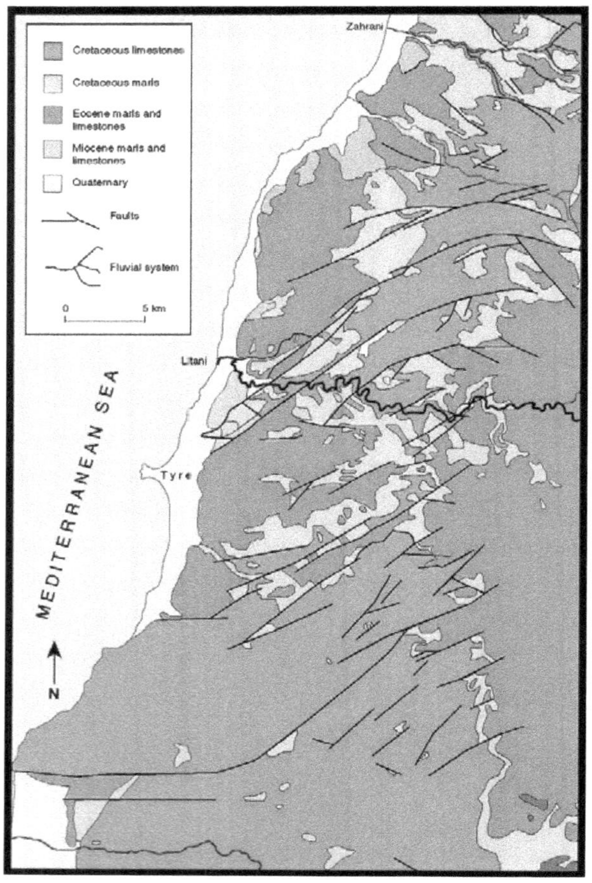

Figure 6.28: Geology of the Tyre vicinity (after Dubertret, 1955).

6.2.2.2 Regional Nile-Levant patterns

Results from Egypt and the southern Levant have been coupled with the Lebanese data to yield a regional map of sediment sources and coastal sinks along the Levantine seaboard (**Figure 6.31**). Nile sediment off the Egyptian coast is dispersed in a large counter-clockwise gyre by geostrophic and wave-driven currents that prevail in the easternmost Mediterranean (Inman and Jenkins, 1984; Frihy *et al.*, 1991; Bergamasco *et al.*, 1992). Sediment dispersal is essentially directed eastwards towards the coasts of Sinai and the southern Levant, before turning north to comb the coasts of Israel. Despite data paucity, it has long been assumed that silt and clay-sized particles were transported beyond the Sinai by bottom currents and suspension. The Israeli and south Lebanese coastline is intersected by ~20 westerly-flowing wadis and rivers. Nir (1984) has calculated that the average amount of sediment reaching the Mediterranean from these rivers is ~1 million tons. Offshore, dispersal of the sediment is affected by the drowned aeolianite ridge topography which creates a series of ridges and sediment troughs (Neev *et al.*, 1987).

(1) The Nile has the lowest saddle values (0.42) on the sedimentary conveyor (Sandler and Herut, 2000). This is consistent with the high expandability (>80 %) of its basaltic derived IS phases. With the exception of the south Lebanon coastal plain, which comprises expandable brown/black soils, fluvial derived samples from Israel and the Lebanon have much higher saddle values between 0.6-0.9. In southern Levant, saddle values of coastal and marine samples are often lower than adjacent continental areas. This pattern attests to a dilution of Levant derived IS with phases from the Nile (**Figure 6.33**). The disparity is less pronounced in northern Israel and southern Lebanon, and leads us to conclude that Nile derived clays are not significant contributors to coastal sedimentation in these areas (**Figure 6.34**). Desert dust derived from the Arabian peninsula also plays a minor, though unquantified, role in Sinai and Levantine sedimentation, especially in the south. Arabian derived desert dust is rich in kaolinite; its contribution to coastal sedimentation is best discernible along the Sinai's Mediterranean façade (**Figure 6.35**).

(2) From the southern Levant to the mouth of the Litani, there is a generally decreasing trend in the Ka/IS of marine samples (Sandler and Herut, 2000). This pattern is not reciprocated in the fluvial clay data, which show no clear geographical trends as a result of the diverse source terrains that fringe the southeastern Mediterranean coast. All the waterways between Gaza and southern Lebanon drain similar geological outcrops. There are five potential sediment sources of note: (i) the Albian-Turonian Judea Group (yields IS, illite, kaolinite and some palygorskite); (ii) the Senonian to Palaeocene outcrops of Mount Scopus, with Eocene geology outcropping in the foothills (yields highly smectitic IS and minor kaolinite [Sandler and Herut, 2000]); (iii) the Oxfordian-Portlandian and Cenomanian-Turonian geology of the Anti-Lebanon range (yields IS, illite and kaolinite, **Figure 6.36** [Dubertret, 1955]); (iv) the Pliocene-Pleistocene fossil sand dunes that fringe the seaboard of both Israel and Lebanon (yield IS and kaolinite); and finally (v) recent Quaternary deposits and palaesols on the region's coastal plains. In Israel these give rise to an IS>kaolinite>illite assemblage, whilst in northern Israel and southern Lebanon the assemblage is dominated by IS>kaolinite.

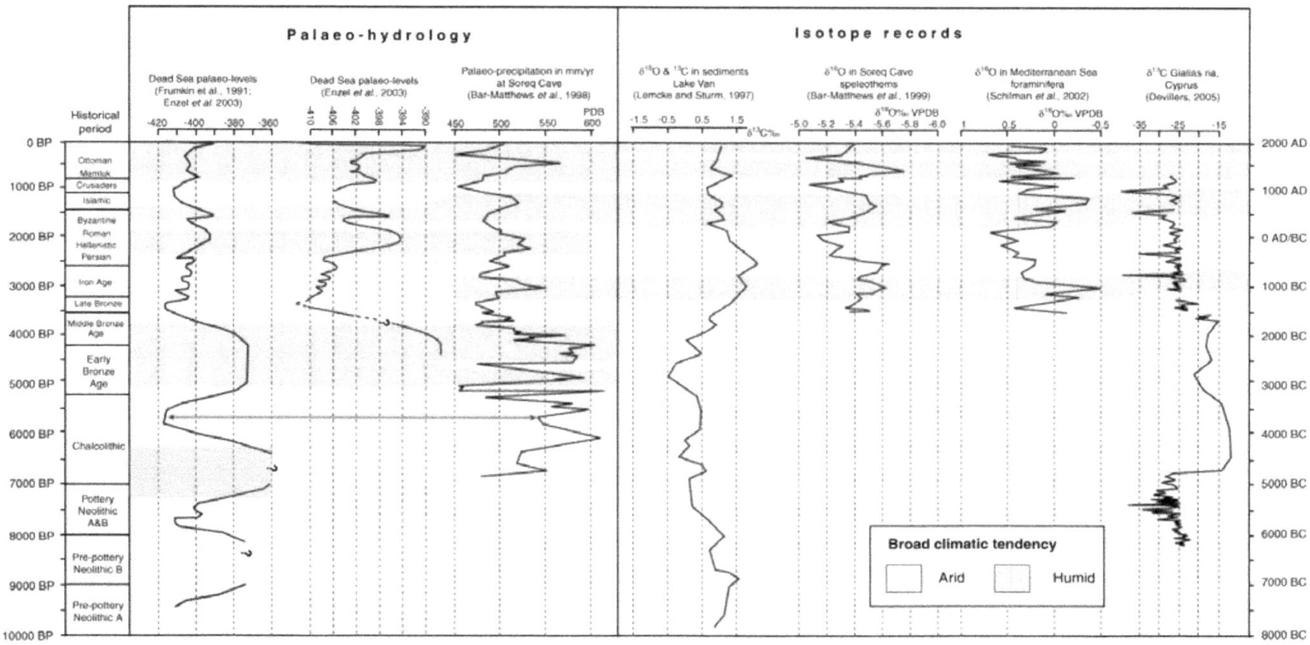

Figure 6.29: Compilation of palaeo-hydrological and isotope climate records from the eastern Mediterranean.

Figure 6.30: Evidence of coastal erosion to the north (lower photograph) and south (upper photograph) of the Litani river. Beach erosion is starting to encroach on agricultural land and is reworking sandy and brown/black soils in the area. This reworking of old sedimentary stocks gives rise to a unique clay signature which is clearly discernable from that of the Litani. Despite proximity to the Litani mouth, the erosion is so pronounced in some areas (lower photograph) that it has necessitated artificial means to entrap sediments (Images: DigitalGlobe).

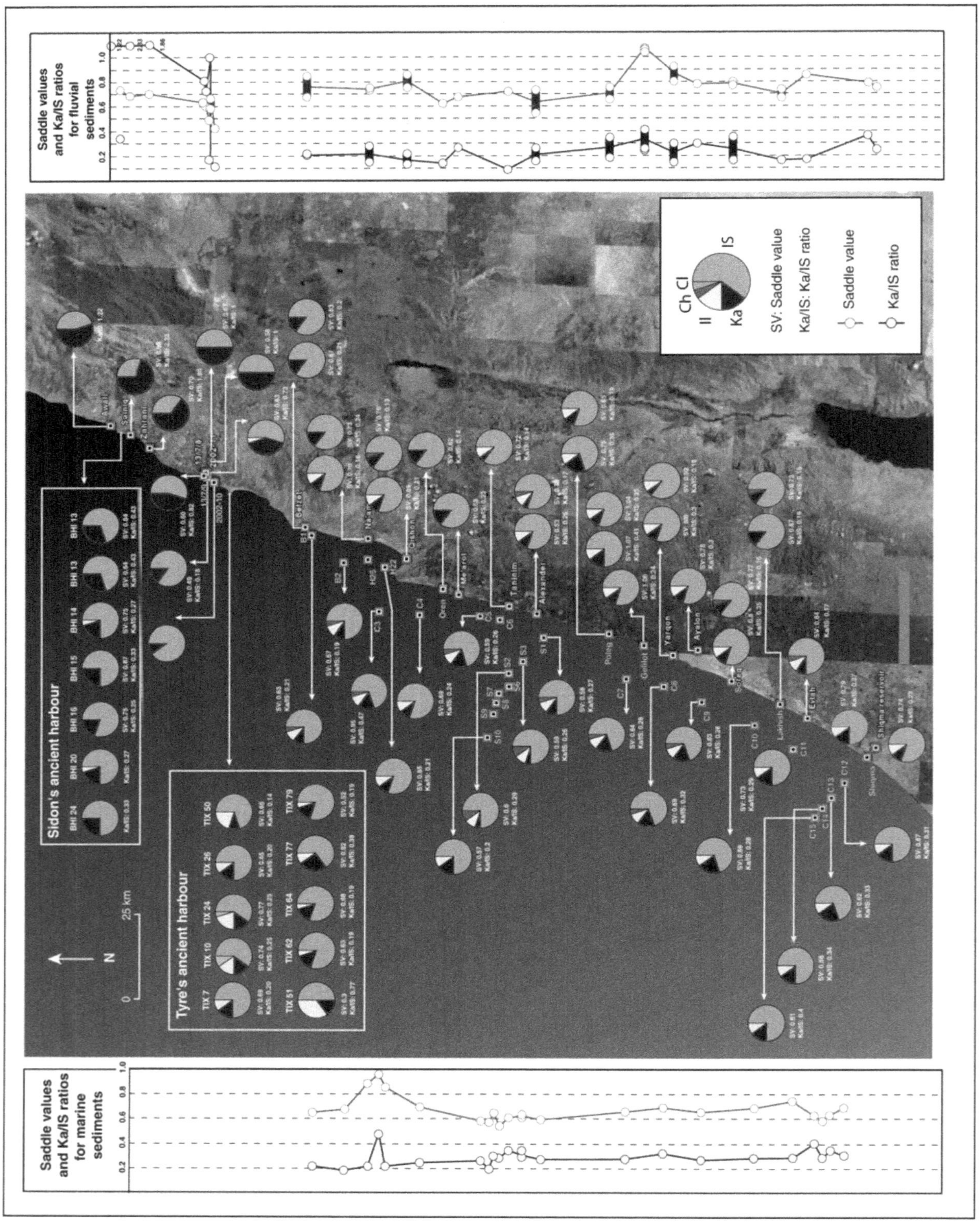

Figure 6.31: Levantine clay signatures from Israel and southern Lebanon (data from present study, Sandler and Herut, 2000 and Ribes *et al.*, 2003).

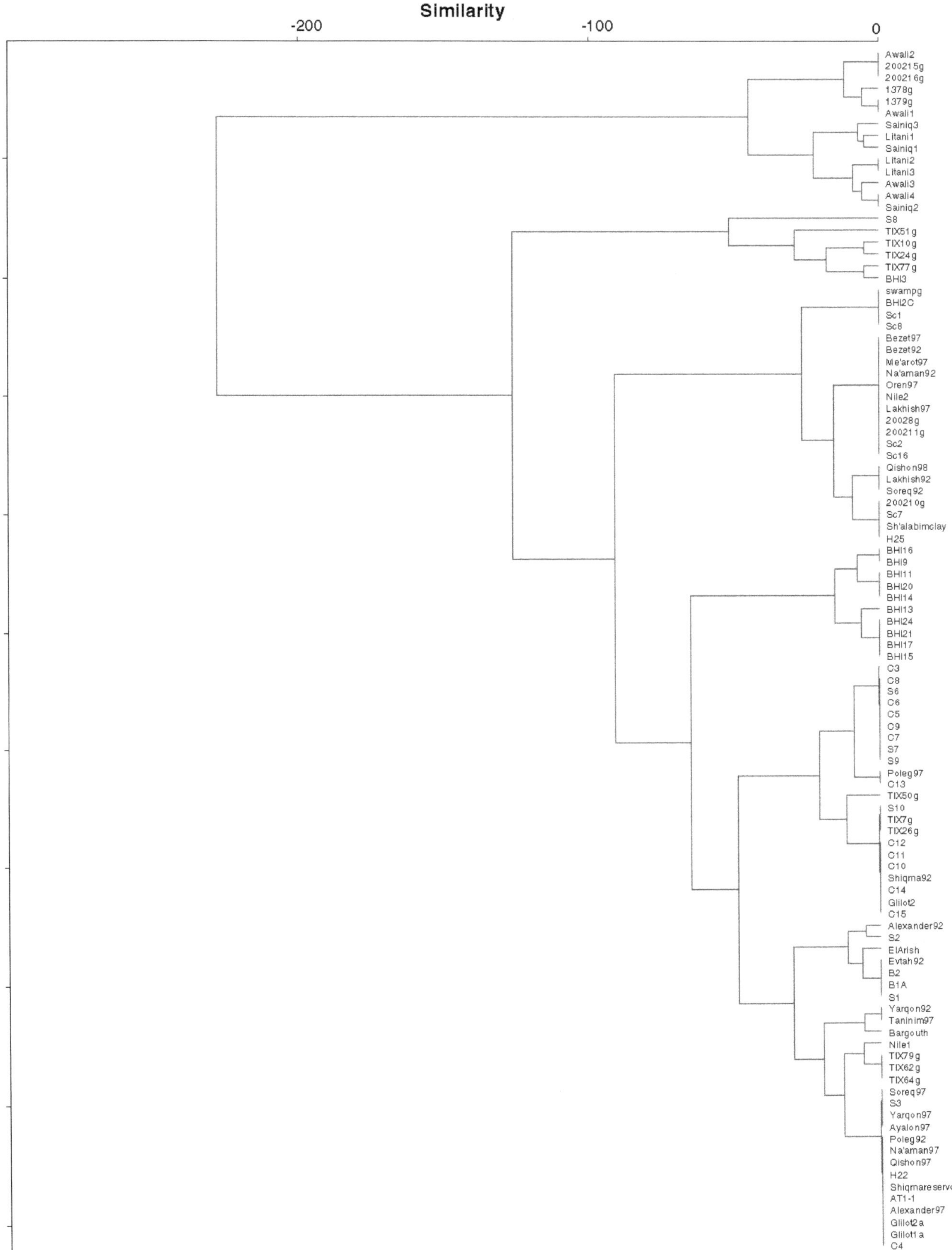

Figure 6.32: Multivariate cluster of the Levantine clay data (algorithm: Ward's method; similarity measure: Euclidean). Despite broad geographical groupings, the analysis is unable to satisfactorily reconcile an overriding spatial or temporal pattern. A number of multivariate statistics were performed with similar results. Note the high degree of scatter in the Lebanese data.

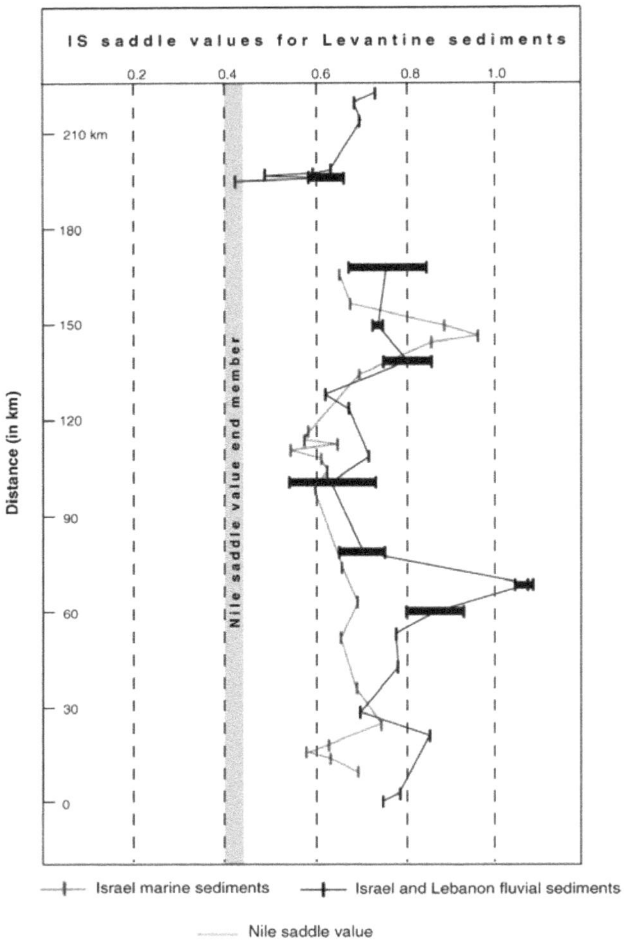

Figure 6.33: North south plot of IS saddle values for fluvial and marine Levantine sediments (Israel data from Sandler and Herut, 2000). The data show a clear offset between marine and adjacent fluvial systems, especially in southern Levant. This suggests a mixing of Nilotic and local sediment sources, which remain dominant, in these areas.

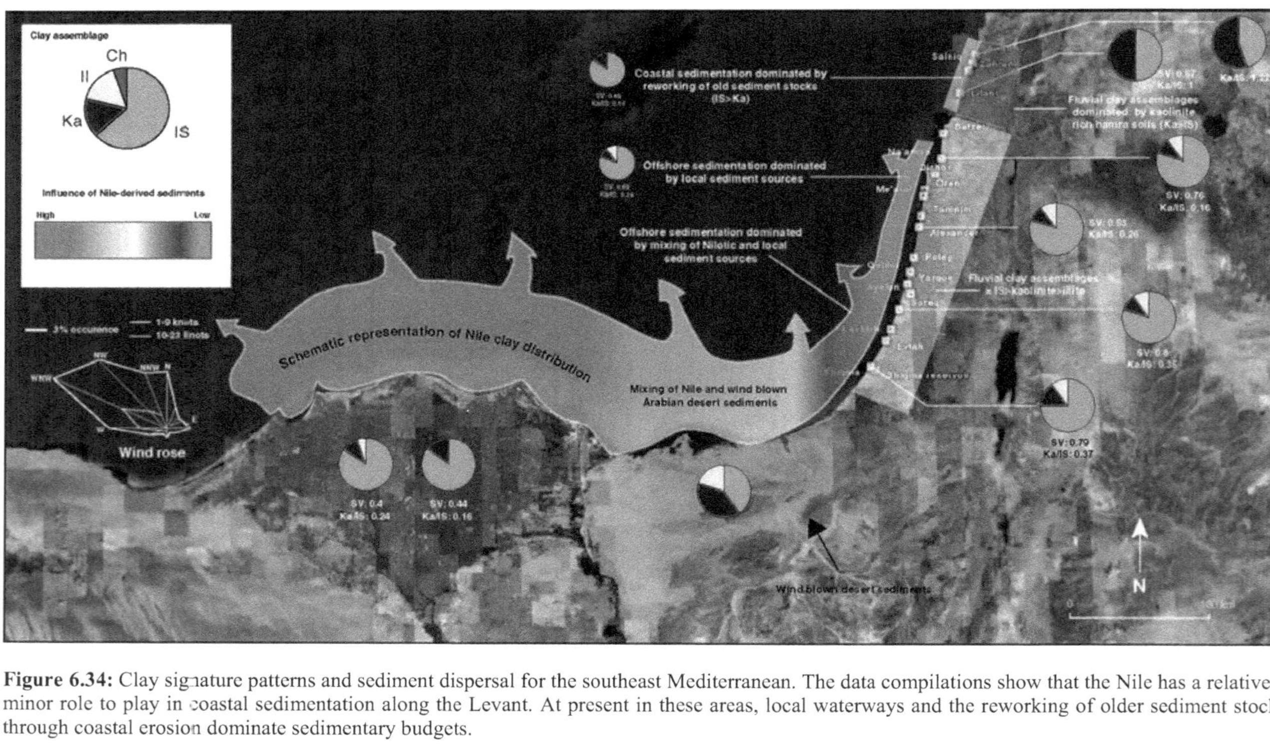

Figure 6.34: Clay signature patterns and sediment dispersal for the southeast Mediterranean. The data compilations show that the Nile has a relatively minor role to play in coastal sedimentation along the Levant. At present in these areas, local waterways and the reworking of older sediment stocks through coastal erosion dominate sedimentary budgets.

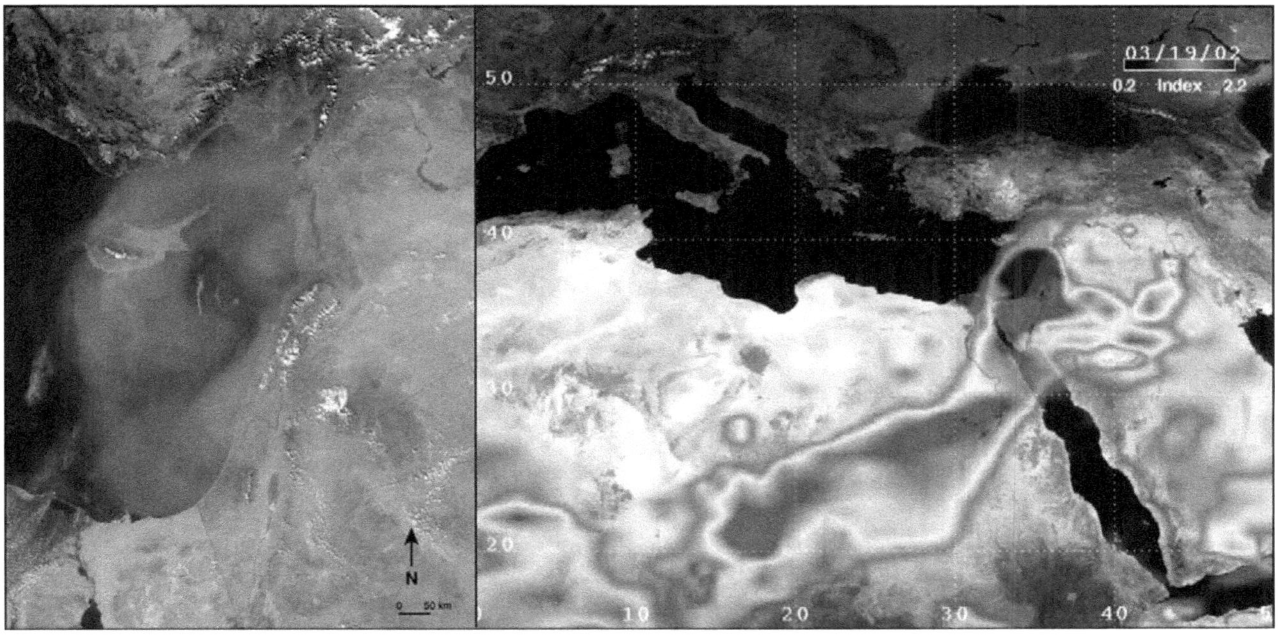

Figure 6.35: Dust storm over the Levant on October 19, 2002. Left: The Terra MODIS instrument captured this image of a large dust storm moving west from the Middle Eastern countries of (clockwise from top) southern Turkey, Syria, Lebanon, Jordan, and Israel, out into the Mediterranean Sea, and over the island of Cyprus. Cyprus lies about 100 miles west of the Syrian coast and about 60 miles south of the Turkish coast. The dust storm stretches about 200 miles out into the Mediterranean, and is about 400 miles across from north to south (credit: Jacques Descloitres, MODIS Land Rapid Response Team, NASA/GSFC). Right: A false-color representation of Total Ozone Mapping Spectrometer (TOMS) measurements of aerosol index was overlain on the true-colour MODIS image to show the extent of the dust cloud. The aerosol index is a measure of how much ultraviolet light is absorbed by the aerosol particles within the atmosphere, and is approximately equal to the optical depth. Red areas indicate high aerosol index values and correspond to the densest portions of the dust cloud. Yellows and greens are moderately high values (credit: Jay Herman, TOMS Aerosol/UV Project Principal Investigator, NASA GSFC).

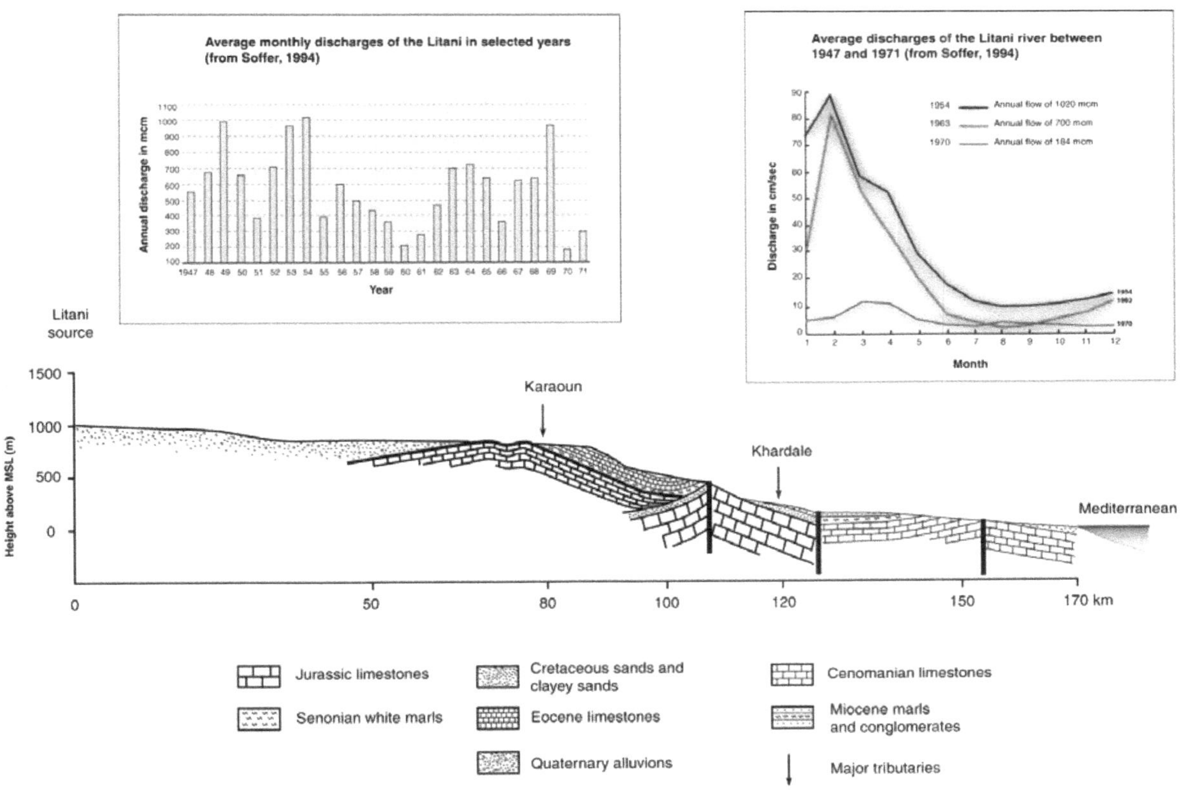

Figure 6.36: Litani long profile and geology (after Abd-El-Al, 1948 and Dubertret, 1955). Based on 25 years of measurements taken between 1947 and 1971 the Litani discharges on average ~580 million cubic metres (MCM) per year, 27 % of this amount during the months of May to October and 73 % between November and April (Soffer, 1994). There are great variations in this 25 year time-series, ranging between a minimum discharge of 184 MCM in 1970 and a maximum of 1020 MCM in 1954.

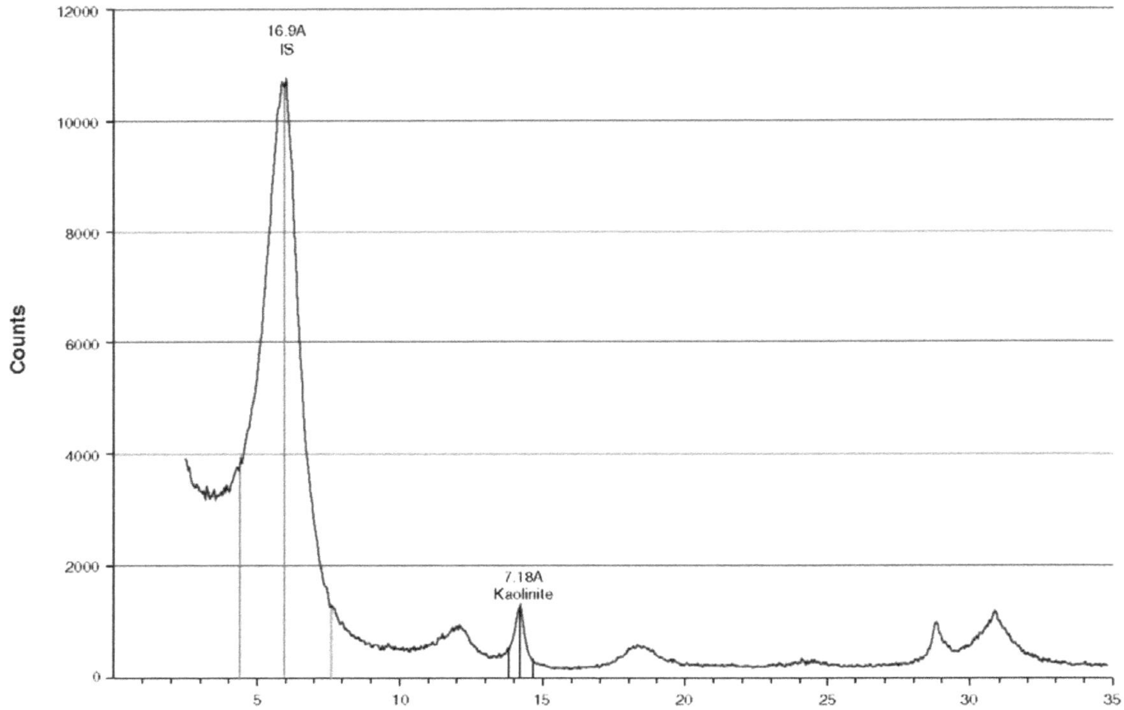

Figure 6.37: Diffractogram of a brown/black soil in the Tyre vicinity. The assemblage comprises dominant smectite (>80%) with low amounts of kaolinite (<20%). The deep brown/black soils are rich in expandable clays (low saddle values) that yield a very compact and prismatic structure.

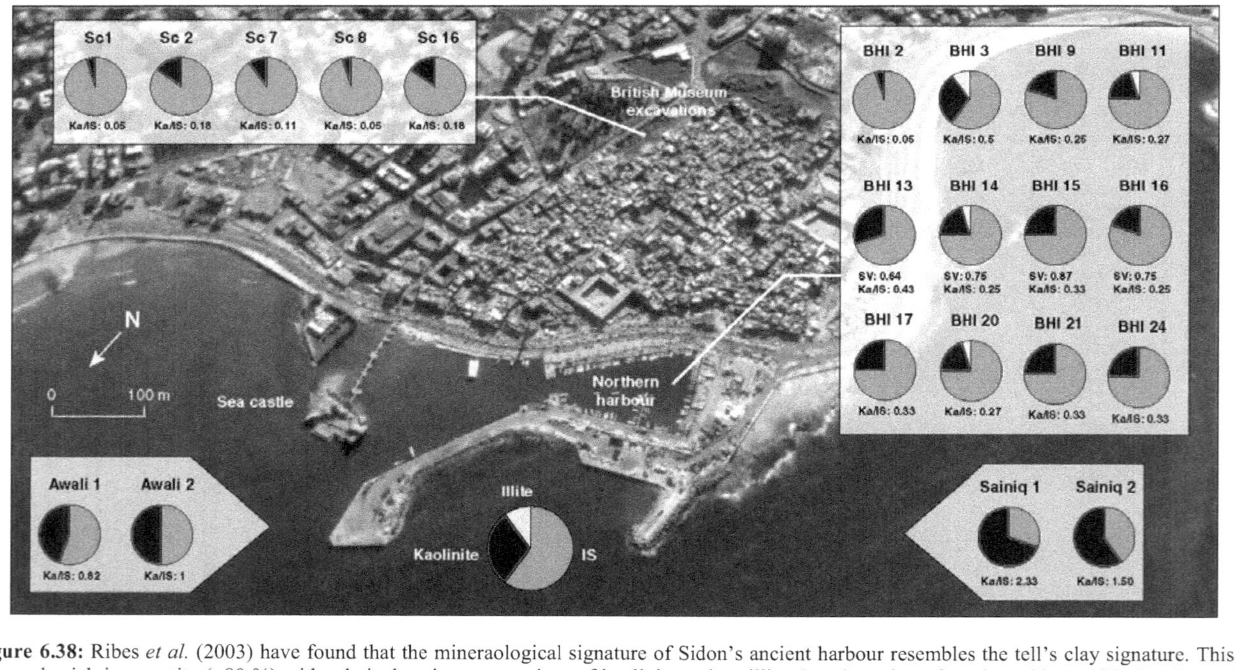

Figure 6.38: Ribes *et al.* (2003) have found that the mineralogical signature of Sidon's ancient harbour resembles the tell's clay signature. This is extremely rich in smectite (>80 %) with relatively minor proportions of kaolinite and no illite. Local erosion of sandy and brown/black soils on the promontory therefore appears to have been the dominant sediment source during the Holocene. The clay rich brown/black soil was also widely used in ceramics and adobe construction.

(3) In Israel, observed changes in the assemblages are relatively minor and remain within a limited frame of IS>kaolinite>illite. A clear shift is observed in the clay composition of fluvial systems in southern Lebanon, characterised by high kaolinite≥IS assemblages. This pattern is replicated by a sharp rise in the Ka/IS ratios from an average of 0.23 south of Betzet to 1.28 between the rivers Litani and Awali. These data are consistent with the erosion of kaolinite-rich hamra soils that dominate the surface geology between Tyre and Sidon. Inversely, the coastal fringe between Tyre and Sidon is characterised by sandy soils and expandable brown/black soils, extremely rich in smectite (>80 %) with relatively minor proportions of kaolinite and no illite (**Figure 6.37**). Ribes *et al.* (2003) have found that the mineraological signature of Sidon's ancient harbour resembles more closely this latter assemblage (**Figure 6.38**). Such a pattern suggests that wave and current-eroded cliffs have a role to play in contributing sediment to the Levantine coast, results corroborated by oceanographic and coastal meteorological data (Goldsmith and Golik, 1980). The role of human societies in affecting harbour sedimentation patterns means that the evidence at Tyre is more variable.

(4) While distal Nile derived sediments have a minor role to play in offshore sedimentation along the Levant (notably in the south), it is not a contributor to inner shelf and coastal sedimentation in this zone. In these areas, sedimentation is dominated by local fluvial systems, storm-wave erosion of coastal cliffs, and strong winds.

6.2.3 Concluding remarks

With the exception of southern Israel, patterns of coastal sedimentation in the Levant are dominated by local fluvial systems. At the scale of Tyre, regional (Litani, erosion of Eocene and Cretaceous outcrops in the Tyre vicinity and reworking of brown/black soils on the Tyrian seaboard) and local sources (erosion of adobe infrastructures, cultural inputs) have been significant in the infilling of Tyre's ancient harbour. The high level of variability observed in the up-core clay signatures evokes Romano-Byzantine deforestation and use of Cretaceous and Eocene geology in local construction. This entrained a clear change in the volume and mineralogy of sediment transported into the basin. At the scale of southern Lebanon, it has been demonstrated that the Nile is not a significant contributor to sedimentation on the inner shelf.

Much of the present research has tended to focus on the study of contemporary samples. At certain sites we question whether this approach accrues the best analogues for sedimentation patterns during the Holocene. As an example, the Nile today yields a mere 15 % of pre-1902 silts and clays (Stanley *et al.*, 1998). Many of the Levant's rivers are also heavily artificialised and there are is an increasing number of anthropogenic infrastructures along the Israeli coast. These are clearly modifying sediment transport paths, notably on the innermost shelf. Such anthropogenic impacts, coupled with sea-level rise reworking old sediment stocks, suggest that scholars must be cautious when extrapolating contemporary sediment dispersal patterns and clay signatures to Holocene sedimentary records. Future work must concentrate on linking fluvial archives with coastal and offshore records to better understand the role and variability of contributing sediment sources over longer centennial and millennial timescales.

6.3 Geochemistry

Geochemist colleagues have undertaken investigations into the trace metal chemistry of sediments at Sidon (Le Roux *et al.*, 2003a-b, 2005). The association of trace metal anomalies are increasingly being used to reconstruct ancient pollution levels during the historical period and antiquity (**Figure 6.39**; Sayre *et al.*, 1992; Renberg *et al.*, 1994; Hong *et al.*, 1994, 1996; Nriagu, 1996; Shotyk *et al.*, 1998, 2005; Martínez-Cortizas *et al.*, 1999; Bränvall *et al.*, 2001). Within this context, it has been demonstrated that lead (Pb) isotopes in harbour sediments are particularly powerful tools in retracing the history of metallurgic activities at coastal sites (Véron *et al.*, 2006). Prior to this research, Alexandria and Marseilles were the only ancient harbour basins to have been probed using geochemical techniques (Le Roux *et al.*, 2005; Véron *et al.*, 2006). There are a number of reasons why harbour sediments are particularly conducive to geochemical analyses: (1) ancient seaports comprise low-energy depocentres. Urban and coastal contaminants are thus concentrated in the basins, and have the potential to yield long time-series of human-environment interactions; (2) trace metals (e.g. copper, lead), have been used by ancient societies since the end of the Neolithic (Mellaart, 1967; Jenkins, 1989; Nriagu, 1998; Aslihan Yener, 2000). Covariation in the geochemical suites of harbour sediments can therefore be linked to changes in metallurgical savoir faire (e.g. copper during the Chalcolithic, copper, tin and lead during the Bronze Age etc.); (3) although small amounts of metals are found in relatively pure form, most must be extracted from more complex ores by removing the 'impurities' (non-metal or other metal) from the combination ore. For example, natural Pb is very rare in the earth's crust, therefore when it is found at an archaeological site its presence can invariably be attributed to a mineral fusion. This renders lead and other trace metals particularly powerful tools in reconstructing anthropogenic contaminations (Shotyk *et al.*, 1998, 2005; Martínez-Cortizas *et al.*, 1999); (4) the majority of trace metals are well-preserved in the fossil record, with relatively low levels of degradation over the Holocene timescale; (5) finally, the geochemical composition of some metals means that they can be provenanced. Lead, for example, is a chemical element that possesses three stable radiogenic isotopes 206Pb, 207Pb and 208Pb. These three isotopes vary as a function of the geological source area which means that lead samples can invariably be linked back to a well-constrained mining area (Goiran, 2001; Véron et al., 2006).

In light of these advantages, the main objectives of Le Roux *et al.*'s (2003a-b, 2005) geochemical analyses at Sidon were to: (1) geochemically quantify levels of pollution during antiquity; (2) attempt to correlate these results with the geoarchaeological data and wider changes in the city's industrial base; and (3) better understand the history of metallurgy at the site (Tylecote, 1992). Details of the methods employed can be reviewed in Le Roux *et al.* (2002) and Véron *et al.* (2006).

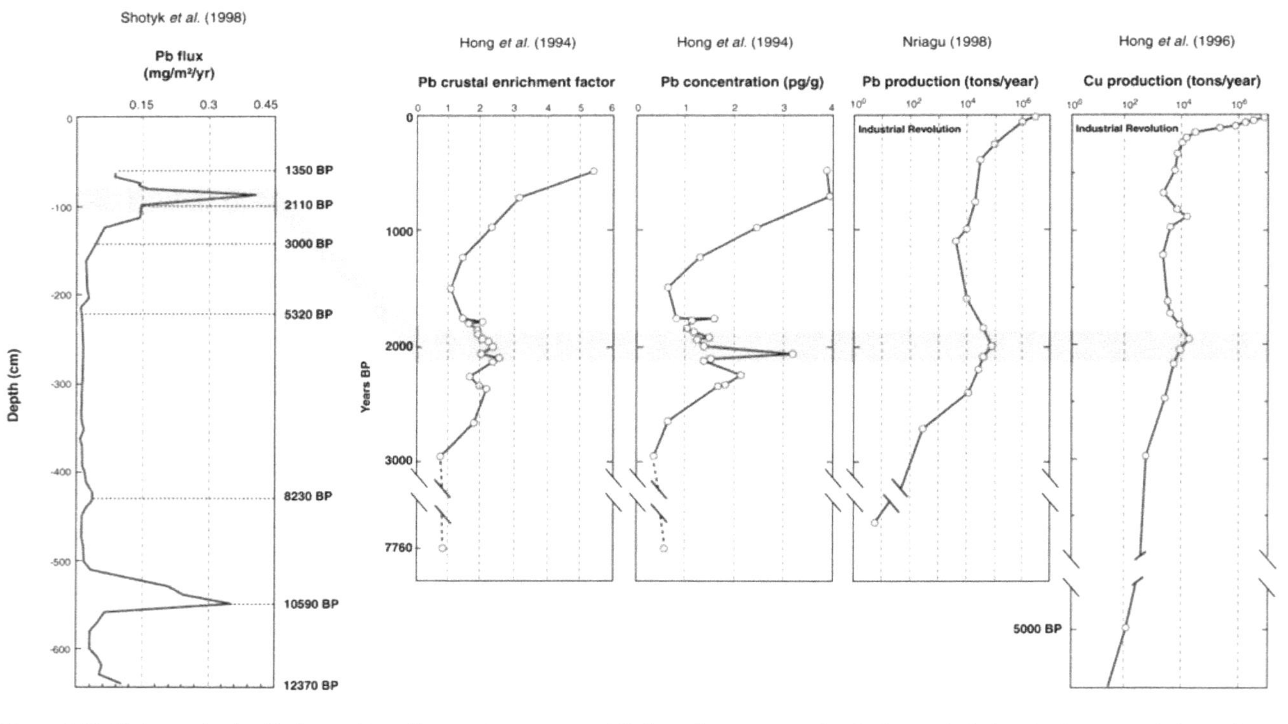

Figure 6.39: Holocene lead pollution archives from various sources. All show the pronounced pollution peak during the Roman period, consistent with findings from Alexandria and Sidon.

6.3.2 What can geochemistry tell us about the occupation history of Sidon?

Le Roux *et al.*'s (2002, 2003a; **Figure 6.40**) studies have concentrated upon the use of lead, and its three radiogenic isotopes 206Pb, 207Pb and 208Pb, as a proxy for pollution and ancient metallurgic activities.

The advantage of using lead to reconstruct ancient anthropogenic impacts is fourfold: (1) first, the element lead is one of the most extensively produced and emitted materials known (Nriagu, 1983, 1998). Its abundance and ease of casting mean that lead use was common in all the ancient civilizations as early as 8,000 years ago. In the Levant, lead beads dating from the seventh millennium BC have been found at Catal Höyük; (2) secondly, it was readily alloyed with other metals from the Bronze Age onwards; (3) third, lead is usually a by-product of the processing of other ores, a universality which renders it ideal as a proxy tracer in long time-series; and (4) finally, the composition of its three radiogenic isotopes 206Pb, 207Pb and 208Pb, means it is possible to differentiate between natural and anthropogenic lead signals. 206Pb, 207Pb and 208Pb are radiogenic and derive respectively from the radioactive decay of 238U, 235U and 232Th (Dickin, 1995). During formation of the ore, the lead is separated from the uranium and thorium. The isotopic composition of the ore is therefore 'fossilised' while the geochemical signature of the surrounding crust continues to evolve. In this way, the 206Pb/207Pb ratio of non-contaminated sediments is generally superior to 1.20 (Ferrand *et al.*, 1999; Bränvall *et al.*, 2001), while anthropogenic lead will generally have a radiogenic ratio inferior to 1.20. On the basis of lead analyses undertaken upon core BH IX, Le Roux *et al.* (2003) identify four phases in Sidon's geochemical data.

(1) Natural lead signal: The base of the core, dated to ~ 2600 BC, shows lead concentrations inferior to 20 ppm. Isotopic ratios of 1.21 are coeval with those of the earth's crust. Le Roux *et al.* (2003) conclude that this lead derives from the natural sediment matrix. There is no anthropogenic signal consistent with the occupation of Sidon during the third millennium BC. This is contradictory with the archaeological evidence unearthed at the site since 1998, which attests to an occupation of the tell from ~3000 BC (Doumet-Serhal, 2003, 2004c). It is possible that the area surrounding the harbour did not have an extensive industrial base at this time, accounting for the absence of lead contaminants in the coastal sediments. Alternatively, it is plausible that the metallurgic centres were located on the periphery of Sidon, around the southern bay for example.

(2) Middle Bronze Age pollution: A shift in crustal 206Pb/207Pb isotopic ratios from 1.21 to 1.19 is consistent with the expansion of metallurgic activities during the Bronze Age. At this time, copper was commonly cast with tin, arsenic, silver and lead. Pb is also commonly a by-product of the processing of other ores. Bronze was in use in Sumer, at Ur, in around 2800 BC, (Crawford, 2004) and in Anatolia shortly afterwards (Aslihan Yener, 2000). From these centres its use spread spasmodically. It appears in the Indus valley in about 2500 BC (Kenoyer, 1998; Possehl, 2002), and progresses westwards through Europe from about 2000 BC.

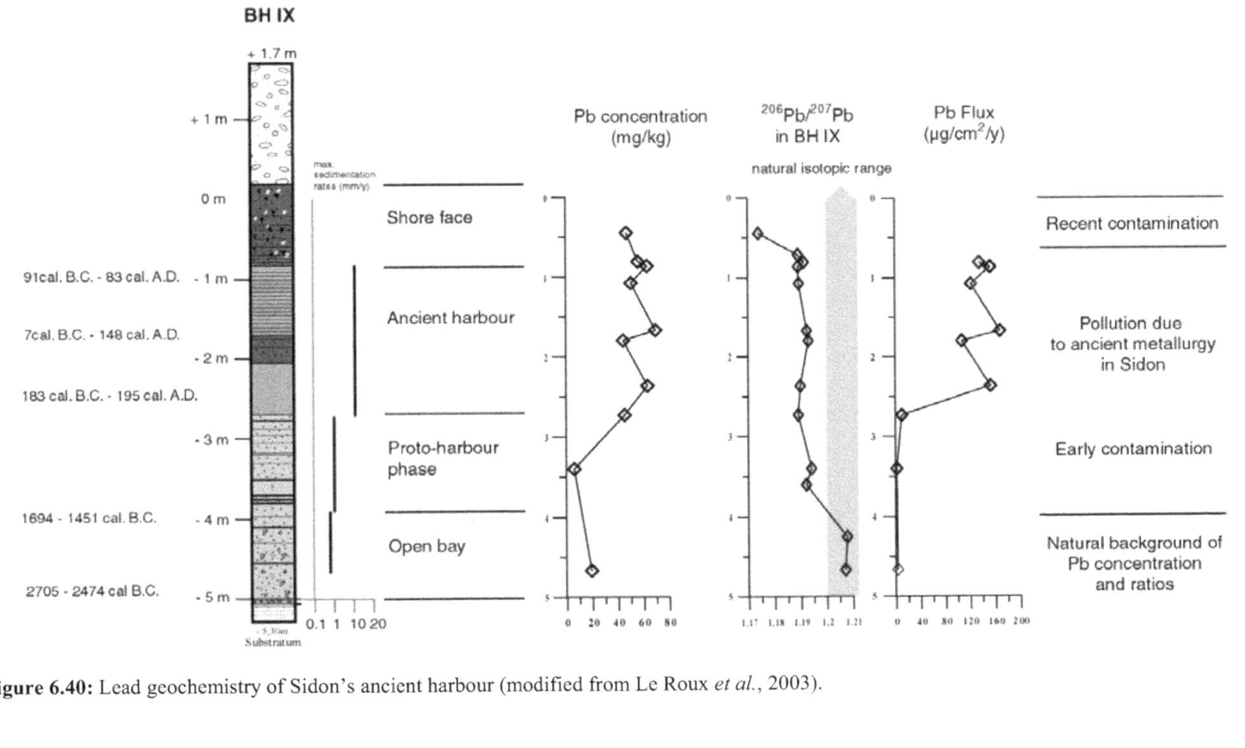

Figure 6.40: Lead geochemistry of Sidon's ancient harbour (modified from Le Roux *et al.*, 2003).

At Sidon, the earliest bronze artefacts have been dated to the early second millennium BC (MB I/II, Doumet-Serhal, 2004d). Low lead concentrations of around 5 ppm are explained by either: (i) the moderate impact of Bronze Age metallurgic activities in harbour pollution; and/or (ii) the relatively high energy nature of the harbour at this time, less conducive to the trapping and archiving of a pollution signal. Le Roux *et al.* (2003a-b) have coupled a study of the harbour pollution with an analysis of Middle Bronze Age weapons found in graves at Sidon (**Figure 6.41**). Their work has revealed the use of arsenic-copper and tin-copper alloys. Using isotope signatures, the copper metals have been provenanced to mines in Turkey, Cyprus and Syria, attesting to Sidon's role in a trade network encompassing much of the eastern Mediterranean (Doumet-Serhal, 2003, 2004a).

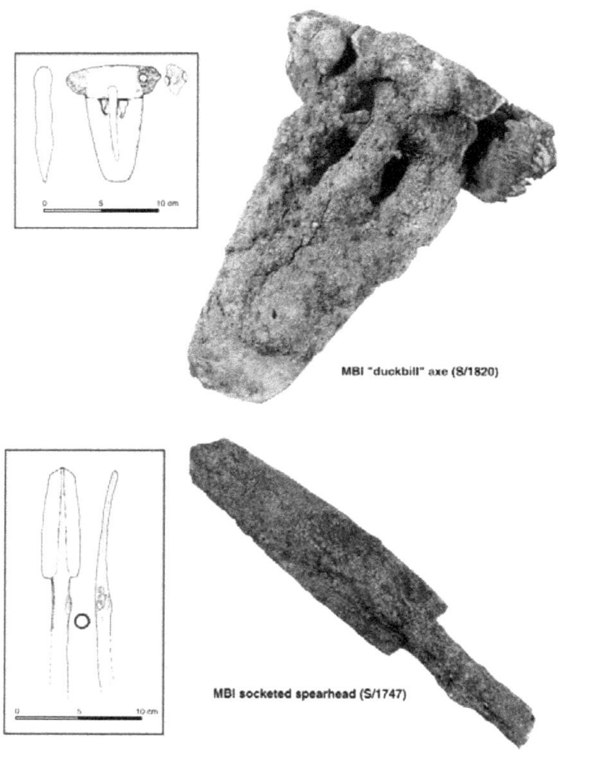

Figure 6.41: Examples of Middle Bronze I (2000-1750 BC) weapons found in graves at Sidon (from Doumet-Serhal, 2004d).

(3) Roman and Byzantine pollution: Harbour dredging means that unfortunately there is a paucity of lead pollution data for the Iron Age. However, constant anthropogenic 206Pb/207Pb ratios of ~1.19 are recorded during the Roman and Byzantine periods. These ratios are accompanied by a sharp rise in lead concentrations to between 40 and 70 ppm. A pronounced rise in pollution levels is consistent with three complementary dynamics (a) an evolution in metallurgic savoir faire, notably the increased use of metal alloys (Craddock, 1976, 1977, 1978); (b) a much larger industrial base compared to the preceding Bronze Age period; and (c) the confined lagoon-like environment created by advanced harbourworks, particularly conducive to the archiving of the city's pollutants. Variability in the lead concentrations, although not pronounced, is explained by fluctuations in harbour energy conditions. For example, higher energy dynamics (coarser grained sediment) are coeval with the 'flushing' of the harbour environment and its lead pollution. The Roman peak in lead pollution is consistent with data from Tyre and Alexandria (Véron *et al.*, 2006). Increased lead and trace metal emissions during ancient Greek and Roman times have been recorded and identified in many long-term archives such as lake sediments in Sweden, ice cores in Greenland, and peat bogs in Spain, Switzerland, and the United Kingdom (Hong *et al.*, 1994; Renberg *et al.*, 1994; Shotyk *et al.*, 1998, 2005; Martínez-Cortizas *et al.*, 1999). Lead is highly malleable, ductile and noncorrosive making it an excellent piping material as demonstrates its Latin name plumbum. Cheap and abundant, the Romans used the metal extensively in building the first sewage and aqueduct systems. At this time it was also used in the sheathing of boats (Casson, 1978b-c; Dell'Orco, 1979). In addition to conduits, lead was also widely used as a container and even in wine and food production (Skovenborg, 1995).

(4) Post-Byzantine demise: Unlike the litho- and biostratigraphical data, the relative demise of Sidon during the post-Byzantine period is not clearly attested to by the geochemical data (**Figure 6.40**). Despite a fall in lead concentrations to 40 ppm it is unclear if this is due to a more exposed harbour or a real fall in pollution levels. 206Pb/207Pb ratios indicate that this Pb remains anthropogenic in origin (Craddock, 1979).

6.3.3 Comparison with Alexandria

The geochemical data from Sidon indicate that the Roman and Byzantine periods were characterised by high levels of coastal pollution. But just how polluted was this Phoenician harbour compared to the Mediterranean's largest coastal metropolis? In an attempt to answer this question we compare and contrast Le Roux *et al.*'s (2003) Lebanese data with geochemical work undertaken at Alexandria (Véron *et al.*, 2006, **Figure 6.42**). Two cores, little or unaffected by Roman dredging, have yielded a continuous lead pollution time series spanning 4000 BC to 1000 AD. This research is particularly interesting as it concurs two pre-Hellenistic pollution peaks. Three pollution phases have been identified:

(1) Around 2500 BC, 206Pb/207Pb ratios around 1.185 and 1.195 suggest the onset of anthropogenic activities during the IVth dynasty. This period, which corresponds to the Old Kingdom, is characterised by a rise in the number of settlements on the Nile delta and an increase in the size of the population to ~1.5 million people (Butzer, 1976; Shaw, 2000). It was also a time of great wealth consolidation and monumental architecture (Djoser's pyramid at Saqqara, Cheops at

Giza etc.). The apparition of bronze in Egypt is slightly posterior to 2500 BC, during the second millennium BC (Leclant, 2005). However, gold, silver, copper and lead were widely used in Egypt in the production of weapons and jewellery from 3500 BC onwards. Whether or not this signal is of local or regional provenance is equivocal. The temptation is to link this geochemical signal to metallurgic activities undertaken by local societies from 2500 BC. It would be paradoxical to think that such a locally unique and endowed site did not attract Bronze Age societies, especially in light of the Levant's trade internationalisation at this time. Significantly, however, no archaeological evidence pertaining to a pre-Hellenistic settlement has yet been unearthed at Alexandria (Baines, 2003). Goiran (2001) favours a local origin for two reasons: (a) there are no lead mines in the lower valley of the Nile that could explain the signal; and (b) studies on the modern distribution of lead in estuaries and deltas indicate that 95 % of the pollutant is deposited at the mouth of rivers (Ferrand *et al.*, 1999). (2) During the Iron Age, a second contamination peak is evidenced by EF of 6-7 and an anthropogenic 206Pb/207Pb ratio of 1.192. This geochemical evidence is consistent with the prosperous Rameses reigns of the XIXth and XXth dynasties and may attest to occupation of the site before the arrival of Alexander the Great in 331 BC. This pollution peak is followed by the absence of a clear lead signature between 800 BC and the establishment of the Hellenistic city, the result of a decline of the Egyptian empire during the Nubian, Assyrian and Persian invasions (Butzer, 1976).

(3) Following the foundation of the Hellenistic city, pollution levels of 200-300 ppm are recorded, rising rapidly to 600 ppm. Greek and Roman lead pollutant concentrations are significantly higher than those measured in Alexandria's modern harbour (90 ± 20 ppm). These levels are also approximately ten times higher than levels recorded at Sidon during the same period, consistent with the city's much larger industrial base. A shift in 206Pb/207Pb ratios is elucidated from crustal derived values of 1.2 prior to the fourth century BC to anthropogenic signatures during the Greek (1.187-1.199) and Roman (1.177-1.185) periods.

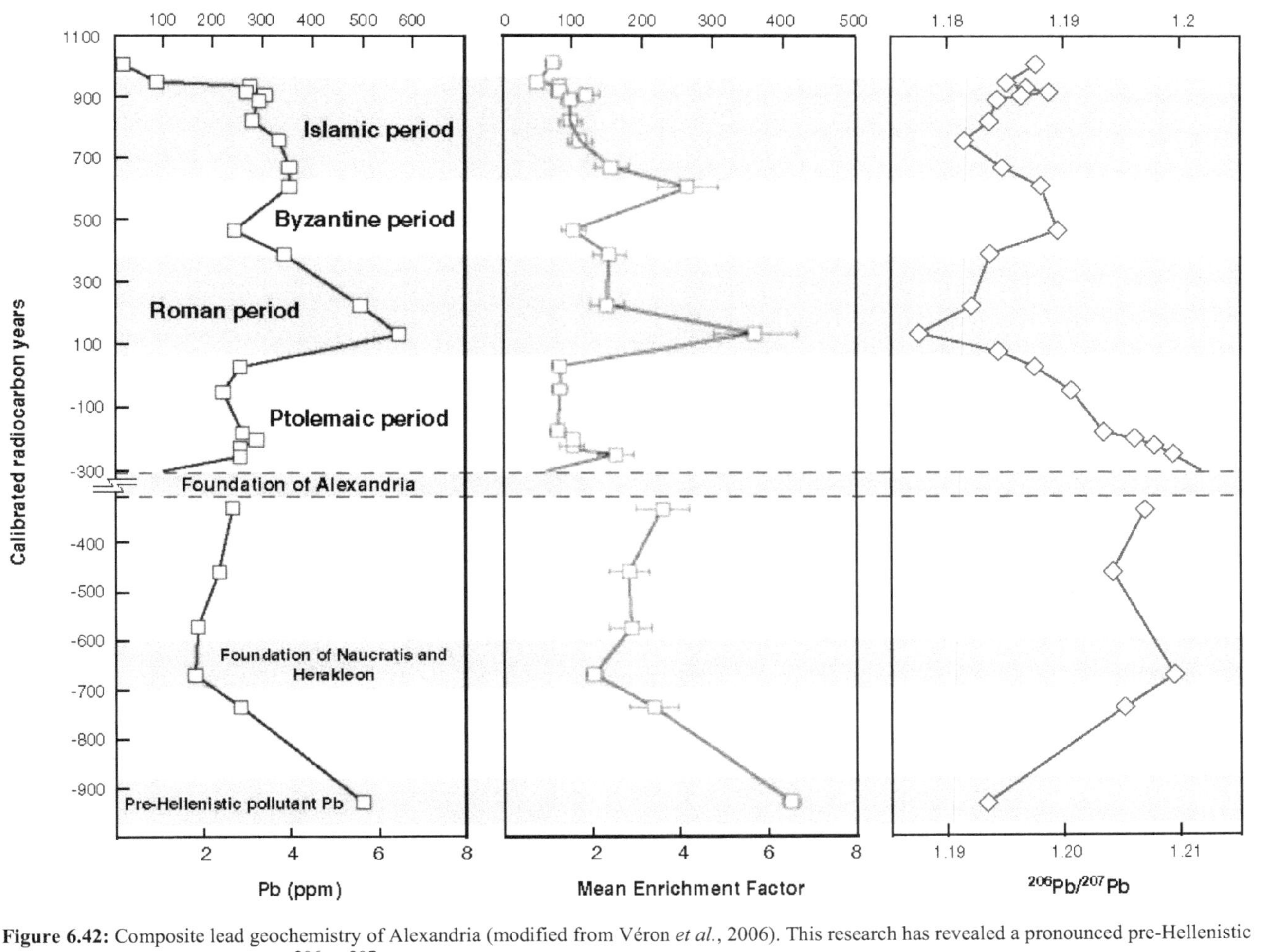

Figure 6.42: Composite lead geochemistry of Alexandria (modified from Véron *et al.*, 2006). This research has revealed a pronounced pre-Hellenistic lead peak of anthropogenic origin ($^{206}Pb/^{207}Pb = \sim 1.19$).

Unlike Sidon, high lead concentration levels of ~300 ppm lend credence to the continued prosperity of Alexandria during the Islamic period. Alexandria was one of three intellectual and technological centres, Carthage and Antioch being the two others, during the Islamic Golden Age (750-1400 AD; Lombard, 2003; Al-Djazairi, 2006).

6.3.4 Concluding remarks

At coastal archaeological sites, trace metal geochemistry of harbour sediments has the potential to significantly advance our understanding of occupation histories and more specifically the evolution of metallurgic activities. While trace metals have long been used in archaeology at terrestrial sites, the high energy nature of coastal archives had long hindered their scope at seaboard locations. Today, much of the debate is centered on differentiating between local autochonous and regional allochthonous signals (Véron, 2004). At present, the evidence suggests that trace metal pollutants are rapidly fixed by marine organisms and/or particles. Attempting to correlate the terrestrial signal of metal artefacts found at the site with the signal obtained in the harbour archives has the potential to yield promising results in this domain. Manifestly, however, more work is required in understanding how pollutants are transported in the source to sink conveyor. Future research must concentrate on investigating long time series in delta and estuarine environments.

With reference to the harbour of Sidon, much weaker pollution signals are recorded during the Bronze Age. This is for three reasons: (1) the smaller nature of the industrial base compared to later Iron Age, Roman and Byzantine periods; (2) a more primitive metallurgic savoir faire; and (3) the dilution and dispersal of a pollution signal due to higher energy coastal processes operating within the harbour at this time. Unfortunately, while Le Roux *et al.*'s study at Sidon underlines the significant scope of trace metal geochemistry to understand human-environment interactions, dredging activities during the Roman period have removed significant tracts of the Iron Age record. Nonetheless significant parallels exist between these coastal records and high-resolution peat bog and ice core pollution archives (**Figure 6.39**; Renberg *et al.*, 1994; Hong *et al.*, 1994, 1996; Shotyk *et al.*, 1998; Martínez-Cortizas *et al.*, 1999; Bränvall *et al.*, 2001).

6.4 Preserving Lebanon's coastal heritage

The sea has played a fundamental role in the history of the Phoenician coast. The Levantine seaboard served as an important communication interface between the ancient Near East and the Mediterranean and attests to a great antiquity of maritime activities (Février, 1935, 1949-1950; Haldane, 1993; Casson, 1994; Katzenstein, 1997; Aubet, 2001). Within this context, Israel and the Lebanon are key regions in understanding the development and evolution of seafaring infrastructure and shipping, providing one of the richest and most continuous maritime archaeological records in the world (Ilan, 1995; Raban, 1997b; Wachsmann, 1998; Marcus, 2002a-b; Pomey and Rieth, 2005). Our research study has elucidated over 5000 years of human-environment interactions, underlining the archaeological importance of Beirut, Sidon and Tyre in comprehending Bronze Age to Islamic period changes in technology, infrastructure and landscapes. The study has facilitated the most sensitive archaeological zones to be identified, with a broad quantification of the archaeological resources available at each site. The idea of establishing a framework for the protection and management of Phoenicia's unparalleled coastal archaeology is one that must be addressed with some degree of urgency. Local (city councils), national (the Department of Antiquities of Lebanon) and international (Association Internationale pour la Sauvegarde de Tyr [AIST], Lebanese British Friends of the National Museum [LBFNM] UNESCO World Heritage Commission) agencies must work together on three areas: (1) to protect Beirut, Sidon and Tyre's exceptional archaeological heritage (see Doumet-Serhal, 2004a); (2) to foster enhanced understanding of the importance of the Lebanon's cultural heritage; and (3) to facilitate participation in archaeology at the local, national and international levels. This includes projects to encourage local participation in archaeological projects, and the development of cultural tourism as a source of durable development (Frost, 1990; Franco, 1996).

In the context of a post-war economy, the Lebanon is presently undergoing rapid economic development exemplified by the urban regeneration and expansion of many of the coast's largest cities. This development needs to be undertaken within the context of a clearly defined cultural management program, and bygone mistakes made in Beirut's city centre should imperatively be avoided (Lauffray, 1995; Karam, 1996; Naccache, 1996, 1998; Seeden, 1999; Raschka, 2006).

The archaeological resources at Beirut, Sidon and Tyre can be divided into two main categories: buried onshore remains (silted up harbours, buried shorelines, buried maritime infrastructure and wrecks) and offshore archaeological contexts (wrecks and drowned land surfaces). Each location has a distinct range of research, management and conservation issues connected to it (Langley and Unger, 1984; Firth and Ferrari, 1992; Woodall, 1993).

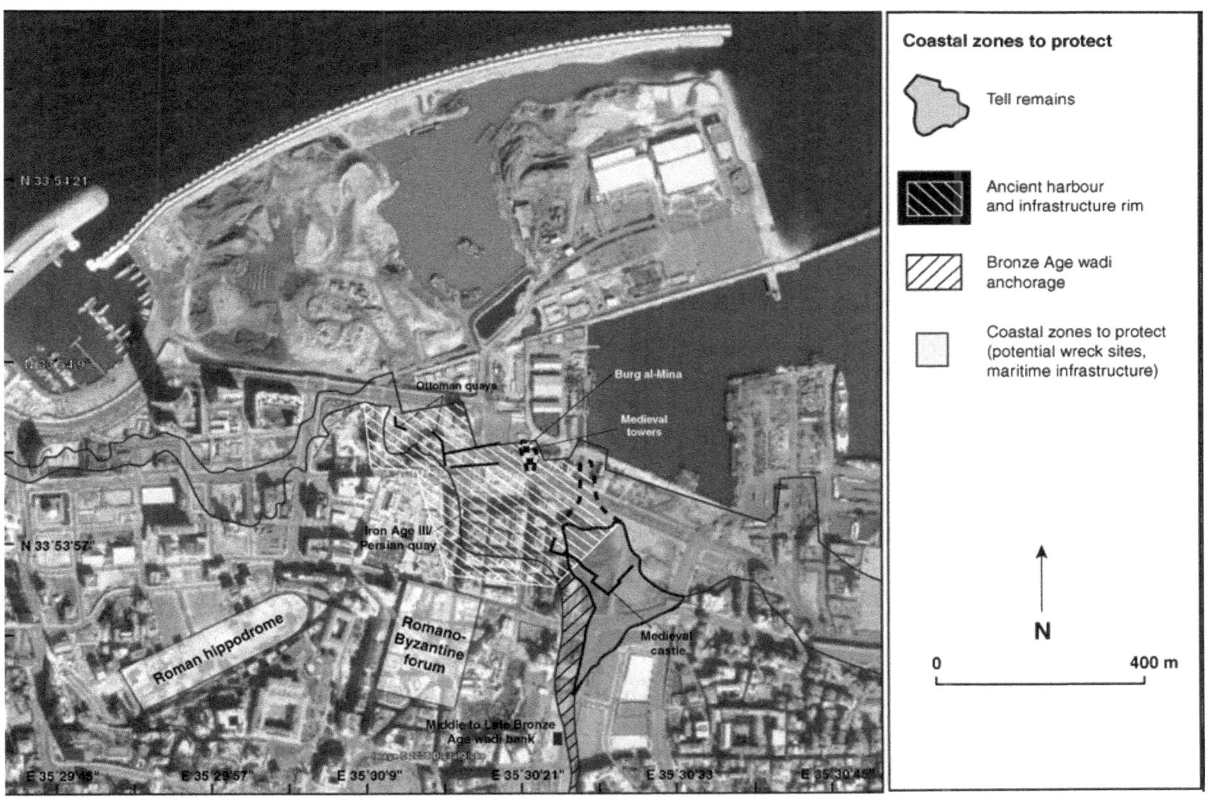

Figure 6.43: Beirut's ancient coastlines and areas of archaeology to imperatively protect (base image: DigitalGlobe, 2006). Archaeological data from Elayi and Sayegh (2000) and Marquis (2004).

6.4.1 Maritime infrastructure

The areas surrounding the present seaports are rich in buried archaeological monuments and relics, bearing witness to a complex history of human occupation spanning 5000 years of history. Using a multi-disciplinary geoarchaeological approach, this study has enhanced understanding of the development of dock and harbour installations. We have identified a number of important seaboard areas (**Figures 6.43-6.45**), significant in understanding the evolution of maritime infrastructure and the expansion of shipping activities along the Levantine seaboard from the Bronze Age onwards. The heart of Beirut, Sidon and Tyre's main anchorage havens today lie landlocked in the city centres, buried beneath tracts of coastal sediment. These sediments are not only rich historical archives key to understanding how ancient societies modified the coastal environment, but the fine-grained silts and sands also offer high preservation potential for the survival of Bronze Age and Iron Age harbourworks. At present, more research is required on installations relating to the construction and provisioning of ships in antiquity, and their evolution through time. Much of the present research is centred on understanding the impact of Bronze Age harbourworks and evolution from natural to artificial seaport infrastructure during the Bronze and Iron Ages. Keys to understanding this include the identification of quays, moles, shipyards, entrepot facilities and so forth. The silted up harbours of all three cities offer exceptional research potential to better comprehend these relationships. While we have clearly linked certain stratigraphic facies to technological changes, there is a relative dearth of data pertaining to the precise nature of these harbourworks at all three sites. At present, only Beirut's silted harbour has yielded Iron Age harbour quays (Elayi and Sayegh, 2000). Offshore at Tyre, Noureddine and Helou (2005) and Descamps (personal communication) have surveyed the northern Roman harbour mole, detailing its archaeological potentiality. Future research should focus on clearly identifying the nature of these archaeological resources.

Figure 6.44: Areas showing rich potential for coastal archaeology at Sidon (base image: DigitalGlobe, 2006).

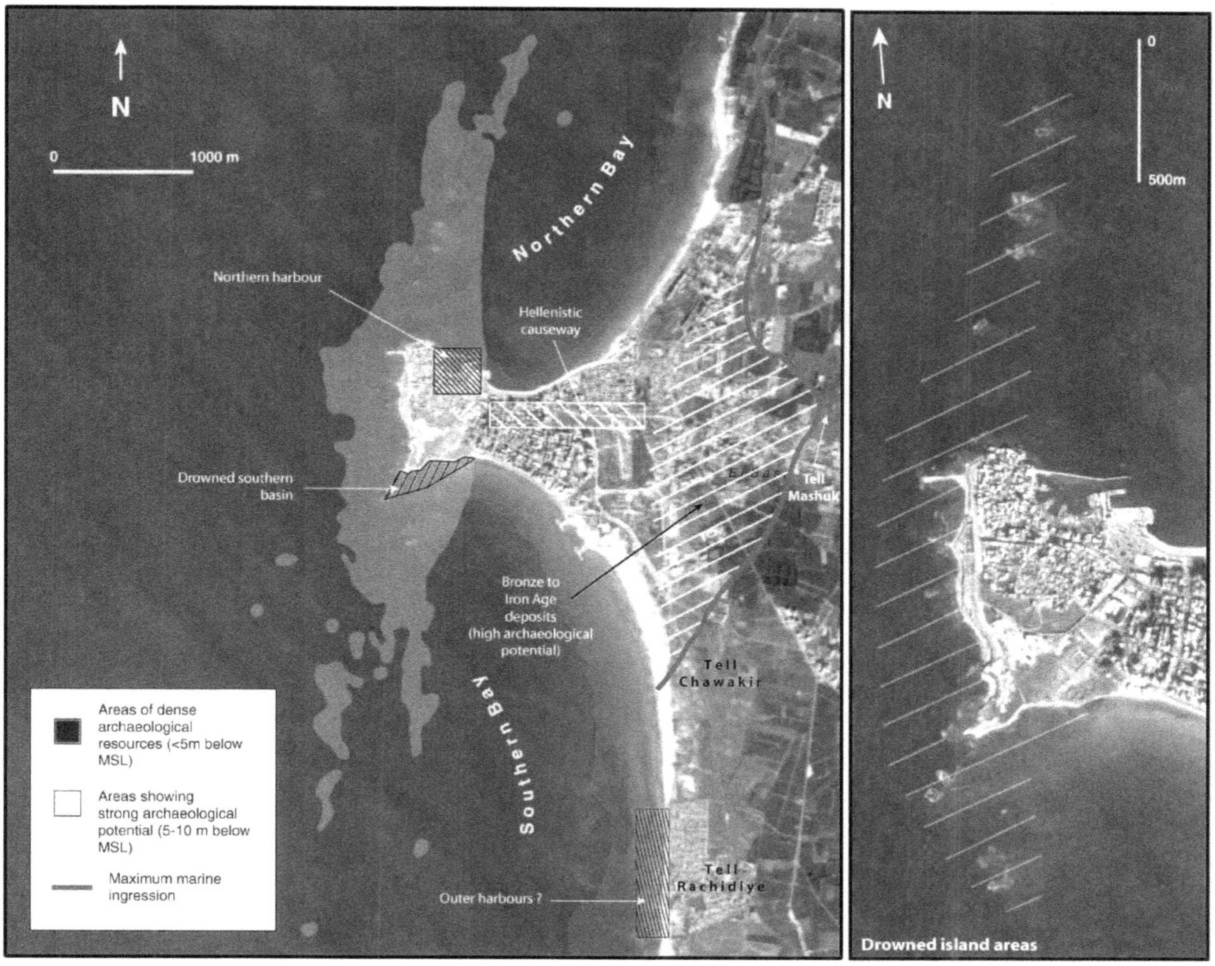

Figure 6.45: Areas showing rich potential for coastal archaeology at Tyre (base image: DigitalGlobe, 2006).

Recommendations:

We propose the heart of all three ancient harbours be protected by national legislation; our work has allowed us to precisely delimit the harbour areas for future protection. One of the chief concerns is to marry urban development with the archaeology, so that the rich historical heritages of Beirut, Sidon and Tyre are integrated into the reconstruction process. New construction permits should not be granted in these sensitive zones until the archaeological scope of the plot/site has been fully investigated. In the event of significant artefacts being unearthed, the zone should be correctly surveyed and recorded. Plots of major scientific importance must imperatively be protected from destruction either by incorporating the archaeology into the proposed construction (feasible solution) or preventing building of the plot. As a counter example, the recent modernisation of Sidon's ancient harbour completely ignored the impressive archaeological remains exposed at the site (**Figure 6.46**). These were amongst the best preserved harbour moles on the Levantine seaboard. Although not completely destroyed, they have been concreted over while a more sustainable solution would have been to expose and accommodate these ancient structures in the new edifice. On a much larger scale, similar mistakes were made in Beirut during the 1990s (Lauffray, 1995; Naccache, 1996, 1998). In reality, conservationists face stiff opposition and must vie with wealthy developers and politicians to protect the archaeological remains in these areas.

6.4.2 Wrecks

As concur other maritime sites on the Levantine coast, the possibility of unearthing Bronze Age to Islamic period wrecks in the area of Beirut, Sidon and Tyre is high (Raveh and Kingsley, 1992; Royal and Kahonov, 2000). Shipwrecks are of great significance in cultural and trading studies, as they comprise sealed and well-dated artefactual assemblages (Croome, 1973; Bass, 1991; Parker, 1995). Recent archaeological work at Istanbul, Naples and Marseilles reveals the high preservation potential of harbour silts in conserving wooden hulls over many millennia. Urban planning policy permitting, there is no immediate threat to these wooden structures in city centres although care must be taken when drilling down for deep foundation works. The question of offshore wrecks is more delicate, as these can be degraded by both erosion and clandestine divers. Even when wreck sites have been identified, it is very difficult to protect and police these areas. Archaeologists must therefore focus upon extracting the maximum amount of data in relatively short time periods (see Galili *et al.*, 1993a).

Figure 6.46: Since 1998, Sidon's northern harbour has been extensively modernised, destroying great tracts of the ancient harbourworks (Marriner and Morhange, 2005). Better urban planning could have saved this unique cultural resource (base image: DigitalGlobe, 2006).

Recommendations:
We recommend government agencies liaise with local fishermen to establish a database of wreck sites along the coasts of Beirut, Sidon and Tyre. There is a clear need for complete appraisal of all known wrecks in the region, and where necessary new protection must be put into place. The silted up ancient harbours offer exceptional opportunities to preserve ancient wrecks and excavate these using classic dig techniques. It is essential that any construction work in these areas be evaluated for the survival of such wrecks. Offshore, attempts should be made to protect known wreck sites and extract the maximum amount of historical and archaeological data. We propose that marine bottoms down to a depth of 5 m be imperatively protected at all three sites (Morhange and Saghieh-Beydoun, 2005; Marriner and Morhange, 2005b). In reality, underwater surveying at offshore sites is very costly and yields much poorer results than similar studies undertaken on terra ferma (see the impressive results obtained at Marseilles and Naples [Pomey, 1995; Hesnard, 2004a-b; Giampaola *et al.*, 2004; Giampaola and Carsana, 2005]).

6.4.3 Submerged landscapes

Our research at Tyre has shown the significant potential of Bronze Age to late Roman land surfaces surviving offshore (**Figure 6.45**). These drowned landscapes have tremendous scope for the preservation of archaeological evidence and can potentially yield important insights into the nature, scale, and pace of coastal change. Tectonic collapse has submerged significant tracts of the ancient island bastion, and we estimate that approximately 50 % (or 470000 m2) of the former island is presently drowned ~3 m below mean sea level. While we have good constraints on the spatial dimensions of the island during antiquity, very little is known about the extent and condition of the archaeological remains. El Amouri *et al.* (2005) have revealed that Poidebard's proposed Egyptian harbour is in reality a drowned urban quarter of the ancient city. This southern basin therefore offers exceptional research and heritage potential. The shallow nature of its archaeology (around 3 m) means the area could be transformed into an underwater archaeology museum. Guided diving tours and the use of glass bottom boats are viable solutions to expose the underwater archaeology to the general public. Not only does this have direct cumulative ramifications for the local economy, but it is also an effective means of policing the area and generating funds for future research.

Recent underwater surveys by Descamps *et al.* suggest that significant Hellenistic, Roman and Byzantine remains also lie to the north of the Sidonian harbour, on the sandstone ridge's presently drowned northern extension (Descamps, personal communication). This ridge runs for some 1400 m from the northern tip of the island. During antiquity, exploitation of the most distal reefs as outer harbours offers scope for the understanding of maritime infrastructure. Unlike archaeology on the present peninsula, these remains have been relatively unaffected by medieval and modern constructions. At present, there are two big threats to the long-term survival of these underwater resources: (1) looting by clandestine divers and treasure hunters: the rich archaeological resources around Tyre have long been known to local divers. It is impossible to estimate the amount of archaeological material that has already been lost or removed from its primary context due to this activity and the sale of artefacts on the black market; (2) erosion: coastal erosion not only undermines the preservation of archaeological resources, but storm and high swell can also expose tracts of archaeology to divers.

Recommendations:
There is a clear need for government and international institutions, in collaboration with local and international academics, to take responsibility for the recording and protection of Tyre's drowned archaeological remains. Similar resources have also been unearthed at Sidon, in the vicinity of Zire island (Frost, 1973; Carayon, 2003; Carayon and Viret, 2004). Greater understanding of the taphonomy of underwater deposits is also required, improving our comprehension of the survival of artefacts and the impacts of erosion and clandestine divers. We have drawn a series of 'sensitivity maps' for all three sites, identifying the cities' areas of high archaeological potential (**Figures 6.43-6.45**). We propose that coastal areas down to a

depth of 5 m be protected at Beirut, Sidon and Tyre. These areas must imperatively be better surveyed for their archaeological resources and correctly policed to avoid the further loss of archaeological resources. The immediate task is to set in place measures that in the long and short term will actively discourage the destruction of underwater cultural heritage, starting at a grassroots local level (in schools, public awareness programs etc.). This does not entail preventing public access to the underwater resources, but rather more correctly policing the most sensitive areas and regularly surveying them to record new areas of archaeology exposed by erosion. Education and public participation must look to change public attitudes and encourage community participation. Although this is already done to a certain extent, national and international agencies must liaise more closely with local fishermen to record any archaeological material they have recovered. A reward system might also encourage new sites to be revealed to the authorities.

6.4.5 Phoenicia's changing coastline

Our coastal reconstructions indicate that Beirut, Sidon and Tyre's coastlines have changed significantly during the past 6000 years. The core networks have detailed the historic pattern of coastal change within all three harbours. At Tyre, we have reconstructed the greater subaerial extension of the island breakwater, its evolution through time and the phases of accretion and progradation of Alexander the Great's isthmus. The presence of this breakwater means that the most pronounced coastal changes have taken place at Tyre. At Beirut and Sidon, research has identified the position of historic coastlines and their evolution since the maximum marine ingression. A broad tendency of silting up and coastline regularisation is observed at all three sites. Within the three cities' silted up harbours, this explains the distribution of archaeological resources. A general seaward shift in the archaeology from the Bronze Age onwards is due to loss of accommodation space.

At Tyre, we have elucidated the probable existence of extensive areas of Bronze Age and Iron Age maritime archaeology around Tell Mashuk, Tell Chawakir and Tell Rachidiye. Electrical resistivity would be a relatively rapid and cost-effective means of better constraining the nature of archaeological resources in this area. Fine-grained deposits to the west of Tell Mashuk offer high preservation potential. Between Tell Chawakir and Tell Rachidiye, a complex system of sand dunes is observed; these aeolian and coastal sediment tracts contain important amounts of archaeology. Sand dunes are complex geomorphological entities prone to instability and sudden large-scale shifts, with implications for recognising, dating and conserving archaeological resources around these tell sites. There is a need for further research on the geomorphology of the sand dunes to better comprehend the distribution of archaeological resources. As demonstrate the exhumed Roman and Byzantine remains on Tyre's tombolo, the archaeological scope of the isthmus is also significant.

Recommendations:
We have elucidated a history of long-term coastal change at all three sites that has greatly advanced understanding of the cities' archaeological resources. Future archaeological work could use this evidence to pinpoint and protect the areas of greatest research potential (Morhange and Saghieh-Beydoun, 2005). As in the silted up harbours, future construction work might attempt to accommodate archaeological findings so as to ensure their long-term protection.

6.4.5 Recommendations for the exhibition of Phoenicia's archaeological heritage

6.4.5.1 Land based

On land, the cultural resources of Beirut, Sidon and Tyre are much easier to manage and present to the public than offshore. We suggest the following ideas: (1) Exhibition. Using museum-based displays can be an important means of translating a message to the public. Some of the material excavated at all three sites is today exposed at the National Museum in Beirut, however neither Sidon nor Tyre possess exhibition centres at present. Examples from Europe show the scope of what can be achieved (Alpözen, 1983; Rosselini, 1984; Basch, 1985; Croome, 1987; Gibbins, 1990; Rodriguez, 1990; Crumlin-Pedersen, 1993). Sites as important as Sidon and Tyre should have their own museums, displaying the most important finds. This is clearly a plan for the future. (2) Data dissemination. Layman books and internet sites can be an effective means of arousing public interest at an international level. At the local level, information pamphlets, brochures and site guides are also important in diffusing site details and serve as introductions to the descriptive history. The recent discoveries at Sidon have, in this respect, been well diffused thanks namely to the efforts of C. Doumet-Serhal in collaboration with the British Museum and the Department of Antiquities. The dissemination of results is an important means of encouraging local and international stakeholders to protect the cities' archaeological heritages. Internet sites can be freely accessed by groups around the globe and serve as useful bridges to further information. Within this context, the archaeological results from Beirut and Sidon have been widely published in both local and international journals (Elayi and Sayegh, 2000; Doumet-Serhal, 2003, 2004a), and at Sidon a book for the lay public is currently in production. This area has not been correctly exploited at Tyre, and information regarding the archaeology of the ancient city and its important cultural resources is not readily accessible. (3) Land based trails. The use of signs and information boards is a practical means of touching a wide cross section of the public. For example, the use of information boards around the ancient harbours of

Beirut, Sidon and Tyre not only informs the general public of the history of the local site, but also allows people to better comprehend the archaeological scope of certain areas. In reality, very few people realise that the heart of Beirut, Sidon and Tyre's ancient harbours are today landlocked beneath the city centres.

6.4.5.2 Marine based

Allowing divers to have access to underwater archaeology can be risky and must be coordinated through education and public relations (Staniforth, 1994). Caesarea Maritima's underwater archaeological park is an example par excellence of what can be achieved. Divers are furnished with waterproof guide books detailing trails which they can follow according to their diving experience. The advantage of diving in ancient harbours is that the archaeology lies in relatively shallow waters. These sheets might provide general information on each site, guidelines on appropriate behaviour and sources for further information. Plinths and site markers can also provide additional information and provide a focal point for coordinating diving on the site (Johnston, 1993). At shallow water sites such as Tyre's southern basin (less than 3-4 m water depth), the scope for such a park is rich. For the non-diving public, glass-bottom boats can also be used to look at the underwater remains.

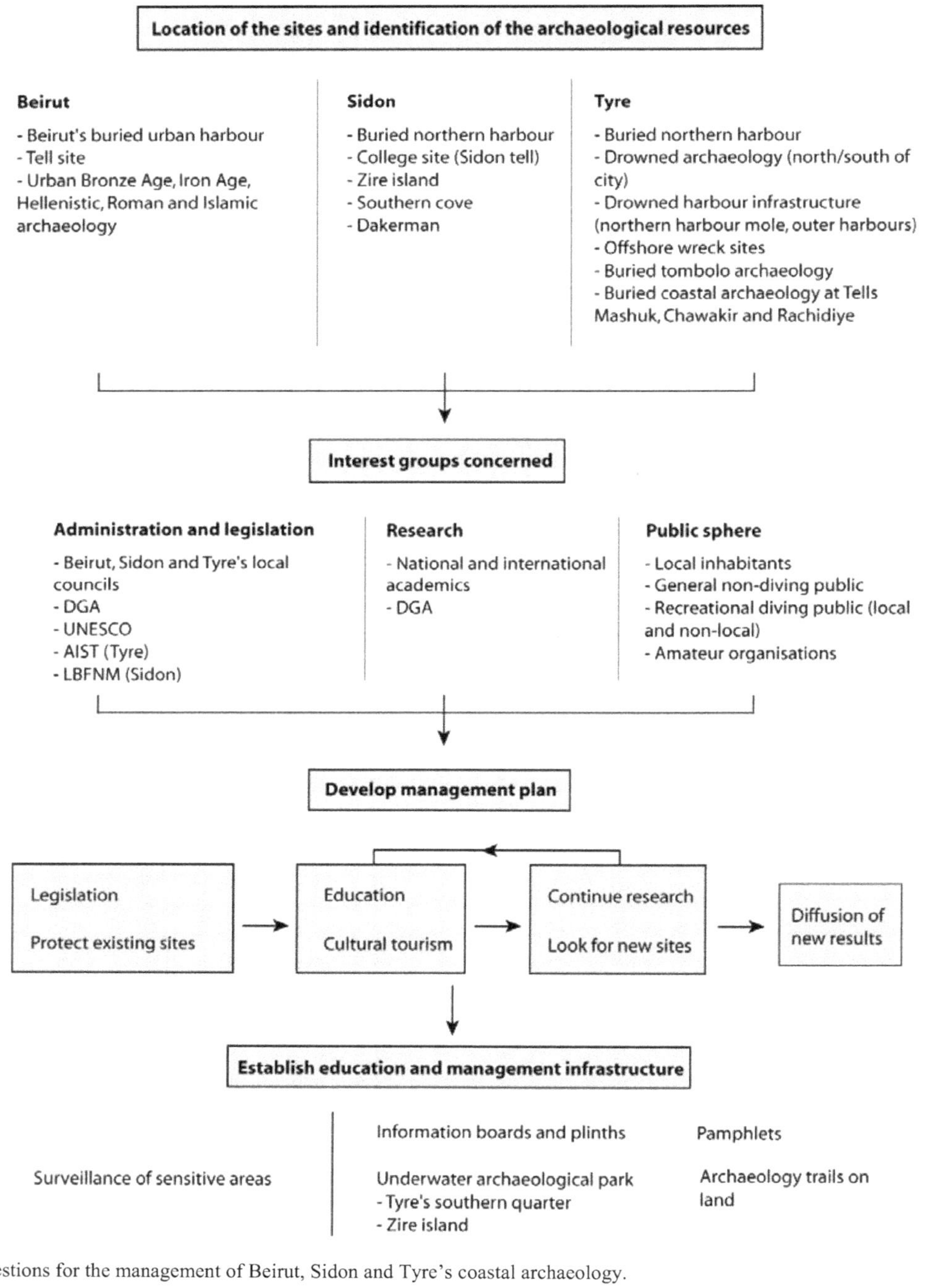

Figure 6.47: Suggestions for the management of Beirut, Sidon and Tyre's coastal archaeology.

6.4.6 Concluding remarks

Beirut, Sidon and Tyre have a rich cultural heritage both on land and at sea (**Figure 6.47**). One of the major problems facing cultural managers at these sites is how to change the attitudes of looters and the construction industry. Although government legislation does not suffice alone (see the counter example of Beirut city centre), there is a clear need for local institutions to implement some form of policy network to ensure the long-term stability of a cultural management process.

At sites such as Sidon and Tyre, the scope to develop a durable development plan based on cultural tourism is rich (Frost, 1990; Elkin and Cafferata, 2001). In Israel, cultural heritage parks have been established at sites such as Caesarea with excellent results. Such parks not only encourage public access to the cultural resource, but also generate funds for preservation of the site and future excavation programs. A management plan to suit the needs of the whole cross-section of society is required. The biggest threat to an underwater management program at Sidon and Tyre is the recreational diving public. While some individuals are extremely interested in archaeology and the preservation of the cities' underwater heritages, a majority of divers remove material out of ignorance or purposefully set out to loot sites for personal financial gain (Eaton, 1980). On the other hand, the local diving and fishing public have an enormous amount of knowledge that can potentially benefit an underwater management program. For example, they are often aware of sites unknown to the authorities. Liaising with these groups can be difficult for a number of reasons, although possible ways around it include the inauguration of a reward system. The Department of Antiquities and Lebanese government must take responsibility for the management of the three sites' cultural heritage, working closely with academics and non-governmental organisations such as UNESCO. Ultimately the preservation of the sites will depend on the effectiveness of the management system, and should be headed by scientists with a good knowledge of their sites.

Conclusions and future research avenues

Ancient harbours comprise excellent sedimentary archives, yielding insights into the magnitude and direction of anthropogenically forced coastal changes during the Holocene (Morhange, 2001; Marriner and Morhange, 2007). In addition to reconstructing the palaeoevironmental evolution of Beirut, Sidon and Tyre, this project has looked to go beyond the site scale of investigation to compare and contrast the now rich geoarchaeological data from around the circum Mediterranean and to formulate a working type stratigraphy of ancient harbours. Traditional disciplinary studies have been shown to be largely inadequate when considered in isolation and, as we have shown here, a geoarchaeological approach is particularly useful in areas of data paucity. An informed earth-science approach can aid in answering three questions imperative to better understanding the maritime archaeological record. (1) **Where?** We have demonstrated that diagnostic litho- and biostratigraphies, consistent with human-modified basins, are clearly translated in the geological record. (2) **When?** The transition from natural to anthropogenic environments can be dated using either radiometric or ceramic dating techniques. (3) **How?** How did local populations modify their coastal environment? Either (a) indirectly, through the production of increased sediment yields; and (b) directly, by modifying their natural environment to produce low energy basins.

The present study has demonstrated that harbour basins are also appropriate for the analysis of archaeological data at three scales. (1) **Basin scale:** An informed geoarchaeological approach can yield insights into the harbour basin topography, its functioning, spatial organisation, and coeval infrastructure through time. (2) **Urban scale:** Information pertaining to the site's occupation history, notably using geochemistry and geophysics, is made possible due to high rates of sedimentation through time. (3) **Regional scale:** Typological data can be derived on how these individual maritime sites evolved on a regional scale. We have also demonstrated that harbour basins are important in better understanding the source to sink sedimentary conveyor.

With reference to this latter point, there is good scope for this Phoenician model to be extrapolated to other sites on the Levantine coast, in addition to Punic harbours in the western Mediterranean. There are, for example, a number of similarities between Sidon and Tyre, and other important seaport settlements on the Levantine coast.

Acre-Sidon: Like Sidon, the ancient city of Acre lies on an easily defendable coastal promontory occupied since the Bronze Age (Briend, 2003). The city's ancient harbour lies east of a south-north trending sandstone ridge that has yielded a well-protected pocket beach since the flooding of the site around 6000 BP (Sivan *et al.*, 2001). Based on our research at Sidon, we hypothesise that Acre's basin underwent a similar palaeogeographical evolution and that the heart of the ancient port presently lies beneath the medieval centre. We hypothesise a rapid silting up of the basin after the Byzantine demise. In **Figure 7.1** a preliminary reconstruction of the harbour's ancient limits is proposed, based on aerial photography and the present urban morphology.

The old city of Acre was designated by UNESCO as a World Heritage Site in 2001. Within this context, earth-science techniques could feasibly be coupled with the present archaeological excavations to better understand the evolution of the site during the past 6000 years BP and precisely pinpoint sites of significant scientific interest.

Arwad-Tyre: Offshore island settlements and harbours are defining traits of both Arwad and Tyre. Arwad lies 2.5 km from the present coastline and has never been linked to the continent, either naturally or artificially. The small dimensions of the island and its distance from the shoreline have not been conducive to the formation of a tombolo, meaning that the eastern fringe of Arwad is important in better understanding the evolution of Tyre's easterly façade during the Bronze and Iron Ages. At Arwad, a large cove - 465 m long by 125 m wide and divided into two discrete basins - lies leeward of the dominant south-westerly swell. We posit that the heart of the ancient Bronze and Iron Age basins lies buried in the urban area fringing the west of the cove and suggest this area be cored to investigate the coastal stratigraphy (**Figure 7.1**). Like Zire island at Sidon, extensive areas of archaeology, including seawalls and mooring bits, are still visible on the island (Frost, 1966; Viret, 2005). It is imperative to attempt to correlate this archaeology with the stratigraphic record.

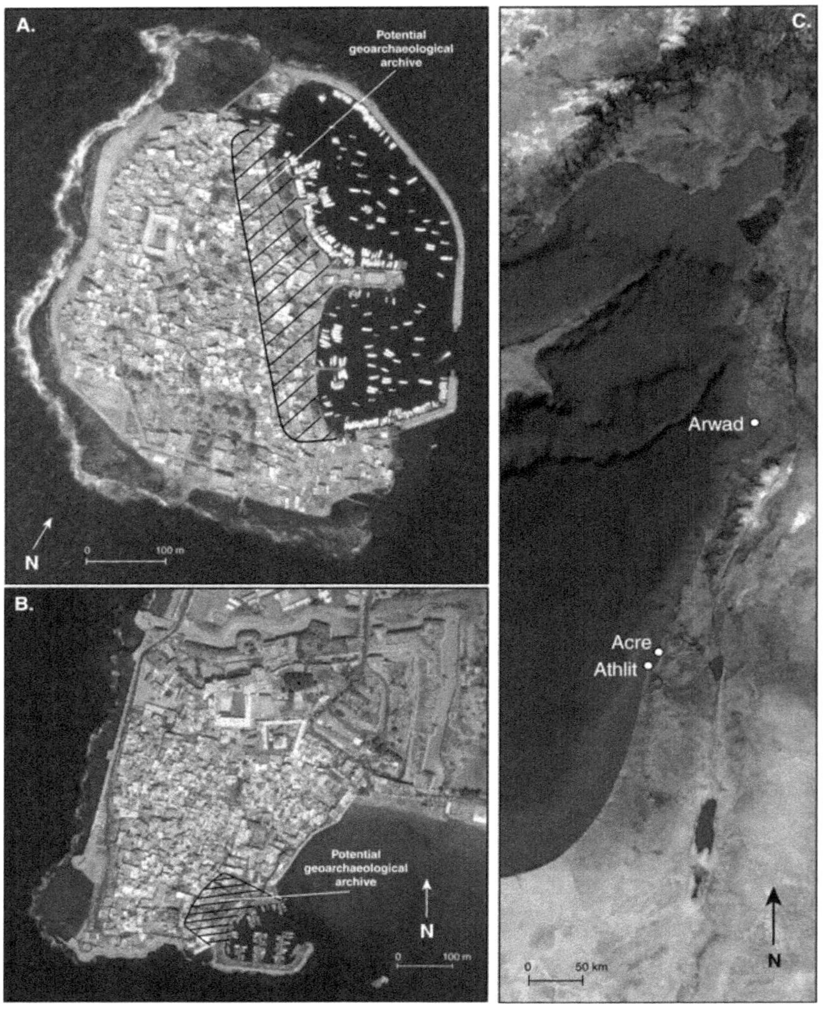

Figure 7.1: Potential geoarchaeological archives to be investigated at Arwad (A) and Acre (B) and their location on the Levantine seaboard (C).

Athlit-Tyre: One of the most promising sites in terms of better comprehending the relationship between harbour technology and coastal stratigraphy is that of Athlit. The harbour at Athlit is typically Phoenician (Lehmann-Hartleben, 1923), comprising two natural basins subsequently modified by human societies from the Iron Age onwards. Haggai has already radiocarbon dated the earliest artificial harbourworks at the site to the ninth to eighth centuries BC (Haggai, 2006). Using our findings in the Lebanon as a workable scientific framework, he has obtained a series of marine cores from inside the harbour and is presently looking to understand the evolution of the basin using coastal stratigraphy (Haggai, personal communication). Unlike Tyre, the archaeology at Athlit has survived in a comparatively conspicuous form and it is hoped that, in collaboration with earth scientists, he will be able to link the artificial infrastructure with diagnostic litho- and biostratigraphies.

Cothons: The research thus far has revealed a significant peculiarity, namely the absence of cothon harbours on the Levantine seaboard. Cothons are artificial basins dug into the substratum and subsequently drowned by marine waters to accommodate maritime vessels. Absence of such basins in Israel, Lebanon and Syria is paradoxical as cothons have been widely attributed to a Phoenician origin (Carayon, 2005). In reality, it seems that the environmental determinism attested to in Bronze Age Levant persisted into the first millennium BC with little need to establish *ex nihilo* basins. Conversely, the rise of new engineering *savoir faire* during the Iron Age means that Phoenician colonists employed the cothon technique more widely when establishing trading outposts, notably in the western Mediterranean (Baurain and Bonnet, 1992; Spanò Giammelario, 1997; Aubet, 2001; Carayon, 2005). The best examples include Mahdia (Zaouali, 1999), Carthage (Baradez, 1958; Hurst and Stager, 1978; Hurst, 1993), Motya (Whitaker, 1921; Isserlin, 1971, 1974), Rachgoun

(Carayon, 2005), Phalasarna (Pirazzoli *et al.*, 1982; Hadjidaki, 1988, 1996) and Jezirat Faraun (Flinder, 1977).

At a general level, we have demonstrated that harbour geoarchaeology is an innovative means of integrating various scales and types of data with potentially profound and pervasive implications for maritime archaeology as a whole. We believe it is fundamental to link wreck archaeology with the coastal, and notably harbour research, to formulate a more holistic picture of maritime landscapes in antiquity. The benefits of a geoscience approach, both in terms of cost effectiveness and data wealth, are arguably unparalleled. It is imperative to develop new scientific arenas, integrating this approach with an interpretative archaeology.

References

Abd-el-Al, I. (1948). Le Litani, étude hydrologique. Service hydrologique de la République Libanaise, Beirut.

Aberg, G., Charalampides, G., Fosse, G. and Hjelmseth, H. (2001). The use of Pb isotopes to differentiate between contemporary and ancient sources of pollution in Greece. Atmos. Environ., 35, 4609-4615.

Acquaro, E. (1988). Gli insediamenti fenici e punici in Italia. Libreria Dello Stato, Roma.

Ager, D. (1989). Lyell's pillars and uniformitarianism. Journal of the Geological Society, 146, 603-605.

Ager, D. (1995). The new catastrophism, the importance of the rare event in geological history. Cambridge University Press, Cambridge.

Aguarod Otal, C. and Erice Lacabe, R. (2003). El Puerto de Caesaraugusta. In: Berlanga, G. P. and Pérez Ballester, J. (Eds.), Puertos Fluviales Antiguos: Ciudad, Desarrollo e Infraestructuras. Federico Morillo, Valencia, pp. 143-155.

Al-Djazairi, S. E. (2006). The Golden Age and Decline of Islamic Civilisation. Bayt Al-Hikma Press, Manchester.

Alpözen, O. (1983). The Bodrum Mueseum of Underwater Archaeology. Museum, 35, 61-63.

Ambert, P. (1995). Le cadre géographique et géologique de Narbonne antique. In : L'Homme préhistorique et la mer, 120e congrès CTHS, Aix-en-Provence, 23-26 oct. 1995, pp. 93-105.

Ambert, M. and Chabal, L. (1992). L'environnement de Lattara (Hérault): potentialités et contraintes. Lattara, 5, 9-26.

Ambraseys, N. N. (1962). Data for the investigation of seismic sea-waves in the Eastern Mediterranean. Bull. Seism. Soc. Am., 52, 895-913.

Ambraseys, N., Melville, C. and Adams, R. (1994). The Seismicity of Egypt, Arabia and the Red Sea: A Historical Review. Cambridge University Press, Cambridge.

Angelo, G. d' and Gargiullo, S. (1978). Guida alle conchiglie Mediterranee. Fabbri Editori, Milano.

Anthony, E. J. (1995). Beach-ridge development and sediment supply: examples from West Africa. Marine Geology, 129, 175-186.

Anthony, E. J. and Blivi, A. B. (1999). Morphosedimentary evolution of a delta-sourced, drift-aligned sand barrier–lagoon complex, western Bight of Benin. Marine Geology, 158, 161-176.

Antonioli, F. and Leoni, G. (1998). Siti archeologici sommersi e loro utilizzazione quali indicatori per lo studio delle variazioni recenti del livello del mare. Quaternario 11, 53-66.

Ardaillon, E. (1896). Rapport sur les fouilles du port de Délos. Bulletin de Correspondance

Hellénique, Ecole française d'Athènes, 20, 428-445.

Arnaud, D. (1992). Les ports de la "Phénicie" à la fin de l'âge du bronze récent (XIV-XIII siècles) d'après les textes cunéiformes de Syrie. Studi micenei ed egeo-Anatolici, 30, 180-181.

Arnaud, D. and Salvini, M. (2000). Une lettre du roi de Beyrouth au roi d'Ougarit de l'époque dite "d'El -Amarna". SMEA, 42/1, 5-17.

Arnaud, P. (2005). Les routes de la navigation antique : Itinéraires en Méditerranée. Errance, Paris.

Arnaud, P. *et al.* (1996). Bey 027, Rapport préliminaire. Bulletin d'Archéologie et d'Architecture Libanaises, 1, 103.

Arnaud-Fassetta, G. and Provansal, M. (1999). High frequency variations of water flux and sediment discharge during the Little Ice Age (1586-1725 AD) in the Rhône Delta (Mediterranean France). Relationship to the catchment basin. Hydrobiologia, 410, 241-250.

Arnaud-Fassetta, G., De Beaulieu, J.-L., Suc, J.-P., Provansal, M., Williamson, D., Leveau, P., Aloïsi, J.-C., Gadel, F., Giresse, P., Oberlin, C. and Duzer, D. (2000). Evidence for an early land use in the Rhône delta (Mediterranean France) as recorded by late Holocene fluvial paleoenvironments (1640-100 BC). Geodinamica Acta, 13, 377-389.

Arnaud-Fassetta, G., Carre, M.-B., Marocco, R., Maselli Scotti, F., Pugliese, N., Zaccaria, C., Bandelli, A., Bresson, V., Manzoni, G., Montenegro, M. E., Morhange, C., Pipan, M., Prizzon, A. and Siché, I. (2003). The site of Aquileia (northeastern Italy): example of fluvial geoarchaeology in a Mediterranean deltaic plain. Géomorphologie, 4, 227-246.

Arteaga, O., Hoffmann, G., Schubart, H. and Schulz, H. D. (1988). Geologisch-archäologische Forschungen zum Verlauf der andalusischen Mittelmeerküste. In: Madrider Beiträge, Band 14, Verlag, pp. 107-126.

Artzy, M. (1985). Merchandise and merchantmen: on ships and shipping in the late Bronze Age Levant. In: Papadopoullos, T. and Chatzestylli, S (Eds.), Proceedings of the Second International Cyprological Congress. Nicosia, pp. 135-140.

Arvieux, L. d' (1735). Mémoires du chevalier d'Arvieux, envoyé extraordinaire du Roi à la Porte, consul d'Alep, d'Alger, de Tripoli, et autres échelles du Levant, contenant ses voyages à Constantinople, dans l'Asie, la Syrie, la Palestine, l'Egypte et la Barbarie, la description de ces pays, les religions, les moeurs, les coutumes, le négoce de ces peuples, et leurs gouvernements. Jean-Baptiste Delespine, Paris.

Aslihan Yener, K. (2000). The Domestication of Metals: The Rise of Complex Metal Industries in Anatolia. Brill Academic Publishers, Leiden.

Atherden, M., Hall, J. and Wright, J. C. (1993). A Pollen Diagram from the Northeast Peloponnese, Greece. Implications for Vegetation History and Archaeology. The Holocene, 3, 351-356.

Athersuch, J., Horne, D. J. and Whittaker, J. E. (1990). Marine and Brackish Water Ostracods (Superfamilies Cypridacea and Cytheracea : Keys and Notes for the Identification of the Species). Brill Academic Publishers, Leiden.

Aubet, M. E. (2001). The Phoenicians and the West: Politics, colonies and trade. Cambridge University Press, Cambridge.

Aubet, M. E. (2004). The Tyre-Al Bass Necropolis. In: Doumet-Serhal, C. (Ed.), Decade: A Decade of Archaeology and History in the Lebanon. Archaeology and History in Lebanon, Beirut, pp. 16-27.

Auriemma, R. (2004). Le strutture sommerse di Egnazia (Br): una rilettura. In: Benini, A. and Giacobelli, M. (Eds.), Atti del II convegno nazionale di archeologia subacquea. Edipuglia, pp. 77-97.

Baedeker, K. (1912). From Palestine and Syria. Handbook for Travellers. 5th Edition. Leipzig.

Baeteman, C., Heyvaert, V. M. A. and Dupin, L. (2005). Geo-environmental Investigation. In: Gasche, H. (Ed), The Persian Gulf Shorelines and the Karheh, Karun, and Jarrahi Rivers, a Geo-archaeological Approach. Akkadica, 125-2, p. 155-215.

Baines, J. (2003). Possible implications of the Egyptian name for Alexandria. Journal of Roman Archaeology, 16, 61-64.

Ballard, R. (1987). The discovery of the Titanic. Guild Publishing, London.

Baradez, J. (1958). Nouvelles recherches sur les ports de Carthage. Karthago, 9, 45-78.

Barash, A. and Danin, Z. (1992). Fauna Palaestina, mollusca I. Annotated list of Mediterranean molluscs of Israel and Sinai. The Israel Academy of Sciences and Humanities, Jerusalem.

Bard, E. (1999). Ice Age Temperatures and Geochemistry. Science, 284, 1133-1134.

Bar-Matthews, M., Ayalon, A. and Kaufman, A. (1997). Late Quaternary paleoclimate in the Eastern Mediterranean Region from stable isotope analysis of speleothems in Soreq cave, Israel. Quaternary Research, 47, 155-168.

Bar-Matthews, M., Ayaolon, A. and Kaufman, A. (1998). Middle to Late Holocene Paleoclimate in the Eastern Mediterranean Region from Stable Isotopic Composition of Speleothems from Soreq Cave, Israel. In: Issar, A. S. and Brown, N. (Eds.) Water, Environment and Society in Times of Climatic Change. Kluwer, Amsterdam, pp. 203-214.

Bar-Matthews, M., Ayalon, A., Kaufman, A. and Wasserburg, G. J. (1999). The eastern Mediterranean paleoclimate as a reflection of regional events: Soreq Cave, Israel. Earth and Planetary Science Letters, 166, 85-95.

Barnett, R. D. (1956). Phoenicia and the Ivory

Trade. Archaeology, 9, 87-97.

Barnett, R. D. (1958). Early Shipping in the Near East. Antiquity, 32, 220-230.

Barnett, R. D. (1969). Ezekiel and Tyre. Eretz-Israel, 9, 6-13.

Basch, L. (1985). The Nautical Museum of the Aegean. Mariner's Mirror, 71, 85.

Basch, L. (1987). Le muse imaginaire de la marine antique. Institut Hellénique pour la preservation de la tradition naitique, Athènes.

Bass, G.A (Ed.) (1974). History of Seafaring. Thames and Hudson, London.

Bass, G. F. (1987). Oldest Known Shipwreck Reveals Splendors of the Bronze Age. National Geographic, 172, 6, 692-733.

Bass, G. F. (1991). Evidence of Trade from Bronze Age Shipwrecks. In: N.H. Gale (Ed.), Bronze Age Trade in the Mediterranean. Paul Åströms Förlag, Jonsered, pp. 69-82.

Basso, D. and Spezzaferri, S. (2000). The distribution of living (stained) benthic foraminifera in Iskenderun Bay (Eastern Turkey): a statistical approach. Boll. Soc. Paleontol., Ital., 39, 207-215.

Battarbee, R. W. (1988). The use of diatom analysis in archaeology: a review. Journal of Archaeological Science, 15, 621-644.

Battarbee, R. W., Jones, V. J., Flower, R. J., Cameron, N. G., Bennion, H., Carvalho, L.

and Juggins, S. (2001). Diatoms. In: Smol, J. P., Last, W., Birks, H. J. B. (Eds.) Tracking Environmental Change using Lake Sediments: Terrestrial, Algal and Siliceous Indicators. Kluwer, Dordrecht, pp. 155-202.

Baurain, C. and Bonnet, C. (1992). Les Phéniciens : marins des trios continents. Armand Colin, Paris.

Bellan-Santini, D., Lacaze, J. C. and Poizat, C. (1994). Les biocénoses marines et littorales de Méditerranée. Coll. Patrimoines Naturels, 19, Muséum National d'Histoire Naturelle.

Bergamasco, A., Malanotte-Rizzoli, P. Long, R. B. and Thacker, W. C. (1992). The seasonal circulation of the eastern Mediterranean investigated with the adjoint method. Earth Science Reviews, 32, 285-309.

Berglund, B. E. (2003). Human impact and climate changes – synchronous events and a causal link? Quaternary International, 105, 7–12.

Bergy, P. A. (1932). Le Paleolithique ancien stratifie a Ras-Beyrouth. Melanges de l'Universite Saint Joseph, 16, 169-216.

Berlanga, G. P. and Pérez Ballester, J. (Eds.) (2003). Puertos Fluviales Antiguos: Ciudad, Desarrollo e Infraestructuras. Federico Morillo, Valencia.

Bertou, J. de (1843). Essai sur topographie de Tyr. Firmin Didot Frères, Paris.

Besson, Le P. J. (1660). Mission de Jésus et des Pères de la Compagnie de Jésus en Syrie par le R. P. Joseph Besson. J. Henault, Paris.

Bikai, P. (1979). The pottery of Tyre (ancient Near East). Aris & Phillips, Oxford.

Bikai, P. and Bikai, P. (1987). Tyre at the end of the twentieth century. Berytus, 35, 67-96.

Billaud, Y. and Marguet, A. (2006). Historique des recherches. In: Dumont, A. (Ed.), Archéologie des lacs et des cours d'eau. Errance, Paris, pp. 9-28.

Bitar, G. and Kouli-Bitar, S. (1998). Inventaire des mollusques marines benthiques du Liban et remarques biogéographiques sur quelques espèces nouvellement signalés. Mesogée, 56, 37-44.

Blackman, D. J. (1973a). The Harbours of Phaselis. International Journal of Nautical Archaeology, 2, 355-364.

Blackman, D. J. (1973b). Evidence of sea level change in ancient harbours and coastal installations. In: Blackman, D. J. (Ed.), Marine Archaeology. Butterworths, London, pp. 115-139.

Blackman, D. J. (1982a). Ancient harbours in the Mediterranean, part 1. International Journal of Nautical Archaeology, 11, 79-104.

Blackman, D. J. (1982b). Ancient harbours in the Mediterranean, part 2. International Journal of Nautical Archaeology, 11, 185-211.

Blackman, D. J. (1996). Further evidence for the use of concrete in ancient harbor construction. In: Raban, A. and Hollum, K. G. (Eds.) Caesarea Maritima: A Retrospective after Two Millennia. Brill, Leiden, pp. 41-49.

Blackman, D. J. (2003). Progress in the Study of Ancient Shipsheds: a Review. In: Beltrame, C. (Ed.), Boats, ships and shipyards: Proceedings of the Ninth International Symposium on Boat and Ship Archaeology, Venice 2000. Oxbow Books, Oxford, pp. 81-90.

Blackman, D. J. (2005). Archaeological evidence for sea level changes. Zeitschrift für Geomorphologie, 137, 61-70.

Blanc, J.-J. (1959). Recherches sédimentologiques littorales et sous-marines en Provence occidentale, PhD thesis. Paris-

Sorbonne, Paris and published by Masson, Paris.

Blanchet, G. (1976). Le temps au Liban - approche d'une climatologie synoptique. PhD thesis, 2 volumes, Université Lyon II, Lyon.

Blivi, A., Anthony, E. J. and Oyédé, L. M. (2002). Sand barrier development in the bight of Benin, West Africa. Ocean and Coastal Management, 45, 185-200.

Bochaca, M., Fauchère, N. and Tranchant, M. (2005). Ports maritimes et ports fluviaux au moyen âge. Publications de la Sorbonne, Série Histoire ancienne et médiévale 81, Paris.

Bonaduce, G., Ciampo, G. and Masoli, M. (1975). Distribution of Ostracoda in the Adriatic sea. Publ. Staz. Zool. Napoli, Napoli, 40, 1-304.

Bonito, M. (1691). Terra tremante, o vero continuatione de' terremoti dalla Creatione del Mondo fino al temp presente. Napoli (reprint Arnaldo Forni Editore, Sala Bolognese, 1980, pp. 834).

Bonner, M. (2005). Arab-Byzantine Relations in Early Islamic Times. Ashgate, Aldershot.

Boomer, I. and Eisenhauer, G. (2002). Ostracod faunas as palaeoenvironmental indicators in marginal marine environments. In: Holmes, J. A. and Chivas, A. R. (Eds.), The Ostracoda: Applications in Quaternary Research, Geophysical Monograph, vol. 131. American Geophysical Union, Washington DC, pp. 135- 149.

Borrut, A. (1999-2000). L'espace maritime syrien au cours des premiers siècles de l'Islam (VIIe-Xe siècle) : le cas de la région entre Acre et Tripoli. Tempora, Annales d'histoire et d'archéologie, 10-11, 1-33.

Borrut, A. (2001). Architecture des espaces portuaires et réseaux défensifs du littoral syro-palestinien dans les sources arabes (VIIe-Xie s.). Archéologie Islamique, 11, 21-46.

Bottema, S. and Woldring, H. (1990). Anthropogenic Indicators in the Pollen Record of the Eastern Mediterranean. In: Bottema, S., Entjes-Nieborg, G. and van Zeist, W. (Eds.), Man's Role in the Shaping of the Eastern Mediterranean Landscape. Balkema, Rotterdam, pp. 231-265.

Bottema, S., Entjes-Nieborg, G. and van Zeist, W. (Eds.) (1990). Man's Role in the Shaping of the Eastern Mediterranean Landscape. Balkema, Rotterdam.

Bouchayer, A. (1931). Marseille ou la mer qui monte. Editions des Portiques, Paris.

Bourguet, M.-N., Lepetit, B., Nordman, D. and Sinarellis, M. (1998). L'invention scientifique de la Méditerranée (Egypte, Morée, Algérie). Editions de l'Ecole des Hautes Etudes en Sciences Sociales, Paris.

Bousquet, B. and Pechoux, P. Y. (1980). Géomorphologie, archéologie, histoire dans le bassin oriental de la Méditerranée: principe, methodes, resultats préliminaires. Méditerranée, 1, 33-45.

Bousquet, B., Dufaure, J. J. and Pechoux, P. Y. (1983). Temps historiques et évolution des paysages égéens. Méditerranée, 2, 3-25.

Bouzek, J. (1996). Bey 069, Sondage A. Bulletin d'Archéologie et d'Architecture Libanaises, 1, 135-147.

Bracco, J.-P. (2005). Archaeological sites. In: Petit-Maire, N. and Vrielinck, B. (Eds.), The Mediterranean basin: the last two climatic cycles - Explanatory notes of the maps. Maison Méditerranéenne des Sciences de l'Homme, Aix-en-Provence, pp. 83-89.

Braidwood, R. J. (1940). Report on two sondages on the coast of Syria, South of Tartous. Syria, 21, 183-221.

Brandon, C. (1996). Cements, Concrete, and Settling Barges at Sebastos. Comparisons with Other Roman Harbor Examples and the Description of Vitruvius. In: Raban, A. and Holum, K.G. (Eds.) Caesarea Maritima - A Retrospective After Two Millennia. Brill, Leiden, pp. 25-40.

Brandon, C. (1997). Techniques d'architecture navale dans la construction des caissons en bois du port du Roi Hérode a Césarée. Cahiers d'Archéologie Subaquatique, 13, 13-33.

Brandon, C. (1999). Pozzolana, lime, and single-mission barges (Area K). In: Holum, K., Raban, A. and Patrich, J. (Eds.), Caesarea Papers, 2, JRA supplement 35, pp. 169-178.

Bränvall, M.-L., Bindler, R., Emteryd, O. and Renberg, I. (2001). Four thousand years of atmospheric lead pollution in northern Europe: a summary from Swedish lake sediments. Journal of Paleolimnology, 25, 421-435.

Bravard, J.-P. and Magny, M. (Eds.) (2002). Les fleuves ont une histoire: paléo-environnement des rivières et des lacs français depuis 15000 ans. Errance, Paris.

Breen, C. and Lane, P. J. (2003). Archaeological approaches to East Africa's changing seascapes. World Archaeology, 35, 469-489.

Breman, E. (1975). The distribution of Ostracodes in the bottom sediments of the Adriatic sea. Diss. Vrije Univ. Amsterdam, Amsterdam.

Briand, F. and Maldonado, A. (Eds.) (1996). Transformations and evolution of the Mediterranean coastline. Bulletin de l'Institut océanographique, numéro special 18, Monaco.

Bridgland, D. R. (2000). River terrace systems in north-west Europe: an archive of environmental change, uplift and early human occupation. Quaternary Science Reviews, 19, 1293-1303.

Briend, J. (2003). La Terre Sainte : cinquante ans d'archéologie. Bayard Centurion, Paris.

Bronk Ramsey, C. (2001). Development of the Radiocarbon Program OxCal. Radiocarbon, 43, 355-363.

Briquel-Chatonnet, F. and Gubel, E. (1999). Les Phéniciens : aux origines du Liban. Gallimard, Paris.

Browder, A. G. and McNinch, J. E. (2006). Linking framework geology and nearshore morphology: Correlation of paleo-channels with shore-oblique sandbars and gravel outcrops. Marine Geology, 231, 141-162.

Brückner, H. (1997). Coastal changes in western Turkey; rapid progradation in historical times. In: Briand, F. and Maldonado, A. (Eds.), Transformations and evolution of the Mediterranean coastline. Bulletin de l'Institut Océanographique, Monaco 18, pp. 63-74.

Brückner, H., Müllenhoff, M., Handl, M., and van der Borg, K. (2002). Holocene landscape evolution of the Büyük Menderes alluvial plain in the environs of Myous and Priene (Western Anatolia, Turkey). Zeitschrift für Geomorphologie, 127, 47-65.

Brückner, H., Vött, A., Schriever, M. and Handl, M. (2005). Holocene delta progradation in the eastern Mediterranean - case studies in their historical context. Méditerranée, 104, 95-106.

Buckley, D. E., Smith, J. N. and Winters, G. V. (1995). Accumulation of contaminant metals in marine sediments of Halifax Harbour, Nova Scotia: environmental factors and historical trends. Applied Geochemistry, 10, 175-195.

Butcher, K. and Thorpe, R. (1997). A note on excavations in central Beirut 1994-96. Journal of Roman Archaeology, 10, 291-306.

Butzer, K. W. (1962). Pleistocene stratigraphy and prehistory in Egypt. Quaternaria, 6, 451-477.

Butzer, K. W. (1976). Early hydraulic civilization in Egypt: a study in cultural ecology. The University of Chicago Press, Chicago.

Butzer, K. W. (1996). 'Sociopolitical Discontinuity in the Near East c. 2200 BCE: Senarios from Palestine and Egypt. In: Dalfes, H. N., Kukla, G. and Weiss, H. (Eds.), Third Millennium B. C. Climate Change and Old World Collapse. Springer, Berlin, New York, pp. 245–96.

Caldara, M., Pennetta, L. and Simone, O. (2002). Holocene Evolution of the Salpi Lagoon (Puglia, Italy). Journal of Coastal Research, SI 36, 124-133.

Caldara, C., Cazzella, A., Fiorentino, G., Lopez, R., Magri, D., Moscoloni, M., Narcisi, B. and Simone, O. (2003). The relationship between the Coppa Nevigata settlement and the wetland area during the Bronze Age (south-eastern Italy). In: Fouache, E. (Ed.), The Mediterranean world environment and history. Elsevier, Paris, pp. 429-437.

Caldwell, J. M. (1939). Sedimentation in harbors. In: Trask, P. D. (Ed.), Applied sedimentation. The American Association of Petroleum Geologists, Tulsa, pp. 291-299.

Canina, L. (1830). Indicazione delle rovine di Ostia e di Porto e della supposizione e dell'intiero loro stato delineata in quattro tavole dall'architetto Luigi Canina. Mercuri e Boraglia, Rome.

Caple, C. (2001). Overview - Degradation, Investigation and Preservation of Archaeological Evidence. In: Brothwell, D. R. and Pollard, A. M. (Eds.), Handbook of Archaeological Sciences. Wiley, Chichester, pp. 587-593.

Carayon, N. (2003). L'île de Ziré à Saïda: nouvelles données archéologiques. Archaeology and History in Lebanon, 18, 95-114.

Carayon, N. (2005). Le cothon ou port artificial creusé. Essai de définition. Méditerranée, 104, 5-13.

Carayon, N. and Viret, J. (2004). L'île de Ziré à Saïda: carrière et port insulaire. In : Denise, F. and Nordiguian, L. (Eds.), Une aventure archéologique : Antoine Poidebard, photographe et aviateur. Editions Parenthèses, Marseille, pp. 314-315.

Carbonel, P. (1980). Les ostracodes et leur interêt dans la définition des écosystèmes estuariens et de plateforme continentale. Essais d'application à des domaines anciens. Mém. Inst. Géol. Bassin d'Aquitaine, 11, 1-378.

Carbonel, P. (1982). Les ostracodes, traceurs des variations hydrologiques dans les systèmes de transition eaux douces-eaux salées. Mém. Soc. Géol. Fr. N.S., 144, 117-128.

Carmel, Z., Inman, D. L., and Golik, A. (1985a). Characteristics of storm waves off the Mediterranean coast of Israel. Coastal Engineering, 9, 1-19.

Carmel, Z., Inman, D. L., and Golik, A. (1985b). Directional wave measurement at Haifa, Israel, and sediment transport along the Nile littoral cell. Coastal Engineering, 9, 21-36.

Carmona, P. and Ruiz, J. M. (2003). Cambios geomorfológicos y puertos históricos en la costa Mediterránea Valenciana. In: Berlanga, G. P. and Pérez Ballester, G. J. (Eds.), Puertos Fluviales Antiguos: Ciudad, Desarrollo e Infraestructuras. Federico Morillo, Valencia, pp. 115-126.

Carmona, P. and Ruiz, J. M. (2004). Geomorphological and geoarchaeological evolution of the coastline of the Tyre tombolo: preliminary results. Bulletin d'Archéologie et d'Architecture Libanaises, Hors-Série 1, 207-219.

Casson, L. (1965). Harbour and river boats of ancient Rome. Journal of Roman Studies, 55, 31-39.

Casson, L. (1975). Bronze Age ships. The evidence of the Thera wall paintings. International Journal of Nautical Archaeology, 4, 3-10.

Casson, L. (1976). Dockside Cranes. International Journal of Nautical Archaeology, 5, 345.

Casson, L. (1978a). The Thera Ships. International Journal of Nautical Archaeology, 7, 232-233.

Casson, L. (1978b). Some evidence of lead sheathing on Roman craft. Mariner's Mirror, 64.

Casson, L. (1978c). More evidence for lead sheathing on Roman craft. Mariner's Mirror, 64, 139-142.

Casson, L. (1994). Ships and Seafaring in Ancient Times. British Museum Press, London.

Catuneanu, O. (2002). Sequence stratigraphy of clastic systems: concepts, merits, and pitfalls. Journal of African Earth Sciences, 35, 1-43.

Catuneanu, O. (2005). Principles of Sequence Stratigraphy. Elsevier, London.

Cayeux, L. (1907). Fixité du niveau de la Méditerranée à l'époque historique. Annales de Géographie, 16, 97-116.

Cayeux, L. (1914). Les déplacements de la mer à l'époque historique. Revue Scientifique, 52, 577-586.

CERC (1975). Shore protection manual: volume 1. US Army Corps of Engineers, Washington D. C.

Cerdan, O., Le Bissonnais, Y., Couturier, A., Bourennane, H. and Souchère, V. (2002). Rill erosion on cultivated hillslopes during two extreme rainfall events in Normandy, France. Soil and Tillage Research, 67, 99-108.

Cervin, G. B. R. de (1977). The Thera ships: other suggestions. Mariner's Mirror, 63, 150-152.

Cervin, G. B. R. de (1978). A further note on the Thera ships. Mariner's Mirror, 64, 150-152.

Chamley, H. (1989). Clay Sedimentology. Springer-Verlag, Berlin.

Chéhab, M. (1939). Tombe phénicienne de Sin el-Fil. In : Mélanges syriens offerts à Monsieur René Dussaud II, Paris, pp. 303-810.

Chen, Z., Warne, A. G., and Stanley, D. J. (1992). Late Quaternary evolution of the northwestern Nile Delta between Rosetta and Alexandria, Egypt. Journal of Coastal Research, 8, 527-561.

Chester, D. K. and James, P. A. (1991). Holocene alluviation in the Algarve, southern Portugal: The case for an anthropogenic cause. Journal of Archaeological Science, 18, 73-87.

Cimerman, F. and Langer, M. R. (1991). Mediterranean Foraminifera. Slovenska Akademija Znanosti in Umetnosti, Ljubljana.

Cinque, A., Russo, F. and Pagano, M. (1991). La successione dei terreni di eta post-romana delle terme di Miseno (Napoli): Nuovi dati per la storia e la stratigrafia del bradisisma puteolano. Bolletino della Societa Geologica Italiana, 110, 231-244.

Clark, A. J. (1990). Seeing Beneath the Soil. Batsford, London.

Clemmensen, L. B., Richardt, N. and Andersen, C. (2001). Holocene sea-level variation and spit development: data from Skagen Odde, Denmark. The Holocene, 11, 323-331.

Coe, A. L. (Ed.) (2003). The Sedimentary Record of Sea-Level Change. Cambridge University Press, Cambridge.

Cohen-Seffer, R., Greenbaum, N., Sivan, D., Jull, T., Barmeir, E., Croitoru, S. and Inbar, M. (2005). Late Pleistocene-Holocene marsh episodes along the Carmel coast, Israel. Quaternary International, 140-141, 103-120.

Coles, J. (1984). The Archaeology of Wetlands. Edinburgh University Press, Edinburgh.

Collina-Girard, J. (1998). Profils littoraux en plongée et niveaux d'érosion eustatiques près de la grotte Cosquer (Marseille) et en Provence. Comptes Rendus de l'Académie des Sciences, 324, 607-615.

Collina-Girard, J. (2001). L'Atlantide devant le dètroit de Gibraltar? Mythe et gèologie. Comptes Rendus de l'Académie des Sciences, 333, 233-240.

Contenau, G. (1920). Mission archéologique à Sidon (1914). Syria, I, 16-55 and 108-154.

Contenau, G. (1924a). Deuxième mission archéologique à Sidon (1920). Syria, IV, 261-281.

Contenau, G. (1924b). Deuxième mission archéologique à Sidon (1920). Syria, V, 9-23 and 123-134.

Conyers, L. B. and Goodman, D. (1997). Ground-Penetrating Radar: an introduction for Archaeologists. Altamira Press, Walnut Creek, Ca.

Copeland, L. and Wescombe, P. J. (1965). Inventory of Stone-Age Sites in Lebanon. Part one: West-Central Lebanon. MUSJ, 41, 61-176.

Copeland, L. and Wescombe, P. J. (1966). Inventory of Stone-Age Sites in Lebanon. Part two: North, South and East-Central Lebanon. MUSJ, 42, 1-174.

Courtaud, J. (2000). Dynamiques géomorphologiques et risques littoraux. Cas du Tombolo de Giens (Var, France méridionale), PhD thesis. Université de Provence, Aix-en-Provence.

Craddock, P. T. (1976). The composition of the copper alloys used by the Greek, Etruscan and Roman civilizations: 1. The Greeks before the archaic period. Journal of Archaeological Science, 3, 93-113.

Craddock, P. T. (1977). The composition of the copper alloys used by the Greek, etruscan and Roman civilisations: 2. The Archaic, Classical and Hellenistic Greeks. Journal of Archaeological Science, 4, 103-123.

Craddock, P. T. (1978). The composition of the copper alloys used by the Greek, Etruscan and Roman civilizations: 3. The origins and early use of brass. Journal of Archaeological Science, 5, 1-16.

Craddock, P. T. (1979). The Copper Alloys of the Medieval Islamic World - Inheritors of the Classical Tradition. World Archaeology, 11, 68-79.

Crawford, H. E. W. (2004). Sumer and the Sumerians. Cambridge University Press, Cambridge.

Croome, A. (1973). Protecting Historic Wrecks. New Scientist, 426.

Croome, A. (1987). The Viking Museum, Roskilde. International Journal of Nautical Archaeology, 16, 348-352.

Croudace, I. W. and Cundy, A. B. (1995). Heavy metal and hydrocarbon pollution in recent sediments from Southampton Water, southern England: a geochemical and isotopic study. Environmental Science and Technology, 29, 1288-1296.

Crumlin-Pedersen, O. (1993). A centre for maritime archaeology in Denmark. International Journal of Nautical Archaeology, 22, 293-296.

Cullen, H. M., deMenocal, P. B., Hemming, S., Hemming, G., Brown, F. H., Guilderson, T. and Sirocko, B. (2000). Climate change and the collapse of the Akkadian empire: Evidence from the deep sea. Geology, 28, 379–382.

Cumberpatch, C. G. (1997). Archaeology in the Beirut Central District. Some notes and observations. Berytus, 42, 157-172.

Cundy, A. B., Sprague, D., Hopkinson, L., Maroukian, H., Gaki-Papanastassiou, K., Papanastassiou, D. and Frogley, M. R. (2006). Geochemical and stratigraphic indicators of late Holocene coastal evolution in the Gythio area, southern Peloponnese, Greece. Marine Geology, 230, 161-177.

Curtis, J. (2000). New excavations at Sidon. National Museum News, 10, 27-40.

Curvers, H. H. (2002). The Lower Town of Beirut (1200-300 BC). A preliminary synthesis. ARAM, 13-14, 51-72.

Curvers, H. H. and Stuart, B. (1996). Bey 008: the 1994 results. Bulletin d'Archéologie et d'Architecture Libanaises, 1, 229.

Curvers, H. H. and Stuart, B. (1997). The BCD infrastructure archaeology project, 1995. Bulletin d'Archéologie et d'Architecture Libanaises, 2, 167-205.

Curvers, H. H. and Stuart, B. (1998-1999). The BCD Archaeology Project 1996-1999. Bulletin d'Archéologie et d'Architecture Libanaises, 3, 13-30.

Curvers, H. and Stuart, B. (2004). Beirut Central Distict Archaeology Project. 1994-2003. In: Doumet-Serhal, C. (Ed.), Decade: A Decade of Archaeology and History in the Lebanon. Archaeology and History of Lebanon, Beirut, pp. 248-260.

Dally, W. R. and Pope, J. (1986). Detached breakwaters for shore protection. Technical Report, Coastal Engineering Research Centre, Waterways Experiment Station, CERC-86-1.

Dalongeville, R. (1975). Les flaques à salinité variable du littoral actuel Libanais, PhD thesis. Université de Brest, Brest.

Dalongeville, R., Lespez, L., Poursoulis, G., Pastre, J.-F., Keraudren, B., Mathieu, R., Prieur, A., Renault-Miskovsky, J., Darmon, F., Kunesh, S., Bernier, P. and Caron, V. (2001). Malia : un marais parle. Bulletin de Correspondance Hellénique, 125, 67-88.

Dalongeville, R. and Müller-Celka, S. (2003). Les variations du niveau marin à Malia (Crète). In: Des chercheurs à la découverte de l'Orient. Maison de l'Orient, Lyon, pp. 30-33.

Darawcheh, R., Sbeinati, M. R., Margottini, C. and Paolini, S. (2000). The 9 July 551 AD Beirut earthquake, eastern Mediterranean region. Journal of Earthquake Engineering, 4, 403-414.

Dark, K. R. (2004). The New Post Office Site in Istanbul and the North-Eastern Harbour of Byzantine Constantinople. International Journal of Nautical Archaeology, 33, 315-319.

Davie, M. F. (1987). Maps and the historical topography of Beirut. Berytus, 35, 141-164.

Davie, M. (2000). Flux mondiaux, expressions locales: Beyrouth et son port au XIXe siècle ottoman. Chronos, 3, 139-172.

Davis, R. A. (Ed.) (1994). Geology of Holocene Barrier Island Systems. Springer-Verlag, Berlin.

Dawson, A. G. (1994). Geomorphological effects of tsunami run-up and backwash. Geomorphology, 10, 83-94.

Dawson, A. G. (1999). Linking tsunami deposits, submarine slides and offshore earthquakes. Quaternary International, 60, 119-126.

Dawson, A. G., Lockett, P. and Shi, S. (2004). Tsunami hazards in Europe. Environment International, 30, 577-585.

De Maria, L. and Turchetti, R. (Eds.) (2004a). Evolución paleoambiental de los puertos y fondeaderos antiguos en el Mediterráneo occidental. Rubbettino Editore, Soveria Mannelli.

De Maria, L. and Turchetti, R. (Eds.) (2004b). Rotte e porti del Mediterraneo dopo la caduta dell'impero romano d'occidente. Rubbettino Editore, Soveria Mannelli.

Del Carmen Berrocal Caparrós, M. (1998). Instalaciones portuarias en Carthago-Nova: la evidencia arqueológica. In: Pérez Ballester, J. and Berlanga, G. P. (Eds.), Reunión Internacional sobre Puertos Antguos y Comercio Marítimo. III Jornades de Arqueología Subacuática: Actas. La Imprenta, Valencia, pp. 98-114.

Dell'Orco, P.G. (1979). The metal sheathing of Roman Craft. Mariner's Mirror, 65.

deMenocal, P.B. (2001). Cultural responses to climate change during the late Holocene. Science, 292, 667-673.

Denise, F. and Nordiguian, L. (Eds.) (2004). Une aventure archéologique: Antoine Poidebard, photographe et aviateur. Editions Parenthèses, Marseilles.

Desjardins, E. E. A. (1876). Géographie historique et administrative de la Gaule romaine. Tome premier, introduction et géographie physique comparée époque romaine - époque actuelle. Hachette, Paris.

Desor, E. and Favre, L. (1874). Le bel âge du bronze lacustre en Suisse. Mémoires de la Société des sciences naturelles de Neuchâtel, IV, 2e partie, 1-111.

Devillers, B. (2005). Morphogenèse et anthropisation holocène d'un bassin versant semi-aride : le Gialias, Chypre, PhD thesis. Université de Provence, Aix-en-Provence.

Dickin, A. P. (1995). Radiogenic Isotope Geology. Cambridge University Press, Cambridge.

Dickinson, O. (1994). The Aegean Bronze Age. Cambridge University Press, Cambridge.

Diodorus Siculus (1967). Diodorus of Sicily: in twelve volumes (Loeb classical library). Heinemann, London.

Dominey-Howes, D., Dawson, A. and Smith, D. (1998). Late Holocene coastal tectonics at Falasarna, western Crete: a sedimentary study. In: Stewart, I. and Vita-Finzi, C. (Eds.), Coastal Tectonics. Geological Society, London, Special Publications 146, pp. 343-352.

Doneddu, M. and Trainito, E. (2005). Conchiglie del Mediterraneo. Il Castello, Trezzano sul Naviglio.

Doumas, C. (1992). The Wall Paintings of Thera. The Thera Foundation, Athens.

Doumet, J. (1980). A Study on the Ancient Purple Colour. Imprimerie Catholique, Beyrouth.

Doumet, J. (2004). Purple Dye. In: Doumet-Serhal, C. (Ed.), Decade: A Decade of Archaeology and History in the Lebanon. Archaeology and History in Lebanon, Beirut, pp. 38-49.

Doumet-Serhal, C. (Ed.) (2003). Sidon - British Museum Excavations 1998-2003. Archaeology and History in Lebanon, 18, 1-144.

Doumet-Serhal, C. (Ed.) (2004a). Decade: A Decade of Archaeology and History in the Lebanon. Archaeology and History in Lebanon, Beirut.

Doumet-Serhal, C. (2004b). Tell Rachidieh: Foreign Relations. In: Doumet-Serhal, C. (Ed.), Decade: A Decade of Archaeology and History in the Lebanon. Archaeology and History in Lebanon, Beirut, pp. 88-99.

Doumet-Serhal, C. (2004c). Excavating Sidon. In: Doumet-Serhal, C. (Ed.), Decade: A Decade of Archaeology and History in the Lebanon. Archaeology and History in Lebanon, Beirut, pp. 102-123.

Doumet-Serhal, C. (2004d). Weapons from the Middle Bronze Age burials at Sidon. In: Doumet-Serhal, C. (Ed.), Decade: A Decade of Archaeology and History in the Lebanon. Archaeology and History in Lebanon, Beirut, pp. 154-177.

Doumet-Serhal, C. (Ed.) (2006a). Sidon: British Museum axcavations in collaboration with the Department of Antiquities of Lebanon. Archaeology and History in Lebanon, 24, 1-160.

Doumet-Serhal, C. (2006b). Sidon: Mediterranean contacts in the Early and Middle Bronze Age, preliminary report. Archaeology and History in Lebanon, 24, 34-47.

Dubar, M. and Anthony, E. J. (1995). Holocene Environmental Change and River-Mouth Sedimentation in the Baie des Anges, French Riviera. Quaternary Research, 43, 329-343.

Dubertret, L. (1940). Sur la structure de la plateforme de Beyrouth et sur ses grès Quaternaires. Compte-Rendu de la Société géologique de France, 8, 83-84.

Dubertret, L. (1955). Carte géologique du Liban au 1:200,000, avec notice explicative. Ministère des Travaux Publics, Beyrouth.

Dubertret, L. (1975). Introduction à la carte géologique au 1:50 000 du Liban. Notes Mémoire Moyen-Orient, 13, 345-403.

Dubois, C. (1907). Pouzzoles antique (histoire et topographie). Ecoles d'Athènes et de Rome, 93.

Dumayne-Peaty, L. (2001). Human Impact on Vegetation. In: Brothwell, D. R. and Pollard, A. M. (Eds.), Handbook of Archaeological Sciences. John Wiley, Chichester, pp. 379-392.

Dumont, A. (Ed.) (2006). Archéologie des lacs et des cours d'eau. Errance, Paris.

Dunand, M. (1936). Fouilles de Byblos I (1926-1932). Paris.

Dunand, M. (1939). Chronique. Bulletin du Musée de Beyrouth, 3, 79-81.

Dunand, M. (1940). Chronique. Bulletin du Musée de Beyrouth, 4, 118.

Dunand, M. (1941). Chronique. Bulletin du Musée de Beyrouth, 5, 88-89.

Dunand, M. (1942-43). Chronique. Bulletin du Musée de Beyrouth, 6, 82-83.

Dunand, M. (1954). Fouilles de Byblos II. Paris.

Dunand, M. (1967). Rapport préliminaire sur les fouilles de Sidon en 1964-1965. Bulletin du Musée de Beyrouth, XX, 27-44.

Duplessy, J.-C., Cortijo, E. and Kallel, N. (2005). Marine records of Holocene climatic variations. Comptes Rendus Geosciences, 337, 87-95.

Duprat, E. (1935). Marseille, l'évolution urbaine. In: Les Bouches du Rhône, Encyclopédie départementale 14, Marseille.

Dussaud, R. (1927). Topographie historique de la Syrie antique et médiévale. Geuthner, Paris.

Eaton, N. (1980). Do you sincerely hope to get rich? Diver, 25.

Edgerton, W. F. (1922-1923). Ancient Egyptian Ships and Shipping. American Journal of Semitic Languages and Literatures, 39, 109-135.

El Amouri, M., El Helou, M., Marquet, M., Noureddine, I. and Seco Alvarez, M. (2005). Mission d'expertise archéologique du port sud de Tyr, sud Liban: résultats préliminaires. Bulletin d'Archéologie et d'Architecture Libanaises, Hors-Série 2, 91-110.

Elayi, J. (1984). Terminologie de la Mer Méditerranée dans les Annales assyriennes. OA, 2, 87-89.

Elayi, J. (1990). Economie des cités phéniciennes sous l'Empire perse. Supplément à Annali dell'Istituto Universitario Orientale di Napoli, Istituti Editoriali e Poligrafici Internazionali, Napoli.

Elayi, J. and Sayegh, H. (2000). Un quartier du port Phénicien de Beyrouth au Fer III/Perse. Archéologie et histoire. Gabalda, Paris.

El Baz, F. (1988). Finding a Pharaoh's Funeral Bark. National Geographic, 173, 512-533.

El-Falaki, M. B. (1872). Mémoire sur l'antique Alexandrie. Copenhagen.

Elkin, D. and Cafferata, H. (2001). Underwater archaeology and cultural tourism - a mutual benefit proposal for Patagonia. Bulletin of the Australian Institute for Maritime Archaeology, 25, 83-88.

Emelyanov, E. M. (1994). Recent bottom sediments of the Levantine sea: their composition and processes of formation. In: Krasheninnikov, V. A. and Hall, J. K. (Eds.), Geological structure of the northeastern Mediterranean. Akademik Nikolaj Strakhov, Jerusalem, pp. 141-158.

Emery, K. O. and Neev, D. (1960). Mediterranean beaches of Israel. Israel Geological Survey Bulletin, 26, 1-24.

Empereur, J.-Y. (1998). Le Phare d'Alexandrie, la merveille retrouvée. Découvertes Gallimard, Paris.

Empereur, J.-Y. and Grimal, N. (1997). Les fouilles sous-marines du phare d'Alexandrie. Compte-rendus de l'Académie des Inscriptions et Belles-Lettres, juillet-octobre, 693-717.

Enzel, Y., Bookman, R., Sharon, D., Gvirtzman, H., Dayan, U., Ziv, B. and Steinc, M. (2003). Late Holocene climates of the Near East deduced from Dead Sea level variations and modern regional winter rainfall. Quaternary Research, 60, 263-273.

Ercilla, G., Díaz, J. I., Alonso, B. and Farran, M. (1995). Late Pleistocene-Holocene sedimentary evolution of the northern Catalonia continental shelf (northwestern Mediterranean Sea). Continental Shelf Research, 15, 1435-1445.

Escoffier, F. F. (1954). Travelling forelands and the shoreline processes associated with them. Bulletin US Beach Erosion Board, 9, 11-14.

Espic, K., Morhange, C., Bourcier, M., Bruzzi, C., Carbonel, P., Nammour, T. and

Doumet-Serhal, C. (2002). Les ports antiques de Sidon: nouvelles données paléo-environnementales. Archaeology and History in Lebanon, 15, 28-36.
Espouy, H. d' (1910). Monuments Antiques. 3 volumes. Paris.

Euzennat, M. (Ed.) (1987). Déplacements des lignes de rivage en Méditerranée. Editions du Centre National de la Recherche Scientifique, Paris.

Fabre, D. (2004/2005). Seafaring in Ancient Egypt. Periplus Publishing, London.

Fahmy, A. M. (1966). Muslim naval organisation in the eastern Mediterranean from the seventh to the tenth century AD. National Publication & Print House, Cairo.

Faraldo Victorica, G. and Curvers, H. H. (2002). Les quartiers de Berytus I: les vestiges ramono-byzantins dans le N-E (Bey 028, 031, 046 et 115). Bulletin d'Archéologie et d'Architecture Libanaises, 6, 259-282.

Felici, E. (1993). Osservazioni sul porto neroniano di Anzio e sulla tecnica romana delle costruzioni portuali in calcestruzzo. Archeologia Subacquea, 1, 71-104.

Felici, E. (2002). Scoperte epigrafici e topografiche sulla costruzione del porto neroniano di Antium. Archeologia Subacquea, 3, 107-122.

Felici, E. and Balderi, G. (1997). Il porto romano di Cosa: appunti per l'interpretazione tecnica di un'opera marittima in cementizio. In: Archeologia Subacquea: studi, richerche e documenti, 2. Libreria dello stato, Roma, pp. 11-19.

Ferrand, J. L., Hamelin, B. and Monaco, A. (1999). Isotopic tracing of anthropogenic Pb inventories and sedimentary fluxes in the Gulf of Lions (NW Mediterranean sea). Continental Shelf Research, 19, 23-47.

Février, J. G. (1935). Les origines de la marine phénicienne. Revue de histoire de la philosophie et d'histoire générale de la civilization, 10, 97-125.

Février, J. G. (1949-1950). L'ancienne marine phénicienne et les découvertes récentes. La Nouvelle Clio, 1-2, 128-143.

Figueiral, I. and Willcox, G. (1999). Archaeobotany: collecting and analytical techniques for sub-fossils. In: Jones, T. P. and Rowe, N P. (Eds.), Fossil plants and spores: modern techniques. The Geological Society, London, pp. 290-294.

Finkbeiner, U. and Sader, H. (1997). Beirut BEY 020: Preliminary report on the 1995 excavations. Bulletin d'Archéologie et d'Architecture Libanaises, 2, 114-166.

Firth, A. and Ferrari, B. (1992). Archaeology and Marine Protected Areas. International Journal of Nautical Archaeology, 21, 67-74.

Fisher, R. V., Heiken, G. and Hulen, J. (1997). Volcanoes: Crucibles of Change. Princeton University Press, Princeton.

Fitchen, J. (1988). Building construction before mechanization. MIT Press, London.

Fleisch, H. (1946). Position de l'Acheuléen à Ras Beyrouth (Liban). Bulletin de la société préhistorique française, 43, 293-299.

Fleming, W.B. (1915). History of Tyre. Columbia University Press, New York.

Flemming, N. C. (1961). Apollonia revisited. Geographical Magazine, 33, 522-530.

Flemming, N. C. (1965). Apollonia. In: du Plat Taylor, J. (Ed.), Marine Archaeology. London, pp. 168-178.

Flemming, N. C. (1969). Archaeological evidence for eustatic change of sea level and earth movements in the Western Mediterranean in the last 2000 years.

Geological Society of America Special Paper, 109, 125.

Flemming, N. C. (1971). Cities in the sea. Doubleday, New York.

Flemming, N. C. (1978). Holocene eustatic changes and coastal tectonics in the northeast Mediterranean: implications for models of crustal consumption. Philosophical Transactions of the Royal Society of London A, 289, 405-458.

Flemming, N. C. (1979-1980). Archaeological indicators of sea level. Oceanis 5, Hors-Série, 149-166.

Flemming, N. C. (1996). Sea level, neotectonics and changes in coastal settlements: threat and response. In: Rice, E. E. (Ed.), The Sea and History. Sutton, pp. 23-52.

Flemming, N. C. (1998). Archaeological evidence for vertical movement on the continental shelf during the Palaeolithic, Neolithic and Bronze Age periods. In: Stewart, I. and Vita-Finzi, C. (Eds.), Coastal Tectonics. Geological Society of London, Special Publications 146, pp. 129-146.

Flemming, N. C. and Webb, C. O. (1986). Tectonic and eustatic coastal changes during the last 10.000 years derived from archeological data. Zeitschrift für Geomorphologie, 62, 1-29.

Fletcher, B. A. (1996). History of Architecture. Architectural Press, Oxford.

Flinder, A. (1977). The island of Jezirat Faraun, its ancient harbour, anchorage and marine defence installations. International Journal of Nautical Archaeology, 6, 127-139.

Flinn, D. (1997). The Role of Wave Diffraction in the Formation of St. Ninian's Ayre (Tombolo) in Shetland, Scotland. Journal of Coastal Research, 13, 202-208.

Foerster Laures, F. (1986). Roman maritime trades. International Journal of Nautical Archaeology, 15, 166-167.

Folk, R. L. (1966). A review of grain-size parameters. Sedimentology, 6, 73-93.

Folk, R. L. and Ward, W. C. (1957). Brazos river bar: a study in the significance of grain size parameters. Journal of Sedimentary Petrology, 27, 3-26.

Forest, C. and Forest, J. D. (1977). Fouilles de la Municipalité de Beyrouth. Geuthner, Paris.

Forstner-Müller, I., Kopetzky, K. and Doumet-Serhal, C. (2006). Egyptian pottery of the late 12th and 13th dynasty from Sidon. Archaeology and History in Lebanon, 24, 52-59.

Fouache, E. (Ed.) (2003). The Mediterranean world environment and history. Elsevier, Paris.

Fouache, E., Dalongeville, R., Kunesch, S., Suc, J.-P., Subally, D., Prieur, A. and Lozouet, P. (2005). The environmental setting of the harbor of the classical site of Oeniades on the Acheloos delta, Greece. Geoarchaeology, 20, 285-302.

Fouache, E. and Pavlopoulos, K. (Eds.) (2005). Sea-level changes in eastern Mediterranean during Holocene: indicators and human impacts. Zeitschrift für Geomorphologie, Supplement Volume 137.

Franco, L. (1996). Ancient Mediterranean harbours: a heritage to preserve. Ocean and Coastal Management, 30, 115-151.

Francou, S. (2002). La mobilité des rivages du port de Beyrouth à l'époque contemporaine, une aide à la recherché géoarchéologique. Archaeology and History in Lebanon, 15, 52-56.

Frechen, M., Dermann, B., Boenigk, W. and Ronen, A. (2001). Luminescence chronology of aeolianites from the section at Givat Olga - Coastal Plain of Israel. Quaternary Science Reviews, 20, 805-809.

Frechen, M., Neber, A., Dermann, B., Tsatskin, A., Boenigk, W. and Ronen, A. (2002). Chronostratigraphy of aeolianites from the Sharon Coastal Plain of Israel. Quaternary International, 89, 31-44.

Frechen, M., Neber, A., Tsatskin, A., Boenigk, W. and Ronen, A. (2004). Chronology of Pleistocene sedimentary cycles in the Carmel Coastal Plain of Israel. Quaternary International, 121, 41-52.

Frihy, O., Fanos, A. M., Khafagy, A. A. and Komar, P. D. (1991). Patterns of nearshore sediment transport along the Nile Delta, Egypt. Coastal Engineering, 15, 409-429.

Frost, H. (1963). Under the Mediterranean. Routledge, London.

Frost, H. (1964). Rouad, ses récifs et mouillages. Prospection sous-marine. Annales Archéologiques de Syrie, 14, 67-74.

Frost, H. (1966). The Arwad plans 1964, a photogrammetric survey of marine installations. Annales Archeologique de Syrie, 16, 13-28.

Frost, H. (1969a). The stone anchors of Byblos. Mélanges Dunand, Mélanges de l'Université Saint Joseph, LXV, 26, 425-442.

Frost, H. (1969b). The stone anchors of Ugarit. Ugaritica, 6, 235-245.

Frost, H. (1970). Bronze-Age Stone-Anchors from the Eastern Mediterranean. Mariner's Mirror, 56, 377-394.

Frost, H. (1971). Recent observations on the submerged harbour works at Tyre. Bulletin du Musée de Beyrouth, 24, 103-111.

Frost, H. (1973). The offshore island harbour at Sidon and other Phoenician sites in the light of new dating evidence. International Journal of Nautical Archaeology, 2, 75-94.

Frost, H. (1990). Museum report - tourism aids archaeology: the Ustica experiment. International Journal of Nautical Archaeology, 19, 341-343.

Frost, H. (1995). Harbours and proto-harbours; early Levantine engineering. In: Karageorghis, V. and D. Michaelides, D. (Eds.), Proceedings of the International Symposium 'Cyprus and the Sea'. University of Cyprus, Nicosia, pp. 1-21.

Frost, H. (2000a). From Byblos to Pharos. In: Some archaeological considerations. Underwater archaeology and coastal management. Focus on Alexandria. UNESCO publishing, coastal management sourcebooks, Paris, pp. 64-68.

Frost, H. (2000b). Installations on the ancient offshore anchorage at Sidon (the rock island of Zire). National Museum News, 10, 69-73.

Frost, H. (2002a). Byblos: the lost temple, the cedars and the sea. A marine archaeological survey. Archaeology and History in Lebanon 15, 57-77.

Frost, H., (2002b). Fourth season of marine investigation: preliminary charting of the offshore shallows. Bulletin d'Archéologie et d'Architecture Libanaises 6, 309-316.

Frost, H. (2004). Byblos and the Sea. In: Doumet-Serhal, C. (Ed.), Decade: A Decade of Archaeology and History in the Lebanon. Archaeology and History in Lebanon, Beirut, pp. 316-347.

Frost, H. (2005). Archaeology, history and the history of archaeology connected with Tyre's harbours. Bulletin d'Archéologie et d'Architecture Libanaises, Hors-Série 2, 45-52.

Frost, H. and Morhange, C. (2000). Proposition de localisation des ports antiques de Byblos (Liban). Méditerranée, 94, 101-104.

Frumkin, A., Magaritz, M., Carmi, I. and Zak I. (1991). The Holocene climatic record of the salt caves of Mount Sedom, Israel. The Holocene, 1, 191-200.

Gale, N. H. and Stos-Gale, Z. A. (1982). Bronze Age copper sources in the Mediterranean: a new approach. Science, 216, 11-19.

Galili, E. and Weinstein-Evron, M. (1985). Prehistory and paleoenvironments of submerged sites along the Carmel coast of Israel. Paléorient, 11, 37-51.

Galili, E., Weinstein-Evron, M. and Ronen, A. (1988). Holocene sea-level changes based on submerged archeological sites off the Northern Carmel coast in Israel. Quaternary Research, 29, 36-42.

Galili, E., Dahari, U. and Sharvit, J. (1993a). Underwater surveys and resue excavations along the Israeli coast. International Journal of Nautical Archaeology, 22, 61-77.

Galili, E., Weinstein-Evron, M., Hershkovitz, I., Gopher, A., Kislev, M., Lernau, O., Kolska-Horwitz, L. and Lernau, H. (1993b). Atlit-Yam: a prehistoric site on the sea floor off the Israeli coast. Journal of Field Archaeology, 20, 133-157.

Galili, E. and Sharvit, J. (1998). Ancient coastal installations and the tectonic stability of the Israeli coast in historical times. In: Stewart, I. S. and Vita-Finzi, C. (Eds.), Coastal tectonics. Geological Society, Special Publications 146, London, pp. 147-163.

Galili, E. and Sharvit, J. (2000). Underwater Archaeological Heritage of Israel: Nature, Policy and Management of Endangered Resource. In: Schutz des Kulturerbes unter Wasser Beiträge zum Internationalen Kongreß für Unterwasserarchäologie (IKUWA 99'), 18-21 Februar 1999. Sassnitz auf Rügen, pp. 555-561.

Galili, E., Raban, A. and Sharvit, J. (2002). Forty years of marine archaeology in Israel. Tropis, 7, 927-961.

Gambin, T. (2004). Islands of the Middle Sea: an archaeology of a coastline. In: De Maria, L. and Turchetti, R. (Eds.), Evolución paleoambiental de los puertos y fondeaderos antiguos en el Mediterráneo occidental. Rubbettino Editore, Soveria Mannelli, pp. 127-146.

Garcia, D. and Vallet, L. (Eds.) (2002). L'espace portuaire de Lattes antique. Lattara, 15, 1-223.

García, J. M. (1998). Portus Carthaginensis. Simbioisis de un emporio y una gran base militar. In: Pérez Ballester, J. and Berlanga, G. P. (Eds.), Reunión Internacional sobre Puertos Antiguos y Comercio Marítimo. III Jornades de Arqueología Subacuática: Actas. La Imprenta, Valencia, pp. 77-97.

Gardner, J. V., Dartnell, P., Mayer, L. A., Hughes Clarke , J. E., Calder, B. R. and Duffy, G. (2005). Shelf-edge deltas and drowned barrier-island complexes on the northwest Florida outer continental shelf. Geomorphology, 64, 133-166.

Garrison, E. G. (1998). A History of Engineering and Technology: Artful Methods. CRC Press, Florida.

Gavin, A. and Maluf, R. (1996). Beirut Reborn: The Restoration and Development of the Central District. Wiley, London.

Gazda, E. K. (2001). Cosa's Contribution to the Study of Roman Hydraulic Concrete: An Historiographic Commentary. In: Goldman, N. W. (Ed.), Classical Studies in Honor of Cleo Rickman Fitch, New York, pp. 145-177.

Gé, T., Migeon, W. and Szepertyski, B. (2005). L'élévation séculaire des berges antiques et médiévales de Bordeaux. Étude géoarchéologique et dendrochronologique. Comptes Rendus Geoscience, 337, 297-303.

Georges, K. (2004). Méthode de traitement statistique appliquée à l'éxoscopie des quartz, environnements fluviaux et littoraux de Provence, PhD thesis. Université de Provence, Aix-en-Provence.

Georges, K. and Prone, A. (2000). Le site littoral de l'anse Saint-Roch à la fin de la montée marine holocène (fouilles de Port Prestige, Antibes, Alpes-Maritimes), résultats préliminaires. Méditerranée, 1-2, 53-59.

Georgiades, A. S. (1907). Les ports de la Grèce dans l'Antiquité qui subsistent encore aujourd'hui. Athènes, Files 1-6.

Geyh, M. A. (1994). The Paleohydrology of the Eastern Mediterranean. In: Bar-Yosef, O. and Kra, R. S. (Eds.), Late Quaternary Chronology and Paleoclimates of the Eastern Mediterranean. Radiocarbon, Tucson, pp. 131-145.

Giampaola, D., Carsana, V. and Boetto, G. (2004). Il mare torna a bagnare Neapolis. Parte II: dalla scoperta del porto al recupero dei relitti. L'archeologo subacqueo, 10, 3, 15-19.

Giampaola, D. and Carsana, V. (2005). Le nuove scoperte: la città, il porto e le machine. In: Lo Sarde, E. (Ed.), Eureka ! il genio degli antichi. Museo archeologico nazionale, Napoli, pp. 116-122.

Gianfrotta, P. A. (1996). Harbor Structures of the Augustan Age in Italy. In: Raban, A. and Holum, K. G. (Eds.), Caesarea Maritima: A Retrospective after Two Millennia. Brill Academic Publishers, Leiden, pp. 65-76.

Gibbins, D. J. L. (1990). The hidden museums of the Mediterranean. New Scientist, 128, 35-40.

Gibbins, D. and Adams, J. (2001). Shipwrecks and maritime archaeology. World Archaeology, 32, 279-291.

Giescke, H.-E. (1983). The Akrotiri Ship Fresco. International Journal of Nautical Archaeology, 12, 123-143.

Gifford, J. A. (1978). Paleogeography of archaeological sites of the Larnaca lowlands, southeastern Cyprus, PhD thesis. University of Minnesota, Minnesota.

Gifford, J., Rapp, G. and Vitali, V. (1992). Paleogeography of Carthage (Tunisia): Coastal change during the first millennium BC. Journal of Archaeological Science, 19, 575-596.

Gill, D. (1999). Non-tectonic settlement of the Herodian harbour in Caesarea. Israel Geological Society Annual Meeting, 24.

Gillmer, T. C. (1975). The Thera Ship. Mariner's Mirror, 61, 321-329.

Gillmer, T. C. (1978). The Thera Ships: A Re-analysis. Mariner's Mirror, 64, 125-133.

Gillmer, T. C. (1985). The Thera Ships as Sailing Vessels. Mariner's Mirror, 71, 401-416.

Ginouves, R. (1998). Espaces architecturaux, bâtiments et ensembles (Dictionnaire méthodique de l'architecture grecque et romaine). Ecole française de Rome, Rome.

Giuliani, C. F. (1996). Note sulla topografia di Portus. In: Mannucci, V. (Ed.), Il parco archeologico naturalistico del porto di Traiano: metodo e progetto. Gangemi Editore, Roma.

Goddio, F., Bernand, A., Bernand, E., Darwish, I., Kiss, Z. and Yoyotte, J. (1998). Alexandria: the Submerged Royal Quarters. Periplus, London.

Goddio, F. and Bernand, A. (2004). Sunken Egypt - Alexandria. Periplus, London.

Goiran, J.-P. (2001). Recherches géomorphologiques dans la région littorale d'Alexandrie, Egypte, PhD thesis. Université de Provence, Aix-en-Provence.

Goiran, J.-P. and Morhange, C. (2003). Géoarchéologie des ports antiques de méditerranée. Topoi, 11, 645-667.

Goiran, J.-P., Marriner, N., Morhange, C., Abd El-Maguib, M., Espic, K., Bourcier, M. and Carbonel, P. (2005). Evolution géomorphologique de la façade maritime d'Alexandrie (Egypte) au cours des six derniers millénaires. Méditerranée, 104, 61-64.

Goldshmidt, V. and Gilboa, M. (1985). Development of an Israeli tidal atlas and comparison with other Mediterranean tidal data. Israel Oceanographic and Limnological Research report H8/85, Haifa.

Goldsmith, V. and Golik, A. (1980). Sediment transport model of the southeastern Mediterranean coast. Marine Geology, 37, 147-175.

Goldsmith, V., and Sofer, S. (1983). Wave climatology of the southeastern Mediterranean. Israel Journal of Earth Science, 32, 1-51.

Golik, A. (1993). Indirect evidence for sediment transport on the continental shelf off Israel. Geo-Marine Letters, 9, 103-108.

Golvin, J.-C. (2003). L'antiquité retrouvée. Errance, Paris.

Good, G. L. (Ed.) (1991). Waterfront Archaeology: International Conference Proceedings. Council for British Archaeology.

Goodfriend, G. A. and Stanley, D. J. (1999). Rapid strand-plain accretion in the northeastern Nile Delta in the 9th century A.D. and the demise of the port of Pelusium. Geology, 27, 147-150.

Gorham, D. L. (2001). Pollen, phytoliths, and other microscopic plant remains in underwater archaeology. International Journal of Nautical Archaeology, 30, 282-298.

Gorham, D. and Bryant, V. M. (2001). The Role of Pollen and Phytoliths in Underwater Archaeology. International Journal of Nautical Archaeology, 30, 299-305.

Gosseaume, E. (1973). Le tombolo triple d'Orbetello (Toscane). Bulletin de la Société Languedocienne de Géographie, 7, 3-11.

Gould, R. A. (2000). Archaeology and the Social History of Ships. Cambridge University Press, Cambridge.

Gouvernet, C. (1948). Une plage ancienne dans le Lacydon à Marseille. Bull. Soc. Linn. de Prov., 16, 13-19.

Green, J. (2004). Maritime Archaeology: A Technical Handbook (Second Edition). Elsevier Academic Press, London.

Grenier, A. (1934). Manuel d'archéologie Gallo-Romaine. Picard, Paris.

Griffiths, D. and Ownby, M. (2006). Assessing the occurrence of Egyptian marl C ceramics in Middle Bronze Age Sidon. Archaeology and History in Lebanon, 24, 63-77.

Guérin, V. (1880). Description geographique, historique et archeologique de la Palestine. Auguste Durand, Paris.

Guidoboni, E., Comastri, A. and Traina, G. (1994). Catalogue of Ancient Earthquakes in the Mediterranean Area up to the 10th Century. Istituto Nazionale di Geofisica, Bologna, Roma.

Guilcher, A. (1958). Coastal and submarine morphology. Methuen and Co. Ltd., London.

Gulliver, F. P. (1896). Cuspate forelands. Geological Society of America Bulletin, 7, 399-422.

Gulliver, F. P. (1899). Shoreline topography. Proceedings of the American Academy of Arts and Science, 34, 149-258.

Gutscher, M. A. (2005). Destruction of Atlantis by a great earthquake and tsunami? A geological analysis of the Spartel Bank hypothesis. Geology, 33, 685-688.

Hadjidaki, E. (1988). Preliminary Report of the Excavations at the Harbor of Phalasarna in West Crete. American Journal of Archaeology, 92, 463-479.

Hadjidaki, E. (1996). The Hellenistic Harbor of Phalasarna in Western Crete : A Comparison with the Hellenistic Inner Harbour of Straton's Tower. In: Raban, A. and Holum, K.G. (Eds.) Caesarea Maritima - A Retrospective After Two Millennia. Brill, Leiden, pp. 53-64.

Haggai, A. (2006). Phoenician Atlit and its newly-excavated harbour: a reassessment. Tel Aviv, Journal of the Institute of Archaeology of Tel Aviv University, 33, 43-60.

Hairy, I. (2006). Alexandrie: nouvelles découvertes autour du Phare. Archéologia, 429, 26-36.

Haldane, C. W. (1990). Egyptian Hulls and Evidence for Caulking. International Journal of Nautical Archaeology, 19, 135-136.

Haldane, D. (1993). At the crossroads of history: nautical archaeology in Syria. Institute for Nautical Archaeology Newsletter, 20, 3-7.

Hall, L. J. (2004). Roman Berytus: Beirut in Late Antiquity. Routledge, London.

Hallager, E. (1988). Aspects of Aegean Long-Distance Trade in the Second Millennium BC. In: Acquaro, E., Godart, L., Mazza, F. and Musti, D. (Eds.), Momenti precoloniali nel mediterraneo antico. Collezione di Studi Fenici 28, Rome, pp. 91-101.

Halliday Saville, L. (1941). Ancient Harbours. Antiquity, 15, 209-232.

Hankey, V. (1993). Egypt, the Aegean and the Levant. Egyptian Archaeology, 3, 27-29.

Harajli, M., Sadek, S. and Asbahan, R. (2002). Evaluation of the seismic hazard of Lebanon. Journal of Seismology, 6, 257-277.

Haslett, S. K. (2002). Quaternary environmental micropalaeontology. Arnold, London.

Hastaoglou-Martinidis, V. (1998). Les villes-ports du basin oriental de la Méditerranée à la fin du XIXe siècle: travaux portuaires et transformations urbaines. In: Petites et grandes villes du bassin Méditerranéen. Etudes autour de l'œuvre d'Etienne Dalmasso. Collection de l'Ecole Française de Rome, Palais Farnèse, pp. 507-525.

Haug, G. H., Gunther, D., Peterson, L. C. D., Sigman, M., Hughen, K. A. and Aeschlimann, B. (2003). Climate and the collapse of Mayan civilization. Science, 299, 1731–1735.

Heinze, M. and Bartl, K. (1997). Bey 024 'Place Debbas', preliminary report. Bulletin d'Archéologie et d'Architecture Libanaises, 2, 236-257.

Hermary, A., Hesnard, A. and Tréziny, H. (1999). Marseille grecque: la cité phocéenne (600-49 av. J.-C.). Errance, Paris.

Hesnard, A. (1994). Une nouvelle fouille du port de Marseille, Place Jules Verne. Compte-rendus de l'Académie des Inscriptions et Belles-Lettres, 1, 195-216.

Hesnard, A. (1995). Les ports antiques de Marseille, Place Jules-Verne. Journal of Roman Archaeology, 8, 65-77.

Hesnard, A. (2004a). Terre submergée, mer enterrée: une "géoarchéologie" du port antique de Marseille. In: De Maria, L. and Turchetti, R. (Eds.), Evolución paleoambiental de los puertos y fondeaderos antiguos en el Mediterráneo occidental. Rubbettino Editore, Soveria Mannelli, pp. 3-29.

Hesnard, A. (2004b). Vitruve, De architectura, V, 12 et le port romain de Marseille. In: Gallina Zevi, A. and Turchetti, R. (Eds.), Le strutture dei porte e degli approdi antichi. Rubbettino Editore, Soveria Mannelli, pp. 175-203.

Hesse, A. (1998). Arguments pour une nouvelle hypothèse de localisation de l'Heptastade d'Alexandrie. Etudes alexandrines, 1, 1-33.

Hesse, A. (2000). Archaeological prospection. In: Ellis, L. (Ed.), Archaeological method and theory: an encyclopedia. Garland, New York, pp. 35-39.

Hine, A. C. (1979). Mechanisms of berm development and resulting beach growth along a barrier spit complex. Sedimentology, 6, 333-351.

Hodge, T. A. (1981). Vitruvius, lead pipes, and lead poisoning. American Journal of Archaeology, 85, 486-491.

Hohlfelder, R. L. (1997). Building harbours in the early Byzantine era: the persistence of Roman technology. Byzantinische Forschungen Internationale Zeitschrift für Byzantinistik, 24, 367-389.

Holden, J., West, L. J., Howard, A. J., Maxfield, E., Panter, I. and Oxley, J. (2006). Hydrological controls of in situ preservation of waterlogged archaeological deposits. Earth-Science Reviews, 78, 59-83.

Holmes, J. A. and Chivas, A. R. (Eds.) (2002). The Ostracoda: Applications in Quaternary Research, Geophysical Monograph, vol. 131. American Geophysical Union, Washington DC.

Holmes, Y. L. (1975). The Foreign Trade of Cyprus During the Late Bronze Age. In: Robertson, N. (Ed.), The Archaeology of Cyprus: Recent Developments. Park Ridge, pp. 90-110.

Holst, S. and Harb, A. (2006). Phoenicians: Lebanon's Epic Heritage. Sierra Sunrise Publishing.

Hong, S., Candelone, J.-P., Patterson, C. C. and Boutron, C. F. (1994). Greenland Ice Evidence of Hemispheric Lead Pollution Two Millennia Ago by Greek and Roman Civilizations. Science, 265, 1841-1843.

Hong, S., Candelone, J.-P., Patterson, C. C. and Boutron, C. F. (1996). History of Ancient Copper Smelting Pollution During Roman and Medieval Times Recorded in Greenland Ice. Science, 272, 246-249.

Horden, P. and Purcell, N. (2000). The Corrupting Sea: A Study of Mediterranean History. Blackwell Publishers, Oxford.

Hornell, J. (1970). Water Transport: Origins and Early Evolution. Newton Abbot.

Horton, B. P. and Edwards, R. J. (2006). Quantifying Holocene Sea Level Change Using Intertidal Foraminifera: Lessons from the British Isles. Journal of Foraminiferal Research, Special Publications 40.

Humphrey, J. W., Oleson, J. P. and Sherwood, A. N. (1998). Greek and Roman Technology: A Sourcebook - Annotated Translations of Greek and Latin Texts and Documents. Routledge, London.

Hurst, H. (1993). Le port militaire de Carthage. Les dossiers d'Archéologie, 183, 42-51.

Hurst, H. and Stager, L. E. (1978). A metropolitan landscape: the Late Punic port of Carthage. World Archaeology, 9, 334-346.

Ilan, D. (1995). The Dawn of Internationalism - the Middle Bronze Age. In: Levy, T. E. (Ed.), The Archaeology of Society in the Holy Land. Leicester University Press, Leicester, pp. 297-319.

Inman, D. L. (1974). Ancient and modern harbors: a repeating phylogeny. In: Proceedings of the 15th International Conference on Coastal Engineering. ASCE, New York, pp. 2049-2067.

Inman, D. L. and Jenkins, S. A. (1984). The Nile littoral cell and man's impact on the coastal zone of the southeastern Mediterranean. Scripps Institution of Oceanography Series, 31, 1-43.

Issar, A. S. (1990). Water Shall Flow from the Rock. Springer, Heidelberg.

Issar, A. S., Govrin, Y., Geyh, M.A., Wakshal, E. and Wolf, M. (1992). Climate changes during the upper Holocene in Israel. Israel Journal of Earth Sciences, 40, 219-223.

Issar, A. S. (2003). Climate changes during the Holocene and their impact on hydrological systems. Cambridge University Press, Cambridge.

Isserlin, B. S. J. (1971). New light on the "cothon" at Motya. Antiquity, 45, 178-186.

Isserlin, B. S. J. (1974). The cothon at Motya: Phoenician harbour works. Archaeology, 27, 188-194.

Izarra, F. de (1993). Le fleuve et les hommes en Gaule romaine. Errance, Paris.

Jameson, M. H. (1973). Halieis at Porto Cheli. In: Blackman, D. J. (Ed.), Marine Archaeology. Colston Papers 23, London.

Jenkins, D. A. (1989). Trace element geochemistry in archaeological sites. Environmental Geochemistry and Health, 11, 57-62.

Jenkins, D. G. (Ed.) (1993). Applied Micropalaeontology. Kluwer Academic Publications, Dordrecht.

Jessup, H. H. (1910). Fifty-three years in Syria. Revell, New York.

Jidejian, N. (1996). Tyre through the ages. Librairie Orientale, Beirut.

Jing, Z. and Rapp, G. (2003). Holocene coastal landscape evolution of the Ambracian embayment in Epirus, Western Greece and its relationships to archaeological settings. In: Wiseman, J. and Zachos, K. (Eds.), Landscape Archaeology in Southern Epirus. American School of Classical Studies, Princeton, pp. 157-198.

Johnson, D. W. (1919). Shore processes and shoreline development. Wiley, New York.

Johnston, P. F. (1993). Treasure salvage, archaeological ethics and maritime museums. International Journal of Nautical Archaeology, 22, 53-60.

Jondet, G. (1916). Les ports submergés de l'ancienne île de Pharos. Mémoires presentés à l'Institut Egyptien, Cairo.

Kaegi, W. E. (2005). Byzantium and the Early Islamic Conquests. Cambridge University Press, Cambridge.

Karageorghis, V. and Michaelides, D. (Eds.) (1995). Proceedings of the International Symposium 'Cyprus and the Sea'. University of Cyprus, Nicosia.

Karam, N. (1996). Beyrouth. L'Histoire qu'on assassine. Self-published, Beirut.

Kassis, A. (2005). Tyr. In: Leclant, J. (Ed.), Dictionnaire de l'Antiquité. Presses Universitaires de France, Paris, pp. 2234-2235.

Katzenstein, H. J. (1997). The History of Tyre. Ben-Gurion University of the Negev Press, Jerusalem.

Kayan, I. (1996). Holocene Coastal Development and Archaeology in Turkey. Zeitschrift für Geomorphologie, Supplement Band 102, 37-59.

Kayan, I. (1999). Holocene stratigraphy and geomorphological evolution of the Aegean coastal plains of Anatolia. Quaternary Science Reviews, 18, 541-548.

Keay, S., Millett, M., Paroli, L. and Strutt, K. (Eds.) (2005a). Portus. Archaeological Monographs of the British School at Rome, 15, 1-360.

Keay, S., Millett, M. and Strutt, K. (2005b). The survey results. In: Keay, S., Millett, M., Paroli, L. and Strutt, K. (Eds.), Portus. Archaeological Monographs of the British School at Rome, 15, London, Oxford, pp. 71-172.

Keller, F. (1854). Die keltischen Pfahlbauten in den Schweizerseen. Mittheilungen der Antiquarischen Gesellschaft in Zürich, 9, 3, 67-101.

Kemp, P. (2001). The History of Ships. Greenwich Editions, London.

Kennedy, H. (1985a). The Last Century of Byzantine Syria: A Reinterpretation. Byzantinische Forschungen, 10, 141-183.

Kennedy, H. (1985b). From polis to madina: urban change in late antique and early Islamic Syria. Past and Present, 106, 3-27.

Kenoyer, J. M. (1998). Ancient cities of the Indus Valley Civilisation. Oxford University Press, Oxford.

Kenrick, J. (1855). History and Antiquities of Phoenicia. B. Fellowes, London.

Kingsley, S. A. (1996). The ancient harbour and anchorage at Dor, Israel: Results of the underwater surveys, 1976-1991. British Archaeological Reports International Series 626, Oxford.

Kiskyras, D. A. (1988). The reasons for the disappearance of the ancient Greek town Helice (Eliki): geological contribution to the search for it. In: Marinos, P. G. and Koukis, G. C. (Eds.), Engineering Geology of Ancient Works, Monuments and Historical Sites. Balkema, Rotterdam, pp. 1301-1306.

Kjerfve, B. and Magill, K. E. (1989). Geographic and hydrodynamic characteristics of shallow coastal lagoons. Marine Geology, 88, 187-199.

Knapp, A. B. (1993). Thalassocracies in Bronze Age Eastern Mediterranean Trade: Making and Breaking a Myth. World Archaeology, 24, 332-347.

Knox, J. C. (2006). Floodplain sedimentation in the Upper Mississippi Valley: natural versus human accelerated. Geomorphology, 79, 286-310.

Kokkalas, S. and Koukouvelas, I. K. (2005). Fault-scarp degradation modeling in central Greece: the Kaparelli and Eliki faults (Gulf of Corinth) as a case study. Journal of Geodynamics, 40, 200-215.

Koukouvelas, I.K., Stamatopoulos, L., Katsonopoulou, D. and Pavlides, S. (2001). A palaeoseismological and geoarchaeological investigation of the Eliki fault, Gulf of Corinth, Greece. Journal of Structural Geology, 23, 531-543.

Kraft, J. C., Rapp, G. and Aschenbrenner, S. E. (1975). Late Holocene Paleogeography of the Coastal Plain of the Gulf of Messina, Greece, and Ist Relationships to Archaeological Settings and Coastal Change. Geological Society of America Bulletin, 86, 1191-1208.

Kraft, J. C., Aschenbrenner, S. E. and Rapp, G. (1977). Paleogeographic reconstructions of coastal Aegean archaeological sites. Science, 195, 941-947.

Kraft, J. C., Kayan, I. and Erol, O. (1980a). Geomorphic reconstructions in the environs of ancient Troy. Science, 209, 1191-1208.

Kraft, J. C., Rapp, G. R. and Aschenbrenner, S. E. (1980b). Late Holocene Palaeogeomorphic Reconstructions in the Area of the Bay of Navarino: Sandy Pylos. Journal of Archaeological Science, 7, 187-210.

Kraft, J. C., Belknap, D. F. and Kayan, I. (1983). Potentials of discovery of human occupation sites on the continental shelves and nearshore coastal zone. In: Masters, P. M. and Flemming, N. C. (Eds.), Quaternary Coastlines and Maritime Archaeology. Academic Press, London, pp. 87-120.

Kraft, J. C., Rapp, G. R., Kayan, I. and Luce, J. V. (2003). Harbor areas at ancient Troy: sedimentology and geomorphology complement Homer's Iliad. Geology, 31, 163-166.

Kraft, J. C., Rapp, G. R., Aschenbrenner, S. E. and Gifford, J. A. (2005). Coastal Change and Archaeological Settings in Elis. Hesperia, 74, 1-39.

Kristiansen, K. and Larsson, T. B. (2005). The Rise of Bronze Age Society: Travels, Transmissions and Transformations. Cambridge University Press, Cambridge.

Laborel, J. and Laborel-Deguen, F. (1994). Biological indicators of relative sea-level variations and co-seismic displacements in the Mediterranean region. Journal of Coastal Research, 10, 395-415.

Laborel, J., Morhange, C., Lafont, R., Le Campion, J., Laborel-Deguen, F. and Sartoretto, S. (1994). Biological evidence of sea-level rise during the last 4500 years on the rocky coasts of continental southwestern France and Corsica. Marine Geology, 120, 203-223.

Lambeck, K. and Bard, E. (2000). Sea-level change along the French Mediterranean coast for the past 30 000 years. Earth and Planetary Science Letters, 175, 203-222.

Lambeck, K., Anzidei, M., Antonioli, F., Benini, A. and Esposito, A. (2004). Sea level in Roman time in the Central Mediterranean and implications for recent change. Earth and Planetary Science Letters, 224, 563-575.

Lamouroux, M. (1967). Contribution à l'étude de la pédogénèse en sols rouges méditerranéens. Sciences du Sol, 2, 55-85.

Lamouroux, M., Paquet, H., Pinta, M. and Millot, G. (1967). Notes preliminaries sur les minéraux argileux des alterations et des sols méditerranéens du Liban. Bulletin du Service de la Carte Géologique d'Alsace Lorraine, 20, 277-292.

Lancaster, L. (2005). Concrete Vaulted Construction in Imperial Rome: Innovations in Context. Cambridge University Press, Cambridge.

Langley, S. B. M. and Unger, R. W. (1984). Nautical archaeology: progress and public responsibility. British Archaeological Report International Series 220, Oxford.

Laorty-Hadji, R. P. (1855). La Syrie, la Palestine et la Judée. Pélerinage à Jérusalem. 6e édition. Bolle-Lasalle, Paris.

Lauffray, J. (1944-45). Forums et monuments de Béryte. Bulletin du Musée de Beyrouth, 7, 13-80.

Lauffray, J. (1946-48). Forums et monuments de Béryte. Bulletin du Musée de Beyrouth, 8, 7-16.

Lauffray, J. (1995). Beyrouth : ce qui n'a pas

été dit. Archéologia, 317, 4-11.

Le Roux, G., Véron, A. and Morhange, C. (2002). Caractérisation géochimique de l'anthropisation dans le port antique de Sidon. Archaeology and History in Lebanon, 15, 37-41.

Le Roux, G., Véron, A. and Morhange, C. (2003a). Geochemical evidences of early anthropogenic activity in harbour sediments from Sidon. Archaeology and History in Lebanon, 18, 115-119.

Le Roux, G., Véron, A., Scholz, C. and Doumet-Serhal, C. (2003b). Chemical and isotopical analyses from the Middle Bronze Age in Sidon. Archaeology and History in Lebanon, 18, 58-61.

Le Roux, G., Véron, A. and Morhange, C. (2005). Lead pollution in the ancient harbours of Marseilles. Méditerranée, 104, 31-35.

Leclant, J. (Ed.) (2005). Dictionnaire de l'Antiquité. Presses Universitaires de France, Paris.

Lefèvre, A. (1995a). Le plus grand chantier d'archéologie urbaine au monde. Archéologia, 316, 14-33.

Lefèvre, A. (1995b). Beyrouth, l'archéologie par le vide. Archéologia, 317, 4-33.

Lehmann-Hartleben, K. (1923). Die Antiken Hafenanlagen im Östlichen Mittelmeer. Dietrich, Leipzig.

Lemcke, G. and Sturm, M. (1997). $\delta^{18}O$ and trace element measurements as proxy for the reconstruction of climate changes at Lake Van (Turkey): preliminary results. In: Dalfes, H. N. *et al.* (Eds.), Third millennium B.C. climate change and Old World collapse, volume 49. Berlin, Springer, Berlin, pp. 178-196.

Lena, G. and Medaglia, S. (2002). Variazioni della linea di costa antica fra Monasterace Marina e la Foce della Fiumara Assi (RC). Geologia dell'Ambiente, 4, 19-22.

Leoni, G. and Dai Pra, G. (1997). Variazioni del livello del mare nel tardo Olocene (ultimi 2500 anni), lungo la costa del Lazio, in base ad indicatori geo-archeologici, interazioni fra neotettonica, eustatismo e clima. ENEA, Dipartimento Ambiente, Centro Ricerche Casaccia RT/AMB/97/8, Roma.

Lespez, L., Dalongeville, R., Pastre, J.-F., Darmon, F., Mathieu, R. and Poursoulis, G. (2003a). Late-Middle-Holocene palaeo-environmental evolution and coastline changes of Malia (Crete). In: Fouache, E. (Ed.), The Mediterranean World Environment and History. Elsevier, London, Paris, pp. 439-452.

Lespez, L., Dalongeville, R., Pastre, J.-F. and Müller-Celka, S. (2003b). Le site de Malia et la mer, approche paléoenvironnementale. Résultats préliminaires : l'analyse sédimentologique du sondage VI. Topoi, 11, 613-633.

Leung Tack, K. D. (1971-72). Etude d'un milieu pollué: le Vieux Port de Marseille. Influence des conditions physiques et chimiques sur la physionomie du peuplement de quai. Téthys, 3, 767-825.

Leveau, P. (2005). L'archéologie du paysage et l'antiquité classique. Agri Centuriati, 2, 9-24.

Leveau, P., Trément, F., Walsh, K. and Barker, G. (Eds.) (1999). Environmental Reconstruction in Mediterranean Landscape Archaeology. Oxbow Books, Oxford.

Lewis, A. R. (1951). Naval power and trade in the Mediterranean AD 500-1100. Princeton University Press, Princeton New Jersey.

Lewis, J. D. (1973). Cosa: an early Roman Harbour. In: Blackman, D. J. (Ed.), Marine Archaeology. Colston Papers 23, London, pp. 219-231.

Linder, E. (1967). Athlit, ville phénicienne d'Israël a-t-elle eu le premier port artificiel de Méditerranée? Archéologia, 17, 25-29.

Lipinski, E. (2004). Itineraria Phoenicia. Peeters Publishers, Leuven.

Lombard, M. (2003). The Golden Age of Islam. Markus Wiener Publishers, Princeton.

Lortet, L. (1884). La Syrie d'aujourd'hui. Voyages dans la Phénicie, le Liban et la Judée, 1875-1880. Hachette, Paris.

Löytved, J. (1876). Plan de Beyrouth dédié à S. M. J. le Sultan Abdel Hamid II, scale 1:12500.

Lyell, C. (1830). Principles of Geology, Volume 1. John Murray, London.

Lyell, C. (1863). Geological Evidence of the Antiquity of Man. John Murray, London.

MacGillivray, J. A. (2003). A Middle Minoan cup from Sidon. Archaeology and History in Lebanon, 18, 20-24.

Maldonado, A. and Stanley, D. J. (1981). Clay mineral patterns as influenced by depositional processes in southeastern Levantine Sea. Sedimentology, 28, 21-32.

Malhotra, V. M. and Mehta, P. K. (1996). Pozzolanic and Cementitious Materials. Gordon & Breach Science Publishers Ltd, Emmaplein.

Mann, D. G. (1999). The species concept in diatoms. Phycologia, 38, 437-495.

Mannucci, V. (1996). Il parco archeologico naturalistico del porto di Traiano. Gangemi editore, Roma.

Marchetti, M. (2002). Environmental changes in the central Po Plain (northern Italy) due to fluvial modifications and anthropogenic activities. Geomorphology, 44, 361-373.

Marcus, E. (2002a). Early seafaring and maritime activity in the southern Levant from prehistory through the third millennium BCE. In: Van den Brink, E. C. M. and Levy, T. E. (Eds.), Egypt and the Levant: Interrelations from the 4th through early 3rd millennium BCE. New Studies in Anthropological Archaeology, Leicester University Press/Continuum International Publishing Group, UK, pp. 403-417.

Marcus, E. (2002b). The southern Levant and maritime trade during the Middle Bronze IIa period. In: Oren, E. and Ahituv, S. (Eds.), Aharon Kempinski Memorial Volume: Studies in Archaeology and Related Disciplines. Beer Sheva, Studies by the Department of Bible and Ancient Near East, vol. 15, Ben-Gurion University of the Negev Press, pp. 241-263.

Marinatos, S. (1971). Excavations at Thera IV (1970 Season). Athens.

Marinatos, S. (1974). Excavations at Thera IV (1972 Season). Athens.

Marinatos, S. (1976). Excavations at Thera VII (1973 Season). Athens.

Markoe, G. E. (2002). Phoenicians. British Museum Press, London.

Marquis, P. (2004). Les fouilles du centre-ville. In: Doumet-Serhal, C. (Ed.), Decade: A Decade of Archaeology and History in the Lebanon. Archaeology and History in Lebanon, Beirut, pp. 266-279.

Marriner, N. (2000). Geoarchaeology and late Holocene coastal evolution of Alexandria's ancient harbours, Egypt, Eastern Mediterranean: a multi-proxy sedimentological and palaeoecological approach, Bachelor of Arts dissertation. University of Durham, Durham.

Marriner, N., de Beaulieu, J.-L. and Morhange, C. (2004). Note on the vegetation landscapes of Sidon and Tyre during antiquity. Archaeology and History in Lebanon, 19, 86-91.

Marriner, N. and Morhange, C. (2005a). Under the city centre, the ancient harbour. Tyre and Sidon: heritages to preserve. Journal of Cultural Heritage, 6, 183-189.

Marriner, N. and Morhange, C. (2005b). Save Tyre. Méditerranée, 104, 129-131.

Marriner, N., Morhange, C., Boudagher-Fadel, M., Bourcier, M. and Carbonel, P. (2005). Geoarchaeology of Tyre's ancient northern harbour, Phoenicia. Journal of Archaeological Science, 32, 1302-1327.

Marriner, N. and Morhange, C. (2006a). Geoarchaeological evidence for dredging in Tyre's ancient harbour, Levant. Quaternary Research, 65, 64-171.

Marriner, N. and Morhange, C. (2006b). The 'Ancient Harbour Parasequence': anthropogenic forcing of the stratigraphic highstand record. Sedimentary Geology, 186, 13-17.

Marriner, N., Morhange, C., Doumet-Serhal, C. and Carbonel, P. (2006a). Geoscience rediscovers Phoenicia's buried harbors. Geology, 34, 1-4.

Marriner, N., Morhange, C. and Doumet-Serhal, C. (2006b). Geoarchaeology of Sidon's ancient harbours, Phoenicia. Journal of Archaeological Science, 33, 1514-1535.

Marriner, N. and Morhange, C. (2007). Geoscience of ancient Mediterranean harbours. Earth-Science Reviews, 80, 137-194.

Mart, Y. and Perecman, I. (1996). Neotectonic activity in Caesarea, the Mediterranean coast of central Israel. Tectonophysics, 254, 139-153.

Martínez-Cortizas, A., Pontevedra-Pombal, X., García-Rodeja, E., Nóvoa-Muñoz, J. C. and Shotyk, W. (1999). Mercury in a Spanish Peat Bog: Archive of Climate Change and Atmospheric Metal Deposition. Science, 284, 939-942.

Masters, P. M. and Flemming, N. C. (Eds.) (1983). Quaternary Coastlines and Maritime Archaeology. Academic Press, London.

Maundrell, H. (1703). A Journey From Aleppo to Jerusalem At Easter A.D. 1697.

McCann, A. M. (1979). The Harbor and Fishery Remains at Cosa, Italy. Journal of Field Archaeology, 6, 391-411.

McCann, A. M. (2002). The Roman Port and Fishery of Cosa. The American Academy in Rome, Rome.

McCann, A. M., Bourgeois, J., Gazda, E. K., Oleson, J. P. and Will, E. L. (1987). The Roman Port and Fishery of Cosa: a Center of Ancient Trade. Princeton University Press, Princeton.

McCaslin, D. E. (1980). Stone anchors in antiquity: coastal settlements and maritime trade-routes in the Eastern Mediterranean ca. 1600-1050 BC. Åström.

McGarry, S., Bar-Matthews, M., Matthews, A., Vaks, A., Schilman, B. and Ayalon, A. (2004). Constraints on hydrological and paleotemperature variations in the Eastern Mediterranean region in the last 140 ka given by the δD values of speleothem fluid inclusions. Quaternary Science Reviews, 23, 919-934.

McKee Smith, J., Sherlock, A. R. and Donald, T. (2001). STWAVE: Steady-State Spectral Wave Model User's Manual for STWAVE Version 3.0. U.S. Army Corps of Engineers, Washington DC.

Mehta, P. K. (1991). Concrete in the Marine Environment. Elsevier, Barking.

Melis, S. (2000). Variations des lignes de rivage aux environs de la ville antique de Nora (Sardaigne, Sud-Ouest-Italie) d'après les données géoarchéologiques. In: Vermeulen, F. and De Dapper, M. (Eds.), Geoarchaeology of the landscapes of Classical Antiquity. Peeters Publishers, Leiden, pp. 127-135.

Mellaart, J. (1967). Catal Hütük: a Neolithic town in Anatolia. Thames and Hudson, London.

Ménanteau, L. and Geffray, O. (2003). Géoarchéologie de la chora de Nymphaion (presqu'île et Détroit de Kertch, Ukraine): apport de la télédétection spatiale et corrélation avec les données archéologiques et historiques. Archeologia, 54, 17-27.

Mendleson, C. (1996). Beirut: Uncovering the Past. National Museum News, 3, 8-9.

Mesnil du Buisson, R. du (1921). Les anciennes défenses de Beyrouth. Syria, 21, 235- 57 and 317-327.

Migowski, C., Stein, M., Prasad, S., Negendank, J. F. W. and Agnon, A. (in press). Holocene climate variability and cultural evolution in the Near East from the Dead Sea sedimentary record. Quaternary Research.

Mikati, R. and Perring, D. (2006). From metropolis to ribat: some recent work on Beirut at the end of antiquity. Archaeology and History in Lebanon, 23, 42-55.

Millet, B. (1989). Fonctionnement hydrodynamique du Bassin de Thau. Validation écologique d'un modèle numérique de circulation. Oceanologica Acta, 12, 37-46.

Millet, B., Blanc, F. and Morhange, C. (2000). Modélisation numérique de la circulation des eaux dans le Vieux-Port de Marseille vers 600 ans avant J.-C. Méditerranée, 94, 61-64.

Millet, B. and Goiran, J.-P. (in press). Impacts of Alexandria's Heptastadion on coastal hydro-sedimentary dynamics during the Hellenistic period: a digital modelling approach. International Journal of Nautical Archaeology.

Milne, G. (1982). Recent work on London's Roman harbour. International Journal of Nautical Archaeology, 11, 163.

Milne, G. (1985). The Port of Roman London. Batsford, London.

Milne, G. and Hobley, B. (Eds.) (1981). Waterfront Archaeology in Britain and Northern Europe (CBA Research Reports). Council for British Archaeology, York.

Milne, G. and Bateman, N. (1983). A Roman harbour in London: excavations and observations near Pudding Lane. Britannia, 15, 207-226.

Ming, D. and Chiew, Y.-M. (2000). Shoreline changes behind detached breakwater. Journal of Waterway, Port, Coastal, and Ocean Engineering, March/April, 63-70.

Mitchell, S. (2006). A History of the Later Roman Empire: AD 284-641. Blackwell, London.

Monicault, J. (1936). Le port de Beyrouth et l'économie des pays du Levant sous le mandat français, PhD thesis. Librairie technique et économique, Paris.

Moore, D. M. and Reynolds, R. C. (1989). X-ray diffraction and the identification and analysis of clay minerals. Oxford University Press, Oxford.

Moore, P. D., Webb, J. A. and Collinson, M. E. (1991). Pollen analysis. Blackwell Scientific Publications, Oxford.

Morgan, M. H. (2000). Vitruvius - The Ten Books on Architecture. Dover Publications, New York.

Morhange, C. (1994). La mobilité des littoraux provençaux: Eléments d'analyse géomorphologiques, PhD thesis. Université de Provence, Aix-en-Provence.

Morhange, C. (Ed.) (2000). Ports antiques et paléoenvironnements littoraux. Méditerranée, 94, 1-112.

Morhange, C. (2001). Mobilité littorale de quelques sites portuaires antiques de Méditerranée : Marseille, Pouzzoes, Cumes, Kition et Sidon, HDR thesis. Université de Provence, Aix-en-Provence.

Morhange, C., Vella, C., Provansal, M., Hesnard, A. and Laborel, J. (1999). Human Impact and Natural Characteristics of the Ancient Ports of Marseille and Fos in Provence, Southern France. In: Barker, G. and Mattingly, D. (Eds.), The Archaeology of Mediterranean Landscapes. Oxbox Books, Oxford, pp. 145-153.

Morhange, C., Goiran, J.-P., Bourcier, M., Carbonel, P., Le Campion, J., Rouchy, J.-M. and Yon, M. (2000). Recent Holocene paleo-environmental evolution and coastline changes of Kition, Larnaca, Cyprus, Mediterranean Sea. Marine Geology, 26, 205-230.

Morhange, C., Laborel, J. and Hesnard, A. (2001). Changes of relative sea level during the past 5000 years in the ancient harbor of Marseilles, Southern France. Palaeogeography, Palaeoclimatology, Palaeoecology, 166, 319-329.

Morhange, C., Blanc, F., Bourcier, M., Carbonel, P., Prone, A., Schmitt-Mercury, S., Vivent, D. and Hesnard, A. (2003a). Bio-sedimentology of the late Holocene deposits of the ancient harbor of Marseilles (Southern France, Mediterranean sea). The Holocene, 13, 593-604.

Morhange, C., Espic, K., Doumet-Serhal, C., Bourcier, M., Carbonel, P. (2003b). The ancient harbours of Sidon, attempt at a synthesis (1998-2002). Archaeology and History in Lebanon, 18, 71-81.

Morhange, C. and Saghieh-Beydoun, M. (Eds.) (2005). Etude géoarchéologique de quatre ports antiques du Liban (Byblos, Beyrouth, Sidon et Tyr). Bulletin d'Archéologie et d'Architecture Libanaises, Hors-Série 2.

Morhange, C., Goiran, J.-P. and Marriner, N. (Eds.) (2005a). Environnements littoraux méditerranéens, heritages et mobilité/Coastal geoarchaeology of the Mediterranean. Méditerranée, 104, 1-140.

Morhange, C., Hamdan Taha, M., Humbert, J. B. and Marriner, N. (2005b). Human settlement and coastal change in Gaza since the Bronze Age. Méditerranée, 104, 75-78.

Morhange, C., Marriner, N., Laborel, J., Todesco, M. and Oberlin, C. (2006a). Rapid sea-level movements and noneruptive crustal deformations in the Phlegrean Fields caldera, Italy. Geology, 34, 93-96.

Morhange, C., Pirazzoli, P. A., Marriner, N., Montaggioni, L. F. and Nammour, T. (2006b). Late Holocene relative sea-level changes in Lebanon, Eastern Mediterranean. Marine Geology, 230, 99-114.

Moscati, S. (Ed.) (1997). Les Phéniciens. Stock, Paris.

Mouterde, R. (1942-43). Monuments et inscriptions de Syrie et du Liban. MUSJ, 25, 1-23.

Mouterde, R. (1966). Regards sur Beyrouth, Phénicienne, Hellénistique et Romaine. Imprimerie Catholique, Beyrouth.

Mouterde, R. and Lauffray, J. (1952). Beyrouth ville romaine. Beyrouth.

Muckelroy, K. (1978). Maritime Archaeology. Cambridge University Press, Cambridge.

Müllenhoff, M., Handl, M., Knipping, M. and Brückner, H. (2004). The evolution of Lake Bafa (Western Turkey) - Sedimentological, microfaunal and palynological results. In: Schernewski, G. and Dolch, T. (Eds.), Geographie der Meere und Küsten. Coastline Reports 1, pp. 55-66.

Müller, G. W. (1894). Ostracoden des Golfes von Neapel un der angrenzenden Meerabschnitte. Flora and fauna Neapel, 21, 1-413.

Murray, J. W. (1991). Ecology and Palaeoecology of Benthic Foraminifera. Longman Scientific and Technical, Harlow.

Naccache, A. (1996). The price of progress. Archaeology, 49, 51-56.

Naccache, A. (1998). Beirut's memorycide: hear no evil, see no evil. In: Meskell, L. (Ed.), Archaeology under fire: nationalism, politics and heritage in the Eastern Mediterranean and Middle East. Routledge, London, pp. 140-158.

Nair, P. K. (1985). Essentials of Palynology. Scholarly Publications.

Neal, A. (2004). Ground-penetrating radar and its use in sedimentology: principles, problems and progress. Earth-Science Reviews, 66, 261-330.

Neev, D., Bakler, N. and Emery, K. O. (1987). Mediterranean Coasts of Israel and Sinai: Holocene Tectonism from Geology, Geophysics, and Archaeology. Taylor and Francis, New York.

Negris, P. (1903a). Observations concernant les variations du niveau de la mer depuis les temps historiques et préhistoriques. Comptes Rendus de l'Académie des Sciences, 137, 222-224.

Negris, P. (1903b). Régression et transgression de la mer depuis l'époque glaciaire jusqu'à nos jours. Revue universelle des mines, de la métallurgie, des travaux publics, des sciences et arts appliqués à l'industrie, 3, 249-281.

Negris, P. (1904a). Vestiges antiques submergés. Mitteilungen des Deutschen Archeologischen Instituts, 29, 340-363.

Negris, P. (1904b). Nouvelles observations sur la dernière transgression de la Méditerranée. Comptes Rendus de l'Académie des Sciences, 2, 379-381.

Negris, P. (1921). Les ports submergés de l'ancienne île de Pharos. Bulletin de la Société Géologique de France, 21, 161-164.

Nibbi, A. (1979). Some Remarks on the Assumption of Ancient Egyptian Sea-Going. Mariner's Mirror, 65, 201, 208.

Nicolaou, K. (1976). The historical topography of Kition. Studies in Mediterranean Archaeology, 153, 1-373.

Nieto, X. and Raurich, X. (1998). La infraestructura portuaria ampuritana. III Jornadas de Arqueología Subacuática, Valencia, pp. 56-76.

Nieto, X., Revil, A., Morhange, C., Vivar, G., Rizzo, E. and Aguelo, X. (2005). La fachada marítima de Ampurias: estudios geofísicos y datos archueológicos. Empúries, 54, 71-100.

Nir, Y. (1984). Recent sediments of the Mediterranean continental shelf and slope, PhD thesis. University of Gothenburg, Gothenburg.

Nir, Y. (1996). The city of Tyre, Lebanon and its semi-artificial tombolo. Geoarchaeology, 11, 235-250.

Nishimura, Y. (2001). Geophysical Prospection in Archaeology. In: Brothwell, D. R. and Pollard, A. M. (Eds.), Handbook of Archaeological Sciences. Wiley, Chichester, pp. 543-553.

Nogueras, P., Burjachs, F., Gallart, F. and Puigdefàbregas, J. (2000). Recent gully erosion in the El Cautivo badlands (Tabernas, SE Spain). CATENA, 40, 203-215.

Nordiguian, L. and Salles, J.-F. (Eds.) (2000). Aux origines de l'archéologie aérienne: A. Poidebard (1878-1955). Presses de l'Université Saint-Joseph, Beyrouth.

Norwich, J. J. (1993). Byzantium: The Apogee. Penguin Books, London.

Noureddine, I. and Helou, M. (2005). Underwater archaeological survey in the northern harbour at Tyre. Bulletin d'Archéologie et d'Architecture Libanaises, Hors-Série 2, 111-128.

Nriagu, J. O. (1983). Lead and lead poisoning in Antiquity. John Wiley, New York.

Nriagu, J. O. (1996). A history of global metal pollution. Science, 272, 223-224.

Nriagu, J. O. (1998). Tales Told in Lead. Science, 281, 1622-1623.

Nuzhet, D., Kukla, G. and Weiss, H. (1997). Third Millennium B.C. Climate Change and Old World Collapse, NATO ASI Series I: Global Environmental Change. Springer-Verlag, Berlin.

Oleson, J. P. (1977). Underwater survey and excavation in the port of Pyrgi (Santa Severa) 1974. Journal of Field Archaeology, 4, 297-308.

Oleson, J. P. (1988). The technology of Roman harbours. International Journal of Nautical Archaeology, 17, 147-157.

Oleson, J. P. and Branton, G. (1992). The Harbour of Caesarea Palaestinae: A Case Study of Technology Transfer in the Roman Empire. Mitteilungen, Leichtweiß-Institut für Wasserbau, 117, 387-421.

Oleson, J. P., Brandon, C., Cramer, S. M., Cucitore, R., Gotti, E. and Hohlfelder, R. L. (2004a). The ROMACONS Project: a Contribution to the Historical and Engineering Analysis of Hydraulic Concrete in Roman Maritime Structures. International Journal of Nautical Archaeology, 33, 199-229.

Oleson, J. P., Brandon, C. and Hohlfelder, R. L. (2004b). The Roman Maritime Concrete Study (ROMACONS): Fieldwork at Portus, Anzio, Santa Liberata, Cosa, 2002-2003. In: Maniscalco, F. (Ed.), Mediterraneum: Tutela e valorizzazione dei beni culturali ed ambientali, vol. 4. Tutela del partimnio culturale sommerso, Naples, pp. 185-194.

Ordóñez Agulla, S. (2003). El Puerto romano de Hispalis. In: Berlanga, G. P. and Pérez Ballester, J. (Eds.), Puertos Fluviales Antiguos: Ciudad, Desarrollo e Infraestructuras. Federico Morillo, Valencia, pp. 59-79.

Ormsby, H. A. (1839). Plan of Beirut, and its East and West Bay, the Site of Ancient Berytus (scale: 1:14,591). Hydrographic Office, London.

Otvos, E. G. (2000). Beach ridges - definitions and significance. Geomorphology, 32, 83-108.

Otvos, E. G. and Giardino, M. J. (2004). Interlinked barrier chain and delta lobe development, northern Gulf of Mexico. Sedimentary Geology, 169, 47-73.

Papadopoulos, G., Vassilopoulou, A. and Plessa, A. (2000). A new catalogue of historical earthquakes in the Corinth rift central Greece: 480 BC - 1910 AD. In: Papadopoulos, G. (Ed.), Historical Earthquakes and Tsunamis in the Corinth rift, Central Greece, Publication No. 12. National Observatory of Athens, Institute of Geodynamics, Athens, pp. 9-119.

Papageorgiou, S., Arnold, M., Laborel, J. and Stiros, S. C. (1993). Seismic uplift of the harbour of ancient Aigeira, Central Greece. International Journal of Nautical Archaeology, 22, 275-281.

Paret, O. (1958). Le mythe des cités lacustres. Dunod, Paris.

Paris, J. (1915). Contributions à l'étude des ports antiques du monde grec, note sur Léchaion. Bulletin de Correspondance hellénique, 29, 6-16.

Paris, J. (1916). Contributions à l'étude des ports antiques du monde grec, les établissements maritimes de Délos. Bulletin de Correspondance hellénique, 40, 5-73.

Parker, A. J. (1995). Maritime cultures and wreck assemblages in the Graeco-Roman world. International Journal if Nautical Archaeology, 24, 87-96.

Paroli, L. (2005). History of Past Research at Portus. In: Keay, S., Millett, M., Paroli, L. and Strutt, K. (Eds.) Portus. Archaeological Monographs of the British School at Rome 15, London, Oxford, pp. 43-59.

Pascual, A., Rodríguez Lázaro, J., Weber, O. and Jouanneau, J. M. (2002). Late Holocene pollution in the Gernika estuary (southern Bay of Biscay) evidenced by the study of Foraminifera and Ostracoda. Hydrobiologia, 475/476, 477- 491.

Paskoff, R., Hurst, H. and Rakob, F. (1985). Position du niveau de la mer et déplacement de la ligne de rivage à Carthage (Tunisie). Compte Rendus de l'Académie des Sciences, 300, II, 13, 613-618.

Pasquinucci, M. (2004). Paleogeografia costiera, porti e approdi in Toscana. In: De Maria, L. and Turchetti, R. (Eds.), Evolución paleoambiental de los puertos y fondeaderos antiguos en el Mediterráneo occidental. Rubbettino Editore, Soveria Mannelli, pp. 61-86.

Pavlopoulos, K., Karkanas, P., Triantaphyllou, M. and Karymbalis, E. (2003). Climate and sea-level changes recorded during late Holocene in the coastal plain of Marathon, Greece. In: Fouache, E. (Ed.), The Mediterranean World Environment and History. Elsevier, London; Paris, pp. 453-465.

Peiser, B. J. (1998). Comparative Analysis of Late Holocene Environmental and Social Upheaval. In: Peiser, B. J., Palmer, T. and Bailey, M. E. (Eds.), Natural Catastrophes during Bronze Age Civilisations. British Archaeological Reports S728, Oxford, pp. 117–139.

Pennington, H. L. and Weber, S. A. (2004). Paleoethnobotany: Modern research connecting ancient plants and ancient peoples. Critical Reviews in Plant Sciences, 23, 3-20.

Péres, J.-M. (1982). Major benthic assemblages. In: Kinne, O. (Ed.), Marine Ecology, vol. 5, part 1. Wiley, Chichester, pp. 373-522.

Péres, J.-M. and Picard, J. (1964). Nouveau manuel de bionomie benthique de la mer Méditerranée. Rec. Trav. Station Marine Endoume 31, Marseille.

Pérez Ballester, J. and Berlanga, G. P. (Eds.) (1998). Reunión Internacional sobre Puertos Antiguos y Comercio Marítimo. III Jornades de Arqueología Subacuática: Actas. La Imprenta, Valencia.

Perrey, A. (1850). Mémoire sur les tremblements de terre ressentis dans la peninsule Turco-hellenique et en Syrie. Acad. R. Sci. de Belgique, Bruxelles, 23, 1-50.

Perring, D. (1999). Excavations in the Souks of Beirut: An Introduction to the Work of the Anglo-Lebanese Team and Summary Report. Berytus, 43, 9-34.

Perring, D., Seeden, H., Sheehan, P. and Williams, T. D. (1996). Archaeological excavations in the souks area of downtown Beirut. Interim report of the AUB project Bey 006 1994-1995. Bulletin d'Archéologie et d'Architecture Libanaises, 1, 176-227.

Petit-Maire, N. and Vrielinck, B. (Eds.) (2005). The Mediterranean basin: the last two climatic cycles - Explanatory notes of the maps. Maison Méditerranéenne des Sciences de l'Homme, Aix-en-Provence.

Pirazzoli, P. A. (1976). Sea level variations in the Northwest Mediterranean during Roman times. Science, 194, 519-521.

Pirazzoli, P. A. (1979-1980). Les viviers à poissons romains en Méditerranée. Oceanis, 5, 191-201.

Pirazzoli, P. A. (1980). Formes de corrosion marine et vestiges archéologiques submergés : interprétation néotectonique de quelques exemples en Grèce et en Yougoslavie. Ann. Inst. océanogr., 56, 101-111.

Pirazzoli, P. A. (1986). The Early Byzantine Tectonic Paroxysm. Zeitschrift für Geomorphologie, 62, 31-49.

Pirazzoli, P. A. (1987a). Submerged remains of ancient Megisti in Castellorizo island (Greece): a preliminary survey. International Journal of Nautical Archaeology, 16, 57-66.

Pirazzoli, P. A. (1987b). Sea-level changes in the Mediterranean. In: Tooley, M. J. and Shennan, I. (Eds.), Sea-level changes. Blackwell Publishers, Oxford, pp. 152-181.

Pirazzoli, P. A. (1988). Sea-level changes and crustal movements in the Hellenic arc (Greece), the contribution of archaeological and historical data. In: Raban, A. (Ed.), Archaeology of coastal changes, Proceedings of the first international symposium "Cities on the sea - past and present". BAR International Series 404, Haifa, pp. 157-184.

Pirazzoli, P. A. (2005). A review of possible eustatic, isostatic and tectonic contributions in eight late-Holocene relative sea-level histories from the Mediterranean area. Quaternary Science Reviews, 24, 18-19, 1989-2001.

Pirazzoli, P. and Thommeret, J. (1973). Une donnée nouvelle sur le niveau marin à Marseille à l'époque romaine. Comptes Rendus de l'Académie des Sciences, 277 D, 2125-2128.

Pirazzoli, P. A., Ausseil-Badie, J., Giresse, P., Hadjidaki, E. and Arnold, M. (1982). Historical Environmental Changes at Phalasarna Harbour, West Crete. Geoarchaeology, 7, 371-392.

Pirazzoli, P. A., Laborel, J., Saliege, J. F., Erol , O. Kayan, I. and Person, A. (1991). Holocene raised shorelines on the Hatay coasts (Turkey): palaeoecological and tectonic implications. Marine Geology, 96, 295-311.

Pirazzoli, P. A., Ausseil-Badie, J., Giresse, P., Hadjidaki, E. and Arnold, M. (1992). Historical environmental changes at Phalasarna Harbor, west Crete. Geoarcheology, 7, 371-392.

Pirazzoli, P. A., Laborel, J. and Stiros, S. C. (1996). Earthquake clustering in the Eastern Mediterranean during historical times. Journal of Geophysical Research, 101, B3, 6083-6097.

Pirenne, H. (1937). Mahomet et Charlemagne. Les Presses Universitaires de France, Paris.

Plassard, J. (1968). Crise séismique au Liban du IV° au VI° siècle. Mélanges de l'Université Saint-Joseph, 44, 10-20.

Plassard, J. and Kogol, B. (1981). Seismicité du Liban. Conseil National de la Recherche Scientifique, Beyrouth.

Pococke, R. (1745) Description of the East and Some Other Countries, vol. II, part I. W. Bowyer, London.

Poidebard, A. (1939). Un grand port disparu, Tyr. Recherches aériennes et sous-marines, 1934-1936. Librairie Orientaliste Paul Geuthner, Paris.

Poidebard, A. and Lauffray, J. (1951). Sidon, aménagements antiques du port de Saïda. Etude aérienne au sol et sous-marine. Imprimerie Catholique, Beyrouth.

Pomerancblum, M. (1966). The distribution of heavy minerals and hydraulic equivalents in sediments of the Mediterranean continental shelf of Israel. Journal of Sedimentary Petrology, 36, 162-174.

Pomey, P. (1995). Les épaves grecques et romaines de la place Jules Verne à Marseille. Compte-Rendus Académie Inscriptions et Belles Lettres, avril-juin, 459-484.

Pomey, P. and Rieth, E. (2005). L'archéologie navale. Errance, Paris.

Poppe, G. T. and Goto, Y. (1991). European seashells, Vol. I. Verlag Christa Hemmen, Wiesbaden.

Poppe, G. T. and Goto, Y. (1993). European seashellsn, Vol. II. Verlag Christa Hemmen, Wiesbaden.

Possehl, G. (2002). The Indus Civilisation. Alta Mira Press, Walnut Creek.

Poulain de Bossay, P. A. (1861). Recherches sur la topographie de Tyr. Madame Maire-Nyon, Paris.

Poulain de Bossay, P. A. (1863). Recherches sur Tyr et Palaetyr. Extrait du recueil de géographie, Paris.

Prone, A. (2003). L'analyse texturale et microstructurale des sols. Exemple pédologique du Nord-Est de la Thailande. Publications de l'Université de Provence, Aix-en-Provence.

Provansal, M., Morhange, C. and Vella, C. (1995). Impacts anthropiques et contraintes naturelles sur les sites portuaires antiques de Marseille et Fos: acquis méthodologique. Méditerranée, 82, 93-100.

Pyatt, F. B., Gilmore, G., Grattan, J. P., Hunt, C. O. and McLaren, S. (2000). An imperial legacy? An exploration of the environmental impact of ancient metal mining and smelting in southern Jordan. Journal of Archaeological Science, 27, 771-778.

Raban, A. (1981). Recent maritime archaeological research in Israel. International Journal of Nautical Archaeology, 10, 287-308.

Raban, A. (1984). Dor: installations maritimes. Revue Biblique, 91, 252-256.

Raban, A. (1985a). Recent Maritime Archaeological Research in Israel. International Journal of Nautical Archaeology, 14, 332-349.

Raban, A. (Ed.) (1985b). Harbour Archaeology. Proceedings of the first international workshop on ancient Mediterranean harbours. Caesarea Maritima, BAR International Series 257, Haifa.

Raban, A. (1985c). The Ancient Harbours of Israel in Biblical Times. In: Raban, A. (Ed.) Harbour Archaeology, Oxford, pp. 11-44.

Raban, A. (1987a). Alternated river courses during the Bronze Age along the Israeli coastline. In: Déplacements des lignes de rivage en Méditerranée d'après les données de l'archéologie. CNRS, Paris.

Raban, A. (1987b). The Harbor of the Sea People at Dor. Biblical Archaeologist, 50, 118-126.
Raban, A. (Ed.) (1988). Archaeology of coastal changes, Proceedings of the first international symposium "Cities on the sea - past and present". BAR International Series 404, Haifa.

Raban, A. (1990). Man instigated coastal changes along the Israeli shore of the Mediterranean in ancient times. In: Entjes-Nieborg, G. and van Zeist W. (Eds.), Man's role in the shaping of the Eastern Mediterranean landscape. Balkema, Rotterdam, pp. 101-112.

Raban, A. (1991). Minoan and Canaanite Harbours. In: Laffineur, R. and Basch, L. (Eds.), Thalassa: L'Egée Préhistorique et la Mer. Aegaeum, 7, pp. 129-146.

Raban, A. (1992a). Sebastos: the royal harbour at Caesarea Maritima - a short-lived giant. International Journal of Nautical Archaeology, 21, 111-124.

Raban, A. (1992b). Archaeological park for divers at Sebastos and other submerged remnants in Caesarea Maritima, Israel. International Journal of Nautical Archaeology, 21, 27-35.

Raban, A. (1995). The heritage of ancient harbour engineering in Cyprus and the Levant. In: Karageorghis, V. and Michaelides, D. (Eds.), Proceedings of Third International Symposium 'Cyprus and the Sea'. University of Cyprus, Nicosia, pp. 139-189.

Raban, A. (1997a). The Phoenician Harbour and Fisherman Village at Atlit. Eretz Israel, 25, 490-508.

Raban, A. (1997b). Phoenician Harbors along the Levantine Coast. Michmanim, 11, 7-27.

Raban, A. and Galili, E. (1985). Recent maritime archaeological research in Israel, a preliminary report. International Journal for Nautical Archaeology, 14, 321-356.

Raban, A. and Holum, K. G. (Eds.) (1996). Caesarea Maritima: A Retrospective after Two Millennia. Brill Academic Publishers, Leiden.

Rapp, G. and Kraft, J. C. (1994). Holocene Coastal Change in Greece and Aegean Turkey. In: Kardulias, P. N. (Ed.), Beyond the Site - Regional Studies in the Aegean Area. University Press of America, Lanham, New York, London, pp. 69-90.

Rapp, G. and Hill, C. L. (1998). Geoarchaeology: The Earth-Science Approach to Archaeological Interpretation. Yale University Press, New Haven.

Raschka, M. (2006). Beirut digs out. In: Vitelli, K. D. (Ed.), Archaeological Ethics. AltaMira Press, pp. 96-102.

Raulin, V. (1869). Description physique de l'ile de Crète. Bertrand, Paris.

Raveh, K. and Kingsley, S. A. (1992). The wreck complex at the entrance to Dor harbour, Israel: preliminary details. International Journal of Nautical Archaeology, 21, 309-315.

Rawlinson, G. (1889). The History of Phoenicia. Longmans, Green, and Co., London.

Reimer, P. J. and McCormac, F. G. (2002). Marine radiocarbon reservoir corrections for the Mediterranean and Aegean seas. Radiocarbon, 44, 159-166.
Reinhardt, E. G., Patterson, R. T. and Schröder-Adams, C. J. (1994). Geoarchaeology of the ancient harbor site of Caesarea Maritima, Israel: Evidence from Sedimentology and Paleoecology of Benthic Foraminifera. Journal of Foraminiferal Research, 24, 37-48.

Reinhardt, E. G., Patterson, R. T., Blenkinsop, J. and Raban, A. (1998). Paleoenvironmental evolution of the inner basin of the ancient harbor at Caesarea Maritima, Israel; foraminiferal and Sr isotopic evidence. Revue de Paleobiologie, 17, 1-21.

Reinhardt, E. G. and Raban, A. (1999). Destruction of Herod the Great's harbor Caesarea Maritima, Israel, geoarchaeological evidence. Geology, 27, 811-814.

Renan, E. (1864). La Mission de Phénicie. Imprimerie Nationale, Paris.

Renberg, I., Persson, M. W. and Emteryd, O. (1994). Pre-industrial atmospheric lead contamination detected in Swedish lake sediments. Nature, 368, 323-326.

Reynolds, P. (1999). Pottery production and economic exchange in 2nd century Berytus: some preliminary observations of ceramic trends from quantified ceramic deposits from the souk excavations in Beirut. Berytus, 43, 35-110.

Ribes, E., Borschneck, D., Morhange, C. and Sandler, A. (2003). Recherche de l'origine des argiles du bassin portuaire antique de Sidon. Archaeology and History in Lebanon, 18, 82-94.

Rickman, G. E. (1988). The archaeology and history of Roman ports. International Journal of Nautical Archaeology, 17, 257-267.

Riedel, H. (1995). Die spätholozäne Entwicklung des Dalyan-Delats (SW-Türkei) - ein Beitrag zur Paläogeographie Kariens. Kölner Geographische Arbeiten, 66, 83-96.

Roberts, D. (2000). Holy Land and Egypt and Nubia. Rizzoli International Publications, New York.

Robinson, S. A., Black, S., Sellwood, B. W. and Valdes, P. J. (2006). A review of palaeoclimates and palaeoenvironments in the Levant and Eastern Mediterranean from 25,000 to 5000 years BP: setting the environmental background for the evolution of human civilisation. Quaternary Science Reviews, 25, 1517-1541.

Rodriguez, A. (1990). A new maritime museum for Malta. International Journal of Nautical Archaeology, 19, 250.

Roller, D. W. (1998). The Building Program of Herod the Great. University of California Press, Los Angeles.

Rosada, G. (2003). Il porto di Aquileia nel sistema degli scali fluvio-lagunari della Decima Regio. In: Berlanga, G. P. and Pérez Ballester, J. (Eds.), Puertos Fluviales Antiguos: Ciudad, Desarrollo e Infraestructuras. Federico Morillo, Valencia, pp. 277-297.

Rosen, A. M. (1986). Cities of Clay: The Geoarchaeology of Tells. The University of Chicago Press, Chicago.

Rosen, A. (1999). Phytolith analysis in Near Eastern Archaeology. In: Pike, S. and Gitin, S. (Eds.), The Practical Impact of Science on Aegean and Near Eastern Archaeology. Archetype Press, London, pp. 86-92.

Rosen, A. M. (2006). Civilizing Climate: Social Responses to Climate Change in the Ancient Near East. AltaMira Press, Lanham.

Rosen, A. and Rosen, S. (2001). Determinist or not determinist? : Climate, environment, and archaeological explanation in the Levant. In: Wolff, S. (Ed.), Studies in the Archaeology of Israel and Neighboring Lands in Memory of Douglas L. Esse. Oriental Institute, University of Chicago, Chicago, pp. 535-554.

Rosen, D. and Kit, E. (1981). Evaluation of the wave characteristics at the Mediterranean coast of Israel. Israel Journal of Earth Science, 30, 120-134.

Rosselini, S. R. (1984). The Cesenatico Maritime Museum. Mariner's Mirror, 7, 119-128.

Rothé, M.-P. and Tréziny, H. (2005). Carte archéologique de la Gaule. Marseille et ses alentours, 13-3. Académie des Inscriptions et Belles-Lettres, Paris.

Rougé, J. (1966). Recherches sur l'organisation du commerce maritime en Méditerranée sous l'empire Romain. S. E. V. P. E. N., Paris.

Round, F. E. (1991). Use of diatoms for monitoring rivers. In: Whitton, B.A., Rott, E. and Friedrich, G. (Eds.), Use of Algae for Monitoring Rivers. Institut für Botanik, Universitat, Innsbruck, pp. 25-32.

Round, F. E. and Crawford, R. M. (1990). The Diatoms: Biology and Morphology of the Genera. Cambridge University Press, Cambridge.

Royal Engineers (1841). Plan of the environs of Beirut (scale: 1:21,120). Royal Engineers.

Royal, J. G. and Kahonov, Y. (2000). An Arab period merchant vessel at Tantura Lagoon, Israel. International Journal of Nautical Archaeology, 29, 151-153.

Ruegg, B. S. D. (1988). Minturnae: A Roman River Seaport on the Garigliano River, Italy. In: Raban, A. (Ed.), Archaeology of coastal changes, Proceedings of the first international symposium "Cities on the sea - past and present". BAR International Series 404, Haifa, pp. 209-228.

Ruiz, F., Abad, M., Bodergat, A. M., Carbonel, P., Rodríguez-Lázaro, J. and Yasuhara, M. (2005). Marine and brackish-water ostracods as sentinels of anthropogenic impacts. Earth-Science Reviews, 72, 89-111.

Russell, K. W. (1985). The earthquake chronology of Palestine and Northwest Arabia from the 2nd through the mid-8th century A.D. Bulletin of the American School of Oriental Research, 260, 37-59.

Saadé, G. (1995). Le port d'Ougarit. In: Yon, M., Sznycer, M. and Bordreuil, P. (Eds.), Le pays d'Ougarit autour de 1200 av. J.-C. Ras Shamra-Ougarit IX, Actes du colloque international, Paris, 28 juin - 1er juillet 1993. Paris.

Sader, H. (1999). Ancient Beirut: Urban Growth in the Light of Recent Excavations. In: Rowe, P. and Sarkis, H. (Eds.), Projecting Beirut: Episodes in the Construction and Reconstruction of a Modern City. Prestel, München, London, New York, pp. 122-134.

Saghieh Beidoun, M. (1997). Evidence of Earthquakes in the Current Excavations of Beirut City Centre. National Museum News, 5, 15-19.

Saghieh-Beydoun, M. (2005). Urban planning in a seaport city: Beirut from Hellenistic to Byzantine periods. Bulletin d'Archéologie et d'Architecture Libanaises, Hor-Série 2, 147-184.

Saghieh-Beydoun, M., 'Allam, M., 'Al'Eddine, A. and Abulhosn, S. (1998-1999). The monumental street 'Cardo Maximus' and the replanning of Roman Berytus. Bulletin d'Archéologie et d'Architecture Libanaises, 3, 95-126.

Saghieh, M. (1996). Bey 001 and 004 preliminary report. Bulletin d'Archéologie et d'Architecture Libanaises, 1, 23-59.

Said, R. (1993). The River Nile: geology, hydrology and utilization. Pergamon Press, New York.

Saidah, R. (1979). Fouilles de Sidon-Dakerman: l'agglomération chalcholithique. Berytus, 27, 29-55.

Samir, A. M. (2000). The response of benthic foraminifera and ostracods to various pollution sources: a study from two lagoons in

Egypt. Journal of Foraminiferal Research, 30, 83-98.

Sanderson, P. and Eliot, I. G. (1996). Shoreline salients on the coast of Western Australia, Journal of Coastal Research, 12, 761-773.

Sandler, A. and Herut, B. (2000). Composition of clays along the continental shelf off Israel: contribution of the Nile versus local sources. Marine Geology, 167, 339-354.

Sanlaville, P. (1977). Étude géomorphologique de la région littorale du Liban. Publications de l'Université Libanaise, Beyrouth.

Sanlaville, P., Dalongeville, R., Bernier, P., and Evin, J. (1997). The Syrian coast: a model of Holocene coastal evolution. Journal of Coastal Research, 13, 385-396.

Sartoretto, S., Collina-Girard, J., Laborel, J. and Morhange, C. (1995). Quand la grotte Cosquer a-t-elle été fermée par la montée des eaux? Méditerranée, 82, 21-24.

Säve-Söderbergh, T. (1946). The Navy of the Eighteenth Egyptian Dynasty. Lundequistka, Uppsala.

Sayegh, H. and Elayi, J. (2000). Rapport préliminaire sur le port de Beyrouth au Fer III/Perse (Bey 039). Transeuphratène, 19, 65-74.

Sayre, E. V., Yener, K. A., Joel, E. C. and Barnes, I. L. (1992). Statistical evaluation of the presently accumulated lead isotope data from Anatolia and surrounding regions. Archaeometry, 34, 73-105.

Schaeffer, C. F. A. (1935). Les fouilles de Ras Shamra-Ugarit, 6ème campagne (printemps 1934), rapport sommaire. Syria, 16, 141-176.

Schilman, B., Bar-Matthews, M., Almogi-Labin, A. and Luz, B. (2001). Global climate instability reflected by Eastern Mediterranean marine records during the late Holocene. Palaeogeography, Palaeoclimatology, Palaeoecology, 176, 157-176.

Schilman, B., Ayalon, A., Bar-Matthews, M., Kagan, E.J. and Almogi-Labin, A. (2002). Sea-land paleoclimate correlation in the Eastern Mediterranean region during the late Holocene. Israel Journal of Earth Sciences, 51, 181-190.

Schmiedt, G. (1970). Atlante aerofotografico delle sedi umane in Italia, parte II, le sedi antiche. Firenze.

Schmiedt, G. (1972). Il livello antico del mar Tirreno: Testimonianze dei resti archeologici. Leo S. Olschki Editore, Florence.

Schmiedt, G. (1975). Antichi Porti d'Italia. Gli Scali Fenico-Punici i porti della Magna-Grecia. Coi Tipi Dell'Istituto Geografico Militare, Firenze.

Schröder, B. and Bochum, B. B. (1996). Late Holocene Rapid Coastal Change in Western Anatolia - Büyük Menderes Plain as a Case Study. Zeitschrift für Geomorphologie, Supplement Band 102, 61-70.

Schulte, L. (2002). Climatic and human influence on river systems and glacier fluctuations in southeast Spain since the Last Glacial Maximum. Quaternary International, 93-94, 85-100.

Schwartz, R. K. and Birkemeier, W. A. (2004). Sedimentology and morphodynamics of a barrier island shoreface related to engineering concerns, Outer Banks, NC, USA. Marine Geology, 211, 215-255.

Scognamiglio, E. (1997). Aggiornamenti per la topografia di Baia sommersa. Archeologia Subacquea, 2, 35-45.

Scollar, I., Tabbagh, A., Hesse, A. and Herzog, I. (1990). Archaeological Prospecting and Remote Sensing. Cambridge University Press, Cambridge.

Scott, R. (1841). Plan of Beyrut, the ancient Berytus (scale: 1:5280). Royal Engineers.

Scranton, R. L. and Ramage, E. S. (1967). Investigations at Corinthian Kenchreai. Hesperia, 36, 124-186.

Seeden, H. (1999). Lebanon's archaeological heritage on trial in Beirut: what future for the capital's past? In: McManamon, F. P. and Hatton, A. (Eds.), Heritage conservation in modern society. One World Archaeology, 33, London, pp. 168-187.

Seeden, H. and Thorpe, R. (1997-1998). Beirut from Ottoman sea walls and landfill to a twelfth century BC burial: Interim report on the archaeological excavations of the souk's northern area (BEY 007). Berytus, 43, 221-254.

Segarra Lagunes, M. M. (2004). Il Tevere e Roma: storia di una simbiosi. Gangemi Editore, Roma.

Sen Gupta, B. K. (Ed.) (2002). Modern Foraminifera. Springer, London.

Service Hydrographique et Océanographique de la Marine (1998). Ports du Liban, 7514 (scale: 1:25000). Brest.

Sgarrella, F. and Moncharmont Zei, M. (1993). Benthic foraminifera of the Gulf of Naples (Italy): Systematics and autoecology. Boll. Soc. Paleontol. Ital., 32, 145-264.

Shackleton, J., van Andel, T. and Runnels, C. (1984). Coastal paleogeography of the central and western Mediterranean during the last 125,000 years and its archaeological implications. Journal of Field Archaeology, 11, 307-314.

Shaw, I. (Ed.) (2000). The Oxford History of Ancient Egypt. Oxford University Press, Oxford.

Shaw, J. W. (1990). Bronze Age Aegean Harboursides. In: Hardy, D. A., Doumas, C. G., Sakellarakis, J. A. and Warren, P. M. (Eds.), Thera and the Aegean World. Thera Foundation, London, pp. 420-436.

Shaw, M. D. (1743).Voyages de Monsieur Shaw, M. D. dans plusieurs provinces de la Barbarie et du Levant, contenant des observations géographiques, physiques, philologiques et mêlées sur les royaumes d'Alger et de Tunis, sur la Syrie, L'Egypte et l'Arabie pétrée, avec des cartes et des figures, ouvrage traduit de l'anglais. Jean Neaulme, La Haye, 2 vols.

Shepard, F. P. (1964). Sea-level changes in the past 6000 years: Possible archaeological significance. Science, 143, 574-576.

Sherwood Illsley, J. (1996). An indexed bibliography of underwater archaeology and related topics. International Maritime Archaeology Series, University of Oxford, Oxford.

Shotyk, W., Weiss, D., Appleby, P. G., Cheburkin, A. K., Frei, R., Gloor, M., Kramers, J. D., Reese, S. and Van Der Knaap, W. O. (1998). History of Atmospheric Lead Deposition Since 12,370 14C yr BP from a Peat Bog, Jura Mountains, Switzerland. Science, 281, 1635-1640.

Shotyk, W., Goodsite, M. E., Roos-Barraclough, F., Givelet, N., Le Roux, G., Weiss, D. J., Cheburkin, A. K., Knudsen, K., Heinemeier, J., Van der Knaap, W. O., Norton, S. A., and Lohse, C. (2005). Accumulation and predominant atmospheric sources of natural and anthropogenic Hg and Pb on the Faroe Islands since 5420 14C yr BP recorded by a peat core from a blanket bog. Geochimica et Cosmochimica Acta, 69, 1-17.

Siani, G., Paterne, M., Arnold, M., Bard, E., Métivier, B., Tisnerat, N. and Bassinot, F. (2000). Radiocarbon reservoir ages in the Mediterranean Sea and Black Sea. Radiocarbon, 42, 271-280.

Silvester, R. and Hsu, J. R. C. (1993). Coastal stabilization: innovative concepts. Prentice Hall, New Jersey.

Simms, A. R., Anderson, J. B. and Blum, M. (2006). Barrier-island aggradation via inlet migration: Mustang Island, Texas. Sedimentary Geology, 187, 105-125.

Sivan, D., Wdowinski, S., Lambeck, K., Galili, E. and Raban, A. (2001). Holocene sea-level changes along the Mediterranean coast of Israel, based on archaeological observations and numerical model. Palaeogeography, Palaeoclimatology, Palaeooecology, 167, 101-117.

Sivan, D. and Porat, N. (2004). Evidence from luminescence for Late Pleistocene formation of calcareous aeolianite (kurkar) and paleosol (hamra) in the Carmel Coast, Israel. Palaeogeography, Palaeoclimatology, Palaeoecology, 211, 95-106.

Skovenborg, E. (1995). Lead in wine through the ages. Journal of Wine Research, 6, 49-64.

Skyring, T. F. (1841). Plan of the Town and defences of Beyrut and its vicinity (scale: 1:4800). Royal Engineers.

Soffer, A. (1994). The Litani River: Fact and Fiction. Middle Eastern Studies, 30, 963-974.

Soloviev, S. L., Solovieva, O. N., Go, C. N., Kim, K. S. and Shchetnikov, N. A. (2000). Tsunamis in the Mediterranean Sea 2000 B.C.-2000 A.D. Kluwer Academic Publishers, Dordrecht.

Soter, S. (1998). Holocene uplift and subsidence of the Helike Delta, Gulf of Corinth, Greece. In: Stewart, I. and Vita-Finzi, C. (Eds.) Coastal Tectonics. Geological Society, London, Special Publications 146, pp. 41-56.

Soter, S. and Katsonopoulou, D. (1998). The search for ancient Helike, 1988-1995. Geological, sonar and bore hole studies. In: Katsonopoulou, D., Soter, S. and Schilardi, D. (Eds.), Ancient Helike and Aigialeia. Athens, pp. 68-114.

Sourdel, D. and J. Sourdel-Thomine, J. (1989). In: E. Poleggi (Ed.), Città portuali del Mediterraneo: storia e archeologia. Sagep Editrice, Genova.

Spanò Giammelario, A. (1997). Les Phéniciens et les Puniques en Sicile. Les Dossiers d'Archéologie, 225, 22-32.

Spratt, T. S. (1865). Travels and Researches in Crete, Vol. 2. J. van Voorst, London.

Staniforth, M. (1994). Public access to maritime archaeology. Bulletin of the Australian Institute for Maritime Archaeology, 18, 13-16.

Stanley, D. J. (1999). Evaluating Use of Rock-Hewn Features for Sea Level Measurement, Israeli Coast. Journal of Coastal Research, 15, 326-331.

Stanley, D. J. (2002). Configuration of the Egypt-to-Canaan coastal margin and north Sinai byway in the Bronze Age. In: Van den Brink, E. C. M. and Levy, T. E. (Eds.), Egypt and the Levant. Leicster University Press, London, pp. 98-117.

Stanley, D. J. (2005). Submergence and burial of ancient coastal sites on the subsiding Nile delta margin, Egypt. Méditerranée, 104, 65-73.

Stanley, D. J. and Warne, A. G. (1994). Worldwide Initiation of Holocene Marine Deltas by Deceleration of Sea-Level Rise. Science, 265, 228-231.

Stanley, D. J. and Galili, E. (1996). Sediment dispersal along northern Israel coast during the early Holocene: geological and archaeological evidence. Marine Geology, 130, 11-17.

Stanley, D. J. and Wingerath, J. G. (1996). Clay mineral distributions to interpret Nile

Cell provenance and dispersal: I. Lower River Nile to delta. Journal of Coastal Research, 12, 911-929.

Stanley, D. J., Mart, Y. and Nir, Y. (1997). Clay mineral distributions to interpret Nile Cell provenance and dispersal: II. Coastal plain from Nile delta to northern Israel. Journal of Coastal Research, 13, 506-533.

Stanley, D. J. and Warne, A. G. (1998). Nile Delta in its destruction phase. Journal of Coastal Research, 14, 794-825.

Stanley, D. J., Nir, Y. and Galili, E. (1998). Clay Mineral Distributions to Interpret Nile Cell Provenance and Dispersal: III. Offshore Margin between Nile Delta and Northern Israel. Journal of Coastal Research, 14, 196-217.

Stanley, D. J., Goddio, F. and Schnepp, G. (2001). Nile flooding sank two ancient cities. Nature, 412, 293-294.

Stanley, D. J., Goddio, F., Jorstad, T. F., and Schnepp, G. (2004a). Submergence of Ancient Greek Cities Off Egypt's Nile Delta - A Cautionary Tale. GSA Today, 14, 4-10.

Stanley, J.-D., Warne, A. G. and Schnepp, G. (2004b). Geoarchaeological Interpretation of the Canopic, Largest of the Relict Nile Delta Distributaries, Egypt. Journal of Coastal Research, 20, 920-930.

Stanley, D. J. and Bernasconi, M. P. (2006). Holocene Depositional Patterns and Evolution in Alexandria's Eastern Harbor, Egypt. Journal of Coastal Research, 22, 283-297.

Stanley, J.-D. and Jorstad, T. F. (2006). Buried Canopic channel identified near Egypt's Nile delta coast with radar (SRTM) imagery. Geoarchaeology, 21, 503-514.

Stapor, F. W. and Stone, G. W. (2004). A new depositional model for the buried 4000 yr BP New Orleans barrier: implications for sea-level fluctuations and onshore transport from a nearshore shelf source. Marine Geology, 204, 215-234.

Stefani, M. and Vincenzi, S. (2005). The interplay of eustasy, climate and human activity in the late Quaternary depositional evolution and sedimentary architecture of the Po Delta system. Marine Geology, 222-223, 19-48.

Stefaniuk, L., Brun, J. P., Munzi, P. and Morhange, C. (2003). L'evoluzione dell'ambiente nei Campi Flegrei e le sue implicazioni storiche: Il caso di Cuma e le ricerche del Centre Jean Bérard nella laguna di Licola. In: Grecia, M. (Ed.), Ambiente e Paesaggio nella Magna Grecia. Istituto per la storia e l'archeologia della Magna Grecia, Taranto, 5-8 ottobre 2002, pp. 397-435.

Stefaniuk, L. and Morhange, C. (2005). Evolution des paysages littoraux dans la depression sud-ouest de Cumes depuis 4000 ans. Méditerranée, 104, 49-59.

Stefaniuk L., Morhange C., Saghieh-Beydoun M., Frost H., Boudagher-Fadel M., Bourcier M. and Noujaim-Clark G. (2005). Localisation et étude paléoenvironnementale des ports antiques de Byblos. Bulletin d'Archéologie et d'Architecture Libanaises, Hors-Série 2, 19-41.

Stewart, A. (1987). Diodorius, Curtius and Arrian on Alexander's mole at Tyre. Berytus, 35, 97-99.

Stiller, M., Kaushansky, P. and Carmi, I. (1983). Recent climatic changes recorded by the salinity of pore waters in the Dead Sea sediments. Hydrobiologia, 103, 75-79.

Stiros, S. (1998). Archaeological evidence for unusually rapid Holocene uplift rates in an active normal faulting terrain: Roman harbour of Aigeira, Gulf of Corinth, Greece. Geoarchaeology, 13, 731-741.

Stiros, S. C. (2001). The AD 365 Crete earthquake and possible seismic clustering during the fourth to sixth centuries in the Eastern Mediterranean: a review of historical and archaeological data. Journal of Structural Geology, 23, 545-562.

Stiros, S. (2005). Social and historical impacts of earthquake-related sea-level changes on ancient (prehistoric to Roman) coastal sites. Zeitschrift für Geomorphologie, Supplement Volume 137, 79-89.

Stiros, S., Arnold, M., Pirazzoli, P. A., Laborel, J., Laborel, F. and Papageorgiou, S. (1992). Historical coseismic uplift on Euboea island, Greece. Earth and Planetary Science Letters, 108, 109-117.

Stiros, S., Pirazzoli, P., Rothaus, R., Papageorgiou, S., Laborel, J. and Arnold, M. (1996). On the date of construction of Lechaion, western harbour of ancient Corinth, Greece. Geoarchaeology, 11, 251-263.

Stochove, V. de (1650). Voyage du Levant du Sr de Stochove. H. A. Velpius, Bruxelles.

Stone, G. W., Liu, B., Pepper, D. A. and Wang, P. (2004). The importance of extratropical and tropical cyclones on the short-term evolution of barrier islands along the northern Gulf of Mexico, USA. Marine Geology, 210, 63-78.

Strong, A. (2002). The Phoenicians in History and Legend. Authorhouse, Bloomington.

Sunamura, T. and Mizuzo, O. (1987). A study on depositional shoreline forms behind an island. Annual Report of the Institute of Geosciences, University of Tsukuba, 13, 71-73.

Tainter, J. A. (1998). The Collapse of Complex Societies. Cambridge University Press, Cambridge.

Tayara, Z. (1991). Etude hydro-pluviométrique comparative des basins versants de la zone côtière intermédiaire du Liban (le Damour, l'Awali-Bisri, le Sainïq et le Zahrani), PhD thesis. Université des sciences et techniques de Lille, Lille.

Televantou, C. (1990). New Light on the West House Wall-paintings. In: Hardy, D. A., Doumas, C. G., Sakellarakis, J. A. and Warren, P. M. (Eds.), Thera and the Aegean World. Thera Foundation, London, pp. 309-324.

Testaguzza, O. (1964). The port of Rome. Archaeology, 14, 285-290.

Thomas, M. A. and van Schalkwyk, A. (1993). Geological hazards associated with intense rain and flooding in Natal. Journal of African Earth Sciences, 16, 193-204.

Thorpe, R. (1998-1999). Bey 045, preliminary report on the excavations. Bulletin d'Archéologie et d'Architecture Libanaises, 3, 57-83.

Thorpe, R., Reuben, A., Beyhum, S., Kouly, S. and Beayno, F. (1998-1999). Bey 007 the Souks area. Preliminary report of the AUB/Acre project. Bulletin d'Archéologie et d'Architecture Libanaises, 3, 31-55.

Todesco, M., Rutqvist, J., Chiodini, G., Pruess, K. and Oldenburg, C. M. (2004). Modelling of recent volcanic episodes at Phlegrean Fields (Italy): Geochemical variations and ground deformation. Geothermics, 33, 531-547.

Treadgold, W. (2000). A Concise History of Byzantium, 285-1461 (European History in Perspective). Palgrave Macmillan, Basingstoke.

Tylecote, R. F. (1992). A History of Metallurgy. The Institute of Materials, London.

van Andel, T. (1989). Late Quaternary sea-level and archaeology. Antiquity, 63, 733-745.

van Andel, T. H. and Shackleton, J. C. (1982). Late Paleolithic and Mesolithic Coastlines of Greece and the Aegean. Journal of Field Archaeology, 9, 445-454.

Van Cotvyck, J. (1620). Itinerarium Hierosolymitanum et Syriacum in quo variarum gentium mores et instituta, insularum, Regionum urbium situs, una ex prisci recentiorisque seaculi usu; una cum eventis quae Auctoritera morique acciderunt, dilucide recensentur. Accessit Synopsis Reipublicae Veneto. Antwerp.

Vandaele, K. and Poesen, J. (1995). Spatial and temporal patterns of soil erosion rates in an agricultural catchment, central Belgium. CATENA, 25, 213-226.

Vasseur, G. (1911). Nouvelles découvertes et observations relatives à Massalia. Bull. Soc. Arch. de Provence, 17, 207-218.

Vasseur, G. (1914). L'origine de Marseille. Annales du Museum d'histoire naturelle de Marseille, 13, 201-224.

Vaumas, E. de (1944). Les terrasses d'abrasion marine de la côte libanaise. Bull. Soc. Geogr. Egypte, 22, 21-85.

Vaumas, E. de (1946). Le relief de Beyrouth et son influence sur le développement de la ville. Publications Techniques et Scientifiques de l'école Française d'Ingénieurs de Beyrouth, 11, Beirut.

Vecchi, L., Morhange, C., Blanc, P. F., Goiran, J. P., Thi Maï, B., Bourcier, M., Carbonel, P., Demant, A., Gasse, F., Girard, M. and Verrecchia, E. (2000). La mobilité des milieux littoraux de Cumes, Champs Phlégréens, Campanie, Italie du Sud. Méditerranée, 94, 71-82.

Vella, C. and Provansal, M. (2000). Relative sea-level rise and neotectonic events during the last 6500 yr on the southern eastern Rhône delta, France. Marine Geology, 170, 27-39.

Vella, C., Fleury, T.-J., Raccasi, G., Provansal, M., Sabatier, F. and Bourcier, M. (2005). Evolution of the Rhône delta plain in the Holocene. Marine Geology, 222-223, 235-265.

Venkatarathnam, K., Biscaye, P. E. and Ryan, W. B. F. (1971). Dispersal patterns of clay minerals in the sediments of the eastern Mediterranean Sea. Marine Geology, 11, 261-282.

Vermeulen, F. and De Dapper, M. (Eds.) (2000). Geoarchaeology of the Landscapes of Classical Antiquity. Peeters Presse, Leuven.

Véron, A. (2004). Le plomb : un pison pour l'homme, de l'or pour le géochimie, HDR thesis. Univsersité Aix-Marseille III, Marseille.

Véron, A., Goiran, J. P., Morhange, C., Marriner, N. and Empereur, J. Y. (2006). Pollutant lead reveals the pre-Hellenistic occupation and ancient growth of Alexandria, Egypt. Geophysical Research Letters, 33, L06409.

Villamont, J. de (1596). Les Voyages du Seigneur de Villamont, Chevalier de L'Ordre de Hierusalem, Gentil-Homme du Pays de Bretaigne: divisez en trois livres. C. de Montr'oeil et J. Richer, Paris.

Villas, C. (1996). Geological invesitagtions. In: Coulson, W. D. E. (Ed.), Ancient Naukratis: Volume II, The Survey at Naukratis and Environs. Oxbow Monograph 60, Oxford, pp. 163-175.

Viret, J. (1999-2000). Nouvelles données sur le port de Tripoli, Tempora. Annales d'histoire et d'archéologie, 10-11, 117-138.

Viret, J. (2000). L'apport d'Antoine Poidebard à l'archéologie des ports antique du Levant: vers une méthode d'investigation moderne. In: Nordiguian, L. and Salles, J. F. (Eds.), Aux origines de l'archéologie moderne: A. Poidebard (1878-1955). Presses de l'Université Saint-Joseph, Beyrouth, pp. 151-163.

Viret, J. (2004). Aux origines de l'archéologie sous-marine. In: Denise, F. and Nordiguian, L. (Eds.), Une aventure archéologique: Antoine Poidebard, photographe et aviateur. Editions Parenthèses, Marseille, pp. 146-165.

Viret, J. (2005). Les " murs de mer " de la côte levantine. Méditerranée, 104, 15-23.

Vlag, D. P. (1992). A model predicting waves and suspended silt concentration in a shallow lake. Hydrobiologia, 235-236, 119-131.

Volney, C. F. (1791). Les Ruines Ou Meditations Sur Les Revolutions Des Empires. Desenne, Volland, Plassan, Paris.

Volney, C. F. (1792). The Ruins, or Meditations on the Revolutions of Empires. J. Johnson, London.

Vött, A., Brückner, H., Schriever, A., Luther, J., Handl, M. and van der Borg, K. (2006a). Holocene Paleogeographies of the Palairos Coastal Plain (Akarnania, Northwest Greece) and Their Geoarchaeological Implications. Geoarchaeology, 21, 649-664

Vött, A., Brückner, H., Handl, M. and Schriever, A. (2006b). Holocene palaeogeographies of the Astakos coastal plain (Akarnania, NW Greece). Palaeogeography, Palaeoclimatology, Palaeoecology, 239, 126-146.

Wachsmann, S. (1998). Seagoing ships and seamanship in the Bronze Age Levant. A & M University Press, Texas.

Ward, C. and Ballard, R. D. (2004). Deep-water Archaeological Survey in the Black Sea: 2000 Season. International Journal of Nautical Archaeology, 33, 1.

Warne, A.G., and Stanley, D. J. (1993a). Late Quaternary evolution of the northwest Nile Delta and adjacent coast in the Alexandria region, Egypt. Journal of Coastal Research, 9, 26-64.

Warne, A. G. and Stanley, D. J. (1993b). Archaeology to Refine Holocene Subsidence Rates Along the Nile Delta Margin, Egypt. Geology, 21, 715-718.

Webster, D. (2002). The Fall of the Ancient Maya. Thames & Hudson, London.

Weinstein, E. N. (1996). Pollen analysis of underwater sites. In: Jansonius, J. and McGregor, D. C. (Eds.), Palynology: Principles and Applications. American Association of Stratigraphic Palynologists Foundation 3, pp. 919-925.

Weiss, H., Courty, M.-A., Wetterstrom, W., Guichard, F., Seniro, L., Meadow, R. and Curnow, A. (1993). The Genesis and Collapse of Third Millennium North Mesopotamian Civilization. Science, 261, 995-1004.

Weiss, H. and Bradley, R.S. (2001). What drives societal collapse? Science, 291, 609-610.

Westaway, R., Bridgland, D. and White, M. (in press). The Quaternary uplift history of central southern England: evidence from the terraces of the Solent River system and nearby raised beaches. Quaternary Science Reviews.

Weydert, P. (1973). Morphologie et sédimentologie de la partie mériodionale du grand récif de Tuléar, PhD thesis. Université Aix-Marseille II, Marseille.

White, V. S. (2002). The Ships of Tarshish: The Phoenicians. Unknown.

Whitaker, J. I. S. (1921). Motya: a Phoenician colony in Sicily. G. Bells & Sons, London.

Willcox, G. (2003). Les Macrorestes Végétaux. In: Geyer, B. and Lefort, J. (Eds.), La Bithynie au Moyen Age. P. Lethielleux, Paris, pp. 201-205.

Wittenberg, L., Kutiel, H., Greenbaum, N. and Inbar, M. (in press). Short-term changes in the magnitude, frequency and temporal distribution of floods in the Eastern Mediterranean region during the last 45 years - Nahal Oren, Mt. Carmel, Israel. Geomorphology.

Woodall, J. N. (Ed.) (1993). Predicaments, Pragmatics and Professionalism: Ethical Conduct in Archaeology. Society of Professional Archaeologists, A & M University, Texas.

Wunderlich, J. (1988). Investigations on the development of the western Nile delta in Holocene times. In: Van der Brink, E. C. M. (Ed.), The archaeology of the Nile delta, Egypt: problems and priorities. Netherlands Foundation for Archaeological Research in Egypt, Amsterdam, pp. 251-257.

Wyld, J. (1840). Plan of the Town and Harbour of Beirut, ancient Berytus (scale: 1:35,000). London.

Yener, K., Sayre, E., Joel, E., Osbal, H., Barnes, I. and Brill, R. (1991). Stable lead isotope studies of Central Taurus ore sources and related artifacts from Eastern Mediterranean chalcolithic and bronze age sites. Journal of Archaeogical Science, 18, 541-577.

Yoffee, N. and Cowgill, G. L. (1988). The Collapse of Ancient States and Civilizations. The University of Arizona Press, Tucson, London.

Yorke, R. A. (1976). Search for submerged Carthage. Geographical Magazine, 49, 24-29.

Yorke, R. A. and Little, J. H. (1975). Offshore survey at Carthage, Tunisia. International Journal of Nautical Archaeology, 4, 85-101.

Yorke, R. A., Little, J. H. and Davidson, D. P. (1976). Offshore survey of the harbours of Carthage. International Journal of Nautical Archaeology, 5, 173-176.

Zaccaria, C. (Ed.) (2001). Strutture portuali e rotte maritime nell' Adriatico di et à Romana. Ecole française de Rome, Rome.

Zaouali, L. (1999). Mahdia, port et arsenal. In: La Méditerranée : l'homme et la mer, dans le cadre du Projet National Mobilisateur : "Les villes-ports en Tunisie". Cahiers du CERES, 21, Tunis, pp. 219-239.

Zenkovich, V. P. (1967). Processes of Coastal development. Oliver and Boyd, London.

Zevi, A. G. and Turchetti, R. (Eds.) (2004a). Le strutture dei porti e degli approdi antichi. Rubbettino Editore, Soveria Mannelli.

Zevi, A. G. and Turchetti, R. (Eds.) (2004b). Méditerranée occidentale antique : les échanges. Rubbettino Editore, Soveria Mannelli.

www.ingramcontent.com/pod-product-compliance
Ingram Content Group UK Ltd.
Pitfield, Milton Keynes, MK11 3LW, UK
UKHW061845041026
15591UKWH00015B/917

9 781407 304366